To Rosemary

FROM THE PREFACE TO THE FIRST EDITION

This book is an outgrowth of a year course in statistical mechanics that I have been giving at the Massachusetts Institute of Technology. It is directed mainly to graduate students in physics.

In this book the starting point of statistical mechanics is taken to be certain phenomenological postulates, whose relation to quantum mechanics I try to state as clearly as I can, and whose physical consequences I try to derive as simply and directly as I can.

Before the subject of statistical mechanics proper is presented, a brief but self-contained discussion of thermodynamics and the classical kinetic theory of gases is given. The order of this development is imperative, from a pedagogical point of view, for two reasons. First, thermodynamics has successfully described a large part of macroscopic experience, which is the concern of statistical mechanics. It has done so not on the basis of molecular dynamics but on the basis of a few simple and intuitive postulates stated in everyday terms. If we first familiarize ourselves with thermodynamics, the task of statistical mechanics reduces to the explanation of thermodynamics. Second, the classical kinetic theory of gases is the only known special case in which thermodynamics can be derived nearly from first principles, i.e., molecular dynamics. A study of this special case will help us understand why statistical mechanics works.

A large part of this book is devoted to selected applications of statistical mechanics. The selection is guided by the interest of the topic to physicists, its value as an illustration of calculating techniques, and my personal taste.

To read the first half of the book the reader needs a good knowledge of classical mechanics and some intuitive feeling for thermodynamics and kinetic theory. To read the second half of the book he needs to have a working knowledge of quantum mechanics. The mathematical knowledge required of the reader does not exceed what he should have acquired in his study of classical mechanics and quantum mechanics.

KERSON HUANG

Cambridge, Massachusetts
February 1963

PREFACE

This book is intended for use as a textbook for a one-year graduate course in statistical mechanics, and as a reference book for workers in the field.

Significant changes and additions have been made at various places in the text and in the problems to make the book more instructive. The general approach remains the same as that stated in the preface to the first edition.

The most substantive change is the addition of the last three chapters on the Landau-Wilson approach to critical phenomena. I hope that these chapters will provide the uninitiated reader with an introduction to this fascinating and important subject, which has developed since the first edition of this book.

To make room for the additions, I have omitted or abridged some of the original material, notably the sections on many-body theory and "rigorous" statistical mechanics, which by now have become separate fields.

All the material in this book probably cannot be covered in a one-year course (even if that were deemed desirable). It might be helpful, therefore, to list the chapters that form the "hard core" of the book. They are the following: Chapters 3, 4, 6, 7, 8, 11, 12 (excepting 3.5, 7.6, 7.7, 7.8).

KERSON HUANG

Marblehead, Massachusetts

CONTENTS

A

THERMODYNAMICS AND KINETIC THEORY

1

THE LAWS OF THERMODYNAMICS

1.1 PRELIMINARIES

Thermodynamics is a phenomenological theory of matter. As such, it draws its concepts directly from experiments. The following is a list of some working concepts which the physicist, through experience, has found it convenient to introduce. We shall be extremely brief, as the reader is assumed to be familiar with these concepts.

(a) A *thermodynamic system* is any macroscopic system.

(b) *Thermodynamic parameters* are measurable macroscopic quantities associated with the system, such as the pressure P, the volume V, the temperature T, and the magnetic field H. They are defined experimentally.

(c) A *thermodynamic state* is specified by a set of values of all the thermodynamic parameters necessary for the description of the system.

(d) *Thermodynamic equilibrium* prevails when the thermodynamic state of the system does not change with time.

(e) The *equation of state* is a functional relationship among the thermodynamic parameters for a system in equilibrium. If P, V, and T are the thermodynamic parameters of the system, the equation of state takes the form

$$f(P, V, T) = 0$$

which reduces the number of independent variables of the system from three to two. The function f is assumed to be given as part of the specification of the system. It is customary to represent the state of such a system by a point in the three-dimensional P-V-T space. The equation of state then defines a surface in this space, as shown in Fig. 1.1. Any point lying on this surface represents a state in equilibrium. In

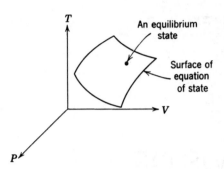

Fig. 1.1 Geometrical representation of the equation of state.

thermodynamics a state automatically means a state in equilibrium unless otherwise specified.

(f) A *thermodynamic transformation* is a change of state. If the initial state is an equilibrium state, the transformation can be brought about only by changes in the external condition of the system. The transformation is *quasi-static* if the external condition changes so slowly that at any moment the system is approximately in equilibrium. It is *reversible* if the transformation retraces its history in time when the external condition retraces its history in time. A reversible transformation is quasi-static, but the converge is not necessarily true. For example, a gas that freely expands into successive infinitesimal volume elements undergoes a quasi-static transformation but not a reversible one.

(g) The *P-V diagram* of a system is the projection of the surface of the equation of state onto the *P-V* plane. Every point on the *P-V* diagram therefore represents an equilibrium state. A reversible transformation is a continuous path on the *P-V* diagram. Reversible transformations of specific types give rise to paths with specific names, such as *isotherms*, *adiabatics*, etc. A transformation that is not reversible cannot be so represented.

(h) The concept of *work* is taken over from mechanics. For example, for a system whose parameters are *P*, *V*, and *T*, the work *dW* done by a system in an infinitesimal transformation in which the volume increases by *dV* is given by

$$dW = P\,dV$$

(i) *Heat* is what is absorbed by a homogeneous system if its temperature increases while no work is done. If ΔQ is a small amount of the heat absorbed, and ΔT is the small change in temperature accompanying the absorption of heat, the *heat capacity C* is defined by

$$\Delta Q = C\Delta T$$

The heat capacity depends on the detailed nature of the system and is given as a part of the specification of the system. It is an experimental fact that, for the same ΔT, ΔQ is different for different ways of heating up the system. Correspondingly, the heat capacity depends on the manner of heating. Commonly considered heat capacities are C_V and C_P, which respectively correspond to heating at constants V and P. Heat capacities per unit mass or per mole of a substance are called *specific heats*.

(*j*)   A *heat reservoir*, or simply *reservoir*, is a system so large that the gain or loss of any finite amount of heat does not change its temperature.

(*k*)   A system is *thermally isolated* if no heat exchange can take place between it and the external world. Thermal isolation may be achieved by surrounding a system with an *adiabatic wall*. Any transformation the system can undergo in thermal isolation is said to take place adiabatically.

(*l*)   A thermodynamic quantity is said to be *extensive* if it is proportional to the amount of substance in the system under consideration and is said to be *intensive* if it is independent of the amount of substance in the system under consideration. It is an important empirical fact that to a good approximation thermodynamic quantities are either extensive or intensive.

(*m*)   The *ideal gas* is an important idealized thermodynamic system. Experimentally all gases behave in a universal way when they are sufficiently dilute. The ideal gas is an idealization of this limiting behavior. The parameters for an ideal gas are pressure P, volume V, temperature T, and number of molecules N. The equation of state is given by Boyle's law:

$$\frac{PV}{N} = \text{constant} \qquad \text{(for constant temperature)}$$

The value of this constant depends on the experimental scale of temperature used.

(*n*)   The equation of state of an ideal gas in fact defines a temperature scale, the *ideal-gas temperature T*:

$$PV = NkT$$

where

$$k = 1.38 \times 10^{-16} \, \text{erg/deg}$$

which is called Boltzmann's constant. Its value is determined by the conventional choice of temperature intervals, namely, the Centigrade degree. This scale has a universal character because the ideal gas has a universal character. The origin $T = 0$ is here arbitrarily chosen. Later we see that it actually has an absolute meaning according to the second law of thermodynamics.

To construct the ideal-gas temperature scale we may proceed as follows. Measure PV/Nk of an ideal gas at the temperature at which water boils and at

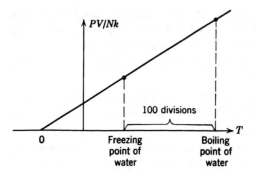

Fig. 1.2 The ideal-gas temperature scale.

which water freezes. Plot these two points and draw a straight line through them, as shown in Fig. 1.2. The intercept of this line with the abscissa is chosen to be the origin of the scale. The intervals of the temperature scale are so chosen that there are 100 equal divisions between the boiling and the freezing points of water. The resulting scale is the Kelvin scale (K). To use the scale, bring anything whose temperature is to be measured into thermal contact with an ideal gas (e.g., helium gas at sufficiently low density), measure PV/Nk of the ideal gas, and read off the temperature from Fig. 1.2. An equivalent form of the equation of state of an ideal gas is

$$PV = nRT$$

where n is the number of moles of the gas and R is the gas constant:

$$R = 8.315 \text{ joule/deg}$$
$$= 1.986 \text{ cal/deg}$$
$$= 0.0821 \text{ liter-atm/deg}$$

Its value follows from the value of Boltzmann's constant and Avogadro's number:

$$\text{Avogadro's number} = 6.023 \times 10^{23} \text{ atoms/mol}$$

Most of these concepts are properly understood only in molecular terms. Here we have to be satisfied with empirical definitions.

In the following we introduce thermodynamic laws, which may be regarded as mathematical axioms defining a mathematical model. It is possible to deduce rigorous consequences of these axioms, but it is important to remember that this model does not rigorously correspond to the physical world, for it ignores the atomic structure of matter, and will thus inevitably fail in the atomic domain. In the macroscopic domain, however, thermodynamics is both powerful and useful. It enables one to draw rather precise and far-reaching conclusions from a few seemingly commonplace observations. This power comes from the implicit assumption that the equation of state is a regular function, for which the thermodynamic laws, which appear to be simple and naive at first sight, are in fact enormously restrictive.

1.2 THE FIRST LAW OF THERMODYNAMICS

In an *arbitrary* thermodynamic transformation let ΔQ denote the net amount of heat absorbed by the system and ΔW the net amount of work done by the system. The first law of thermodynamics states that the quantity ΔU, defined by

$$\Delta U = \Delta Q - \Delta W \tag{1.1}$$

is the same for all transformations leading from a given initial state to a given final state.

This immediately defines a state function U, called the internal energy. Its value for any state may be found as follows. Choose an arbitrary fixed state as reference. Then the internal energy of any state is $\Delta Q - \Delta W$ in *any* transformation which leads from the reference state to the state in question. It is defined only up to an arbitrary additive constant. Empirically U is an extensive quantity. This follows from the saturation property of molecular forces, namely, that the energy of a substance is doubled if its mass is doubled.

The experimental foundation of the first law is Joule's demonstration of the equivalence between heat and mechanical energy—the feasiblity of converting mechanical work completely into heat. The inclusion of heat as a form of energy leads naturally to the inclusion of heat in the statement of the conservation of energy. The first law is precisely such a statement.

In an infinitesimal transformation, the first law reduces to the statement that the differential

$$dU = dQ - dW \tag{1.2}$$

is exact. That is, there exists a function U whose differential is dU; or, the integral $\int dU$ is independent of the path of the integration and depends only on the limits of integration. This property is obviously not shared by dQ or dW.

Given a differential of the form $df = g(A, B)\, dA + h(A, B)\, dB$, the condition that df be exact is $\partial g / \partial B = \partial h / \partial A$. Let us explore some of the consequences of the exactness of dU. Consider a system whose parameters are P, V, T. Any pair of these three parameters may be chosen to be the independent variables that completely specify the state of the system. The other parameter is then determined by the equation of state. We may, for example, consider $U = U(P, V)$. Then*

$$dU = \left(\frac{\partial U}{\partial P}\right)_V dP + \left(\frac{\partial U}{\partial V}\right)_P dV \tag{1.3}$$

The requirement that dU be exact immediately leads to the result

$$\frac{\partial}{\partial V}\left[\left(\frac{\partial U}{\partial P}\right)_V\right]_P = \frac{\partial}{\partial P}\left[\left(\frac{\partial U}{\partial V}\right)_P\right]_V \tag{1.4}$$

The following equations, expressing the heat absorbed by a system during an infinitesimal reversible transformation (in which $dW = P\, dV$), are easily ob-

---

*The symbol $(\partial U / \partial P)_V$ denotes the partial derivative of U with respect to P, with V held constant.

tained by successively choosing as independent variables the pairs $(P, V), (P, T)$, and (V, T):

$$dQ = \left(\frac{\partial U}{\partial P}\right)_V dP + \left[\left(\frac{\partial U}{\partial V}\right)_P + P\right] dV \qquad (1.5)$$

$$dQ = \left[\left(\frac{\partial U}{\partial T}\right)_P + P\left(\frac{\partial V}{\partial T}\right)_P\right] dT + \left[\left(\frac{\partial U}{\partial P}\right)_T + P\left(\frac{\partial V}{\partial P}\right)_T\right] dP \qquad (1.6)$$

$$dQ = \left(\frac{\partial U}{\partial T}\right)_V dT + \left[\left(\frac{\partial U}{\partial V}\right)_T + P\right] dV \qquad (1.7)$$

Called dQ equations, these are of little practical use in their present form, because the partial derivatives that appear are usually unknown and inaccessible to direct measurement. They will be transformed to more useful forms when we come to the second law of thermodynamics.

It can be immediated deduced from the dQ equations that

$$C_V \equiv \left(\frac{\Delta Q}{\Delta T}\right)_V = \left(\frac{\partial U}{\partial T}\right)_V \qquad (1.8)$$

$$C_P \equiv \left(\frac{\Delta Q}{\Delta T}\right)_P = \left(\frac{\partial H}{\partial T}\right)_P \qquad (1.9)$$

where $H = U + PV$ is called the enthalpy of the system.

We consider the following examples of the application of the first law.

(a) *Analysis of Joule's Free-Expansion Experiment.* The experiment in question concerns the free expansion of an ideal gas into a vacuum. The initial and final situations are illustrated in Fig. 1.3.

Experimental Finding. $T_1 = T_2$.

Deductions. $\Delta W = 0$, since the gas performs no work on its external surrounding. $\Delta Q = 0$, since $\Delta T = 0$. Therefore, $\Delta U = 0$ by the first law.

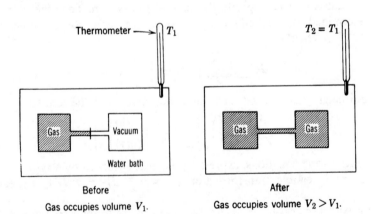

Before
Gas occupies volume V_1.

After
Gas occupies volume $V_2 > V_1$.

Fig. 1.3 Joule's free-expansion experiment.

Thus two states with the same temperature but different volumes have the same internal energy. Since temperature and volume may be taken to be the independent parameters, and since U is a state function, we conclude that for an ideal gas U is a function of the temperature alone. This conclusions can also be reached theoretically, without reference to a specific experiment, with the help of the second law of thermodynamics.

(b) *Internal Energy of Ideal Gas.* Since U depends only on T, (1.8) yields

$$C_V = \left(\frac{\partial U}{\partial T}\right)_V = \frac{dU}{dT}$$

Assuming C_V to be independent of the temperature, we obtain

$$U = C_V T + \text{constant}$$

The additive constant may be arbitrarily set equal to zero.

(c) *The Quantity $C_P - C_V$ for an Ideal Gas.* The enthalpy of an ideal gas is a function of T only:

$$H = U + PV = (C_V + Nk)T$$

Hence, from (1.9),

$$C_P = \left(\frac{\partial H}{\partial T}\right)_P = \frac{dH}{dT} = C_V + Nk$$

or

$$C_P - C_V = Nk$$

Thus it is more efficient to heat an ideal gas by keeping the volume constant than to heat it by keeping the pressure constant. This is intuitively obvious; at constant volume no work is done, so all the heat energy goes into increasing the internal energy.

1.3 THE SECOND LAW OF THERMODYNAMICS

Statement of the Second Law

From experience we know that there are processes that satisfy the law of conservation of energy yet never occur. For example, a piece of stone resting on the floor is never seen to cool itself spontaneously and jump up to the ceiling, thereby converting the heat energy given off into potential energy. The purpose of the second law of thermodynamics is to incorporate such experimental facts into thermodynamics. Its experimental foundation is common sense, as the following equivalent statements of the second law will testify.

Kelvin Statement. There exists no thermodynamic transformation whose *sole* effect is to extract a quantity of heat from a given heat reservoir and to convert it entirely into work.

Clausius Statement. There exists no thermodynamic transformation whose *sole* effect is to extract a quantity of heat from a colder reservoir and to deliver it to a hotter reservoir.

In both statements the key word is "sole." An example suffices to illustrate the point. If an ideal gas is expanded reversibly and isothermally, work is done by the gas. Since $\Delta U = 0$ in this process, the work done is equal to the heat absorbed by the gas during the expansion. Hence a certain quantity of heat is converted entirely into work. This is not the sole effect of the transformation, however, because the gas occupies a larger volume in the final state. This process is allowed by the second law.

The Kelvin statement K and the Clausius statement C are equivalent. To prove this we prove that if the Kelvin statement is false, the Clausius statement is false, and vice versa.

Proof that *K* False $\Rightarrow$ *C* False Suppose K is false. Then we can extract heat from a reservoir at temperature T_1 and convert it entirely into work, with no other effect. Now we can convert this work into heat and deliver it to a reservoir at temperature $T_2 > T_1$ with no other effect. (A practical way of carrying out this particular step is illustrated by Joule's experiment on the equivalence of heat and energy.) The net result of this two-step process is the transfer of an amount of heat from a colder reservoir to a hotter one with no other effect. Hence C is false.　■

Proof that *C* False $\Rightarrow$ *K* False First define an *engine* to be a thermodynamic system that can undergo a cyclic transformation (i.e., a transformation whose final state is identical with the initial state), in which system does the following things, and only the following things:

(a) absorbs an amount of heat $Q_2 > 0$ from reservoir T_2;

(b) rejects an amount of heat $Q_1 > 0$ to reservoir T_1, with $T_1 < T_2$;

(c) performs an amount of work $W > 0$.

Suppose C is false. Extract Q_2 from reservoir T_1 and deliver it to reservoir T_2, with $T_2 > T_1$. Operate an engine between T_2 and T_1 for one cycle, and arrange the engine so that the amount of heat extracted by the engine from T_2 is exactly Q_2. The net result is that an amount of heat is extracted from T_1 and entirely converted into work, with no other effect. Hence K is false.　■

The Carnot Engine

An engine that does all the things required by the definition in a reversible way is called a *Carnot engine*. A Carnot engine consists of any substance that is made to go through the reversible cyclic transformation illustrated in the *P-V* diagram of Fig. 1.4, where *ab* is isothermal at temperature T_2, during which the system absorbs heat Q_2; *bc* is adiabatic; *cd* is isothermal at temperature T_1, with

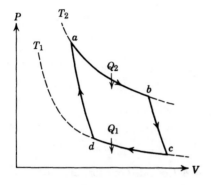

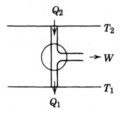

Fig. 1.4 The Carnot engine.

$T_1 < T_2$, during which time the system rejects heat Q_1; and da is adiabatic. It may also be represented schematically as in the lower part of Fig. 1.4. The work done by the system in one cycle is, according to the first law,

$$W = Q_2 - Q_1$$

since $\Delta U = 0$ in any cyclic transformation. The efficiency of the engine is defined to be

$$\eta = \frac{W}{Q_2} = 1 - \frac{Q_1}{Q_2}$$

We show that if $W > 0$, then $Q_1 > 0$ and $Q_2 > 0$. The proof is as follows. It is obvious that $Q_1 \neq 0$, for otherwise we have an immediate violation of the Kelvin statement. Suppose $Q_1 < 0$. This means that the engine absorbs the amount of heat Q_2 from T_2 and the amount of heat $-Q_1$ from T_1 and converts the net amount of heat $Q_2 - Q_1$ into work. Now we may convert this amount of work, which by assumption is positive, into heat and deliver it to the reservoir at T_2, with no other effect. The net result is the transfer of the positive amount of heat $-Q_1$ from T_1 to T_2 with no other effect. Since $T_2 > T_1$ by assumption, this is impossible by the Clausius statement. Therefore $Q_1 > 0$. From $W = Q_2 - Q_1$ and $W > 0$ it follows immediately that $Q_2 > 0$.   ∎

In the same way we can show that if $W < 0$ and $Q_1 < 0$, then $Q_2 < 0$. In this case the engine operates in reverse and becomes a "refrigerator."

The importance of the Carnot engine lies in the following theorem.

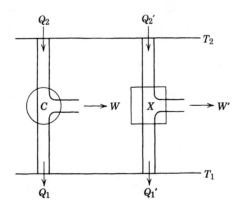

Fig. 1.5 Construction for the proof of Carnot's theorem.

CARNOT'S THEOREM

No engine operating between two given temperatures is more efficient than a Carnot engine.

Proof Operate a Carnot engine C and an arbitrary engine X between the reservoirs T_2 and T_1 ($T_2 > T_1$), as shown in Fig. 1.5. We have, by the first law,

$$W = Q_2 - Q_1$$
$$W' = Q_2' - Q_1'$$

Let

$$\frac{Q_2}{Q_2'} = \frac{N'}{N}$$

where N' and N are two integers. This equality can be satisfied to any degree of accuracy by making N', N sufficiently large. Now operate the C engine N cycles in reverse, and the X engine N' cycles. At the end of this operation we have

$$W_{\text{total}} = N'W' - NW$$
$$(Q_2)_{\text{total}} = N'Q_2' - NQ_2 = 0$$
$$(Q_1)_{\text{total}} = N'Q_1' - NQ_1$$

On the other hand we can also write

$$W_{\text{total}} = (Q_2)_{\text{total}} - (Q_1)_{\text{total}} = -(Q_1)_{\text{total}}$$

The net result of the operation is thus a violation of the Kelvin statement, unless

$$W_{\text{total}} \leq 0$$

which implies that

$$(Q_1)_{\text{total}} \geq 0$$

In other words we must have

$$N'Q_1' - NQ_1 \geq 0$$
$$Q_2Q_1' - Q_2'Q_1 \geq 0$$
$$\frac{Q_1}{Q_2} \leq \frac{Q_1'}{Q_2'}$$

Therefore

$$\left(1 - \frac{Q_1}{Q_2}\right) \geq \left(1 - \frac{Q_1'}{Q_2'}\right)$$ ■

Since X is arbitrary, it can be a Carnot engine. Thus we have the trivial

COROLLARY

All Carnot engines operating between two given temperatures have the same efficiency.

Absolute Scale of Temperature

The corollary to Carnot's theorem furnishes a definition of a scale of temperature, the absolute scale. It is defined as follows. If the efficiency of a Carnot engine operating between two reservoirs of respective absolute temperatures θ_1 and θ_2 $(\theta_2 > \theta_1)$ is η, then

$$\frac{\theta_1}{\theta_2} = 1 - \eta$$

Since $0 \leq \eta \leq 1$, the absolute temperature of any reservoir is always greater than zero. To obtain a uniform temperature scale arrange a series of ordered Carnot engines, all performing the same amount of work W, so that the heat rejected by any Carnot engine is absorbed by the next one, as shown in Fig. 1.6. Obviously, for all n, we have

$$Q_{n+1} - Q_n = W$$
$$\frac{Q_n}{Q_{n+1}} = \frac{\theta_n}{\theta_{n+1}}$$

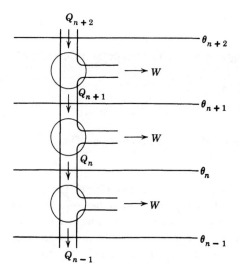

Fig. 1.6 A series of Carnot engines used to define the absolute scale of temperature.

The last equation can be rewritten in the form

$$\frac{\theta_n}{Q_n} = \frac{\theta_{n+1}}{Q_{n+1}} = x$$

Hence $x \equiv \theta_n/Q_n$ is independent of n. It is easily seen that

$$\theta_{n+1} - \theta_n = xW$$

which is independent of n. Hence a temperature scale of equal intervals results. Choosing $xW = 1$ K results in the absolute Kelvin scale of temperature. It is to be noted that

(a) The definition of the absolute scale of temperature is independent of the specific properties of any substance. It depends only on a property that is common to all substances, the second law of thermodynamics.

(b) The limit $\theta = 0$ is the greatest lower bound of the temperature scale and is called the absolute zero. Actually no Carnot engine exists with absolute zero as the temperature of the lower reservoir, for that would violate the second law. The absolute zero exists only in a limiting sense.

(c) The absolute Kelvin scale θ is identical with the ideal-gas temperature scale T, if $T > 0$. This is easily proved by using an ideal gas to form a Carnot engine. From now on we do not distinguish between the two and denote the absolute temperature by T.

1.4 ENTROPY

The second law of thermodynamics enables us to define a state function S, the entropy, which we shall find useful. We owe this possibility to the following theorem.

CLAUSIUS' THEOREM

In *any* cyclic transformation throughout which the temperature is defined, the following inequality holds:

$$\oint \frac{dQ}{T} \leq 0$$

where the integral extends over one cycle of the transformation. The equality holds if the cyclic transformation is reversible.

Proof Let the cyclic transformation in question be denoted by $\mathcal{O}$. Divide the cyclic into n infinitesimal steps for which the temperature may be considered to be constant in each step. The system is imagined to be brought successively into contact with heat reservoirs at temperatures $T_1, T_2, \ldots, T_n$. Let Q_i be the amount of heat absorbed by the system during the ith step from the heat reservoir of temperature T_i. We shall prove that

$$\sum_{i=1}^{n} \left(\frac{Q_i}{T_i} \right) \leq 0$$

The theorem is obtained as we let $n \to \infty$. Construct a set of n Carnot engines $\{C_1, C_2, \ldots, C_n\}$ such that C_i

(a) operates between T_i and T_o $(T_o \geq T_i,$ all $i)$,
(b) absorbs amount of heat $Q_i^{(o)}$ from T_o,
(c) rejects amount of heat Q_i to T_i.

We have, by definition of the temperature scale.

$$\frac{Q_i^{(o)}}{Q_i} = \frac{T_o}{T_i}$$

Consider one cycle of the combined operation $\mathcal{O} + \{C_1 + \cdots + C_n\}$. The net result of this cycle is that an amount of heat

$$Q_o = \sum_{i=1}^{n} Q_i^{(o)} = T_o \sum_{i=1}^{n} \left(\frac{Q_i}{T_i} \right)$$

is absorbed from the reservoir T_o and converted entirely into work, with no other effect. According to the second law this is impossible unless $Q_o \leq 0$. Therefore

$$\sum_{i=1}^{n} \left(\frac{Q_i}{T_i} \right) \leq 0$$

This proves the first part of the theorem.

If $\mathcal{O}$ is reversible, we reverse it. Going through the same arguments, we arrive at the same inequality except that the signs of Q_i are reversed:

$$- \sum_{i=1}^{n} \left(\frac{Q_i}{T_i} \right) \leq 0$$

Combining this with the previous inequality (which of course still holds for a reversible $\mathcal{O}$) we obtain

$$\sum_{i=1}^{n} \left(\frac{Q_i}{T_i} \right) = 0 \qquad\blacksquare$$

COROLLARY
For a reversible transformation, the integral

$$\int \frac{dQ}{T}$$

is independent of the path and depends only on the initial and final states of the transformation.

Proof Let the initial state be A and the final state be B. Let I, II denote two arbitrary reversible paths joining A to B, and let II′ be the reverse of II. Clausius' theorem implies that

$$\int_{\text{I}} \frac{dQ}{T} + \int_{\text{II}'} \frac{dQ}{T} = 0$$

But

$$\int_{\text{II}'} \frac{dQ}{T} = -\int_{\text{II}} \frac{dQ}{T}$$

Hence,

$$\int_{\text{I}} \frac{dQ}{T} = \int_{\text{II}} \frac{dQ}{T} \qquad\blacksquare$$

This corollary enables us to define a state function, the entropy S. It is defined as follows. Choose an arbitrary fixed state O as reference state. The entropy $S(A)$ for any state A is defined by

$$S(A) \equiv \int_{O}^{A} \frac{dQ}{T}$$

where the path of integration is any reversible path joining O to A. Thus the entropy is defined only up to an arbitrary additive constant.* The difference in the entropy of two states, however, is completely defined:

$$S(A) - S(B) = \int_{B}^{A} \frac{dQ}{T}$$

where the path of integration is any reversible path jointing B to A. It follows from this formula that in any infinitesimal *reversible* transformation the change in S is given by

$$dS = \frac{dQ}{T}$$

which is an exact differential.

We note the following properties of the entropy:

(a) For an arbitrary transformation,

$$\int_{A}^{B} \frac{dQ}{T} \leq S(B) - S(A)$$

The equality holds if the transformation is reversible.

---

*This definition depends on the assumption that every equilibrium state A is accessible from any state O through a reversible transformation. In other words, the surface of the equation of state consists of one continuous sheet. If the surface of the equation of state consists of two disjoint sheets, our definition only defines S for each sheet up to an additive constant. The absolute value of this constant, which becomes relevant when we go from one sheet to another, is the subject of the third law of thermodynamics, which states that at absolute zero $S = 0$ for all sheets.

Fig. 1.7 Reversible path R and irreversible path I connecting states A and B.

Proof Let R and I denote respectively any reversible and any irreversible path joining A to B, as shown in Fig. 1.7. For path R the assertion holds by definition of S. Now consider the cyclic transformation made up of I plus the reverse of R. From Clausius' theorem we have

$$\int_I \frac{dQ}{T} - \int_R \frac{dQ}{T} \leq 0$$

or

$$\int_I \frac{dQ}{T} \leq \int_R \frac{dQ}{T} \equiv S(B) - S(A) \qquad \blacksquare$$

(*b*) The entropy of a thermally isolated system never decreases.

Proof A thermally isolated system cannot exchange heat with the external world. Therefore $dQ = 0$ for any transformation. By the previous property we immediately have

$$S(B) - S(A) \geq 0$$

The equality holds if the transformation is reversible. $\qquad \blacksquare$

An immediate consequence of this is that for a thermally isolated system the state of equilibrium is the state of maximum entropy consistent with external constraints.

For a physical interpretation of the entropy, we consider the following example. One mole of ideal gas expands isothermally from volume V_1 to V_2 by two routes: Reversible isothermal expansion and irreversible free expansion. Let us calculate the change of entropy of the gas and of the external surroundings.

Reversible Isothermal Expansion. The arrangement is illustrated in Fig. 1.8. In the P-V diagram the states of the gas (and not its surroundings) are represented. Since the gas is ideal, $U = U(T)$. Hence $\Delta U = 0$. The amount of heat absorbed is equal to the work done, which is the shaded area in the P-V diagram:

$$\Delta Q = RT \log \frac{V_2}{V_1}$$

Hence,

$$(\Delta S)_{\text{gas}} = \int \frac{dQ}{T} = \frac{\Delta Q}{T} = R \log \frac{V_2}{V_1}$$

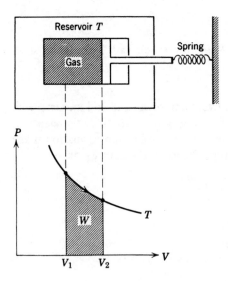

Fig. 1.8 Reversible isothermal expansion of an ideal gas.

The reservoir supplies the amount of heat $-\Delta Q$. Hence,

$$(\Delta S)_{\text{reservoir}} = -\frac{\Delta Q}{T} = -R \log \frac{V_2}{V_1}$$

The change in entropy of the whole arrangement is zero. An amount of work

$$W = \Delta Q = RT \log \frac{V_2}{V_1}$$

is stored in the spring connected to the piston. This can be used to compress the gas, reversing the transformation.

Free Expansion. This process is illustrated in Fig. 1.3. The initial and final states are the same as in the reversible isothermal expansion. Therefore $(\Delta S)_{\text{gas}}$ is the same as in that case because S is a state function. Thus

$$(\Delta S)_{\text{gas}} = R \log \frac{V_2}{V_1}$$

Since no heat is supplied by the reservoir we have

$$(\Delta S)_{\text{reservoir}} = 0$$

which leads to an increase of entropy of the entire system of gas plus reservoir:

$$(\Delta S)_{\text{total}} = R \log \frac{V_2}{V_1}$$

In comparison with the previous case, an amount of useful energy

$$W = T(\Delta S)_{\text{total}}$$

is "wasted," for it could have been extracted by expanding the gas reversibly. This example illustrates the fact that irreversibility is generally "wasteful," and is marked by an increase of entropy of the total system under consideration. For this reason the entropy of a state may be viewed as a measure of the unavailability of useful energy in that state.

It may be noted in passing that heat conduction is an irreversible process and thus increases the total entropy. Suppose a metal bar conducts heat from reservoir T_2 to reservoir T_1 at the rate of Q per second. The net increase in entropy per second of the entire system under consideration is

$$Q\left(\frac{1}{T_1} - \frac{1}{T_2}\right) > 0$$

The only reversible way to transfer heat is to operate a Carnot engine between the two reservoirs.

We might indulge in the following thought. The entropy of the entire universe, which is as isolated a system as exists, can never decrease. Furthermore, we have ample evidence, by just looking around us, that the universe is not unchanging, and that most changes are irreversible. It follows that the entropy of the universe constantly increases, and will lead relentlessly to a "heat death" of the universe—a state of maximum entropy. Is this the fate of the universe? In a universe in which the second law of thermodynamics is rigorously correct, the affirmative answer is inescapable. In fact, however, ours is not such a universe, although this conclusion cannot be arrived at within thermodynamics.

Our universe is governed by molecular laws, whose invariance under time reversal denies the existence of any natural phenomenon that absolutely distinguishes between the past and the future. The proper answer to the question we posed is no. The reason is that the second law of thermodynamics cannot be a rigorous law of nature.

This leads to the new question, "In what sense, and to what extent, is the second law of thermodynamics correct?" We examine this question in our discussion of kinetic theory (see Section 4.4) where we see that the second law of thermodynamics is correct "on the average," and that in macroscopic phenomena deviations from this law are so rare that for all practical purposes they never occur.

1.5 SOME IMMEDIATE CONSEQUENCES OF THE SECOND LAW

The consequences of the second law that we discuss here are based on the fact that dS is an exact differential. Let us first recall one of the dQ equations, (1.7):

$$dQ = C_V\, dT + \left[\left(\frac{\partial U}{\partial V}\right)_T + P\right] dV$$

Putting $dQ = T\,dS$, we obtain

$$dS = \left(\frac{C_V}{T}\right) dT + \frac{1}{T}\left[\left(\frac{\partial U}{\partial V}\right)_T + P\right] dV \tag{1.10}$$

Since dS is an exact differential, we must have

$$\left(\frac{\partial}{\partial V}\right)_T \left(\frac{C_V}{T}\right) = \left(\frac{\partial}{\partial T}\right)_V \left[\frac{1}{T}\left(\frac{\partial U}{\partial V}\right)_T + \frac{P}{T}\right] \tag{1.11}$$

Putting $C_V = (\partial U/\partial T)_V$ and carrying out the differentiations on the right-hand side, we obtain, after some algebra,

$$\left(\frac{\partial U}{\partial V}\right)_T = T\left(\frac{\partial P}{\partial T}\right)_V - P \tag{1.12}$$

It is now possible to calculate $(\partial U/\partial V)_T$ for an ideal gas. Using the equation of state $P = NkT/V$, we obtain

$$\left(\frac{\partial U}{\partial V}\right)_T = \frac{NkT}{V} - P = 0$$

Hence U is a function of T alone. This was earlier deduced from Joule's free expansion experiment with the help of the first law. We now see that it is a logical consequence of the second law.

Substitution of (1.12) into (1.10) yields

$$T\,dS = C_V\,dT + T\left(\frac{\partial P}{\partial T}\right)_V dV \tag{1.13}$$

Going through similar steps for another form of the dQ equation, equation (1.6), we obtain

$$T\,dS = C_P\,dT - T\left(\frac{\partial V}{\partial T}\right)_P dP \tag{1.14}$$

It is possible to re-express (1.13) and (1.14) in such a way that only quantities that are conveniently measurable appear in these equations. To do this we need the following mathematical lemma, which we state without proof.

LEMMA

Let x, y, z be quantities satisfying a functional relation $f(x, y, z) = 0$. Let w be a function of any two of x, y, z. Then

(a)
$$\left(\frac{\partial x}{\partial y}\right)_w \left(\frac{\partial y}{\partial z}\right)_w = \left(\frac{\partial x}{\partial z}\right)_w$$

(b)
$$\left(\frac{\partial x}{\partial y}\right)_z = \frac{1}{\left(\dfrac{\partial y}{\partial x}\right)_z}$$

(c)
$$\left(\frac{\partial x}{\partial y}\right)_z \left(\frac{\partial y}{\partial z}\right)_x \left(\frac{\partial z}{\partial x}\right)_y = -1 \quad \text{(chain relation)}$$

Let us define the following quantities, which are experimentally measurable.

$$\alpha \equiv \frac{1}{V}\left(\frac{\partial V}{\partial T}\right)_P \qquad \text{(coefficient of thermal expansion)} \qquad (1.15)$$

$$\kappa_T \equiv -\frac{1}{V}\left(\frac{\partial V}{\partial P}\right)_T \qquad \text{(isothermal compressibility)} \qquad (1.16)$$

$$\kappa_S \equiv -\frac{1}{V}\left(\frac{\partial V}{\partial P}\right)_S \qquad \text{(adiabatic compressibility)} \qquad (1.17)$$

Using the lemma we have

$$\left(\frac{\partial P}{\partial T}\right)_V = -\frac{1}{(\partial T/\partial V)_P(\partial V/\partial P)_T} = \frac{(\partial V/\partial T)_P}{-(\partial V/\partial P)_T} = \frac{\alpha}{\kappa_T} \qquad (1.18)$$

Equations (1.13) and (1.14) can now be written in the desired forms:

$$T\,dS = C_V\,dT + \frac{\alpha T}{\kappa_T}\,dV \qquad (1.19)$$

$$T\,dS = C_P\,dT - \alpha TV\,dP \qquad (1.20)$$

These are known as the $T\,dS$ equations.

Next we try to express $C_P - C_V$ for any substance in terms of other experimental quantities. Equating the right-hand side of (1.13) to that of (1.14), we have

$$C_V\,dT + T\left(\frac{\partial P}{\partial T}\right)_V dV = C_P\,dT - T\left(\frac{\partial V}{\partial T}\right)_P dP$$

Choosing P and V to be the independent variables, we write

$$dT = \left(\frac{\partial T}{\partial V}\right)_P dV + \left(\frac{\partial T}{\partial P}\right)_V dP$$

Substitution of this into the previous equation yields

$$\left[(C_P - C_V)\left(\frac{\partial T}{\partial V}\right)_P - T\left(\frac{\partial P}{\partial T}\right)_V\right] dV$$

$$+ \left[(C_P - C_V)\left(\frac{\partial T}{\partial P}\right)_V - T\left(\frac{\partial V}{\partial T}\right)_P\right] dP = 0$$

Since dV and dP are independent of each other, the coefficients of dV and dP must separately vanish. From the first of these we obtain

$$C_P - C_V = \frac{T(\partial P/\partial T)_V}{(\partial T/\partial V)_P} = -T\left[\left(\frac{\partial V}{\partial T}\right)_P\right]^2\left(\frac{\partial P}{\partial V}\right)_T$$

where the chain relation was used. Therefore

$$C_P - C_V = \frac{TV\alpha^2}{\kappa_T} \qquad (1.21)$$

This shows that $(C_P - C_V) > 0$ if $\kappa_T \geq 0$. From experience we know that $\kappa_T \geq 0$ for most substances, but this fact is implied by neither the first law nor the second law. It can be proven in statistical mechanics, where use is made of the nature of intermolecular forces, and where it is known as Van Hove's theorem.

Finally we consider $\gamma \equiv C_P/C_V$. Equations (1.13) and (1.14) remain valid for adiabatic transformations for which $dS = 0$. The following expressions for C_V and C_P are therefore true:

$$C_V = -T\left(\frac{\partial P}{\partial T}\right)_V \left(\frac{\partial V}{\partial T}\right)_S$$

$$C_P = T\left(\frac{\partial V}{\partial T}\right)_P \left(\frac{\partial P}{\partial T}\right)_S$$

dividing one by the other, we find

$$\frac{C_P}{C_V} = -\frac{(\partial V/\partial T)_P(\partial P/\partial T)_S}{(\partial P/\partial T)_V(\partial V/\partial T)_S} = -\frac{(\partial V/\partial T)_P}{(\partial P/\partial T)_V}\left(\frac{\partial P}{\partial V}\right)_S = \frac{(\partial V/\partial P)_T}{(\partial V/\partial P)_S}$$

(1.22)

Combining (1.21) and (1.22), we can solve for C_V and C_P separately:

$$C_V = \frac{TV\alpha^2 \kappa_S}{(\kappa_T - \kappa_S)\kappa_T}$$

(1.23)

$$C_P = \frac{TV\alpha^2}{\kappa_T - \kappa_S}$$

(1.24)

1.6 THERMODYNAMIC POTENTIALS

We introduce two auxiliary state functions, the Helmholtz free energy A and the Gibbs thermodynamic potential G (or Gibbs free energy). They are defined as follows:

$$A = U - TS \tag{1.25}$$

$$G = A + PV \tag{1.26}$$

They are useful in determining the equilibrium state of a system that is not isolated. We discuss their significances separately.

The physical meaning of the free energy A is furnished by the fact that in an isothermal transformation the change of the free energy is the negative of the maximum possible work done by the system. To see this, let a system undergo an arbitrary isothermal transformation from state A to state B. We have from the second law that

$$\int_A^B \frac{dQ}{T} \leq S(B) - S(A)$$

or, since T is constant,

$$\frac{\Delta Q}{T} \leq \Delta S$$

where ΔQ is the amount of heat absorbed during the transformation and $\Delta S = S(B) - S(A)$. By use of the first law we can rewrite the inequality just given as follows:

$$W \leq -\Delta U + T\Delta S \tag{1.27}$$

where W is the work done by the system. Since the right side is none other than $-\Delta A$, we have

$$W \leq -\Delta A \tag{1.28}$$

The equality holds if the transformation is reversible.   ∎

Suppose $W = 0$; then (1.28) reduces to a useful theorem.

THEOREM

For a mechanically isolated system kept at constant temperature the Helmholtz free energy never increases.

COROLLARY

For a mechanically isolated system kept at constant temperature the state of equilibrium is the state of minimum Helmholtz free energy.

In an infinitesimal reversible transformation it is easily verified that

$$dA = -P\,dV - S\,dT \tag{1.29}$$

From this follow the relations

$$P = -\left(\frac{\partial A}{\partial V}\right)_T \tag{1.30}$$

$$S = -\left(\frac{\partial A}{\partial T}\right)_V \tag{1.31}$$

which are members of a class of relations known as Maxwell relations. If the function $A(V, T)$ is known, then P and S are calculable by (1.30) and (1.31).

As an example of the principle of minimization of free energy, consider a gas in a cylinder kept at constant temperature. A sliding piston divides the total volume V into two parts V_1 and V_2, in which the pressures are respectively P_1 and P_2. If the piston is released and allowed to slide freely, what is its equilibrium position? By the principle just stated the position of the piston must minimize the free energy of the total system. Suppose equilibrium has been established. Then a slight change in the position of the piston should not change the free energy, since it is at a minimum. That is, $\delta A = 0$. Now A is a function of V_1, V_2, and T. Hence

$$0 = \delta A = \left(\frac{\partial A}{\partial V_1}\right)_T \delta V_1 + \left(\frac{\partial A}{\partial V_2}\right)_T \delta V_2$$

Since $V_1 + V_2 = V$ remains constant, we must have $\delta V_1 = -\delta V_2$. Therefore

$$0 = \left[\left(\frac{\partial A}{\partial V_1} \right)_T - \left(\frac{\partial A}{\partial V_2} \right)_T \right] \partial V_1$$

As δV_1 is arbitrary, its coefficient must vanish. We thus obtain the equilibrium condition

$$\left(\frac{\partial A}{\partial V_1} \right)_T = \left(\frac{\partial A}{\partial V_2} \right)_T$$

which becomes, through use of (1.30),

$$P_1 = P_2$$

a result that is intuitively obvious.

We consider now the Gibbs potential. Its importance lies with the following theorem.

THEOREM

For a system kept at constant temperature and pressure the Gibbs potential never increases.

COROLLARY

For a system kept at constant temperature and pressure the state of equilibrium is the state of minimum Gibbs potential.

Proof If T is kept constant, then in any transformation

$$W \le -\Delta A$$

Now specialize the situation further by keeping the pressure constant, thereby making $W = P\Delta V$. We then have

$$P\Delta V + \Delta A \le 0$$
$$\Delta G \le 0 \qquad\qquad \blacksquare$$

In a infinitesimal reversible transformation

$$dG = -S\,dT + V\,dP \qquad (1.32)$$

From this we immediately obtain more Maxwell relations:

$$S = -\left(\frac{\partial G}{\partial T} \right)_P \qquad (1.33)$$

$$V = \left(\frac{\partial G}{\partial P} \right)_T \qquad (1.34)$$

Still two more Maxwell relations may be obtained by considering the differential changes of the enthalpy:

$$H = U + PV$$
$$dH = T\,dS + V\,dP \qquad (1.35)$$

from which follow

$$V = \left(\frac{\partial H}{\partial P} \right)_S \tag{1.36}$$

$$T = \left(\frac{\partial H}{\partial S} \right)_P \tag{1.37}$$

Further Maxwell relations are

$$T = \left(\frac{\partial U}{\partial S} \right)_V \tag{1.38}$$

and

$$P = -\left(\frac{\partial U}{\partial V} \right)_S \tag{1.39}$$

which follow from the first law, $dU = -P\,dV + T\,dS$.

The eight Maxwell relations may be conveniently summarized by the following diagram:

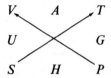

The functions A, G, H, U are flanked by their respective natural arguments, for example, $A = A(V, T)$. The derivative with respect to one argument, with the other held fixed, may be found by going along a diagonal line, either with or against the direction of the arrow. Going against the arrow yields a minus sign; for example, $(\partial A/\partial T)_V = -S$, $(\partial A/\partial V)_T = -P$.

1.7 THE THIRD LAW OF THERMODYNAMICS

The second law of thermodynamics enables us to define the entropy of a substance up to an arbitrary additive constant. The definition of the entropy depends on the existence of a reversible transformation connecting an arbitrarily chosen reference state O to the state A under consideration. Such a reversible transformation always exists if both O and A lie on one sheet of the equation of state surface. If we consider two different substances, or metastable phases of the same substance, the equation of state surface may consist of more than one disjoint sheets. In such cases the kind of reversible path we have mentioned may not exist. Therefore the second law does not uniquely determine the difference in entropy of two states A and B, if A refers to one substance and B to another. In 1905 Nernst supplied a rule for such a determination. This rule has since been called the *third law of thermodynamics*. It states:

The entropy of a system at absolute zero is a universal constant, which may be taken to be zero.

The generality of this statement rests in the facts that (a) it refers to any system, and that (b) it states that $S = 0$ at $T = 0$, regardless of the values of any other parameter of which S may be a function. It is obvious that the third law renders the entropy of any state of any system unique.

The third law immediately implies that any heat capacity of a system must vanish at absolute zero. Let R be any reversible path connecting a state of the system at absolute zero to the state A, whose entropy is to be calculated. Let $C_R(T)$ be the heat capacity of the system along the path R. Then, by the second law,

$$S(A) = \int_0^{T_A} C_R(T) \frac{dT}{T} \tag{1.40}$$

But according to the third law

$$S(A) \underset{T_A \to 0}{\to} 0 \tag{1.41}$$

Hence we must have

$$C_R(T) \underset{T \to 0}{\to} 0 \tag{1.42}$$

In particular, C_R may be C_V or C_P. The statement (1.42) is experimentally verified for all substances so far examined.

A less obvious consequence of the third law is that at absolute zero the coefficient of thermal expansion of any substance vanishes. This may be shown as follows. From the $T\,dS$ equations we can deduce the equalities

$$\left(\frac{\partial S}{\partial T}\right)_P = \frac{C_P}{T} \tag{1.43}$$

$$\left(\frac{\partial S}{\partial P}\right)_T = -\left(\frac{\partial V}{\partial T}\right)_P \tag{1.44}$$

Combining these we arrive at

$$\left(\frac{\partial C_P}{\partial P}\right)_T = -T\left(\frac{\partial^2 V}{\partial T^2}\right)_P \tag{1.45}$$

From (1.44) and (1.40) we have, for the coefficient of thermal expansion α, the expression

$$V\alpha \equiv \left(\frac{\partial V}{\partial T}\right)_P = -\left(\frac{\partial S}{\partial P}\right)_T = -\frac{\partial}{\partial P}\int_0^T C_P\frac{dT}{T} = -\int_0^T \left(\frac{\partial C_P}{\partial P}\right)_T \frac{dT}{T} \tag{1.46}$$

where the integrations proceed along a path of constant P. Using (1.45), we rewrite this as

$$V\alpha = \int_0^T \left(\frac{\partial^2 V}{\partial T^2}\right)_P dT = \left(\frac{\partial V}{\partial T}\right)_P - \left[\left(\frac{\partial V}{\partial T}\right)_P\right]_{T=0} \tag{1.47}$$

Therefore

$$\alpha \underset{T \to 0}{\to} 0 \tag{1.48}$$

In a similar fashion we can show that

$$\left(\frac{\partial P}{\partial T}\right)_V \underset{T\to 0}{\to} 0 \tag{1.49}$$

Combined with (1.48), this implies that on the P-T diagram the melting curve has zero tangent at $T = 0$.

It is experimentally found that C_P can be represented by the following series expansion at low temperatures:

$$C_P = T^x(a + bT + cT^2 + \cdots) \tag{1.50}$$

where x is a positive constant, and $a, b, c, \ldots$ are functions of P. Differentiating (1.50) with respect to P, we find that

$$\left(\frac{\partial C_P}{\partial P}\right)_T = T^x(a' + b'T + c'T^2 + \cdots) \tag{1.51}$$

Substituting this into (1.46), we have

$$V\alpha = -\int_0^T dT(a'T^{x-1} + b'T^x + \cdots) = -T^x\left(\frac{a'}{x} + \frac{b'T}{x+1} + \frac{c'T^2}{x+2} + \cdots\right)$$

Hence

$$\frac{V\alpha}{C_P} \underset{T\to 0}{\to} \text{ finite constant} \tag{1.52}$$

This has the consequence that a system cannot be cooled to absolute zero by a finite change of the thermodynamic parameters. For example, from one of the $T\,dS$ equations we find that through an adiabatic change dP of the pressure, the temperature changes by

$$dT = \left(\frac{V\alpha}{C_P}\right)T\,dP \tag{1.53}$$

By virtue of (1.52), the change of P required to produce a finite change in the temperature is unbounded as $T \to 0$.

The unattainability of absolute zero is sometimes stated as an alternative formulation of the third law. This statement is independent of the second law, for the latter only implies that there exists no Carnot engine whose lower reservoir is at absolute zero.* Whether it is possible to make a system approach absolute zero from a higher temperature is an independent question. According to (1.53), it depends on the behavior of the specific heat, of which the second law says nothing.

Before experimental techniques were well developed for low temperatures, it was generally believed that heat capacities of substances remain constant down

---

*The second law requires that in any reversible transformation we must have $dQ = T\,dS$, where dS is an exact differential. Hence $dQ = 0$ when $T = 0$. That is, all transformations at absolute zero are adiabatic. Even if we had a system at absolute zero, there would still be no reversible way to heat it to a higher temperature. Thus we cannot construct a Carnot engine whose lower reservoir is at absolute zero.

to absolute zero, as classical kinetic theory predicts (i.e., $x = 0$ in (1.50)). If this were so, we see directly from (1.53) that the unattainability of absolute zero would be automatic. This is why the question did not receive attention until the turn of the century, when it was discovered that heat capacities tend to vanish at low temperatures. We now see that even if the heat capacities vanish at absolute zero, the absolute zero is still unattainable.

We see, when we come to quantum statistical mechanics, that the third law of thermodynamics is a macroscopic manifestation of quantum effects (see Section 8.4). The foregoing discussions, which are somewhat abstract, become concrete and physical when they are presented in the context of quantum statistical mechanics. The importance of the third law of thermodynamics, therefore, does not lie in these abstract considerations, but in its practical usefulness. We end our discussion of the third law with one of its applications.

The free energy of a system is defined as

$$A = U - TS \tag{1.54}$$

which, according to the third law, can be written in the form

$$A = U - T \int_0^T \frac{C_V}{T'} \, dT' \tag{1.55}$$

where the integral in the second term extends over a path of constant volume. There is no arbitrary additive constant except the one already contained in U. This formula, together with

$$U = \int_0^T C_V \, dT' + \text{constant} \tag{1.56}$$

enables us to determine both U and A up to the same arbitrary additive constant from measurements of C_V.

To illustrate the practical use of these formulas we consider the melting point of solid quartz. The stable phase of quartz at low temperatures is a crystalline solid. The liquid phase (glass), however, can be supercooled and can exist in metastable equilibrium far below the melting point. Hence a direct measurement of the melting point is difficult. It can, however, be determined indirectly through the use of (1.55). Let the specific heat c_V of both solid and liquid quartz be measured through a range of temperatures at a fixed volume V. Let Δc_V denote their difference, which is a function of temperature. Then the difference in internal energy per unit mass of the two phases is obtained by numerically integrating Δc_V at constant V:

$$\Delta u = \int \Delta c_V \, dT'$$

Using (1.56) we have, for the difference in free energy per unit mass of the two phases,

$$\Delta a = \Delta u - T \int_0^T \frac{\Delta c_V}{T'} \, dT'$$

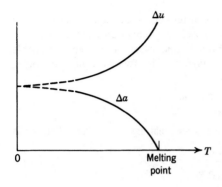

Fig. 1.9 Determination of the melting point through use of the third law.

Plotting Δu and Δa as a function of T at a fixed V, we should obtain a graph that looks qualitatively like that shown in Fig. 1.9. The melting point is the temperature at which $\Delta a = 0$, since the condition of phase equilibrium at fixed T and V is the equality of the free energies per unit mass. In practice the point at which $\Delta a = 0$ may be obtained either by direct integration up to that point, or by extrapolation.

PROBLEMS

1.1 Find the equations governing an adiabatic transformation of an ideal gas.

1.2 (*a*) An engine is represented by the cyclic transformation shown in the accompanying T-S diagram, where A denotes the area of the shaded region and B the area of the region below it. Show that this engine is not as efficient as a Carnot engine operating between the highest and the lowest available temperatures.
(*b*) Show that an arbitrary reversible engine cannot be more efficient than a Carnot engine operating between the highest and the lowest available temperatures.

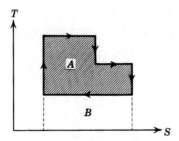

1.3 Let a real gas undergo the free-expansion process. Supply all relevant arguments to show that the change in temperature ΔT is related to the change in the volume ΔV by the formula

$$\Delta T = \left(\frac{\partial T}{\partial V} \right)_U \Delta V$$

where ΔT, ΔV are small quantities. In particular, explain why this formula looks the same as that for a reversible process, although the process under consideration is irreversible.

1.4 Two isotherms of 1 mol of a substance that can undergo a gas-liquid transition are shown in the accompanying P-V diagram. The absolute temperatures are T_2 and T_1, respectively. The substance is made to go through one cycle of a cyclic reversible transformation $ABCDEF$, as indicated in the diagram. The following information is given:

   (i) ABC and DEF are isothermal transformations.

   (ii) FA and CD are adiabatic transformations.

   (iii) In the gas phase ($BCDE$) the substance is an ideal gas. At A the substance is pure liquid.

   (iv) Latent heat along AB: $L = 200$ cal/mol

$$T_2 = 300 \text{ K}$$
$$T_1 = 150 \text{ K}$$
$$V_A = 0.5 \text{ liter}$$
$$V_B = 1 \text{ liter}$$
$$V_C = 2.71828 \text{ liter}$$

Calculate the net amount of work done by the substance.

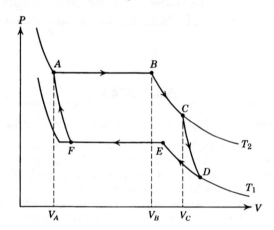

1.5 A substance has the following properties:

   (i) At a constant temperature T_0 the work done by it on expansion from V_0 to V is

$$W = RT_0 \log \frac{V}{V_0}$$

   (ii) The entropy is given by

$$S = R\frac{V}{V_0}\left(\frac{T}{T_0}\right)^a$$

   where V_0, T_0, and a are fixed constants.

(**a**) Calculate the Helmholtz free energy.

(**b**) Find the equation of state.

(**c**) Find the work done at an arbitrary constant temperature T.

2

SOME APPLICATIONS
OF THERMODYNAMICS

2.1 THERMODYNAMIC DESCRIPTION OF PHASE TRANSITIONS

The surface of the equation of state of a typical substance is shown in Fig. 2.1, where the shaded areas are cylindrical surfaces, representing regions of phase transition. The P-V and P-T diagrams are shown in Fig. 2.2. We study here the implications of the second law for these phase transitions.

Let us consider the transition between the gas phase and the liquid phase. The transition takes place at a constant temperature and pressure, as shown in Fig. 2.3. This pressure $P(T)$ is called the vapor pressure at the temperature T. Let the system be initially in state 1, where it is all liquid. When heat is added to the system, some of the liquid will be converted into gas, and so on until we reach state 2, where the system is all gas, as schematically shown in Fig. 2.4. The important facts are that

(a) during the phase transition both P and T remain constant;

(b) in the gas-liquid mixture the liquid exists in the same state as at 1 and the gas exists in the same state as at 2.

As a result, knowing the properties of the states 1 and 2 suffices for a complete description of the phase transition. The isotherm in the P-V diagram is horizontal during the phase transition, because the gas phase has a smaller density than the liquid phase. Consequently, when a certain mass of liquid is converted into gas, the total volume of the system expands, although P and T remain unchanged. Such a transition is known as a "first-order transition."

The dependence of the vapor pressure $P(T)$ on the temperature may be found by applying the second law. Consider a gas-liquid mixture in equilibrium at temperature T and vapor pressure $P(T)$. Let the mass of the liquid be m_1 and the mass of the gas be m_2. If the system is in equilibrium with the given T and

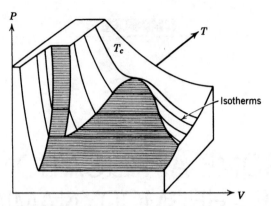

Fig. 2.1 Surface of equation of state of a typical substance (not to scale).

$P(T)$, the Gibbs potential of this state must be at a minimum. That is, if any parameters other than T and P are varied slightly, we must have $\delta G = 0$. Let us vary the composition of the mixture by converting an amount δm of liquid to gas, so that

$$-\delta m_1 = \delta m_2 = \delta m \qquad (2.1)$$

The total Gibbs potential of the gas-liquid mixture may be represented, with neglect of surface effects, as

$$G = m_1 g_1 + m_2 g_2 \qquad (2.2)$$

where g_1 is the Gibbs potential per unit mass of the liquid in state 1 and g_2 is that for the gas in state 2. They are also called *chemical potentials*. They are independent of the total mass of the phases but may depend on the density of the phases (which, however, are not altered when we transfer mass from one phase to the other). Thus

$$\delta G = 0 = -(g_1 - g_2)\,\delta m$$

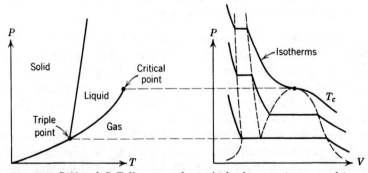

Fig. 2.2 $P\text{-}V$ and $P\text{-}T$ diagrams of a typical substance (not to scale).

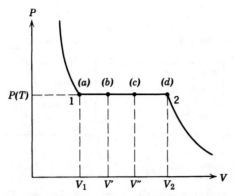

Fig. 2.3 An isotherm exhibiting a phase transition.

The condition for equilibrium is then

$$g_1 = g_2 \tag{2.3}$$

This condition determines the vapor pressure, as we shall see.

The chemical potentials $g_1(P, T)$ and $g_2(P, T)$ are two state functions of the liquid and gas respectively. Recall that in each phase we have

$$\left(\frac{\partial g}{\partial T}\right)_P = -s \quad \text{(entropy per unit mass)} \tag{2.4}$$

$$\left(\frac{\partial g}{\partial P}\right)_T = v \quad \text{(volume per unit mass)} \tag{2.5}$$

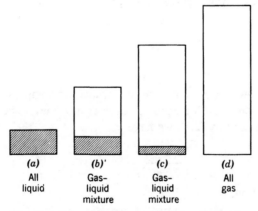

Fig. 2.4 Schematic illustration of a first-order phase transition. The temperature and the pressure of the system remain constant throughout the transition. The total volume of the system changes as the relative amount of the substance in the two phases changes, because the two phases have different densities.

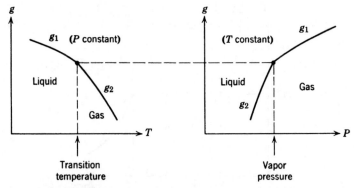

Fig. 2.5 Chemical potentials g_1, g_2 for the two phases in a first-order phase transition.

We see that the first derivative of g_1 is different from that of g_2 at the transition temperature and pressure:

$$\left[\frac{\partial (g_2 - g_1)}{\partial T} \right]_P = -(s_2 - s_1) < 0 \tag{2.6}$$

$$\left[\frac{\partial (g_2 - g_1)}{\partial P} \right]_T = v_2 - v_1 > 0 \tag{2.7}$$

This is why the transition is called "first-order." The behavior of $g_1(P, T)$ and $g_2(P, T)$ are qualitatively sketched in Fig. 2.5.

To determine the vapor pressure we proceed as follows. Let

$$\begin{aligned}
\Delta g &= g_2 - g_1 \\
\Delta s &= s_2 - s_1 \\
\Delta v &= v_2 - v_1
\end{aligned} \tag{2.8}$$

where all quantities are evaluated at the transition temperature T and vapor pressure P. The condition for equilibrium is that T and P be such as to make $\Delta g = 0$. Dividing (2.6) by (2.7), we have

$$\frac{(\partial \Delta g / \partial T)_P}{(\partial \Delta g / \partial P)_T} = -\frac{\Delta s}{\Delta v} \tag{2.9}$$

By the chain relation,

$$\left(\frac{\partial \Delta g}{\partial T} \right)_P \left(\frac{\partial T}{\partial P} \right)_{\Delta g} \left(\frac{\partial P}{\partial \Delta g} \right)_T = -1$$

or

$$\frac{(\partial \Delta g / \partial T)_P}{(\partial \Delta g / \partial P)_T} = -\left(\frac{\partial P}{\partial T} \right)_{\Delta g} \tag{2.10}$$

The reason the chain relation is valid here is that Δg is a function of T and P, and hence there must exist a relation of the form $f(T, P, \Delta g) = 0$. The derivative

$$\frac{dP(T)}{dT} = \left(\frac{\partial P}{\partial T}\right)_{\Delta g = 0} \tag{2.11}$$

is precisely the derivative of the vapor pressure with respect to temperature under equilibrium conditions, for Δg is held fixed at the value zero. Combining (2.11), (2.10), and (2.9), we obtain

$$\frac{dP(T)}{dT} = \frac{\Delta s}{\Delta v} \tag{2.12}$$

The quantity

$$l = T\Delta s \tag{2.13}$$

is called the *latent heat of transition*. Thus

$$\frac{dP(T)}{dT} = \frac{l}{T\Delta v} \tag{2.14}$$

This is known as the *Clapeyron equation*. It governs the vapor pressure in any first-order transition.

It may happen in a phase transition that $s_2 - s_1 = 0$ and $v_2 - v_1 = 0$. When this is so the first derivatives of the chemical potentials are continuous across the transition point. Such a transition is not of the first order and would not be governed by the Clapeyron equation, and its isotherm would not have a horizontal part in the P-V diagram. Ehrenfest defines a phase transition to be an nth-order transition if, at the transition point,

$$\frac{\partial^n g_1}{\partial T^n} \neq \frac{\partial^n g_2}{\partial T^n} \quad \text{and} \quad \frac{\partial^n g_1}{\partial P^n} \neq \frac{\partial^n g_2}{\partial P^n}$$

whereas all lower derivatives are equal. A well-known example is the second-order transition in superconductivity. On the other hand many examples of phase transitions cannot be described by this scheme. Notable among these are the Curie point transition in ferromagnets, the order-disorder transition in binary alloys, and the λ transition in liquid helium. In these cases the specific heat diverges logarithmically at the transition point. Since the specific heat is related to the second derivative of g these examples cannot be characterized by the behaviors of the higher derivatives of g, because they do not exist. Modern usuage distinguishes only between first-order and higher-order transitions, and the latter are usually indiscriminately called "second-order" transitions.

2.2 SURFACE EFFECTS IN CONDENSATION

If we compress a gas isothermally, it is supposed to start condensing at a point O, as shown in the P-V diagram in Fig. 2.6. If we compress the system further, the pressure is supposed to remain constant. Actually the pressure will sometimes follow the dotted line shown in Fig. 2.6; along this dotted line the system is not

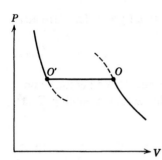

Fig. 2.6 Supersaturation and supercooling.

in stable equilibrium, however, because the slightest jar will abruptly reduce the pressure to the correct vapor pressure. Similarly, if a liquid is expanded beyond the point O', it will sometimes follow the dotted curve shown, but this too would not be a situation of stable equilibrium. These phenomena are respectively known as supersaturation and supercooling. They owe their existence to surface effects, which we have previously ignored. We give a qualitative discussion of the surface effects responsible for supersaturation.

Vapor pressure as we have defined it is the pressure at which a gas can coexist in equilibrium with an infinitely large body of its own liquid. We now denote it by $P_\infty(T)$. On the other hand, the pressure at which a gas can coexist in equilibrium with a finite droplet (of radius r) of its own liquid is not the vapor pressure $P_\infty(T)$ but a higher pressure $P_r(T)$. The difference between $P_\infty(T)$ and $P_r(T)$ is due to the surface tension of the droplet. Before we try to give a qualitative description of the mechanism of condensation, we calculate $P_r(T)$.

Suppose a droplet of liquid is placed in an external medium that exerts a pressure P on the droplet. Then the work done by the droplet on expansion is empirically given by

$$dW = P\,dV - \sigma\,da \tag{2.15}$$

where da is the increase in the surface area of the droplet and σ the coefficient of surface tension. The first law now takes the form

$$dU = dQ - P\,dV + \sigma\,da \tag{2.16}$$

Integrating this, we obtain for the internal energy of a droplet of radius r the expression

$$U = \tfrac{4}{3}\pi r^3 u_\infty + 4\pi\sigma r^2 \tag{2.17}$$

where u_∞ is the internal energy per unit volume of an infinite droplet. Correspondingly the Gibbs potential takes the form

$$G = \tfrac{4}{3}\pi r^3 g_\infty + 4\pi\sigma r^2$$

Consider a droplet of radius r in equilibrium with a gas of temperature T and pressure P. For given T and P, r must be such that the total Gibbs potential of the entire system is at a minimum. This condition determines a relation between P and r for a given T. Let the mass of the droplet be M_1 and the mass of the gas be M_2. The total Gibbs potential is of the form

$$G_{\text{total}} = M_2 g_2 + M_1 g_1 + 4\pi\sigma r^2 \qquad (2.18)$$

where g_2 and g_1 are respectively the chemical potential of an infinite body of gas and liquid. We now imagine the radius of the drop changed slightly by evaporation, so that $\delta M_1 = -\delta M_2$. The equilibrium condition is

$$\delta G_{\text{total}} = 0 = \delta M_1 \left(-g_2 + g_1 + 8\pi\sigma r \frac{\partial r}{\partial M_1} \right) \qquad (2.19)$$

Since

$$\frac{\partial r}{\partial M_1} = \frac{1}{4\pi\rho r^2} \qquad (2.20)$$

where ρ is the mass density of the droplet, we have

$$g_2 - g_1 = \frac{2\sigma}{\rho r} \qquad (2.21)$$

as the condition for equilibrium. Differentiating both sides with respect to the pressure P of the gas at constant temperature, and remembering the Maxwell relation $(\partial g/\partial P)_T = 1/\rho$, we obtain

$$\frac{1}{\rho'} - \frac{1}{\rho} = -\frac{2\sigma}{\rho r^2}\left(\frac{\partial r}{\partial P} \right)_T - \frac{2\sigma}{\rho^2 r}\left(\frac{\partial \rho}{\partial P} \right)_T \qquad (2.22)$$

where ρ' is the mass density of the gas. We assume that the gas is sufficiently dilute to be considered as ideal gas. Hence

$$\rho' = \left(\frac{m}{kT} \right) P \qquad (2.23)$$

where m is the mass of a gas atom. Further, $1/\rho$ may be neglected compared to $1/\rho'$, and $(\partial\rho/\partial P) \approx 0$. Therefore

$$\left(\frac{\partial r}{\partial P} \right)_T = -\left(\frac{kT}{m} \right)\frac{\rho r^2}{2\sigma P} \qquad (2.24)$$

Integrating both sides of this equation we find P as a function of r for a given temperature:

$$P_r(T) = P_\infty(T)\exp\left(\frac{2\sigma m}{\rho kT}\frac{1}{r} \right) \qquad (2.25)$$

which is the expression we seek. A graph of $P_r(T)$ is shown in Fig. 2.7.

Now we can give a qualitative description of what happens when a gas starts to condense. According to (2.25) only liquid droplets of a given radius r can exist in equilibrium with the gas at a given T and P. The droplets that are too large find the external pressure too high. They attempt to reduce the external pressure by gathering vapor, but this makes them grow still larger. The droplets that are too small, on the other hand, find the pressure too low, and tend to evaporate, but this makes them still smaller, and they eventually disappear. Thus, unless all

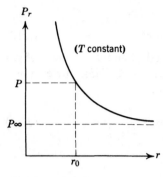

Fig. 2.7 The pressure at which a liquid droplet of radius r can exist in equilibrium with its own vapor.

droplets initially have exactly the same radius r_0 (which is unlikely), the average size will shift towards a larger value. This can only be achieved through a net condensation of vapor onto the droplets, thereby lowering pressure of the vapor. The whole process then repeats. Thus there is a self-sustaining process favoring the formation of larger and larger droplets until all the liquid forms a single body essentially infinite in size. At this point the pressure of the vapor drops to P_∞, the equilibrium vapor pressure.

The instability, as we noted, is triggered by the presence of droplets larger than a critical size. This critical size decreases as the degree of supersaturation increases (since P_r increases with the degree of supersaturation), making the presence of too-large droplets more and more likely. Finally, when P_r becomes so large that the critical size is of the order of the molecular radius, the presence of large droplets becomes a certainty, through the momentary formation of bound states of a few molecules in random collision. Supersaturation cannot be pushed beyond this point.

It is clear that supercooling can be discussed in a similar way by considering bubbles instead of droplets.

2.3 VAN DER WAALS EQUATION OF STATE

Van der Waals attempted to find a simple qualitative way to improve the equation of state of a dilute gas by incorporating the effects of molecular interaction. The result is the Van der Waals equation of state.

In most substances the potential energy between two molecules as a function of the intermolecular separation has the qualitative shape shown in Fig. 2.8. The attractive part of the potential energy originates from the mutual electric polarization of the two molecules and the repulsive part from the Coulomb repulsion of the overlapping electronic clouds of the molecules. Van der Waals idealized the situation by approximating the repulsive part by an infinite hard-sphere repulsion, so that the potential energy looks like that illustrated in Fig. 2.9. Thus each molecule is imagined to be a hard elastic sphere of diameter d surrounded by an attractive force field. The effects of the repulsive and attractive parts are then discussed separately.

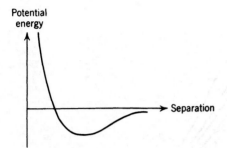

Fig. 2.8 Typical intermolecular potential.

The main effect of the hard core would be to forbid the presence of any other molecule in a certain volume centered about a molecule. If V is the total volume occupied by a substance, the effective volume available to one of its molecules would be smaller than V by the totality of such excluded volumes, which is a constant depending on the molecular diameter and the number of molecules present:

$$V_{\text{eff}} = V - b \tag{2.26}$$

where b is a constant characteristic of the substance under discussion.

The qualitative effect of the attractive part of the potential energy is a tendency for the system to form a bound state. If the attraction is sufficiently strong, the system will exist in an N-body bound state, which requires no external wall to contain it. Thus we may assume that the attraction produces a decrease in the pressure that the system exerts on an external wall. The amount of decrease is proportional to the number of pairs of molecules, within interaction range, in a layer near the wall. This in turn is roughly proportional to N^2/V^2. Since N and the range of interaction are constants, the true pressure P of the system may be decomposed into two parts:

$$P = P_{\text{kinetic}} - \frac{a}{V^2} \tag{2.27}$$

where a is another constant characteristic of the system and P_{kinetic} is defined by the equation itself.

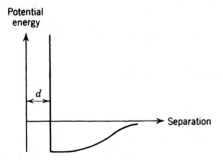

Fig. 2.9 Idealized intermolecular potential.

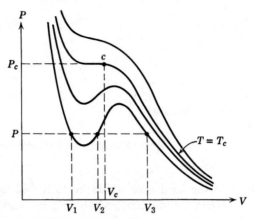

Fig. 2.10 P-V diagram of the Van der Waals equation of state.

The hypothesis of Van der Waals is that for 1 mol of the substance

$$V_{\text{eff}} P_{\text{kinetic}} = RT$$

where R is the gas constant. Therefore the equation of state is

$$(V - b)\left(P + \frac{a}{V^2}\right) = RT \tag{2.28}$$

This is the Van der Waals equation of state. Some isotherms corresponding to this equation of state are shown in Fig. 2.10. There exists a temperature T_c, called the critical temperature, at which the "kink" in the isotherm disappears. The point of inflection c is called the critical point. Its pressure P_c, volume V_c, and temperature T_c can be expressed in terms of a and b as follows. For a given T and P, (2.28) generally has three roots in V (e.g., the values V_1, V_2, V_3 shown in Fig. 2.10). As T increases these roots move together, and at $T = T_c$ they merge into V_c. Thus in the neighborhood of the critical point the equation of state must read

$$(V - V_c)^3 = 0$$

or

$$V^3 - 3V_c V^2 + 3V_c^2 V - V_c^3 = 0 \tag{2.29}$$

This is to be compared with (2.28) when we put $T = T_c$ and $P = P_c$:

$$(V - b)\left(P_c + \frac{a}{V^2}\right) = RT_c$$

or

$$V^3 - \left(b + \frac{RT_c}{P_c}\right)V^2 + \frac{a}{P_c}V - \frac{ab}{P_c} = 0 \tag{2.30}$$

We thus obtain the simultaneous equations

$$3V_c = b + \frac{RT_c}{P_c}$$

$$3V_c^2 = \frac{a}{P_c}$$

$$V_c^3 = \frac{ab}{P_c}$$

which may be solved to yield

$$RT_c = \frac{8a}{27b}$$

$$P_c = \frac{a}{27b^2} \tag{2.31}$$

$$V_c = 3b$$

Therefore the Van der Waals constants a and b may be fitted to experiments by measuring any two of T_c, P_c, and V_c.

Let us measure P in units of P_c, T in units of T_c, and V in units of V_c:

$$\bar{P} = \frac{P}{P_c}, \qquad \bar{T} = \frac{T}{T_c}, \qquad \bar{V} = \frac{V}{V_c} \tag{2.32}$$

Then the Van der Waals equation of state becomes

$$\left(\bar{P} + \frac{3}{\bar{V}^2}\right)\left(\bar{V} - \tfrac{1}{3}\right) = \tfrac{8}{3}\bar{T} \tag{2.33}$$

This is a remarkable equation because it does not explicitly contain any constant characteristic of the substance. If the Van der Waals hypothesis were correct, (2.33) would hold for all substances. The assertion that the equation of state when expressed in terms of $\bar{P}$, $\bar{T}$, and $\bar{V}$ is a universal equation valid for all substances is called the *law of corresponding states.**

Looking at the isotherms of the Van der Waals equation of state we notice that the "kink" in a typical isotherm is unphysical, for it implies a negative compressibility. Its occurrence may be attributed to the implicit assumption that the system is homogeneous, with no allowance made for the possible coexistence of two phases. The situation may be improved by making what is called a *Maxwell construction* in the following manner. We ask whether it is possible to have two different states of the Van der Waals system coexisting in equilibrium. It is immediately obvious that for this to be possible the two states must have the same P and T. Therefore only states like those at volumes V_1, V_2, V_3 in Fig. 2.10 need be considered as candidates. The further principle we apply is the minimiza-

---

*For experimental evidence for the law of corresponding states, see Chapter 16, especially Fig. 16.2.

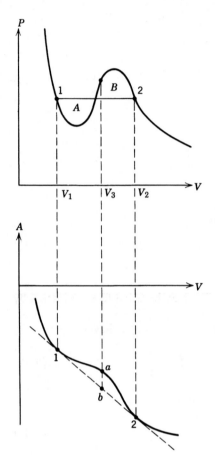

Fig. 2.11 The Maxwell construction.

tion of free energy. Let the temperature and the total volume of the system be fixed. Then we assume that the system is either in one homogeneous phase, or is composed of more than one phase. The situation that has the lower free energy is the equilibrium situation.

The free energy may be calculated by integrating $-P\,dV$ along an isotherm:

$$A(T, V) = -\int_{\text{isotherm}} P\,dV \qquad (2.34)$$

This may be done graphically, as shown in Fig. 2.11. It is seen that the states 1 and 2 can coexist because they have the same T and P. Further, the point b, which lies between 1 and 2 on the common tangent passing through 1 and 2, represents a state in which part of the system is in state 1 and part in state 2, because the free energy of this state is obviously a linear combination of those of 1 and 2. We note that point b lies lower than point a, which represents the free energy of a homogeneous system at the same T and V. Hence b, the phase separation case, is the equilibrium situation. Thus between the points 1 and 2 on the isotherm the system breaks up into two phases, with the pressure remaining constant. In other words the system undergoes a first-order phase transition. In the P-V diagram the points 1 and 2 are so located that the areas A and B are equal. To show this, let us write down all the conditions determining 1 and 2:

$$-\frac{\partial A}{\partial V_1} = -\frac{\partial A}{\partial V_2} \qquad \text{(equal pressure)}$$

$$\frac{A_2 - A_1}{V_2 - V_1} = \frac{\partial A}{\partial V_1} \qquad \text{(common tangent)}$$

Combining these we can write

$$\left(-\frac{\partial A}{\partial V_1}\right)(V_2 - V_1) = -(A_2 - A_1)$$

or

$$P_1(V_2 - V_1) = \int_{V_1}^{V_2} P \, dV$$

whose geometrical meaning is precisely $A = B$. This geometrical construction is known as the Maxwell construction.

The Maxwell construction shows that we obtain a much more sensible equation of state when we incorporate into our scheme of description the possibility that the system can separate into two phases. A more formal way to see this is given in Section 7.8, in the context of the choice of "ensembles" in statistical mechanics.

2.4 OSMOTIC PRESSURE

If we cover one end of an open glass tube with a "semipermeable membrane" that is permeable to water but not to sugar in solution, fill the tube with a sugar solution, and then dip this end of the tube into a beaker of water, we find that the sugar solution rises to a height h above the level of the water, as illustrated in Fig. 2.12. This indicates that the sugar solution has a pressure $\rho g h$ higher than that of pure water at the same temperature. This pressure must be due to the presence of the sugar, and is called the *osmotic pressure* exerted by the sugar in solution. It is by virtue of this pressure that a living cell, which is mostly water, can absorb sugar when it is immersed in a sugar solution. The osmotic pressure P' exerted by n_1 moles of solute in a very dilute solution of temperature T and volume V is experimentally given by

$$P' = \frac{n_1 R T}{V} \tag{2.35}$$

We derive this result with the help of the second law of thermodynamics.

Consider a solution containing n_0 moles of solvent and n_1 moles of solute, with

$$n_1/n_0 \ll 1$$

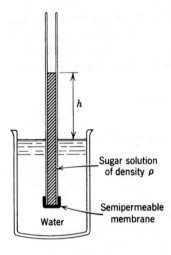

Sugar solution of density ρ

Semipermeable membrane

Water

Fig. 2.12 Osmotic pressure.

The free energy of the solution can be obtained from the definition $A = U - TS$. To this end we first discuss the internal energy of the solution. It is a function of T, P, n_0, n_1. Further, it is assumed to be a homogeneous function of n_0 and n_1. That is, if n_0 and n_1 are simultaneously increased by a certain factor, U increases by the same factor. By making a Taylor series expansion we can write

$$U(T, P, n_0, n_1) = U(T, P, n_0, 0) + n_1(\partial U/\partial n_1)_{n_1=0} + \cdots$$

Up to first-order terms, with the assumption of homogeneity in mind, we rewrite this in the form

$$U(T, P, n_0, n_1) = n_0 u_0(T, P) + n_1 u_1(T, P) \qquad (2.36)$$

Similarly the volume occupied by the solution is

$$V(T, P, n_0, n_1) = n_0 v_0(T, P) + n_1 v_1(T, P) \qquad (2.37)$$

where u_0 and v_0 are the internal energy per mole and volume per mole of the pure solvent respectively. On the other hand, u_1 and v_1 have no comparably simple interpretation.

Let us now consider the entropy of the solution. Imagine that the solution undergoes an infinitesimal reversible transformation, with n_0 and n_1 held fixed. The change in entropy is the exact differential

$$dS = \frac{dQ}{T} = \frac{1}{T}(dU + P\,dV) = n_0\left[\frac{1}{T}(du_0 + P\,dv_0) + \frac{n_1}{n_0}\frac{1}{T}(du_1 + P\,dv_1)\right]$$

Since n_1/n_0 is arbitrary, the two differentials

$$ds_0 = \frac{1}{T}(du_0 + P\,dv_0), \qquad ds_1 = \frac{1}{T}(du_1 + P\,dv_1)$$

must be separately exact. Therefore the entropy has the form

$$S(T, P, n_0, n_1) = n_0 s_0(T, P) + n_1 s_1(T, P) + \lambda(n_0, n_1) \qquad (2.38)$$

where the constant of integration $\lambda(n_0, n_1)$ does not depend on P, T. Accordingly we can find $\lambda(n_0, n_1)$ by making T so high and P so low that the solution completely evaporates and becomes a mixture of two ideal gases, the entropy of which we can explicitly calculate. It will be seen that the osmotic pressure arises solely from the term $\lambda(n_0, n_1)$.

The entropy of 1 mol of ideal gas at given T and P is

$$s(T, P) = c_P \log T - R \log P + K \tag{2.39}$$

where K is a numerical constant. The entropy of a mixture of two ideal gases of n_0 and n_1 moles respectively is

$$
\begin{aligned}
S_{\text{ideal}}&(T, P, n_0, n_1)\\
&= (n_0 c_{P_0} + n_1 c_{P_1}) \log T - n_0 R \log P_0 - n_1 R \log P_1 + n_0 K_0 + n_1 K_1
\end{aligned} \tag{2.40}
$$

where P_0 and P_1 are the partial pressures of the two gases. To express the entropy in terms of the total pressure P, we make use of the facts

$$
\begin{aligned}
P &= P_0 + P_1\\
P_0/n_0 &= P_1/n_1
\end{aligned} \tag{2.41}
$$

which imply

$$
\begin{aligned}
P_0 &= \frac{n_0 P}{n_0 + n_1} \approx P\\[2mm]
P_1 &= \frac{n_1 P}{n_0 + n_1} \approx \frac{n_1}{n_0} P
\end{aligned} \tag{2.42}
$$

Thus

$$
\begin{aligned}
S_{\text{ideal}}(T, P, n_0, n_1) = {}&(n_0 c_{P_0} + n_1 c_{P_1}) \log T - (n_0 + n_1) R \log P\\
&- n_1 R \log(n_1/n_0) + n_0 K_0 + n_1 K_1
\end{aligned} \tag{2.43}
$$

Comparison of this with (2.38) yields

$$\lambda(n_0, n_1) = -n_1 R \log(n_1/n_0) + n_0 K_0 + n_1 K_1 \tag{2.44}$$

The first term on the right side is known as the *entropy of mixing*, which arises directly from the mixing of the two gases with additive partial pressures. It is the entropy of mixing that gives rise to osmotic pressure. The free energy of the solution can now be written in the form

$$
\begin{aligned}
A(T, P, n_0, n_1) = {}&n_0 a_0(T, P)\\
&+ n_1 a_1(T, P) + n_1 RT \log(n_1/n_0)
\end{aligned} \tag{2.45}
$$

where a_0 is the free energy per mole of the pure solvent. The explicit forms of a_1 and a_0 are irrelevant to our purpose.

To find the osmotic pressure, consider a solution separated from the pure solvent by a semipermeable membrane, as shown in Fig. 2.13. The pressure of the solution is by definition higher than that of the pure solvent by P', the osmotic pressure. The total free energy of this composite system is

$$A = (n_0 + n_0') a_0 + n_1 a_1 + n_1 RT \log(n_1/n_0) \tag{2.46}$$

Suppose the semipermeable membrane is displaced reversibly, with the tempera-

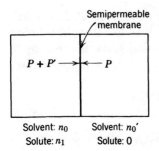

Solvent: n_0 Solvent: n_0' **Fig. 2.13** Thought experiment in the derivation of the
Solute: n_1 Solute: 0   osmotic pressure.

ture and the total volume of the composite system held fixed. Then n_0 suffers a change dn_0, and n_0' suffers a change $-dn_0$, for, as far as the pure solvent is concerned, the membrane is nonexistent. The volume of the *solution* changes by the amount $v_0\, dn_0$. The work done by the entire composite system is

$$dW = P'v_0\, dn_0 \tag{2.47}$$

According to the second law this is equal to the negative of the change in free energy dA, given by

$$-dA = \frac{n_1}{n_0} RT\, dn_0 \tag{2.48}$$

Therefore

$$P' = \frac{n_1 RT}{n_0 v_0}$$

Since $n_1/n_0 \ll 1$, however, $V = n_0 v_0$ is just the volume occupied by the solution. Hence

$$P' = \frac{n_1 RT}{V} \tag{2.49}$$

It is easy to see that the boiling point of a solution is higher than that of the pure solvent, on account of osmotic pressure. To deduce the change in boiling point, let us first find the difference between the vapor pressure of a solution and that of a pure solvent. This can be done by considering the arrangement shown in Fig. 2.14, which is self-explanatory. Under equilibrium conditions the difference between the pressures is the difference between the pressures of the vapor at B and at C. The pressure at C, however, is the same as that at A, because the vapor is at rest. Hence

$$\Delta P_{\text{vapor}} = P_B - P_C = P_B - P_A = \rho'gh \tag{2.50}$$

where ρ' is the mass density of the vapor. On the other hand, the osmotic pressure is by definition equal to the pressure exerted by the column of solution of height h:

$$P' = \rho gh \tag{2.51}$$

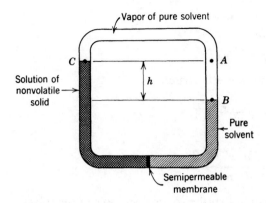

Fig. 2.14 Aid in the derivation of the difference in vapor pressure of a solution and the pure solvent.

where ρ is the density of the solution. Dividing (2.50) by (2.51) we have

$$\frac{\Delta P_{\text{vapor}}}{P'} = \frac{\rho'}{\rho} \tag{2.52}$$

which reduces, on using (2.49), to

$$\Delta P_{\text{vapor}} = \frac{\rho'}{\rho} \frac{n_1}{n_0} \frac{RT}{v_0} \tag{2.53}$$

where v_0 is the volume per mole of the solvent. Thus, at a given temperature, the solution has a lower vapor pressure than the pure solvent by the amount (2.53). The meaning of this formula may be made vivid by the qualitative plot of the vapor pressures in Fig. 2.15, from which we immediately see that the solution has a higher boiling point. The rise in boiling point ΔT can be deduced from the Clapeyron equation, which gives us the slope of the vapor pressure curve of either the solution or the solvent. In the approximation that we are using, these two

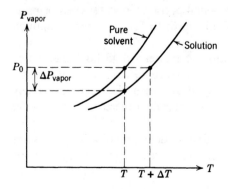

Fig. 2.15 Difference in boiling point between a solution and the pure solvent.

slopes may be taken to be the same and given by

$$\frac{dP}{dT} = \frac{l}{T \Delta v}$$

where l and Δv both refer to the pure solvent. Therefore

$$\Delta T = \frac{\Delta P_{vapor}}{(dP/dT)}$$

or

$$\frac{\Delta T}{T} = \frac{\Delta v}{l} \frac{\rho'}{\rho} \frac{n_1}{n_0} \frac{RT}{v_0}$$

We may further make the approximation that the volume per mole of the solvent is negligible compared to that of its vapor, and that the vapor is an ideal gas.

$$\Delta v \approx \frac{RT}{P}$$

$$\frac{\rho'}{\rho} = \frac{P v_0}{RT}$$

We obtain, with these approximations,

$$\frac{\Delta T}{T} = \frac{n_1}{n_0} \frac{RT}{l} \tag{2.54}$$

where l is the heat of vaporization per mole of the pure solvent.

2.5 THE LIMIT OF THERMODYNAMICS

We have seen that a substance in solution exerts osmotic pressure. A solution, in the derivation we have given, is any mixture of substances for which the entropy is greater than the sum of the entropies of the individual substances before they were mixed. Thermodynamics itself does not tell us what entropy really is, however, and therefore it does not tell us what constitutes a solution and what does not. For example, on purely thermodynamic grounds there is no way to answer the question, "Does a suspension of small particles in water exert osmotic pressure?" To answer this question we would have to form a definite opinion concerning the constitution of matter.

A nonbeliever in the atomic constitution of matter would hold that a fine suspension of particles does not exert osmotic pressure because it is not qualitatively different from having rocks in water. Even when the rocks are finely pulverized, they are still rocks. Would you think placing a piece of rock in water changes the pressure of the water?

Atomic theory says that a suspension of fine particle is qualitatively the same as a solution, which is a suspension of atoms. Thus there should be an osmotic pressure.

The question is not a philosophical one, but an experimental one. To settle it, all one has to do is to measure the osmotic pressure of a suspension.

It is, of course, difficult to measure directly the osmotic pressure exerted by a suspension of particles, even if this pressure exists. For example, a suspension of 5×10^{10} particles/ml at room temperature would exert an osmotic pressure of 10^{-9} atm. Accordingly, indirect methods of detection have to be used. In 1905 Einstein proposed to measure the density $n(x)$ of a vertical column of suspended particles as a function of height x. If there were no osmotic pressure, all the suspended particles would eventually sink to the bottom. Assuming that there is osmotic pressure, we can deduce $n(x)$ as follows. The osmotic pressure at height x is

$$P'(x) = n(x)kT \tag{2.55}$$

If $n(x)$ is not a constant, there will be a net force per unit volume acting on the particles at height x, given by

$$F_{\text{osmotic}}(x) = -kT\frac{dn(x)}{dx} \tag{2.56}$$

The force per unit volume due to gravity, on the other hand, is

$$F_{\text{gravity}}(x) = -mgn(x) \tag{2.57}$$

where m is the mass of a suspended particle. Under equilibrium conditions these two forces must cancel.* Therefore

$$\frac{dn(x)}{dx} + \frac{mg}{kT}n(x) = 0 \tag{2.58}$$

from which follows immediately

$$n(x) = n(0)e^{-mgx/kT} \tag{2.59}$$

This formula was verified by experiments.

Equation (2.59) can also be derived from purely kinetic considerations. If the viscosity of the medium is η and the radius of a suspended particle is r, a suspended particle falling under gravity will eventually reach the terminal velocity $mg/6\pi r\eta$, according to Stokes' law. The flux of particles falling because of gravity is therefore

$$-\frac{n(x)mg}{6\pi r\eta}$$

On the other hand, when $n(x)$ is not a constant, these particles are expected from kinetic theory to diffuse, giving rise to a net upward flux of

$$D\frac{dn(x)}{dx}$$

where D is the coefficient of diffusion. In equilibrium these two fluxes must be

---

*We assume that the force of buoyancy can be neglected.

equal. Therefore

$$\frac{dn(x)}{dx} + \frac{mg}{6\pi r \eta D} n(x) = 0 \tag{2.60}$$

Comparison of this with (2.58) yields

$$D = \frac{kT}{6\pi r \eta} \tag{2.61}$$

Experimental verification of this relation also constitutes a demonstration of the existence of osmotic pressure in a suspension.

Finally, from experiments involving suspensions, we can deduce atomic constants such as Avogadro's number. These experiments have to do with the motion of a single suspended particle (Brownian motion) and are beyond the scope of the present discussion.

PROBLEMS

2.1 What is the boiling point of water on Mt. Evans, Colorado, where the atmospheric pressure is two-thirds that at sea level?

2.2 A substance whose state is specified by P, V, T can exist in two distinct phases. At a given temperature T the two phases can coexist if the pressure is $P(T)$. The following information is known about the two phases. At the temperatures and pressures where they can coexist in equilibrium,

   (i) there is no difference in the specific volume of the two phases;
   (ii) there is no difference in the specific entropy of the two phases;
   (iii) the specific heat c_P and the volume expansion coefficient α are different for the two phases.

(a) Find $dP(T)/dT$ as a function of T.

(b) What is the qualitative shape of the transition region in the P-V diagram? In what way is it different from that of an ordinary gas-liquid transition?

The phase transition we have described is a *second-order transition*.

2.3 A cloud chamber contains water vapor at its equilibrium vapor pressure $P_\infty(T_0)$ corresponding to an absolute temperature T_0. Assume that

   (i) the water vapor may be treated as an ideal gas;
   (ii) the specific volume of water may be neglected compared to that of the vapor;
   (iii) the latent heat l of condensation and $\gamma = c_P/c_V$ may be taken to be constants: $l = 540$ cal/g, $\gamma = \frac{3}{2}$.

(a) Calculate the equilibrium vapor pressure $P_\infty(T)$ as a function of the absolute temperature T.

(b) The water vapor is expanded adiabatically until the temperature is T, $T < T_0$. Assume the vapor is now supersaturated. If a small number of droplets of water is formed (catalyzed, e.g., by the presence of ions produced by the passage of an α particle), what is the equilibrium radius of these droplets?

(*c*) In the approximations considered, does adiabatic expansion always lead to super-saturation?

2.4 Show that the heat capacity at constant volume C_V of a Van der Waals gas is a function of the temperature alone.

2.5 Derive the Maxwell construction by considerations involving the minimization of the Gibbs potential instead of the Helmholtz free energy.

2.6 Consider an open tank partitioned in two by a vertical semipermeable membrane that is permeable to water but not to sugar in solution. Fill the tank with water and dissolve sugar on the left side of the partition. The level of the sugar solution will be higher than that of the pure water because of osmotic pressure. Since the partition is permeable to water, will the water in the sugar solution leak out through the partition?

3

THE PROBLEM
OF KINETIC THEORY

3.1 FORMULATION OF THE PROBLEM

The system under consideration in the classical kinetic theory of gases is a dilute gas of N molecules enclosed in a box of volume V. The temperature is sufficiently high and the density is sufficiently low for the molecules to be localized wave packets whose extensions are small compared to the average intermolecular distance. For this to be realized the average de Broglie wavelength of a molecule must be much smaller than the average interparticle separation:

$$\frac{\hbar}{\sqrt{2mkT}} \left(\frac{N}{V} \right)^{1/3} \ll 1 \qquad (3.1)$$

Under such conditions each molecule may be considered a classical particle with a rather well-defined position and momentum. Furthermore, two molecules may be considered to be distinguishable from each other. The molecules interact with each other through collisions whose nature is specified through a given differential scattering cross section σ. Throughout our discussion of kinetic theory only the special case of a system of one kind of molecule will be considered.

An important simplification of the problem is made by ignoring the atomic structure of the walls containing the gas under consideration. That is, the physical walls of the container are replaced by idealized surfaces which act on an impinging gas molecule in a simple way, e.g., reflecting it elastically.

We are not interested in the motion of each molecule in detail. Rather, we are interested in the distribution function $f(\mathbf{r}, \mathbf{p}, t)$, so defined that

$$f(\mathbf{r}, \mathbf{p}, t) \, d^3r \, d^3p \qquad (3.2)$$

is the number of molecules which, at time t, have positions lying within a volume element d^3r about $\mathbf{r}$ and momenta lying within a momentum-space element d^3p about p. The volume elements d^3r and d^3p are not to be taken literally as

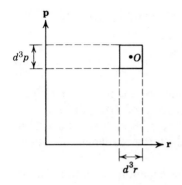

Fig. 3.1 The six-dimensional μ space of a molecule.

mathematically infinitesimal quantities. They are finite volume elements which are large enough to contain a very large number of molecules and yet small enough so that compared to macroscopic dimensions they are essentially points. That such a choice is possible can be seen by an example. Under standard conditions there are about 3×10^{19} molecules/cm^3 in a gas. If we choose $d^3r \sim 10^{-10}$ cm^3, which to us is small enough to be called a point, there are still on the order of 3×10^9 molecules in d^3r.

To make the definition of $f(\mathbf{r}, \mathbf{p}, t)$ more precise, let us consider the six-dimensional space, called the μ space, spanned by the coordinates* $(\mathbf{r}, \mathbf{p})$ of a molecule. The μ space is schematically represented in Fig. 3.1. A point in this space represents a state of a molecule. At any instant of time, the state of the entire system of N molecules is represented by N points in μ space. Let a volume element $d^3r\, d^3p$ be constructed about each point in μ space, such as that shown about the point O in Fig. 3.1. If we count the number of points in this volume element, the result is by definition $f(\mathbf{r}, \mathbf{p}, t)\, d^3r\, d^3p$. If the sizes of these volume elements are chosen so that each of them contains a very large number of points, such as 10^9, and if the density of these points does not vary rapidly from one element to a neighboring element, then $f(\mathbf{r}, \mathbf{p}, t)$ may be regarded as a continuous function of its arguments. If we cover the entire μ space with such volume elements, we can make the approximation

$$\sum f(\mathbf{r}, \mathbf{p}, t)\, d^3p\, d^3r \approx \int f(\mathbf{r}, \mathbf{p}, t)\, d^3p\, d^3r \qquad (3.3)$$

where the sum on the left extends over all the centers of the volume elements, and the integral on the right side is taken in the sense of calculus. Such an approximation will always be understood.

Having defined the distribution function, we can express the information that there are N molecules in the volume V through the normalization condition

$$\int f(\mathbf{r}, \mathbf{p}, t)\, d^3r\, d^3p = N \qquad (3.4)$$

---

*For brevity, the collection of spatial and momentum coordinates $(\mathbf{r}, \mathbf{p})$ is referred to as the coordinates of a molecule.

If the molecules are uniformly distributed in space, so that f is independent of $\mathbf{r}$, then

$$\int f(\mathbf{r}, \mathbf{p}, t)\, d^3p = \frac{N}{V} \qquad (3.5)$$

The aim of kinetic theory is to find the distribution function $f(\mathbf{r}, \mathbf{p}, t)$ for a given form of molecular interaction. The limiting form of $f(\mathbf{r}, \mathbf{p}, t)$ as $t \to \infty$ would then contain all the equilibrium properties of the system. The aim of kinetic theory therefore includes the derivation of the thermodynamics of a dilute gas.

To fulfill this aim, our first task is to find the equation of motion for the distribution function. The distribution function changes with time, because molecules constantly enter and leave a given volume element in μ space. Suppose there were no molecular collisions (i.e., $\sigma = 0$). Then a molecule with the coordinates $(\mathbf{r}, \mathbf{p})$ at the instant t will have the coordinates $(\mathbf{r} + \mathbf{v}\,\delta t, \mathbf{p} + \mathbf{F}\,\delta t)$ at the instant $t + \delta t$, where $\mathbf{F}$ is the external force acting on a molecule, and $\mathbf{v} = \mathbf{p}/m$ is the velocity. We may take δt to be a truly infinitesimal quantity. Thus all the molecules contained in a μ-space element $d^3r\, d^3p$, at $(\mathbf{r}, \mathbf{p})$, at the instant t, will all be found in an element $d^3r'\, d^3p'$, at $(\mathbf{r} + \mathbf{v}\,\delta t, \mathbf{p} + \mathbf{F}\,\delta t)$, at the instant $t + \delta t$. Hence in the absence of collisions we have the equality

$$f(\mathbf{r} + \mathbf{v}\,\delta t, \mathbf{p} + \mathbf{F}\,\delta t, t + \delta t)\, d^3r'\, d^3v' = f(\mathbf{r}, \mathbf{v}, t)\, d^3r\, d^3v$$

which reduces to

$$f(\mathbf{r} + \mathbf{v}\,\delta t, \mathbf{p} + \mathbf{F}\,\delta t, t + \delta t) = f(\mathbf{r}, \mathbf{p}, t) \qquad (3.6)$$

because $d^3r\, d^3p = d^3r'\, d^3p'$. The last fact is easily established if we assume that the external force $\mathbf{F}$ depends on position only. At any instant t, we may choose $d^3r\, d^3p$ to be a six-dimensional cube. It is sufficient to show that the area of any projection of this cube, say, $dx\, dp_x$, does not change. A simple calculation will show that this projection, originally a square, becomes a parallelogram of the same area in the time δt, as illustrated in Fig. 3.2. This invariance is valid as long as $(\mathbf{r}, \mathbf{p})$ are canonically conjugate generalized coordinates.

When there are collisions (i.e., $\sigma > 0$), equality (3.6) must be modified. We write

$$f(\mathbf{r} + \mathbf{v}\,\delta t, \mathbf{p} + \mathbf{F}\,\delta t, t + \delta t) = f(\mathbf{r}, \mathbf{p}, t) + \left(\frac{\partial f}{\partial t}\right)_{\text{coll}} \delta t \qquad (3.7)$$

which *defines* $(\partial f / \partial t)_{\text{coll}}$. Expanding the left-hand side to the first order in δt, we obtain the equation of motion for the distribution function as we let $\delta t \to 0$:

$$\left(\frac{\partial}{\partial t} + \frac{\mathbf{p}}{m} \cdot \nabla_{\mathbf{r}} + \mathbf{F} \cdot \nabla_{\mathbf{p}}\right) f(\mathbf{r}, \mathbf{p}, t) = \left(\frac{\partial f}{\partial t}\right)_{\text{coll}} \qquad (3.8)$$

where $\nabla_{\mathbf{r}}, \nabla_{\mathbf{p}}$ are, respectively, the gradient operators with respect to $\mathbf{r}$ and $\mathbf{p}$. This equation is not meaningful until we explicitly specify $(\partial f / \partial t)_{\text{coll}}$. It is in specifying this term that the assumption that the system is a dilute gas becomes relevant.

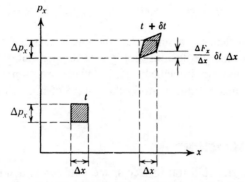

Fig. 3.2 The invariance of the volume element in μ space under dynamical evolution in time.

An explicit form for $(\partial f/\partial t)_{\text{coll}}$ can be obtained by going back to its definition (3.7). Consider Fig. 3.3, where the square labeled A represents the μ-space volume element at $\{\mathbf{r}, \mathbf{p}, t\}$ and the one labeled B represents that at $\{\mathbf{r} + \mathbf{v}\,\delta t, \mathbf{p} + \mathbf{F}\,\delta t, t + \delta t\}$, where δt eventually tends to zero. During the time interval δt, some molecules in A will be removed from A by collision. We regard A as so small that *any* collision that a molecule in A suffers will knock it out of A. Such a molecule with not reach B. On the other hand, there are molecules outside A which, through collisions, will get into A during the time interval δt. These will be in B. Therefore the number of molecules in B at $t + \delta t$, as $\delta t \to 0$, equals the original number of molecules in A at time t plus the *net* gain of molecules in A due to collisions during the time interval δt. This statement is the content of (3.7), and may be expressed in the form

$$\left(\frac{\partial f}{\partial t}\right)_{\text{coll}} \delta t = (\overline{R} - R)\,\delta t \tag{3.9}$$

where $R\,\delta t\, d^3r\, d^3p$ = no. of collisions occurring during the time
between t and $t + \delta t$ in which one of the
initial molecules is in $d^3r\, d^3p$ about $(\mathbf{r}, \mathbf{p})$ (3.10)
$\overline{R}\,\delta t\, d^3r\, d^3p$ = no. of collisions occurring during the time
between t and $t + \delta t$, in which one of the
final molecules is in $d^3r\, d^3p$ about $(\mathbf{r}, \mathbf{p})$ (3.11)

Strictly speaking, we make a small error here. For example, in (3.10), we are implicitly assuming that if a molecule qualifies under the description, none of its

Fig. 3.3 A volume element in μ space at the times t and $t + \delta t$.

partners in collision qualifies. This error is negligible because of the smallness of d^3p.

To proceed further, we assume that the gas is extremely dilute, so that we may consider only binary collisions and ignore the possibility that three or more molecules may collide simultaneously. This considerably simplifies the evaluation of R and $\bar{R}$. It is thus natural to study the nature of binary collisions next.

3.2 BINARY COLLISIONS

We consider an elastic collision in free space between two spinless molecules of respective masses m_1 and m_2. The momenta of the molecules in the initial state are denoted by $\mathbf{p}_1$ and $\mathbf{p}_2$, respectively, and the energies by ϵ_1 and ϵ_2, with $\epsilon_i = p_i^2/2m_i$. The corresponding quantities in the final state are indicated by a prime. Momentum and energy conservation require that

$$\mathbf{p}_1 + \mathbf{p}_2 = \mathbf{p}_1' + \mathbf{p}_2'$$
$$E = \epsilon_1 + \epsilon_2 = \epsilon_1' + \epsilon_2' \tag{3.12}$$

where E is the total energy. We define the total mass M and reduced mass μ by

$$M \equiv m_1 + m_2, \qquad \mu \equiv \frac{m_1 m_2}{m_1 + m_2} \tag{3.13}$$

and the total momentum $\mathbf{P}$ and relative momentum $\mathbf{p}$ by

$$\mathbf{P} \equiv \mathbf{p}_1 + \mathbf{p}_2$$
$$\mathbf{p} \equiv \frac{m_2 \mathbf{p}_1 - m_1 \mathbf{p}_2}{m_1 + m_2} = \mu(\mathbf{v}_1 - \mathbf{v}_2) \tag{3.14}$$

where $\mathbf{v}_i = \mathbf{p}_i/m_i$ is the velocity. Solving these for $\mathbf{p}_1$ and $\mathbf{p}_2$ gives

$$\mathbf{p}_1 = \frac{m_1}{M}\mathbf{P} - \mathbf{p}, \qquad \mathbf{p}_2 = \frac{m_2}{M}\mathbf{P} + \mathbf{p} \tag{3.15}$$

It can be easily verified that the total energy is given by

$$E = \frac{P^2}{2M} + \frac{p^2}{2\mu} \tag{3.16}$$

The conditions for momentum-energy conservation become simply

$$\mathbf{P} = \mathbf{P}', \qquad |\mathbf{p}| = |\mathbf{p}'| \tag{3.17}$$

That is, the collision merely rotates the relative momentum without changing its magnitude. Let the angle between $\mathbf{p}'$ and $\mathbf{p}$ be θ, and the azimuthal angle of $\mathbf{p}'$ about $\mathbf{p}$ be ϕ. These angles completely specify the kinematics of the collision. They are collectively denoted by Ω, and are called the scattering angles. We depict the kinematics geometrically in Fig. 3.4. If the potential responsible for the scattering is a central potential (i.e., dependent only on the magnitude of the distance between the molecules), then the scattering is independent of ϕ.

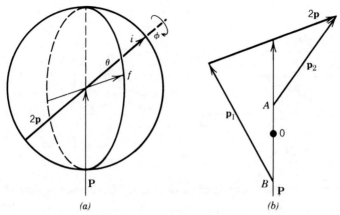

Fig. 3.4 Geometrical representation of the kinematics of a binary collision. (a) The total momentum **P** is unchanged (momentum conservation). The relative momentum is rotated from the initial direction i to the final direction f, without change in magnitude (energy conservation). The scattering angles are the angles θ, ϕ of f relative to i. (b) The individual momenta of the colliding partners p_1 and p_2 may be constructed from **P** and **2p**, as shown. The points A and B will coincide at O, the midpoint of **P**, if the colliding molecules have equal mass.

The dynamical aspects of the collision are contained in the differential cross section $d\sigma/d\Omega$, which is defined experimentally as follows. Consider a beam of particle 2 incident on particle 1, regarded as the target. The incident flux I is defined as the number of incident particles crossing unit area per second, from the viewpoint of the target:

$$I = n|\mathbf{v}_1 - \mathbf{v}_2| \qquad (3.18)$$

where n is the density of particles in the incident beam. The differential cross section $d\sigma/d\Omega$ is defined by the statement

$$I(d\sigma/d\Omega)\, d\Omega \equiv \text{Number of incident molecules scattered per second} \atop \text{into the solid angle element } d\Omega \text{ about the direction } \Omega \qquad (3.19)$$

The differential cross section has the dimension of area. The number of molecules scattered into $d\Omega$ per second is equal to the number of molecules in the incident beam crossing an area $d\sigma/d\Omega$ per second. The total cross section is the number of molecules scattered per second, regardless of scattering angle:

$$\sigma_{\text{tot}} = \int d\Omega \, \frac{d\sigma}{d\Omega} \qquad (3.20)$$

In classical mechanics the differential cross section can be calculated from the intermolecular potential as follows. First we transform the coordinate system to the center-of-mass system, in which the total momentum is zero. Since we are considering only the nonrelativistic domain, this involves a trivial translation of

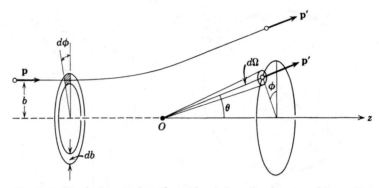

Fig. 3.5 Classical scattering of a molecule by a fixed center of force O.

all velocities by a constant amount. We only need to follow the trajectory of one of the particles, which will move along an orbit, as if it were scattered by a fixed center of force O, as illustrated in Fig. 3.5. It approaches O with momentum **p**, the relative momentum, and will recede from O with momentum **p′**, the rotated relative momentum. The normal distance between the line of approach and O is called the impact parameter b. By conservation of angular momentum, it is also the normal distance between the line of recession and O. This is indicated in Fig. 3.5, together with the scattering angles. From the geometry it is clear that

$$I \frac{d\sigma}{d\Omega} d\Omega = I b \, db \, d\phi \tag{3.21}$$

We can find the relation between b and the scattering angles from the classical orbit equation, thereby obtaining $d\sigma/d\Omega$ as a function of the scattering angles.*

The use of classical mechanics to calculate the differential cross section in this problem is an old tradition (started by Maxwell) predating quantum mechanics. To be correct, however, we must use quantum mechanics, notwithstanding the fact that between collisions we regard the molecules as classical particles. The reason is that when the molecules collide their wave functions necessarily overlap, and they see each other as plane waves of definite momenta rather than wave packets of well-defined positions. Furthermore, formulating the scattering problem quantum mechanically makes the kinematics and symmetries of the problem more obvious.

In quantum mechanics the basic quantity in a scattering problem is the transition matrix (T matrix), whose elements are the matrix elements of a certain operator $T(E)$ between the initial (i) and final (f) state:

$$T_{\text{fi}} = \langle 1', 2' | T(E) | 1, 2 \rangle$$
$$T(E) = \mathcal{H}' + \mathcal{H}'(E - \mathcal{H}_0 + i\epsilon)^{-1} \mathcal{H}' + \cdots \tag{3.22}$$

where $\mathcal{H}_0$ is the unperturbed Hamiltonian, $\mathcal{H}'$ the potential, and $\epsilon \to 0^+$. A

*See any book on classical mechanics, for example, L. D. Landau and E. M. Lifshitz, *Mechanics* (Pergamon, Oxford, 1960), Chapter IV.

collision is a transition from the initial state to a set of final states. For final states in the infinitesimal momentum-space element $d^3p_1' \, d^3p_2'$, the rate is*

$$dP_{12 \to 1'2'} = I \, d\sigma = d^3p_1' \, d^3p_2' \, \delta^4(P_f - P_i)|T_{fi}|^2$$
$$\delta^4(P_f - P_i) \equiv \delta^3(\mathbf{P} - \mathbf{P}')\delta(E - E') \tag{3.23}$$

The transition rate into any region of momentum space can be obtained by integrating the above over the appropriate region. To obtain the differential cross section, integrate over the recoil momentum $\mathbf{p}_1$ (which is fixed by momentum conservation) and the magnitude p_2' (which is determined by energy conservation) to obtain

$$I\frac{d\sigma}{d\Omega} = \int dp_2' \, p_2'^2 \int d^3p_1' \, \delta^4(P_f - P_i)|T_{fi}|^2 \tag{3.24}$$

The integrations are trivial, and yield a factor representing the density of final states because of the δ functions that enforce momentum-energy conservation. For the formal manipulations that we are going to do, however, it is best to leave the integrations undone.

The T matrix is invariant under spatial rotations and reflections and under time reversal, because all molecular interactions originate in the electromagnetic interaction, which have these invariance properties. Explicitly, we have

$$\langle \mathbf{p}_2', \mathbf{p}_1'|T|\mathbf{p}_1, \mathbf{p}_2 \rangle = \langle R\mathbf{p}_2', R\mathbf{p}_1'|T|R\mathbf{p}_1, R\mathbf{p}_2 \rangle$$
$$\langle \mathbf{p}_2', \mathbf{p}_1'|T|\mathbf{p}_1, \mathbf{p}_2 \rangle = \langle -\mathbf{p}_2, -\mathbf{p}_1|T| -\mathbf{p}_1', -\mathbf{p}_2' \rangle \tag{3.25}$$

where $R\mathbf{p}$ is the vector obtained from $\mathbf{p}$ after performing a spatial rotation about an axis, and/or a reflection with respect to a plane. For elastic scattering the density of states are the same in the initial and final states. Thus invariances of the T matrix directly implies corresponding invariances of the differential cross section.

If the molecules have spin, (3.25) remains valid provided we interpret $\mathbf{p}$ to include the spin coordinate. A rotation rotates both the momentum and the spin, but a reflection does not affect the spin. Under time reversal both the momentum and the spin change sign.

From (3.25) we can deduce that the inverse collision, defined as the collision with initial and final states interchanged, has the same T matrix (and hence the same cross section):

$$\langle \mathbf{p}_2', \mathbf{p}_1'|T|\mathbf{p}_1, \mathbf{p}_2 \rangle = \langle \mathbf{p}_2, \mathbf{p}_1|T|\mathbf{p}_1', \mathbf{p}_2' \rangle \quad \text{or} \quad T_{fi} = T_{if} \tag{3.26}$$

To show this we represent the collision by the schematic drawing A of Fig. 3.6, which is self-explanatory. The diagram A' beneath it has the same meaning as

---

*A word about normalization. With the definition (3.22), and with single-particle states normalized to one particle per unit volume, there should be a factor $(2\pi\hbar)^3$ multiplying the three-dimensional δ function for momentum conservation, a factor $2\pi/\hbar$ multiplying the δ function for energy conservation, and a factor $(2\pi\hbar)^{-3}$ multiplying each volume element d^3p of momentum space. Since we are not going to calculate cross sections, these factors are a nuisance to write out. We redefine the T matrix appropriately to absorb these factors.

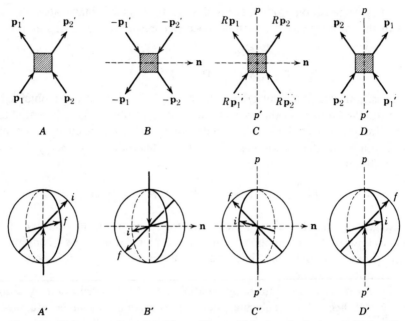

Fig. 3.6 Symmetry operations that take a collision to the inverse collision. In A', B', C', and D', i and f, respectively, denote initial and final relative momenta.

Fig. 3.4. The T matrix for this collision is the same as the time-reversed collision represented by B. Now rotate the coordinate system through $180°$ about a suitable axis **n** perpendicular to the total momentum, and then reflect with respect to a plane pp' perpendicular to **n**. As a result we obtain the collision D, which is the inverse of the original collision, and which has the same T matrix because of (3.25).

If collisions were treated classically, the inverse collision could be very different from the original collision. As a concrete example consider the classical collision between a sphere and a wedge.* A glance at Fig. 3.7 proves the point. But this is irrelevant for molecular scattering, because molecules are not describable as "wedges" or the like. A nonspherically symmetric molecule is one with nonzero spin, and exists in an eigenstate of the spin. The angular orientation, being conjugate to the angular momentum, is completely uncertain. The symmetry between collision and inverse collision remains valid when the spin is taken into account, as stated previously.

3.3 THE BOLTZMANN TRANSPORT EQUATION

To derive an explicit formula for $(\partial f/\partial t)_{\text{coll}}$, we assume that the gas is sufficiently dilute that only binary collisions need be taken into account. The effect of

---

*Both made of concrete.

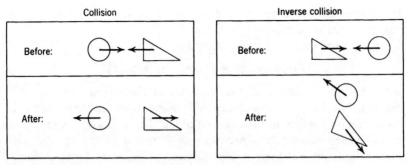

Fig. 3.7 Classical collision between macroscopic objects.

external forces on collisions are ignored on the assumption that these forces, if present, would vary little over the range of the intermolecular potential.

The number of transitions $12 \to 1'2'$ in a volume element d^3r at r, owing to collisions during the time interval δt is

$$dN_{12} \, dP_{12 \to 1'2'} \, \delta t$$

where dN_{12} is the initial number of colliding pairs $(\mathbf{p}_1, \mathbf{p}_2)$. We introduce the two-particle correlation function F by

$$dN_{12} = F(\mathbf{r}, \mathbf{p}_1, \mathbf{p}_2, t) \, d^3r \, d^3p_1 \, d^3p_2 \tag{3.27}$$

Thus, in the notation of (3.10), we have

$$R \, \delta t \, d^3r \, d^3p_1 = \delta t \, d^3r \, d^3p_1 \int d^3p_2 \, dP_{12 \to 1'2'} \, F(\mathbf{r}, \mathbf{p}_1, \mathbf{p}_2, t) \tag{3.28}$$

Using (3.23), we obtain

$$R = \int d^3p_2 \, d^3p_1' \, d^3p_2' \, \delta^4(P_f - P_i) |T_{fi}|^2 F(\mathbf{r}, \mathbf{p}_1, \mathbf{p}_2, t) \tag{3.29}$$

Similarly, we find

$$\bar{R} = \int d^3p_2 \, d^3p_1' \, d^3p_2' \, \delta^4(P_i - P_f) |T_{if}|^2 F(\mathbf{r}, \mathbf{p}_1', \mathbf{p}_2', t) \tag{3.30}$$

The δ functions in (3.29) and (3.30) are identical, and $T_{fi} = T_{if}$ by (3.25). Hence

$$\left(\frac{\partial f}{\partial t} \right)_{\text{coll}} = \bar{R} - R = \int d^3p_2 \, d^3p_1' \, d^3p_2' \, \delta^4(P_f - P_i) |T_{fi}|^2 (F_{1'2'} - F_{12}) \tag{3.31}$$

where $F_{12} = F(\mathbf{r}, \mathbf{p}_1, \mathbf{p}_2, t)$. Note that we can integrate over the vector $\mathbf{p}_1'$ and the magnitude p_2, so that the differential cross section appears in the integrand of (3.30):

$$\left(\frac{\partial f}{\partial t} \right)_{\text{coll}} = \int d^3p_2 \, d\Omega \, |\mathbf{v}_1 - \mathbf{v}_2| (d\sigma/d\Omega)(F_{1'2'} - F_{12}) \tag{3.32}$$

For formal manipulations, however, it is more convenient to leave it in the form (3.31).

The expression we obtained is exact for a sufficiently dilute gas. But it contains the unknown correlation function F.

We now introduce the crucial assumption

$$F(\mathbf{r}, \mathbf{p}_1, \mathbf{p}_2, t) \approx f(\mathbf{r}, \mathbf{p}_1, t) f(\mathbf{r}, \mathbf{p}_2, t) \tag{3.33}$$

This says that the momenta of two particles in the volume element d^3r are uncorrelated, so that the probability of finding them simultaneously is the product of the probability of finding each alone. This is known as the "assumption of molecular chaos." It is necessary to obtain a closed equation for the distribution function, but there is otherwise no justification for it at this point. We shall come back to analyze its meaning later.

With the assumption of molecular chaos, we have

$$\left(\frac{\partial f_1}{\partial t} \right)_{\text{coll}} = \int d^3p_2 \, d\Omega |\mathbf{v}_1 - \mathbf{v}_2| (d\sigma/d\Omega)(f_1'f_2' - f_1 f_2) \tag{3.34}$$

where the following abbreviations have been used:

$$\begin{aligned}
f_1 &\equiv f(\mathbf{r}, \mathbf{p}_1, t) \\
f_2 &\equiv f(\mathbf{r}, \mathbf{p}_2, t) \\
f_1' &\equiv f(\mathbf{r}, \mathbf{p}_1', t) \\
f_2' &\equiv f(\mathbf{r}, \mathbf{p}_2', t)
\end{aligned} \tag{3.35}$$

Substituting (3.34) into (3.8) we obtain the Boltzmann transport equation

$$\left(\frac{\partial}{\partial t} + \frac{\mathbf{p}_1}{m} \cdot \nabla_r + \mathbf{F} \cdot \nabla_{p_1} \right) f_1 = \int d^3p_2 \, d^3p_1' \, d^3p_2' \, \delta^4(P_f - P_i) |T_{fi}|^2 (f_2'f_1' - f_2 f_1) \tag{3.36}$$

which is a nonlinear integro-differential equation for the distribution function.

We have considered only the case of a single species of spinless molecules. If we consider different types of molecules, then we have to introduce a separate distribution function for each type, and the collision term will couple them if different types of molecules can scatter each other. If the molecules have spin, or if we consider excitation of the molecules through scattering, then the different spin states or excited states should be considered as different species of molecules.

3.4 THE GIBBSIAN ENSEMBLE

Gibbs introduced the idea of a statistical ensemble to describe a macroscopic system, which has proved to be a very important concept. We shall use it here to present another approach to the Boltzmann transport equation.

A state of the gas under consideration can be specified by the $3N$ canonical coordinates $q_1, \ldots, q_{3N}$ and their conjugate momenta $p_1, \ldots, p_{3N}$. The $6N$-dimensional space spanned by $\{p_i, q_i\}$ is called the Γ space, or phase space, of the system. A point in Γ space represents a state of the entire N-particle system, and is referred to as the *representative point*. This is in contrast to the μ space introduced earlier, which refers to only one particle.

It is obvious that a very large (in fact, infinite) number of states of the gas corresponds to a given macroscopic condition of the gas. For example, the condition that the gas is contained in a box of volume 1 cm^3 is consistent with an infinite number of ways to distribute the molecules in space. Through macroscopic measurements we would not be able to distinguish between two gases existing in different states (thus corresponding to two distinct representative points) but satisfying the same macroscopic conditions. Thus when we speak of a gas under certain macroscopic conditions, we are in fact referring not to a single state, but to an infinite number of states. In other words, we refer not to a single system, but to a collection of systems, identical in composition and macroscopic condition but existing in different states. With Gibbs, we call such a collection of systems an *ensemble*, which is geometrically represented by a distribution of representative points in Γ space, usually a continuous distribution. It may be conveniently described by a density function $\rho(p, q, t)$, where (p, q) is an abbreviation for $(p_1, \ldots, p_{3N}; q_1, \ldots, q_{3N})$, so defined that

$$\rho(p, q, t) \, d^{3N}p \, d^{3N}q \tag{3.37}$$

is the number of representative points that at time t are contained in the infinitesimal volume element $d^{3N}p \, d^{3N}q$ of Γ space centered about the point (p, q). An ensemble is completely specified by $\rho(p, q, t)$. It is to be emphasized that members of an ensemble are mental copies of a system and do not interact with one another.

Given $\rho(p, q, t)$ at any time t, its subsequent values are determined by the dynamics of molecular motion. Let the Hamiltonian of a system in the ensemble be $\mathscr{H}(p_1, \ldots, p_{3N}; q_1, \ldots, q_{3N})$. The equations of motion for a system are given by

$$\dot{p}_i = -\frac{\partial \mathscr{H}}{\partial q_i} \qquad (i = 1, \ldots, 3N)$$

$$\dot{q}_i = \frac{\partial \mathscr{H}}{\partial p_i} \qquad (i = 1, \ldots, 3N) \tag{3.38}$$

These will tell us how a representative point moves in Γ space as time evolves. We assume that the Hamiltonian does not depend on any time derivative of p and q. It is then clear that (3.38) is invariant under time reversal and that (3.38) uniquely determines the motion of a representative point for all times, when the position of the representative point is given at any time. It follows immediately from these observations that the locus of a representative point is either a simple closed curve or a curve that never intersects itself. Furthermore, the loci of two distinct representative points never intersect.

We now prove the following theorem.

LIOUVILLE'S THEOREM

$$\frac{\partial \rho}{\partial t} + \sum_{i=1}^{3N} \left(\frac{\partial \rho}{\partial p_i} \dot{p}_i + \frac{\partial \rho}{\partial q_i} \dot{q}_i \right) = 0 \qquad (3.39)$$

Proof Since the total number of systems in an ensemble is conserved, the number of representative points leaving any volume in Γ space per second must be equal to the rate of decrease of the number of representative points in the same volume. Let ω be an arbitrary volume in Γ space and let S be its surface. If we denote by $\mathbf{v}$ the $6N$-dimensional vector whose components are

$$\mathbf{v} \equiv (\dot{p}_1, \dot{p}_2, \ldots, \dot{p}_{3N}; \dot{q}_1, \dot{q}_2, \ldots, \dot{q}_{3N})$$

and $\mathbf{n}$ the vector locally normal to the surface S, then

$$-\frac{d}{dt} \int_\omega d\omega \, \rho = \int_S dS \, \mathbf{n} \cdot \mathbf{v}\rho$$

With the help of the divergence theorem in $6N$-dimensional space, we convert this to the equation

$$\int_\omega d\omega \left[\frac{\partial \rho}{\partial t} + \nabla \cdot (\mathbf{v}\rho) \right] = 0 \qquad (3.40)$$

where ∇ is the $6N$-dimensional gradient operator:

$$\nabla \equiv \left(\frac{\partial}{\partial p_1}, \frac{\partial}{\partial p_2}, \ldots, \frac{\partial}{\partial p_{3N}}; \frac{\partial}{\partial q_1}, \frac{\partial}{\partial q_2}, \ldots, \frac{\partial}{\partial q_{3N}} \right)$$

Since ω is an arbitrary volume the integrand of (3.40) must identically vanish. Hence

$$-\frac{\partial \rho}{\partial t} = \nabla \cdot (\mathbf{v}\rho) = \sum_{i=1}^{3N} \left[\frac{\partial}{\partial p_i}(\dot{p}_i \rho) + \frac{\partial}{\partial q_i}(\dot{q}_i \rho) \right]$$

$$= \sum_{i=1}^{3N} \left(\frac{\partial \rho}{\partial p_i} \dot{p}_i + \frac{\partial \rho}{\partial q_i} \dot{q}_i \right) + \sum_{i=1}^{3N} \rho \left(\frac{\partial \dot{p}_i}{\partial p_i} + \frac{\partial \dot{q}_i}{\partial q_i} \right)$$

By the equations of motion (3.38) we have

$$\frac{\partial \dot{p}_i}{\partial p_i} + \frac{\partial \dot{q}_i}{\partial q_i} = 0 \qquad (i = 1, \ldots, 3N)$$

Therefore

$$-\frac{\partial \rho}{\partial t} = \sum_{i=1}^{3N}{}' \left(\frac{\partial \rho}{\partial p_i} \dot{p}_i + \frac{\partial \rho}{\partial q_i} \dot{q}_i \right) \qquad \blacksquare$$

Liouville's theorem is equivalent to the statement

$$\frac{d\rho}{dt} = 0 \qquad (3.41)$$

since by virtue of the equations of motion p_i and q_i are functions of the time. Its geometrical interpretation is as follows. If we follow the motion of a representative point in Γ space, we find that the density of representative points in its neighborhood is constant. Hence the distribution of representative points moves in Γ space like an incompressible fluid.

The observed value of a dynamical quantity O of the system, which is generally a function of the coordinates and conjugate momenta, is supposed to be its averaged value taken over a suitably chosen ensemble:

$$\langle O \rangle = \frac{\int d^{3N}p \, d^{3N}q \, O(p, q) \rho(p, q, t)}{\int d^{3N}p \, d^{3N}q \, \rho(p, q, t)} \qquad (3.42)$$

This is called the ensemble average of O. Its time dependence comes from that of ρ, which is governed by Liouville's theorem. In principle, then, this tells us how a quantity approaches equilibrium—the central question of kinetic theory. In the next section we shall derive the Boltzmann transport equation using this approach.

Under certain conditions one can prove an *ergodic theorem*, which says that if one waits a sufficiently long time, the locus of the representative point of a system will cover the entire accessible phase space. More precisely, it says that the representative point comes arbitrarily close to any point in the accessible phase space. This would indicate that the ensemble corresponding to thermodynamic equilibrium is one for which ρ is constant over the accessible phase space. This is actually what we shall assume.*

3.5 THE BBGKY HIERARCHY

One can define correlation functions f_s, which give the probability of finding s particles having specified positions and momenta, in the systems forming an ensemble. The function f_1 is the familiar distribution function. The exact equations of motion for f_s in classical mechanics can be written down. They show that to find f_1 we need to know f_2, which in turns depends on a knowledge of f_3, and so on till we come the full N-body correlation function f_N. This system of equations is known as the BBGKY[†] hierarchy. We shall derive it and show how the chain of equations can be truncated to yield the Boltzmann transport equation. The "derivation" will not be any more rigorous than the one already given, but it will give new insight into the nature of the approximations.

Consider an ensemble of systems, each being a gas of N molecules enclosed in volume V, with Hamiltonian $\mathcal{H}$. Instead of the general notation $\{p_i, q_i\}$

---

*See the remarks about the relevance of the ergodic theorem in Section 4.5.

[†] BBGKY stands for Bogoliubov-Born-Green-Kirkwood-Yvon. For a detailed discussion and references see N. N. Bogoliubov in *Studies in Statistical Mechanics*, J. de Boer and G. E. Uhlenbeck, Eds., Vol. I (North-Holland, Amsterdam, 1962).

$(i = 1, \ldots, 3N)$, we shall denote the coordinates by the Cartesian vectors $\{\mathbf{p}_i, \mathbf{r}_i\}$ $(i = 1, \ldots, N)$, for which we use the abbreviation

$$z_i = (\mathbf{p}_i, \mathbf{r}_i), \qquad \int dz_i = \int d^3 p_i \, d^3 r_i \tag{3.43}$$

The density function characterizing the ensemble is denoted by $\rho(1, \ldots, N, t)$, and assumed to be symmetric in $z_1, \ldots, z_N$. Its integral over all phase space is a constant by Liouville's theorem; hence we can normalized it to unity:

$$\int dz_1 \cdots dz_N \rho(1, \ldots, N, t) = 1 \tag{3.44}$$

Thus the ensemble average of any function $O(1, \ldots, N)$ of molecular coordinates can be written as

$$\langle O \rangle \equiv \int dz_1 \cdots dz_N \rho(1, \ldots, N, t) O(1, \ldots, N) \tag{3.45}$$

Using the Hamiltonian equations of motion (3.38), we rewrite Liouville's theorem in the form

$$\frac{\partial \rho}{\partial t} = \sum_{i=1}^{N} \left(\nabla_{p_i} \rho \cdot \nabla_{r_i} \mathscr{H} - \nabla_{r_i} \rho \cdot \nabla_{p_i} \mathscr{H} \right) \tag{3.46}$$

Assume that the Hamiltonian is of the form

$$\mathscr{H} = \sum_{i=1}^{N} \frac{\mathbf{p}^2}{2m} + \sum_{i=1}^{N} U_i + \sum_{i<j} v_{ij}$$
$$U_i = U(\mathbf{r}_i) \tag{3.47}$$
$$v_{ij} = v_{ji} = v\left(|\mathbf{r}_i - \mathbf{r}_j| \right)$$

Then

$$\nabla_{p_i} \mathscr{H} = \frac{\mathbf{p}_i}{m}$$
$$\nabla_{r_i} \mathscr{H} = -\mathbf{F}_i - \sum_{\substack{j=1 \\ (j \neq i)}}^{N} \mathbf{K}_{ij} \tag{3.48}$$

where

$$\mathbf{F}_i = -\nabla_{r_i} U(\mathbf{r}_i)$$
$$\mathbf{K}_{ij} = -\nabla_{r_i} v\left(|\mathbf{r}_i - \mathbf{r}_j| \right) \tag{3.49}$$

Liouville's theorem can now be cast in the form

$$\left[\frac{\partial}{\partial t} + h_N(1, \ldots, N) \right] \rho(1, \ldots, N) = 0 \tag{3.50}$$

where

$$h_N(1,\ldots,N) = \sum_{i=1}^{N} S_i + \frac{1}{2} \sum_{\substack{i,j=1 \\ (i \neq j)}}^{N} P_{ij}$$

$$S_i \equiv \frac{\mathbf{p}_i}{m} \cdot \nabla_{r_i} + \mathbf{F}_i \cdot \nabla_{p_i} \qquad (3.51)$$

$$P_{ij} \equiv \mathbf{K}_{ij} \cdot \nabla_{p_i} + \mathbf{K}_{ji} \cdot \nabla_{p_j} = \mathbf{K}_{ij} \cdot \left(\nabla_{p_i} - \nabla_{p_j} \right)$$

The single-particle distribution function is defined by

$$f_1(\mathbf{p},\mathbf{r},t) \equiv \left\langle \sum_{i=1}^{N} \delta^3(\mathbf{p} - \mathbf{p}_i)\delta^3(\mathbf{r} - \mathbf{r}_i) \right\rangle = N \int dz_2 \cdots dz_N \rho(1,\ldots,N,t)$$

$$(3.52)$$

The factor N in the last form comes from the fact that all terms in the sum in the preceding term have the same value, owing to the fact that ρ is symmetric in $z_1,\ldots,z_N$. Integrating f_1 over z_1 yields the correct normalization N, by virtue of (3.44).

The general s-particle distribution function, or correlation function, is defined by

$$f_s(1,\ldots,z,t) \equiv \frac{N!}{(N-s)!} \int dz_{s+1} \cdots dz_N \rho(1,\ldots,N,t) \qquad (s = 1,\ldots,N)$$

$$(3.53)$$

The combinatorial factor in front comes from the fact that we do not care which particle is at z_1, which is at z_2, etc. The equation of motion is

$$\frac{\partial}{\partial t} f_s = \frac{N!}{(N-s)!} \int dz_{s+1} \cdots dz_N \frac{\partial}{\partial t}\rho = -\frac{N!}{(N-s)!} \int dz_{s+1} \cdots dz_N h_N \rho$$

$$(3.54)$$

We isolate those terms in h_N involving only the coordinates $z_1,\ldots,z_s$:

$$h_N(1,\ldots,N) = \sum_{i=1}^{s} S_i + \sum_{s+1}^{N} S_i + \frac{1}{2} \sum_{\substack{i,j=1 \\ (i \neq j)}}^{s} P_{ij} + \frac{1}{2} \sum_{\substack{i,j=s+1 \\ (i \neq j)}}^{N} P_{ij} + \sum_{i=1}^{s} \sum_{j=s+1}^{N} P_{ij}$$

$$= h_s(1,\ldots,s) + h_{N-s}(s+1,\ldots,N) + \sum_{i=1}^{s} \sum_{j=s+1}^{N} P_{ij} \qquad (3.55)$$

Note that

$$\int dz_{s+1} \cdots dz_N h_{N-s}(s+1,\ldots,N)\rho(1,\ldots,N) = 0 \qquad (3.56)$$

because h_{N-s} consists of gradient terms in $\mathbf{p}$ with $\mathbf{p}$-independent coefficients, and a gradient term in $\mathbf{r}$ with an $\mathbf{r}$-independent coefficient. Thus the integral evaluates ρ on the boundary of phase space, where we assume ρ to vanish. Substituting (3.55) into (3.54), we obtain

$$
\begin{aligned}
\left(\frac{\partial}{\partial t} + h_s\right) f_s &= -\frac{N!}{(N-s)!} \int dz_{s+1} \cdots dz_N \sum_{i=1}^{s} \sum_{j=s+1}^{N} P_{ij}\rho(1,\ldots,N) \\
&= -\sum_{i=1}^{s} \int dz_{s+1} P_{i,s+1} \frac{N!}{(N-s+1)!} \int dz_{s+2} \cdots dz_N \rho(1,\ldots,N) \\
&= -\sum_{i=1}^{s} \int dz_{s+1} P_{i,s+1} f_{s+1}(1,\ldots,s+1)
\end{aligned}
\tag{3.57}
$$

In passing from the first to the second equation we have used the fact that the sum over j gives $N - s$ identical terms. Now substitute P_{ij} from (3.51), and note that the second term there does not contribute, because it leads to a vanishing surface term. We then arrive at

$$
\left(\frac{\partial}{\partial t} + h_s\right) f_s(1,\ldots,s) = -\sum_{i=1}^{s} \int dz_{s+1} \mathbf{K}_{i,s+1} \cdot \nabla_{p_i} f_{s+1}(1,\ldots,s+1)
$$
$$
(s = 1,\ldots,N) \quad (3.58)
$$

which is the BBGKY hierarchy. The left side of each of the equations above is a "streaming term," involving only the s particles under consideration. For $s > 1$ it includes the effect of intermolecular scattering among the s particles. The right-hand side is the "collision integral," which describes the effect of scattering between the particles under consideration with an "outsider," thus coupling f_s to f_{s+1}.

The first two equations in the hierarchy read

$$
\left(\frac{\partial}{\partial t} + \frac{\mathbf{p}_1}{m} \cdot \nabla_{r_1} + \mathbf{F}_1 \cdot \nabla_{p_1}\right) f_1(z_1, t) = -\int dz_2 \mathbf{K}_{12} \cdot \nabla_{p_1} f_2(z_1, z_2, t) \tag{3.59}
$$

$$
\left[\frac{\partial}{\partial t} + \frac{\mathbf{p}_1}{m} \cdot \nabla_{r_1} + \frac{\mathbf{p}_2}{m} \cdot \nabla_{r_2} + \mathbf{F}_1 \cdot \nabla_{p_1} + \mathbf{F}_2 \cdot \nabla_{p_2} + \tfrac{1}{2}\mathbf{K}_{12} \cdot \left(\nabla_{p_1} - \nabla_{p_2}\right)\right]
$$
$$
\times f_2(z_1, z_2, t)
$$
$$
= -\int dz_3 \left(\mathbf{K}_{13} \cdot \nabla_{p_1} + \mathbf{K}_{23} \cdot \nabla_{p_2}\right) f_3(z_1, z_2, z_3, t) \tag{3.60}
$$

The terms in the equations above have dimensions f_s/time, and different time scales are involved:

$$
\mathbf{K} \cdot \nabla_p \sim \frac{1}{\tau_c}
$$
$$
\mathbf{F} \cdot \nabla_p \sim \frac{1}{\tau_e} \tag{3.61}
$$
$$
\frac{\mathbf{p}}{m} \cdot \nabla_r \sim \frac{1}{\tau_s}
$$

where τ_c is the duration of a collision, τ_e is the time for a molecule to traverse a characteristic distance over which the external potential varies significantly, and τ_s is the time for a molecule to traverse a characteristic distance over which the correlation function varies significantly. The time τ_c is the shortest, and τ_e the longest.

The equation for f_1 is unique in the hierarchy, in that "streaming" sets a rather slow time scale, for it does not involve intermolecular scattering, (there being only one particle present.) The collision integral, which has more rapid variations, sets the time scale of f_1. This is why the equilibrium condition is determined by the vanishing of the collision integral.

In contrast, the equation for f_2 (and higher ones as well) contains a collision term of the order $1/\tau_c$ on the left side. The collision integral on the right side is smaller by a factor of the order nr_0^3 (where n is the density, and r_0 the range of the intermolecular potential) because the integration of $\mathbf{r}_3$ extends only over a volume of radius r_0. Now $r_0 \approx 10^{-8}$ cm and $n \approx 10^{19}$ cm^{-3} under standard conditions. Hence $nr_0^3 \approx 10^{-5}$. Thus for f_2 (and higher equations too) the time scale is set by the streaming terms instead of the collision integral, which we shall neglect.

With neglect of the right side of (3.60), the hierarchy is truncated at f_2, and we have only two coupled equations for f_1 and f_2:

$$\left(\frac{\partial}{\partial t} + \frac{\mathbf{p}_1}{m} \cdot \nabla_{r_1}\right) f_1(z_1, t) = -\int_{r_0} dz_2\, \mathbf{K}_{12} \cdot \nabla_{p_1} f_2(z_1, z_2, t) \equiv \left(\frac{\partial f_1}{\partial t}\right)_{\text{coll}} \quad (3.62)$$

$$\left[\frac{\partial}{\partial t} + \frac{\mathbf{p}_1}{m} \cdot \nabla_{r_1} + \frac{\mathbf{p}_2}{m} \cdot \nabla_{r_2} + \tfrac{1}{2}\mathbf{K}_{12} \cdot \left(\nabla_{p_1} - \nabla_{p_2}\right)\right] f_2(z_1, z_2, t) = 0 \quad (3.63)$$

where we have set all external forces to zero, for simplicity. We shall also assume for simplicity that the force $\mathbf{K}$ vanishes outside a range r_0. To remind us of this fact, we put the subscript r_0 on the integral in the first equation, indicating that the spatial part of the integral is subject to $|\mathbf{r}_1 - \mathbf{r}_2| < r_0$.

The salient qualitative features of (3.62) and (3.63) are that f_2 varies in time with characteristic period τ_c, and in space with characteristic distance r_0, while f_1 varies much less rapidly, by a factor nr_0^3. Thus f_1 measures space and time with much coarser scales than f_2. ·

The correlations in f_2 are produced by collisions between particles 1 and 2. When their positions are so far separated as to be out of molecular interaction range, we expect that there will be no correlation between 1 and 2, and f_2 will assume a product form (neglecting, of course, possible correlations produced by collisions with a third particle):

$$f_2(z_1, z_2, t) \xrightarrow[|\mathbf{r}_1 - \mathbf{r}_2| \gg r_0]{} f_1(z, t) f_1(z_2, t) \quad (3.64)$$

To evaluate $(\partial f_1/\partial t)_{\text{coll}}$, however, we need f_2 not in the uncorrelated region, but in the region where the two particles are colliding. To look at this

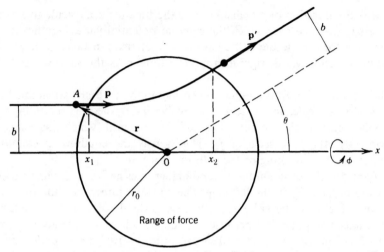

Fig. 3.8 Illustration of behavior of two-particle correlation function. The separation between the two particles is **r**, and the relative momentum **p**. The two particles are correlated only inside the range of the intermolecular force, indicated by the sphere of radius r_0. Outside the sphere, the correlation function is a product of two one-particle distribution functions. In equilibrium there is a steady scattering of beams of particles of all momenta, from all directions, at all impact parameters.

region it is convenient to use total and relative coordinates, defined as follows:

$$\mathbf{P} = \mathbf{p}_2 + \mathbf{p}_1 \qquad \mathbf{p} = \frac{\mathbf{p}_2 - \mathbf{p}_1}{2}$$

$$\mathbf{R} = \frac{\mathbf{r}_2 + \mathbf{r}_1}{2} \qquad \mathbf{r} = \mathbf{r}_2 - \mathbf{r}_1 \tag{3.65}$$

Then (3.63) becomes

$$\left(\frac{\partial}{\partial t} + \frac{\mathbf{p}}{m} \cdot \nabla_R + \frac{\mathbf{p}}{m} \cdot \nabla_r + \mathbf{K}(\mathbf{r}) \cdot \nabla_p \right) f_2(\mathbf{P}, \mathbf{R}, \mathbf{p}, \mathbf{r}, t) = 0 \tag{3.66}$$

$$\mathbf{K}(\mathbf{r}) \equiv -\nabla v(r)$$

Transform to the center-of-mass system by putting $\mathbf{P} = 0$. The above can then be rewritten, to first order in dt, as the streaming condition (with **P** and **R** suppressed for clarity):

$$f_2\left(\mathbf{p} + \mathbf{K}(\mathbf{r}) \, dt, \mathbf{r} + \frac{\mathbf{p}}{m} \, dt, t + dt \right) = f_2(\mathbf{p}, \mathbf{r}, t) \tag{3.67}$$

It traces the classical trajectories in the force field **K** centered at O, as illustrated in Fig. 3.8. If f_2 were peaked at point A initially, then (3.67) says that as time goes on the peak will move along the trajectory for that particular initial condition.

The equilibrium situation, for which $\partial f_2/\partial t = 0$, is a steady-state scattering, by the force field $\mathbf{K}$, of a beam of particles consisting of all momenta, at all impact parameters. Referring to Fig. 3.8, we may describe the steady state as follows: Outside the sphere of interaction the uncorrelated factorized form (3.64) holds. However, boundary values of the momenta are correlated through the fact that momenta entering the sphere at a specific impact parameter must leave the sphere at the correct scattering angle, and vice versa.

To "derive" the Boltzmann transport equation, we assume that, since f_2 has a shorter time scale than f_1, it reaches equilibrium earlier than f_1. Thus we set $\partial f_2/\partial t = 0$, and assume f_2 has attained the equilibrium form described earlier. Similarly, we assume that the range of force r_0 is essentially zero from the point of view of f_1. Thus in the factorized form of f_2 just before and after a collision, we can put $\mathbf{r}_2$ and $\mathbf{r}_1$ both equal to the same value.

With this in mind, we substitute (3.63) into (3.62) to obtain

$$
\left(\frac{\partial f_1}{\partial t} \right)_{\text{coll}} \equiv - \int_{r_0} dz_2 \, \mathbf{K}_{12} \cdot \nabla_{p_1} f_2(z_1, z_2, t)
$$

$$
= - \int_{r_0} dz_2 \, \mathbf{K}_{12} \cdot \left(\nabla_{p_1} - \nabla_{p_2} \right) f_2(z_1, z_2, t)
$$

$$
= \frac{1}{m} \int_{r_0} dz_2 \, \left(\mathbf{p}_1 \cdot \nabla_{r_1} + \mathbf{p}_2 \cdot \nabla_{r_2} \right) f_2(z_1, z_2, t) \tag{3.68}
$$

Using the coordinates defined in (3.66), and neglecting the gradient with respect to $\mathbf{R}$, we have

$$
\left(\frac{\partial f_1}{\partial t} \right)_{\text{coll}} = - \frac{1}{m} \int d^3 p_2 \int_{r < r_0} d^3 r \, (\mathbf{p}_1 - \mathbf{p}_2) \cdot \nabla_r f_2
$$

$$
= \frac{1}{m} \int d^3 p_2 \, |\mathbf{p}_1 - \mathbf{p}_2| \int d\phi \, b \, db \int_{x_1}^{x_2} dx \, \frac{\partial}{\partial x} f_2 \tag{3.69}
$$

where the notation is indicated in Fig. 3.8. Now we set

$$
f_2(x_1) = f_1(\mathbf{p}_1) f_1(\mathbf{p}_2)
$$

$$
f_2(x_2) = f_1(\mathbf{p}_1') f_1(\mathbf{p}_2')
$$

where $\mathbf{p}_1', \mathbf{p}_2'$ are the final momenta in the scattering process, when the initial momenta are $\mathbf{p}_1, \mathbf{p}_2$ and the impact parameter is b. Using the definition (3.21) of the classical cross section, we finally have

$$
\left(\frac{\partial f_1}{\partial t} \right)_{\text{coll}} = \int d^3 p_2 \, d\Omega \, |\mathbf{v}_1 - \mathbf{v}_2| (d\sigma/d\Omega)(f_1' f_2' - f_1 f_2) \tag{3.70}
$$

which is the same as (3.34).

PROBLEMS

3.1 Give a few numerical examples to show that the condition (3.1) is fulfilled for physical gases at room temperatures.

3.2 Explain qualitatively why all molecular interactions are electromagnetic in origin.

3.3 For the collision between perfectly elastic spheres of diameter a,

(*a*) calculate the differential cross section with classical mechanics in the coordinate system in which one of the spheres is initially at rest;

(*b*) compare your answer with the quantum mechanical result. Consider both the low-energy and the high-energy limit. (See, e.g., L. I. Schiff, *Quantum Mechanics*, 2nd ed. (McGraw-Hill, New York, 1955), p. 110).

3.4 Consider a mixture of two gases whose molecules have masses m and M, respectively, and which are subjected to external forces $\mathbf{F}$ and $\mathbf{Q}$, respectively. Denote the respective distribution functions by f and g. Assuming that only binary collisions between molecules are important, derive the Boltzmann transport equation for the system.

3.5 This problem illustrates in a trivial case how the ensemble density tends to a uniform density over the accessible phase space. Consider an ensemble of systems, each of which consists of a single free particle in one dimension with momentum p and coordinate q. The particle is confined to a one-dimensional box with perfectly reflecting walls located at $q = -1$ and $q = 1$ (in arbitrary units.) Draw a square box of unit sides in the pq plane (the phase space). Draw a square of sides $1/2$ in the upper left corner of this box. Let the initial ensemble correspond to filling this corner box uniformly with representative points.

(*a*) What is the accessible part of the phase space? (i.e., the region that the representative points can reach through dynamical evolution from the initial condition.)

(*b*) Consider how the shape of the distribution of representative point changes at regular successive time intervals. How does the distribution look after a long time?

Suggestion: When a particle is being reflected at a wall, its momentum changes sign. Represent what happens in phase space by continuing the locus of the representative point to a fictitious adjacent box in pq space, as if the wall were absent. "Fold" the adjacent box onto the original box properly to get the actual trajectory of the representative point. After a long time, you need many such adjacent boxes. The "folding back" will then give you a picture of the distribution.

4

THE EQUILIBRIUM STATE OF A DILUTE GAS

4.1 BOLTZMANN'S *H* THEOREM

We define the equilibrium distribution function as the solution of the Boltzmann transport equation that is independent of time. We shall see that it is also the limiting form of the distribution function as the time tends to infinity. Assume that there is no external force. It is then consistent to assume further that the distribution function is independent of $\mathbf{r}$ and hence can be denoted by $f(\mathbf{p}, t)$. The equilibrium distribution function, denoted by $f_0(\mathbf{p})$, is the solution to the equation $\partial f(\mathbf{p}, t)/\partial t = 0$. According to the Boltzmann transport equation (3.36), $f_0(\mathbf{p})$ satisfies the integral equation

$$0 = \int d^3p_2 \, d^3p_1' \, d^3p_2' \, \delta^4(P_f - P_i)|T_{fi}|^2(f_2'f_1' - f_2f_1) \tag{4.1}$$

where p_1 is a given momentum.

A sufficient condition for $f_0(\mathbf{p})$ to solve (4.1) is

$$f_0(\mathbf{p}_2')f_0(\mathbf{p}_1') - f_0(\mathbf{p}_2)f_0(\mathbf{p}_1) = 0 \tag{4.2}$$

where $\{\mathbf{p}_1, \mathbf{p}_2\} \rightarrow \{\mathbf{p}_1', \mathbf{p}_2'\}$ is any possible collision (i.e., one with nonvanishing cross section). We show that this condition is also necessary, and we thus arrive at the interesting conclusion that $f_0(p)$ is independent of $d\sigma/d\Omega$, as long as the latter is nonzero.

To show the necessity of (4.2) we define with Boltzmann the functional

$$H(t) \equiv \int d^3v f(\mathbf{p}, t) \log f(\mathbf{p}, t) \tag{4.3}$$

where $f(\mathbf{p}, t)$ is the distribution function at time t, satisfying

$$\frac{\partial f(p_1, t)}{\partial t} = \int d^3p_2 \, d^3p_1' \, d^3p_2' \, \delta^4(P_f - P_i)|T_{fi}|^2(f_2'f_1' - f_2f_1) \tag{4.4}$$

Differentiation of (4.3) yields

$$\frac{dH(t)}{dt} = \int d^3v \, \frac{\partial f(\mathbf{p}, t)}{\partial t} [1 + \log f(\mathbf{p}, t)] \tag{4.5}$$

Therefore $\partial f/\partial t = 0$ implies $dH/dt = 0$. This means that a necessary condition for $\partial f/\partial t = 0$ is $dH/dt = 0$. We now show that the statement

$$\frac{dH}{dt} = 0 \tag{4.6}$$

is the same as (4.2). It would then follow that (4.2) is also a necessary condition for the solution of (4.1). To this end we prove the following theorem.

BOLTZMANN'S H THEOREM
If f satisfies the Boltzmann transport equation, then

$$\frac{dH(t)}{dt} \le 0 \tag{4.7}$$

Proof Substituting (4.4) into the integrand of (4.5) we have*

$$\frac{dH}{dt} = \int d^3p_2 \, d^3p_1' \, d^3p_2' \, \delta^4(P_f - P_i)|T_{fi}|^2 (f_2'f_1' - f_2f_1)(1 + \log f_1) \tag{4.8}$$

Interchanging $\mathbf{p}_1$ and $\mathbf{p}_2$ in this integrand leaves the integral invariant because T_{fi} is invariant under such an interchange. Making this change of variables of integration and taking one-half of the sum of the new expression and (4.8), we obtain

$$\frac{dH}{dt} = \frac{1}{2} \int d^3p_2 \, d^3p_1' \, d^3p_2' \, \delta^4(P_f - P_i)|T_{fi}|^2$$
$$\times (f_2'f_1' - f_2f_1)[2 + \log(f_1f_2)] \tag{4.9}$$

This integral is invariant under the interchange of $\{\mathbf{p}_1, \mathbf{p}_2\}$ and $\{\mathbf{p}_1', \mathbf{p}_2'\}$ because for every collision there is an inverse collision with the same T matrix. Hence

$$\frac{dH}{dt} = -\frac{1}{2} \int d^3p_2 \, d^3p_1' \, d^3p_2' \, \delta^4(P_f - P_i)|T_{fi}|^2 (f_2'f_1' - f_2f_1)$$
$$\times [2 + \log(f_1'f_2')] \tag{4.10}$$

Taking half the sum of (4.9) and (4.10) we obtain

$$\frac{dH}{dt} = \frac{1}{4} \int d^3p_2 \, d^3p_1' \, d^3p_2' \, \delta^4(P_f - P_i)|T_{fi}|^2 (f_2'f_1' - f_2f_1)$$
$$\times [\log(f_1f_2) - \log(f_1'f_2')] \tag{4.11}$$

The integrand of the integral in (4.11) is never positive. ∎

---

*Note that the use of (4.4) presupposes that the state of the system under consideration satisfies the assumption of molecular chaos.

As a by-product of the proof, we deduce from (4.11) that $dH/dt = 0$ if and only if the integrand of (4.11) identically vanishes. This proves that the statement (4.6) is identical with (4.2). It also shows that under an arbitrary initial condition $f(\mathbf{p}, t) \underset{t \to \infty}{\to} f_0(\mathbf{p})$.

4.2 THE MAXWELL-BOLTZMANN DISTRIBUTION

It has been shown that the equilibrium distribution function $f_0(\mathbf{p})$ is a solution of (4.2). It will be called the Maxwell-Boltzmann distribution. To find it, let us take the logarithm of both sides of (4.2):

$$\log f_0(\mathbf{p}_1) + \log f_0(\mathbf{p}_2) = \log f_0(\mathbf{p}_1') + \log f_0(\mathbf{p}_2') \tag{4.12}$$

Since $\{\mathbf{p}_1, \mathbf{p}_2\}$ and $\{\mathbf{p}_1', \mathbf{p}_2'\}$ are, respectively, the initial and final velocities of *any* possible collision, (4.12) has the form of a conservation law. If $\chi(\mathbf{p})$ is any quantity associated with a molecule of velocity $\mathbf{p}$, such that $\chi(\mathbf{p}_1) + \chi(\mathbf{p}_2)$ is conserved in a collision between molecules $\mathbf{p}_1$ and $\mathbf{p}_2$, a solution of (4.12) is

$$\log f_0(\mathbf{p}) = \chi(\mathbf{p})$$

The most general solution of (4.12) is

$$\log f_0(\mathbf{p}) = \chi_1(\mathbf{p}) + \chi_2(\mathbf{p}) + \cdots$$

where the list $\chi_1, \chi_2, \ldots$ exhausts all independently conserved quantities. For spinless molecules, these are the energy and the momentum of a molecule, and, of course, a constant. Hence $\log f$ is a linear combination of $\mathbf{p}^2$ and the three components of $\mathbf{p}$ plus an arbitrary constant:

$$\log f_0(\mathbf{p}) = -A(\mathbf{p} - \mathbf{p}_0)^2 + \log C$$

or

$$f_0(\mathbf{p}) = Ce^{-A(\mathbf{p} - \mathbf{p}_0)^2} \tag{4.13}$$

where C, A, and the three components of $\mathbf{p}_0$ are five arbitrary constants. We can determine these constants in terms of observed properties of the system.

Applying the condition (3.5), and denoting the particle density N/V by n we have

$$n = C\int d^3p\, e^{-A(\mathbf{p} - \mathbf{p}_0)^2} = C\int d^3p\, e^{-Ap^2} = C\left(\frac{\pi}{A}\right)^{3/2}$$

from which we conclude that $A > 0$ and

$$C = \left(\frac{A}{\pi}\right)^{3/2} n \tag{4.14}$$

Let the average momentum $\langle \mathbf{p} \rangle$ of a gas molecule be defined by

$$\langle \mathbf{p} \rangle \equiv \frac{\int d^3p\, \mathbf{p} f_0(\mathbf{p})}{\int d^3p\, f_0(\mathbf{p})} \tag{4.15}$$

Then

$$\langle \mathbf{p} \rangle = \frac{C}{n} \int d^3p \, \mathbf{p} e^{-A(\mathbf{p}-\mathbf{p}_0)^2} = \frac{C}{n} \int d^3p \, (\mathbf{p} + \mathbf{p}_0) e^{-Ap^2} = \mathbf{p}_0 \quad (4.16)$$

Thus we must take $\mathbf{p}_0 = 0$, if the gas has no translational motion as a whole.

Next we calculate the average energy ϵ of a molecule, defined by

$$\epsilon \equiv \frac{\int d^3p \, (p^2/2m) f_0(\mathbf{p})}{\int d^3p \, f_0(\mathbf{p})} \quad (4.17)$$

We have, setting $\mathbf{p}_0 = 0$,

$$\epsilon = \frac{C}{2nm} \int d^3p \, \frac{p^2}{2m} e^{-Ap^2} = \frac{2\pi C}{nm} \int_0^\infty dp \, p^4 e^{-Ap^2} = \frac{3}{4Am}$$

The constant A is therefore related to the average energy by

$$A = \frac{3}{4\epsilon m} \quad (4.18)$$

Substituting this into (4.14) we obtain for the constant C the expression

$$C = n\left(\frac{3}{4\pi\epsilon m}\right)^{3/2} \quad (4.19)$$

To relate the average energy ϵ to a directly measurable quantity, let us find the equation of state corresponding to the equilibrium distribution function. We do this by calculating the pressure, which is defined as the average force per unit area exerted by the gas on one face of a perfectly reflecting plane exposed to the gas. Let the disk shown in Fig. 4.1 represent such a unit area, and let us call the axis normal to it the x axis. A molecule can hit this disk only if the x component of its momentum p_x is positive. Then it loses an amount of momentum $2p_x$ upon reflection from this disk. The number of molecules reflected by the disk per second is the number of molecules contained in the cylinder shown in Fig. 4.1 with $v_x > 0$. This number is $v_x f_0(\mathbf{p}) \, d^3p$, with $v_x > 0$. Therefore the pressure is,

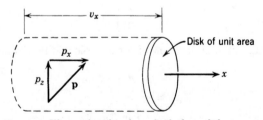

Fig. 4.1 Illustration for the calculation of the pressure.

for a gas with zero average velocity

$$P = \int_{v_x > 0} d^3p \, 2p_x v_x f_0(\mathbf{p}) = \int d^3p \, p_x v_x f_0(\mathbf{p})$$

$$= \frac{C}{m} \int d^3p \, p \, p_x^2 e^{-Ap^2} = \frac{C}{3m} \int d^3p \, p \, p^2 e^{-Ap^2} \qquad (4.20)$$

where the last step comes about because $f_0(\mathbf{p})$ depends only on $|\mathbf{p}|$ so that the average values of p_x^2, p_y^2, and p_z^2 are all equal to one-third of the average of $p^2 = p_x^2 + p_y^2 + p_z^2$. Finally we notice that

$$P = \tfrac{2}{3}C \int d^3p \, \frac{p^2}{2m} e^{-Ap^2} = \tfrac{2}{3}n\epsilon \qquad (4.21)$$

This is the equation of state. Experimentally we define the temperature T by $P = nkT$, where k is Boltzmann's constant. Hence

$$\epsilon = \tfrac{3}{2}kT \qquad (4.22)$$

In terms of the temperature T, the average momentum $\mathbf{p_0}$, and the particle density n the equilibrium distribution function for a dilute gas in the absence of external force is

$$f_0(\mathbf{p}) = \frac{n}{(2\pi mkT)^{3/2}} e^{-(\mathbf{p} - \mathbf{p_0})^2/2mkT} \qquad (4.23)$$

This is the Maxwell-Boltzmann distribution, the probability of finding a molecule with momentum $\mathbf{p}$ in the gas, under equilibrium conditions.*

If a perfectly reflecting wall is introduced into the gas, $f_0(\mathbf{p})$ will remain unchanged because $f_0(\mathbf{p})$ depends only on the magnitude of $\mathbf{p}$, which is unchanged by reflection from the wall.

For a gas with $\mathbf{p_0} = 0$ it is customary to define the most probable speed $\bar{v}$ of a molecule by the value of v at which $4\pi p^2 f(\mathbf{p})$ attains a maximum. We easily find $\bar{p} = \sqrt{2mkT}$. The most probable speed is therefore

$$\bar{v} = \sqrt{\frac{2kT}{m}} \qquad (4.24)$$

The root mean square speed v_{rms} is defined by

$$v_{rms} \equiv \left[\frac{\int d^3p \, v^2 f_0(\mathbf{p})}{\int d^3p \, f_0(\mathbf{p})} \right]^{1/2} = \sqrt{\frac{3kT}{m}} \qquad (4.25)$$

At room temperatures these speeds for an O_2 gas are of the order of magnitude of 10^5 cm/s.

---

*We have assumed, in accordance with experimental facts, that the temperature T is independent of the average momentum $\mathbf{p_0}$.

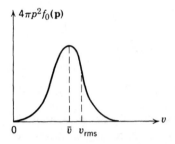

Fig. 4.2 The Maxwell-Boltzmann distribution.

A plot of $4\pi p^2 f_0(\mathbf{p})$ against $v = p/m$ is shown in Fig. 4.2. We notice that $f_0(\mathbf{p})$ does not vanish, as it should, when v exceeds the velocity of light c. This is because we have used Newtonian dynamics for the molecules instead of the more correct relativistic dynamics. The error is negligible at room temperatures, because $\bar{v} \ll c$. The temperature above which relativistic dynamics must be used can be roughly estimated by putting $\bar{v} = c$, from which we obtain $kT \approx mc^2$. Hence $T \approx 10^{13}$ K for H_2.

Let us now consider the equilibrium distribution for a dilute gas in the presence of an external conservative force field given by

$$\mathbf{F} = -\nabla\phi(\mathbf{r}) \tag{4.26}$$

We assert that the equilibrium distribution function is now

$$f(\mathbf{r}, \mathbf{p}) = f_0(\mathbf{p})e^{-\phi(\mathbf{r})/kT} \tag{4.27}$$

where $f_0(\mathbf{p})$ is given by (4.23). To prove this we show that (4.27) satisfies Boltzmann's equation. We see immediately that $\partial f/\partial t = 0$ because (4.27) is independent of the time. Furthermore $(\partial f/\partial t)_{\text{coll}} = 0$ because $\phi(\mathbf{r})$ is independent of $\mathbf{p}$:

$$\left(\frac{\partial f}{\partial t}\right)_{\text{coll}} = e^{-2\phi(\mathbf{r})/kT}\int d^3p_2 \, d^3p_1' \, d^3p_2' \, \delta^4(P_f - P_i)|T_{fi}|^2(f_2'f_1' - f_2 f_1) = 0$$

Hence it is only necessary to verify that

$$\left(\frac{\mathbf{p}}{m}\cdot\nabla_\mathbf{r} + \mathbf{F}\cdot\nabla_\mathbf{p}\right)f(\mathbf{r}, \mathbf{p}) = 0$$

and this is trivial. We may absorb the factor $\exp(-\phi/kT)$ in (4.27) into the density n and write

$$f(\mathbf{r}, \mathbf{p}) = \frac{n(\mathbf{r})}{(2\pi m kT)^{3/2}}e^{-(\mathbf{p} - \mathbf{p}_0)^2/2mkT} \tag{4.28}$$

where

$$n(\mathbf{r}) = \int d^3p f(\mathbf{r}, \mathbf{p}) = n_0 e^{-\phi(\mathbf{r})/kT} \tag{4.29}$$

Finally we derive the thermodynamics of a dilute gas. We have defined the temperature by (4.22) and we have obtained the equation of state. By the very definition of the pressure, the work done by the gas when its volume increases by dV is $P\,dV$. The internal energy is defined by

$$U(T) = N\epsilon = \tfrac{3}{2}NkT \tag{4.30}$$

which is obviously a function of the temperature alone.

The analog of the first law of thermodynamics now takes the form of a definition for the heat absorbed by the system:

$$dQ = dU + P\,dV \tag{4.31}$$

It tells us that heat added to the system goes into the mechanical work $P\,dV$ and the energy of molecular motion dU. From (4.31) and (4.30) we obtain for the heat capacity at constant volume

$$C_V = \tfrac{3}{2}Nk \tag{4.32}$$

The analog of the second law of thermodynamics is Boltzmann's H theorem, where we identify H with the negative of the entropy per unit volume divided by Boltzmann's constant:

$$H = -\frac{S}{Vk} \tag{4.33}$$

Thus the H theorem states that for a fixed volume (i.e., for an isolated gas) the entropy never decreases, which is a statement of the second law.

To justify (4.33) we calculate H in equilibrium:

$$H_0 = \int d^3p\, f_0 \log f_0 = n\left\{ \log\left[n\left(\frac{1}{2\pi mkT}\right)^{3/2}\right] - \frac{3}{2}\right\}$$

Using the equation of state we can rewrite this as

$$-kVH_0 = \tfrac{3}{2}Nk \log\left(PV^{5/3}\right) + \text{constant} \tag{4.34}$$

We recognize that the right-hand side is the entropy of an ideal gas in thermodynamics. It follows from (4.34), (4.33), and (4.31) that $dS = dQ/T$.

Thus we have derived all of classical thermodynamics for a dilute gas; and moreover, we were able to calculate the equation of state and the specific heat. The third law of thermodynamics cannot be derived here because we have used classical mechanics and thus are obliged to confine our considerations to high temperatures.

4.3 THE METHOD OF THE MOST PROBABLE DISTRIBUTION

We have noted the interesting fact that the Maxwell-Boltzmann distribution is independent of the detailed form of molecular interactions, as long as they exist. This fact endows the Maxwell-Boltzmann distribution with universality. We might suspect that as long as we are interested only in the *equilibrium* behavior of

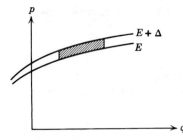

Fig. 4.3 The microcanonical ensemble corresponding to a gas contained in a finite volume with energy between E and $E + \Delta$.

a gas there is a way to derive the Maxwell-Boltzmann distribution without explicitly mentioning molecular interactions. Such a derivation is now supplied. Through it we shall understand better the meaning of the Maxwell-Boltzmann distribution. The conclusion we reach will be the following. If we choose a state of the gas at random from among all its possible states consistent with certain macroscopic conditions, the probability that we shall choose a Maxwell-Boltzmann distribution is overwhelmingly greater than that for any other distribution.

We shall use the approach of the Gibbsian ensemble described in Sec. 3.4. We assume that in equilibrium the system is equally likely to be found in any state consistent with the macroscopic conditions. That is, the density function is a constant over the accessible portion of Γ space.

Specifically we consider a gas of N molecules enclosed in a box of volume V with perfectly reflecting walls. Let the energy of the gas lie between E and $E + \Delta$, with $\Delta \ll E$. The ensemble then consists of a uniform distribution of points in a region of Γ space bounded by the energy surfaces of energies E and $E + \Delta$, and the surfaces corresponding to the boundaries of the containing box, as illustrated schematically in Fig. 4.3. Since the walls are perfectly reflecting, energy is conserved, and a representative point never leaves this region. By Liouville's theorem the distribution of representative points moves likes an incompressible fluid, and hence maintains a constant density at all times. This ensemble is called a *microcanonical ensemble*.

Next consider an arbitrary distribution function of a gas. A molecule in the gas is confined to a finite region of μ space because the values of p and q are restricted by the macroscopic conditions. Cover this finite region of μ space with volume elements of volume $\omega = d^3p \, d^3q$, and number them from 1 to K, where K is a very large number which eventually will be made to approach infinity. We refer to these volume elements as cells. An arbitrary distribution function is defined if we specify the number of molecules n_i found in the ith cell. These are called occupation numbers, and they satisfy the conditions

$$\sum_{i=1}^{K} n_i = N \tag{4.35}$$

$$\sum_{i=1}^{K} \epsilon_i n_i = E \tag{4.36}$$

where ϵ_i is the energy of a molecule in the ith cell:

$$\epsilon_i = \frac{p_i^2}{2m}$$

where $\mathbf{p}_i$ is the momentum of the ith cell. It is in (4.36), and only in (4.36), that the assumption of a dilute gas enters. An arbitrary set of integers $\{n_i\}$ satisfying (4.35) and (4.36) defines an arbitrary distribution function. The value of the distribution function in the ith cell, denoted by f_i, is

$$f_i = \frac{n_i}{\omega} \tag{4.37}$$

This is the distribution function for one member in the ensemble. The equilibrium distribution function is the above averaged over the microcanonical ensemble, which assigns equal weight to all systems satisfying (4.35) and (4.36):

$$\bar{f}_i = \frac{\langle n_i \rangle}{\omega}$$

This is the same definition as (3.52) except that we have replaced the infinitesimal element $d^3r\,d^3p$ by a finite cell of volume ω. Unfortunately this ensemble average is difficult to calculate. So we shall adopt a somewhat different approach, which will yield the same result for a sufficiently large system.

It is clear that if the state of the gas is given, then f is uniquely determined; but if f is given, the state of the gas is not uniquely determined. For example, interchanging the positions of two molecules in the gas leads to a new state of the gas, and hence moves the representative in Γ space; but that does not change the distribution function. Thus a given distribution function f corresponds not to a point, but to a volume in Γ space, which we call the *volume occupied by f*. We shall assume that the equilibrium distribution function is the *most probable distribution function*, i.e., that which occupies the largest volume in Γ space.

The procedure is then as follows:

(a) Choose an arbitrary distribution function by choosing an arbitrary set of allowed occupations numbers. Calculate the volumes it occupies by counting the number of systems in the ensemble that have these occupation numbers.

(b) Vary the distribution function to maximize the volume.

Let us denote by $\Omega\{n_i\}$ the volume in Γ space occupied by the distribution function corresponding to the occupation numbers $\{n_i\}$. It is proportional to the number of ways of distributing N distinguishable molecules among K cells so that there are n_i of them in the ith cell ($i = 1, 2, \ldots, K$). Therefore

$$\Omega\{n_i\} \propto \frac{N!}{n_1! n_2! n_3! \cdots n_K!} g_1^{n_1} g_2^{n_2} \cdots g_K^{n_K} \tag{4.38}$$

where g_i is a number that we will put equal to unity at the end of the calculation but that is introduced here for mathematical convenience. Taking the logarithm of (4.38) we obtain

$$\log \Omega \{ n_i \} = \log N! - \sum_{i=1}^{K} \log n_i! + \sum_{i=1}^{K} n_i \log g_i + \text{constant}$$

Now assume that each n_i is a very large integer, so we can use Stirling's approximation, $\log n_i! \approx n_i \log n_i - 1$. We then have

$$\log \Omega \{ n_i \} = N \log N - \sum_{i=1}^{K} n_i \log n_i! + \sum_{i=1}^{K} n_i \log g_i + \text{constant} \quad (4.39)$$

To find the equilibrium distribution we vary the set of integers $\{ n_i \}$ subject to the conditions (4.35) and (4.36) until $\log \Omega$ attains a maximum. Let $\{ \bar{n}_i \}$ denote the set of occupation numbers that maximizes $\log \Omega$. By the well-known method Lagrange multipliers we have

$$\delta [\log \Omega \{ n_i \}] - \delta \left(\alpha \sum_{i=1}^{K} n_i + \beta \sum_{i=1}^{K} \epsilon_i n_i \right) = 0 \qquad (n_i = \bar{n}_i) \qquad (4.40)$$

where α, β are Lagrange's multipliers. Now the n_i can be considered independent of one another. Substituting (4.39) into (4.40) we obtain

$$\sum_{i=1}^{K} \left[-(\log n_i + 1) + \log g_i - \alpha - \beta \epsilon_i \right] \delta n_i = 0 \qquad (n_i = \bar{n}_i)$$

Since δn_i are independent variations, we obtain the equilibrium condition by setting the summand equal to zero:

$$\log \bar{n}_i = -1 + \log g_i - \alpha - \beta \epsilon_i$$
$$\bar{n}_i = g_i e^{-\alpha - \beta \epsilon_i - 1} \qquad (4.41)$$

The most probable distribution function is, by (4.37) and (4.41),

$$\bar{f}_i = C e^{-\beta \epsilon_i} \qquad (4.42)$$

where C is a constant. The determination of the constants C and β proceeds in the same way as for (4.13). Writing $\bar{f}_i \equiv f(\mathbf{p}_i)$, we see that $f(\mathbf{p})$ is the Maxwell-Boltzmann distribution (4.23) for $\mathbf{p}_0 = 0$. To show that (4.41) actually corresponds to a maximum of $\log \Omega \{ n_i \}$ we calculate the second variation. It is easily shown that the second variation of the quantity on the left side of (4.40), for $n_i = \bar{n}_i$, is

$$- \sum_{i=1}^{K} \frac{1}{n_i} (\delta n_i)^2 < 0$$

We have obtained the Maxwell-Boltzmann distribution as the most probable distribution, in the sense that among all the systems satisfying the macroscopic conditions the Maxwell-Boltzmann distribution is the distribution common to the largest number of them. The question remains: What fraction of these systems have the Maxwell-Boltzmann distributions? In other words, how probable is the most probable distribution? The probability for the occurrence of any set of occupation numbers $\{n_i\}$ is given by

$$P\{n_i\} = \frac{\Omega\{n_i\}}{\sum_{\{n_j'\}} \Omega\{n_j'\}} \tag{4.43}$$

where the sum in the denominator extends over all possible sets of integers $\{n_j'\}$ satisfying (4.35) and (4.36). The probability for finding the system in the Maxwell-Boltzmann distribution is therefore $P\{\bar{n}_i\}$. A direct calculation of $P\{\bar{n}_i\}$ is not easy. We shall be satisfied with an estimate, which, however, becomes an exact evaluation if this probability approaches unity.

The ensemble average of n_i is given by

$$\langle n_i \rangle = \frac{\sum_{\{n_j\}} n_i \Omega\{n_j\}}{\sum_{\{n_j\}} \Omega\{n_j\}} \tag{4.44}$$

It is obvious from (4.38) that

$$\langle n_i \rangle = g_i \frac{\partial}{\partial g_i} \log \left[\sum_{\{n_j\}} \Omega\{n_j\} \right] \tag{4.45}$$

if we let $g_i \to 1$. The deviations from the average value can be estimated by calculating the mean square fluctuation $\langle n_i^2 \rangle - \langle n_i \rangle^2$. We can express $\langle n_i^2 \rangle$ in terms of $\langle n_i \rangle$ as follows:

$$\langle n_i^2 \rangle \equiv \frac{\sum n_i^2 \Omega}{\sum \Omega} = \frac{g_i \dfrac{\partial}{\partial g_i} \left(g_i \dfrac{\partial}{\partial g_i} \sum \Omega \right)}{\sum \Omega} \tag{4.46}$$

where the sum $\sum$ extends over all allowed $\{n_i\}$. Through the series of steps given next we obtain the desired results:

$$
\begin{aligned}
\langle n_i^2 \rangle &= g_i \frac{\partial}{\partial g_i} \left(\frac{1}{\sum \Omega} g_i \frac{\partial}{\partial g_i} \sum \Omega \right) - g_i \left(\frac{\partial}{\partial g_i} \frac{1}{\sum \Omega} \right) g_i \frac{\partial}{\partial g_i} \sum \Omega \\
&= g_i \frac{\partial}{\partial g_i} \left(g_i \frac{\partial}{\partial g_i} \log \sum \Omega \right) + \left(\frac{1}{\sum \Omega} g_i \frac{\partial}{\partial g_i} \sum \Omega \right)^2 \\
&= g_i \frac{\partial}{\partial g_i} \langle n_i \rangle + \langle n_i \rangle^2 \tag{4.47}
\end{aligned}
$$

Therefore the mean square fluctuation is

$$\langle n_i^2 \rangle - \langle n_i \rangle^2 = g_i \frac{\partial}{\partial g_i} \langle n_i \rangle \tag{4.48}$$

where we must let $g_i \to 1$ at the end of the calculation.

If the mean square fluctuation is large compared to $\langle n_i \rangle^2$, then $\langle n_i \rangle$ may differ considerably from $\bar{n}_i$; but then neither of them will be physically meaningful. If the mean square fluctuation is small compared to $\langle n_i \rangle^2$, we may expect $\langle n_i \rangle$ and $\bar{n}_i$ to be almost equal. We assume the latter is so, and we shall see that this is a consistent assumption.* Putting

$$\langle n_i \rangle \approx \bar{n}_i$$

we find from (4.41) and (4.48) that

$$\langle n_i^2 \rangle - \langle n_i \rangle^2 \approx \bar{n}_i$$

or

$$\sqrt{\left\langle \left(\frac{n_i}{N}\right)^2 \right\rangle - \left\langle \frac{n_i}{N} \right\rangle^2} \approx \frac{\sqrt{\bar{n}_i/N}}{\sqrt{N}} \qquad (4.49)$$

Since $\bar{n}_i/N$ is less than one, the right side of (4.49) becomes vanishingly small if N is the number of molecules in 1 mol of gas, namely $N \approx 10^{23}$. This result implies that the probability $P\{n_i\}$ defined by (4.43) has an extremely sharp peak at $\{n_i\} = \{\bar{n}_i\}$. The width of the peak is such that $P\{n_i\}$ is essentially reduced to zero when any n_i/N differ from $\bar{n}_i/N$ by a number of the order of $1/\sqrt{N}$. A schematic plot of $P\{n_i\}$ is shown in Fig. 4.4. We shall call the distributions lying within the peak "essentially Maxwell-Boltzmann" distributions. They are physically indistinguishable from the strict Maxwell-Boltzmann distribution. From these considerations we conclude that in a physical gas any state picked out at random from among all those satisfying the given macroscopic conditions will almost certainly have a distribution function that is Maxwell-Boltzmann.

The meaning of the Maxwell-Boltzmann distribution is therefore as follows. If a dilute gas is prepared in an arbitrary initial state, and if there exist interactions to enable the gas to go into states other than the initial state, the gas will in time almost certainly become Maxwell-Boltzmann, because among all possible states of the gas satisfying the macroscopic conditions (which are conserved by the interactions), almost all of them have the Maxwell-Boltzmann distribution. This, however, does not tell us how long it will take for the gas to reach the equilibrium situation. Nor does it rule out the possibility that the gas may never reach the equilibrium situation, nor that of leaving the equilibrium situation after attaining it. From this point of view, we see that the laws of thermodynamics are not rigorously true but only overwhelmingly probable.

To illustrate these ideas, consider a gas enclosed in a cubical box with perfectly reflecting walls. Suppose initially the gas molecules are distributed in an arbitrary way within the box, and all have exactly the same velocity parallel to one edge of the box. If there are no interactions, this distribution will be maintained indefinitely, and the system never becomes Maxwell-Boltzmann. For such a gas thermodynamics is invalid. If there is molecular interaction, *no matter how small*, the initial distribution will, through collisions, change with time. Since almost any state of the gas will have a Maxwell-Boltzmann distribution, it is

---

*This assumption can be proved by the method described in Chapter 9. The desired result is essentially stated in (9.29).

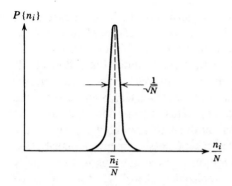

$P\{n_i\}$

$\frac{1}{\sqrt{N}}$

$\frac{\bar{n}_i}{N}$

$\frac{n_i}{N}$

Fig. 4.4 Probability of a gas having the occupation numbers $\{n_i\}$. The most probable occupation numbers $\{\bar{n}_i\}$ correspond to the Maxwell-Boltzmann distribution. Occupation numbers $\{n_i\}$ for which $P\{n_i\}$ lies within the peak are called "essentially Maxwell-Boltzmann" distributions.

reasonable that the distribution after a *sufficiently long time*, depending on the collision cross section, will become Maxwell-Boltzmann. The considerations we have made cannot tell us how long this time must be. They only tell us what the equilibrium situation is, if it is reached.

The derivation of the Maxwell-Boltzmann distribution presented here is independent of the earlier derivation based on the Boltzmann transport equation. Neither of these derivations is rigorous. In the present one there are assumptions that we did not justify, and in the previous one there was the assumption of molecular chaos, which remains unproved and is not related to the assumptions made here. The present method seems to be more satisfactory as a derivation of the Maxwell-Boltzmann distribution because it reveals more clearly the statistical nature of the Maxwell-Boltzmann distribution. The method of the most probable distribution, however, does not furnish information about a gas not in equilibrium, whereas the Boltzmann transport equation does. Hence the main value of the Boltzmann equation lies in its application to nonequilibrium phenomena.

4.4 ANALYSIS OF THE *H* THEOREM

We now discuss the physical implication of Boltzmann's H theorem. For a given distribution function $f(\mathbf{p}, t)$, H is defined by

$$H = \int d^3p\, f(\mathbf{p}, t) \log f(\mathbf{p}, t) \tag{4.50}$$

The time evolution of H is determined by the time evolution of $f(\mathbf{p}, t)$, which does not in general satisfy the Boltzmann transport equation. It satisfies the Boltzmann transport equation only at the instant when the assumption of molecular chaos happens to be valid.

The H theorem states that *if at a given instant t the state of the gas satisfies the assumption of molecular chaos*, then at the instant $t + \epsilon(\epsilon \to 0)$,

(a) $\dfrac{dH}{dt} \leq 0$

(b) $\dfrac{dH}{dt} = 0$ if and only if $f(\mathbf{v}, t)$ is the Maxwell-Boltzmann distribution.

The proof of the theorem given earlier is rigorous in the limiting case of an infinitely dilute gas. Therefore an inquiry into the validity of the H theorem can only be an inquiry into the validity of the assumption of molecular chaos.

We recall that the assumption of molecular chaos states the following: If $f(\mathbf{p}, t)$ is the probability of finding a molecule with velocity $\mathbf{p}$ at time t, the probability of simultaneously finding a molecule with velocity $\mathbf{p}$ and a molecule with velocity $\mathbf{p}'$ at time t is $f(\mathbf{p}, t)f(\mathbf{p}', t)$. This assumption concerns the correlation between two molecules and has nothing to say about a form of the distribution function. Thus a state of the gas possessing a given distribution function may or may not satisfy the assumption of molecular chaos. For brevity we call a state of the gas a state of "molecular chaos" if it satisfies the assumption of molecular chaos.

We now show that when the gas is in a state of "molecular chaos" H is at a local peak. Consider a dilute gas, in the absence of external force, prepared with an initial condition that is invariant under time reversal.* Under these conditions, the distribution function depends on the magnitude but not the direction of $\mathbf{v}$. Let the gas be in a state of "molecular chaos" and be non-Maxwell-Boltzmann at time $t = 0$. According to the H theorem $dH/dt \leq 0$ at $t = 0^+$. Now consider another gas, which at $t = 0$ is precisely the same as the original one except that all molecular velocities are reversed in direction. This gas must have the same H and must also be in a state of "molecular chaos." Therefore for this new gas $dH/dt \leq 0$ at $t = 0^+$. On the other hand, according to the invariance of the equations of motion under time reversal, the future of the new gas is the past of the original gas. Therefore for the original gas we must have

$$\frac{dH}{dt} \leq 0 \qquad \text{at } t = 0^+$$

$$\frac{dH}{dt} \geq 0 \qquad \text{at } t = 0^-$$

Thus H is at a local peak,[†] as illustrated in Fig. 4.5.

When H is not at a local peak, such as at the points a and c in Fig. 4.5, the state of the gas is not a state of "molecular chaos." Hence molecular collisions, which are responsible for the change of H with time, can create "molecular chaos" when there is none and destroy "molecular chaos" once established.

It is important to note that dH/dt is not necessarily a continuous function of time; it can be changed abruptly by molecular collisions. Overlooking this fact might lead us to conclude, erroneously, that the H theorem is inconsistent with the invariance under time reversal. A statement of the H theorem that is manifestly invariant under time reversal is the following. If there is "molecular chaos" now, then $dH/dt \leq 0$ in the next instant. If there will be "molecular chaos" in the next instant, then $dH/dt \geq 0$ now.

---

*These simplifying features are introduced to avoid the irrelevant complications arising from the time reversal properties of the external force and the agent preparing the system.

[†] The foregoing argument is due to F. E. Low (unpublished).

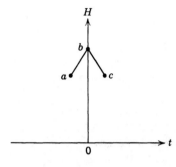

Fig. 4.5 H is at a local peak when the gas is in a state of "molecular chaos."

We now discuss the general behavior of H as a function of time. Our discussion rests on the following premises.

(a) H is at its smallest possible value when the distribution function is strictly Maxwell-Boltzmann. This easily follows from (4.50), and it is independent of the assumption of molecular chaos.*

(b) Molecular collisions happen at random, i.e., the time sequence of the states of a gas is a sequence of states chosen at random from those that satisfy the macroscopic conditions. This assumption is plausible but unproved.

From these premises it follows that the distribution function of the gas is almost always essentially Maxwell-Boltzmann, i.e., a distribution function contained within the peak shown in Fig. 4.4. The curve of H as a function of time consists mostly of microscopic fluctuations above the minimum value. Between two points at which H is at the minimum value there is likely to be a small peak.

If at any instant the gas has a distribution function appreciably different from the Maxwell-Boltzmann distribution, then H is appreciably larger than the minimum value. Since collisions are assumed to happen at random, it is overwhelmingly probable that after the next collision the distribution will become essentially Maxwell-Boltzmann and H will decrease to essentially the minimum value. By time reversal invariance it is overwhelmingly probable that before the last collision H was at essentially the minimum value. Thus H is overwhelmingly likely to be at a sharp peak when the gas is in an improbable state. The more improbable the state, the sharper the peak.

A very crude model of the curve of H as a function of time is shown in Fig. 4.6. The duration of a fluctuation, large or small, should be of the order of the time between two successive collisions of a molecule, i.e., 10^{-11} sec for a gas under ordinary conditions. The large fluctuations, such as that labeled a in Fig. 4.6, almost never occur spontaneously.† We can, of course, prepare a gas in an improbable state, e.g., by suddenly removing a wall of the container of the gas, so that H is initially at a peak. But it is overwhelmingly probable that within a few collision times the distribution would be reduced to an essentially Maxwell-Boltzmann distribution.

*See Problem 4.9.
†See Problems 4.5 and 4.6.

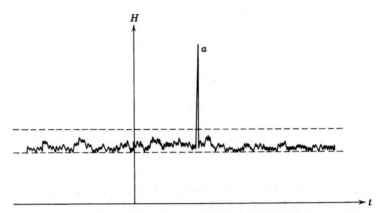

Fig. 4.6 H as a function of time. The range of values of H lying between the two horizontal dashed lines is called the "noise range."

Most of the time the value of H fluctuates within a small range above the minimum value. This range, shown enclosed by dashed lines in Fig. 4.6, corresponds to states of the gas with distribution functions that are essentially Maxwell-Boltzmann, i.e., distribution functions contained within the peak of Fig. 4.4. We call this range the "noise range." These features of the curve of H have been deduced only through plausibility arguments, but they are in accord with experience. We can summarize them as follows.

(a) For all practical purposes H never fluctuates spontaneously above the noise range. This corresponds to the observed fact that a system in thermodynamic equilibrium never spontaneously goes out of equilibrium.

(b) If at an instant H has a value above the noise range, then, for all practical purposes, H always decreases after that instant. In a few collision times its value will be within the noise range. This corresponds to the observed fact that if a system is initially not in equilibrium (the initial state being brought about by external agents), it always tends to equilibrium. In a few collision times it will be in equilibrium. This feature, together with (a), constitutes the second law of thermodynamics.

(c) Most of the time the value of H fluctuates in the noise range, in which dH/dt is as frequently positive as negative. (This is not a contradiction to the H theorem, because the H theorem merely requires that when the system is in a state of "molecular chaos," then $dH/dt \leq 0$ in the next instant.) These small fluctuations produce no observable change in the equation of state and other thermodynamic quantities. When H is in the noise range, the system is, for all practical purposes, in thermodynamic equilibrium. These fluctuations, however, do lead to observable effects,

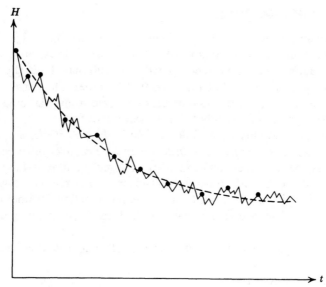

Fig. 4.7 The solid curve is H as a function of time for a gas initially in an improbable state. The dots are the points at which there is "molecular chaos." The dashed curve is that predicted by the Boltzmann transport equation.

e.g., the fluctuation scattering of light. We witness it in the blue of the sky.

We have argued that whenever the distribution function is not strictly Maxwell-Boltzmann, H is likely to be at a local peak. On the other hand, it was shown earlier that in a state of "molecular chaos" H *is* at a local peak. Thus we may regard a state of "molecular chaos" as a convenient mathematical model for a state that does not have a strictly Maxwell-Boltzmann distribution function. Hence the Boltzmann transport equation may be regarded as valid in a statistical sense. To illustrate this, let us imagine that a gas is prepared in an improbable initial state. The curve of H as a function of time might look like the solid curve in Fig. 4.7. Let us mark with a dot a point on this curve at which the gas is in a state of "molecular chaos." All these dots must be at a local peak of H (but not all local peaks are marked with a dot). By assumption of the randomness of the time sequence of states, they are likely to be evenly distributed in time. The distribution of dots might look like that illustrated in Fig. 4.7.

A solution to the Boltzmann transport equation would yield a smooth curve of negative slope that tries to fit these dots, as shown by the dashed curve in Fig. 4.7. It is in this sense that the Boltzmann transport equation provides a description of the approach to equilibrium.

These arguments make it only plausible that the Boltzmann transport equation is useful for the description of the approach to equilibrium. The final test lies in the comparison of results with experiments.

4.5 THE POINCARÉ CYCLE

When Boltzmann announced the H theorem a century ago, objections were raised against it on the ground that it led to "paradoxes." These are the so-called "reversal paradox" and "recurrence paradox," both based on the erroneous statement of the H theorem that $dH/dt \leq 0$ at all times. The correct statement of the H theorem, as given in the last section, is free from such objections. We mention these "paradoxes" purely for historical interest.

The "reversal paradox" is as follows: *The H theorem singles out a preferred direction of time. It is therefore inconsistent with time reversal invariance.* This is not a paradox, because the statement of the alleged paradox is false. We have seen in the last section that time reversal invariance is consistent with the H theorem, because dH/dt need not be a continuous function of time. In fact, we have made use of time reversal invariance to deduce interesting properties of the curve of H.

The "recurrence paradox" is based on the following true theorem.

POINCARÉ'S THEOREM

A system having a finite energy and confined to a finite volume will, after a sufficiently long time, return to an arbitrarily small neighborhood of almost any given initial state.

By "almost any state" is meant any state of the system, except for a set of measure zero (i.e., a set that has no volume, e.g., a discrete point set). A neighborhood of a state has an obvious definition in terms of the Γ space of the system.

A proof of Poincaré's theorem is given at the end of this section. This theorem implies that H is an almost periodic function of time. The "recurrence paradox" arises in an obvious way, if we take the statement of the H theorem to be $dH/dt \leq 0$ at all times. Since this is not the statement of the H theorem, there is no paradox. In fact, Poincaré's theorem furnishes further information concerning the curve of H.

Most of the time H lies in the noise range. Poincaré's theorem implies that the small fluctuations in the noise range repeat themselves. This is only to be expected.

For the rare spontaneous fluctuations above the noise range, Poincaré's theorem requires that if one such fluctuation occurs another one must occur after a sufficiently long time. The time interval between two large fluctuations is called a *Poincaré cycle*. A crude estimate (see Problem 4.7) shows that a Poincaré cycle is of the order of e^N, where N is the total number of molecules in the system. Since $N \approx 10^{23}$, a Poincaré cycle is extremely long. In fact, it is essentially the same number, be it $10^{10^{23}}$ s or $10^{10^{23}}$ ages of the universe, (the age of the universe being a mere 10^{10} years.) Thus it has nothing to do with physics.

We mentioned the ergodic theorem in Section 3.4, but did not use it as a basis for the microcanonical ensemble, even though, on the surface, it seems to be the justification we need. The reason is that existing proofs of the theorem all share a characteristic of the proof the Poincaré theorem given below, i.e., an avoidance of dynamics. For this reason, they cannot provide the true relaxation time for a system to reach local equilibrium, (typically about 10^{-15} s for real systems,) but have a characteristic time scale of the order of the Poincaré cycle. For this reason, the ergodic theorem has so far been an interesting mathematical exercise irrelevant to physics.

Proof of Poincaré's Theorem Let a state of the system be represented by a point in Γ space. As time goes on, any point in Γ space traces out a locus that is uniquely determined by any given point on the locus. Let g_0 be an arbitrary volume element in Γ space of volume ω_0. After time t all the points in g_0 will be in another volume element g_t, of volume ω_t, which is uniquely determined by g_0. By Liouville's theorem, $\omega_t = \omega_0$.

Let Γ_0 denote the subspace that is the union of all g_t for $0 \leq t < \infty$. Let its volume be Ω_0. Similarly, let Γ_τ denote the subspace that is the union of all g_t for $\tau \leq t < \infty$. Let its volume be Ω_τ. The numbers Ω_0 and Ω_τ are finite because, since the energy of the system and the spatial extension of the system are finite, a representative point is confined to a finite region of Γ space. The definitions immediately imply that Γ_0 contains Γ_τ.

We may think of Γ_0 and Γ_τ in a different way. Imagine the region Γ_0 to be filled uniformly with representative points. As time goes on, Γ_0 will evolve into some other regions that are uniquely determined. It is clear, from the definitions, that after a time τ, Γ_0 will become Γ_τ. Hence, by Liouville's theorem,

$$\Omega_0 = \Omega_\tau$$

We recall that Γ_0 contains all the future destinations of the points in g_0, and Γ_τ contains all the future destinations of the points in g_τ, which in turn is evolved from g_0 after the time τ. It has been shown that Γ_0 has the same volume as Γ_τ. Therefore Γ_0 and Γ_τ must contain the same set of points except for a set of measure zero.

In particular, Γ_τ contains all of g_0 except for a set of measure zero. But, by definition, all points in Γ_τ are future destinations of the points in g_0. Therefore all points in g_0, except for a set of measure zero, must return to g_0 after a sufficiently long time. Since g_0 can be made as small as we wish, Poincaré's theorem follows. ∎

PROBLEMS

4.1 Describe an experimental method for the verification of the Maxwell-Boltzmann distribution.

4.2 A cylindrical column of gas of given temperature rotates about a fixed axis with constant angular velocity. Find the equilibrium distribution function.

4.3 (*a*) What fraction of the H_2 gas at sea level and at a temperature of 300 K can escape from the earth's gravitational field?

(*b*) Why do we still have H_2 gas in the atmosphere at sea level?

4.4 Using relativistic dynamics for gas molecules find, for a dilute gas of zero total momentum,

(*a*) the equilibrium distribution function;

(*b*) the equation of state.

> *Answer.* PV is independent of the volume. Hence it is NkT by definition of T.

4.5 (*a*) *Estimate* the probability that a stamp (mass = 0.1 g) resting on a desk top at room temperature (300 K) will spontaneously fly up to a height of 10^{-8} cm above the desk top.

> *Hint.* Think not of one stamp but of an infinite number of noninteracting stamps placed side by side. Formulate an argument showing that these stamps obey the Maxwell-Boltzmann distribution.

> *Answer.* Let m = mass of stamps, h = height, g = acceleration of gravity. Probability $\approx e^{-mgh/kT}$

4.6 A room of volume $3 \times 3 \times 3$ m^3 is under standard conditions (atmospheric pressure and 300 K).

(*a*) *Estimate* the probability that at any instant of time a 1-cm^3 volume anywhere within this room becomes totally devoid of air because of spontaneous statistical fluctuations.

(*b*) estimate the same for a 1-Å^3 volume.

> *Answer.* Let N = total number of air molecules, V = volume of room, v = the volume devoid of air. Probability $\approx e^{-N(v/V)}$

4.7 Suppose the situation referred to in Problem 4.6*a* has occurred. Describe qualitatively the behavior of the distribution function thereafter. Estimate the time it takes for such a situation to occur again, under the assumption that molecular collisions are such that the time sequence of the state of the system is a random sequence of states.

4.8 (*a*) Explain why in (4.42) we arrived at the formula for the Maxwell-Boltzmann distribution for a gas with no average momentum ($p_0 = 0$), although average momentum was not specified as a macroscopic condition in (4.35) and (4.36).

(*b*) Derive the Maxwell-Boltzmann distribution for a gas with average velocity v_0, using the method of the most probable distribution.

4.9 Let

$$H = \int d^3p f(\mathbf{p}, t) \log f(\mathbf{p}, t)$$

where $f(\mathbf{v}, t)$ is arbitrary except for the conditions

$$\int d^3p f(\mathbf{p}, t) = n$$

$$\frac{1}{n} \int d^3p \frac{p^2}{2m} f(\mathbf{p}, t) = \epsilon$$

Show that H is minimum when f is the Maxwell-Boltzmann distribution.

5

TRANSPORT PHENOMENA

5.1 THE MEAN FREE PATH

To begin our discussion on the approach to equilibrium of a gas initially not in equilibrium, we introduce the qualitative concept of the mean free path and related quantities.

A gas is not in equilibrium when the distribution function is different from the Maxwell-Boltzmann distribution. The most common case of a nonequilibrium situation is that in which the temperature, density, and average velocity are not constant throughout the gas. To approach equilibrium, these nonuniformities have to be ironed out through the transport of energy, mass, and momentum from one part of the gas to another. The mechanism of transport is molecular collision, and the average distance over which molecular properties can be transported in one collision is the *mean free path*. It is the average distance traveled by a molecule between successive collisions. We give an estimate of its order of magnitude.

The number of collisions happening per second per unit volume at the point **r** in a gas is given by

$$Z = \int d^3p_1 \, d^3p_2 \, d^3p_1' \, d^3p_2' \, \delta^4(P_f - P_i)|T_{fi}|^2 f(\mathbf{r}, \mathbf{p}_1, t) f(\mathbf{r}, \mathbf{p}_2, t)$$

where $f(\mathbf{r}, \mathbf{p}, t)$ is the distribution function. The integration over $\mathbf{p}_1'$ and $\mathbf{p}_2'$ can be immediately effected to yield

$$Z = \int d^3p_1 \int d^3p_2 \, \sigma_{\text{tot}} \, |\mathbf{v}_1 - \mathbf{v}_2| f(\mathbf{r}, \mathbf{p}_1, t) f(\mathbf{r}, \mathbf{p}_2, t) \tag{5.1}$$

A free path is defined as the distance traveled by a molecule between two successive collisions. Since it takes two molecules to make a collision, every collision terminates two free paths. The total number of free paths occurring per second per unit volume is therefore $2Z$. Since there are n molecules per unit volume, the average number of free paths traveled by a molecule per second is

$2Z/n$. The *mean free path*, which is the average length of a free path, is given by

$$\lambda = \frac{n}{2Z}\bar{v} \qquad (5.2)$$

where $\bar{v} = \sqrt{2kT/m}$ is the most probable speed of a molecule. The average duration of a free path is called the *collision time* and is given by

$$\tau = \frac{\lambda}{\bar{v}} \qquad (5.3)$$

For a gas in equilibrium, $f(\mathbf{r}, \mathbf{p}, t)$ is the Maxwell-Boltzmann distribution. Assume for an order-of-magnitude estimate that σ_{tot} is insensitive to the energy of the colliding molecules and may be replaced by a constant of the order of πa^2 where a is the molecular diameter. Then we have

$$Z = \frac{\sigma_{tot}}{m}\int d^3p_1 \int d^3p_2 |\mathbf{p}_1 - \mathbf{p}_2| f(\mathbf{p}_1) f(\mathbf{p}_2)$$

$$= \frac{\sigma_{tot}\, n^2}{m(2\pi mkT)^3}\int d^3p_1 \int d^3p_2 |\mathbf{p}_1 - \mathbf{p}_2| \exp\left[-\frac{p_1^2 + p_2^2}{2mkT}\right]$$

$$= \frac{\sigma_{tot}n^2}{m(2\pi mkT)^3}\int d^3P \int d^3p |\mathbf{p}| \exp\left[-\frac{1}{kT}\left(\frac{P^2}{4m} + \frac{p^2}{m}\right)\right]$$

where $\mathbf{P} = \mathbf{p}_1 + \mathbf{p}_2$, $\mathbf{p} = \frac{1}{2}(\mathbf{p}_2 - \mathbf{p}_1)$. The integrations are elementary and give

$$Z = 2n^2\sigma_{tot}\sqrt{\frac{kT}{\pi m}} = \sqrt{\frac{2}{\pi}}\, n^2\sigma_{tot}\,\bar{v} \qquad (5.4)$$

Therefore

$$\lambda = \frac{\sqrt{\pi/8}}{n\sigma_{tot}}$$

$$\tau = \frac{\sqrt{\pi/8}}{n\sigma_{tot}\,\bar{v}} \qquad (5.5)$$

We see that the mean free path is independent of the temperature and is inversely proportional to the density times the total cross section.

The following are some numerical estimates. For H_2 gas at its critical point,

$$\lambda \approx 10^{-7}\ \text{cm}$$

$$\tau \approx 10^{-11}\ \text{s}$$

For H_2 gas in interstellar space, where the density is about 1 molecule/cm^3,

$$\lambda \approx 10^{15}\ \text{cm}$$

The diameter of H_2 has been taken to be about 1 Å.

From these qualitative estimates, it is expected that in H_2 gas under normal conditions, for example, any nonuniformity in density or temperature over distances of order 10^{-7} cm will be ironed out in the order of 10^{-11} s. Variations in density or temperature over macroscopic distances may persist for a long time.

5.2 EFFUSION

An important quantity governing the behavior of a gas is the ratio of the mean free path to some other characteristic length, such as

The size of the box containing the gas.

The diameter of a hole through which gas molecules may pass.

The wavelength of density fluctuations.

When the mean free path is large compared to any other length in the problem, the gas is said to be in the *collisionless regime*. A practical example is the process of *effusion*, whereby a gas leaks through a very small hole of diameter much smaller than the mean free path—a phenomenon of great interest to all experimentalists who maintain vacuum systems.

In effusion the gas molecules do not collide as they go through the hole. Therefore the flux I through the hole, defined as the number of molecules crossing the hole per second per unit area of the hole, is just the flux of molecules impinging on the surface area of the hole. The contribution to the flux from molecules of velocity $\mathbf{v}$ is given by

$$dI = d^3p\, v_x f(\mathbf{p})$$

where the x axis is chosen normal to the hole. The total flux is therefore

$$I = \int_{v_x > 0} d^3p\, v_x f(\mathbf{p})$$

Assuming the Maxwell-Boltzmann distribution, we have

$$I = \frac{nm^3}{(2\pi mkT)^{3/2}} \int_0^\infty dv_x\, v_x e^{-mv_x^2/2kT} \int_{-\infty}^\infty dv_y\, e^{-mv_y^2/2kT} \int_{-\infty}^\infty dv_z\, e^{-mv_z^2/2kT}$$

$$= n\sqrt{\frac{kT}{2\pi m}} = \frac{n\bar{v}}{2\sqrt{\pi}}$$

Eliminating n through $P = nkT$, we obtain

$$I = \frac{P}{\sqrt{2\pi mkT}} \tag{5.6}$$

The inverse proportionality to $\sqrt{m}$ makes the process useful as a means of separating isotopes.

The opposite of the collisionless regime is one in which the mean free path is much smaller than the other characteristic lengths of the problem, exemplified by the flow of a gas through a very large hole. In this case the gas molecules will undergo many collisions as they pass through the hole, and will "thermalize" locally. The prevailing condition is known as the *hydrodynamic regime*, and will be the subject of the rest of this chapter.

5.3 THE CONSERVATION LAWS

To investigate nonequilibrium phenomena, we must solve the Boltzmann transport equation, with given initial conditions, to obtain the distribution function as a function of time. Some rigorous properties of any solution to the Boltzmann equation may be obtained from the fact that in any molecular collision there are dynamical quantities that are rigorously conserved.

Let $\chi(\mathbf{r}, \mathbf{p})$ be any quantity associated with a molecule of velocity $\mathbf{p}$ located at $\mathbf{r}$, such that in any collision $\{\mathbf{p}_1, \mathbf{p}_2\} \rightarrow \{\mathbf{p}_1', \mathbf{p}_2'\}$ taking place at $\mathbf{r}$, we have

$$\chi_1 + \chi_2 = \chi_1' + \chi_2' \qquad (5.7)$$

where $\chi_1 = \chi(\mathbf{r}_1, \mathbf{p}_1)$, etc. We call χ a conserved property. The following theorem holds.

THEOREM

$$\int d^3p\, \chi(\mathbf{r}, \mathbf{p}) \left[\frac{\partial f(\mathbf{r}, \mathbf{p}, t)}{\partial t} \right]_{\text{coll}} = 0 \qquad (5.8)$$

where $(\partial f/\partial t)_{\text{coll}}$ is the right side of (3.36).*

Proof By definition of $(\partial f/\partial t)_{\text{coll}}$ we have

$$\int d^3p\, \chi \left(\frac{\partial f}{\partial t} \right)_{\text{coll}} = \int d^3p_1\, d^3p_2\, d^3p_1'\, d^3p_2'\, \delta^4(P_f - P_i)|T_{fi}|^2 (f_2'f_1' - f_2 f_1)\chi_1 \qquad (5.9)$$

Making use of the properties of T_{fi} discussed in Section 3.2, and proceeding in a manner similar to the proof of the H theorem, we make each of the following interchanges of integration variables.

First: $\mathbf{p}_1 \rightleftarrows \mathbf{p}_2$

Next: $\mathbf{p}_1 \rightleftarrows \mathbf{p}_1'$ and $\mathbf{p}_2 \rightleftarrows \mathbf{p}_2'$

Next: $\mathbf{p}_1 \rightleftarrows \mathbf{p}_2'$ and $\mathbf{p}_2 \rightleftarrows \mathbf{p}_1'$

For each case we obtain a different form for the same integral. Adding the three

*Note that it is not required that f be a solution of the Boltzmann transport equation.

new formulas so obtained to (5.9) and dividing the result by 4 we get

$$\int d^3 p\, \chi \left(\frac{\partial f}{\partial t} \right)_{\text{coll}} = \frac{1}{4} \int d^3 p_1\, d^3 p_2\, d^3 p_1'\, d^3 p_2'\, \delta^4 (P_f - P_i) |T_{fi}|^2$$
$$\times (f_2' f_1' - f_2 f_1)(\chi_1 + \chi_2 - \chi_1' - \chi_2') \equiv 0 \quad \blacksquare$$

The conservation theorem relevant to the Boltzmann transport equation is obtained by multiplying the Boltzmann transport equation on both sides by χ and then integrating over $\mathbf{p}$. The collision term vanishes by virtue of (5.8), and we have*

$$\int d^3 p\, \chi(\mathbf{r}, \mathbf{p}) \left(\frac{\partial}{\partial t} + \frac{p_i}{m} \frac{\partial}{\partial x_i} + F_i \frac{\partial}{\partial v_i} \right) f(\mathbf{r}, \mathbf{p}, t) = 0 \tag{5.10}$$

We may rewrite (5.10) in the form

$$\frac{\partial}{\partial t} \int d^3 p\, \chi f + \frac{\partial}{\partial x_i} \int d^3 p\, \chi \frac{p_i}{m} f - \int d^3 p\, \frac{\partial \chi}{\partial x_i} \frac{p_i}{m} f + \int d^3 p\, \frac{\partial}{\partial p_i} (\chi F_i f)$$

$$- \int d^3 p\, \frac{\partial \chi}{\partial p_i} F_i f - \int d^3 p\, \chi \frac{\partial F_i}{\partial p_i} f = 0 \tag{5.11}$$

The fourth term vanishes if $f(\mathbf{r}, \mathbf{p}, t)$ is assumed to vanish when $|\mathbf{p}| \to \infty$. This conservation theorem is most useful in hydrodynamics, where the velocity $\mathbf{v} = \mathbf{p}/m$ rather than the momentum $\mathbf{p}$ is a directly measurable quantity. Accordingly, we shall reexpress $\mathbf{p}$ in terms of $\mathbf{v}$, where convenient. We also define the average value $\langle A \rangle$ by

$$\langle A \rangle \equiv \frac{\int d^3 p\, A f}{\int d^3 p\, f} = \frac{1}{n} \int d^3 p\, A f \tag{5.12}$$

where

$$n(\mathbf{r}, t) \equiv \int d^3 p\, f(\mathbf{r}, \mathbf{p}, t) \tag{5.13}$$

We obtain finally the desired theorem:

CONSERVATION THEOREM

$$\frac{\partial}{\partial t} \langle n\chi \rangle + \frac{\partial}{\partial x_i} \langle n v_i \chi \rangle - n \left\langle v_i \frac{\partial \chi}{\partial x_i} \right\rangle - \frac{n}{m} \left\langle F_i \frac{\partial \chi}{\partial v_i} \right\rangle - \frac{n}{m} \left\langle \frac{\partial F_i}{\partial v_i} \chi \right\rangle = 0 \tag{5.14}$$

where χ is any conserved property. Note that $\langle nA \rangle = n \langle A \rangle$ because n is

---

*The summation convention, whereby a repeated vector index is understood to be summed from 1 to 3, is used.

independent of **v**. From now on we restrict our attention to velocity-independent external forces so that the last term of (5.14) may be dropped.

For simple molecules the independent conserved properties are mass, momentum, and energy. For charged molecules we also include the charge, but this extension is trivial. Accordingly we set successively

$$\chi = m \qquad \text{(mass)}$$

$$\chi = mv_i \qquad (i = 1, 2, 3) \qquad \text{(momentum)}$$

$$\chi = \tfrac{1}{2}m|\mathbf{v} - \mathbf{u}(\mathbf{r}, t)|^2 \qquad \text{(thermal energy)}$$

where

$$\mathbf{u}(\mathbf{r}, t) \equiv \langle \mathbf{v} \rangle$$

We should then have three independent conservation theorems.

For $\chi = m$ we have immediately

$$\frac{\partial}{\partial t}(mn) + \frac{\partial}{\partial x_i}\langle mnv_i \rangle = 0$$

or, introducing the mass density

$$\rho(\mathbf{r}, t) \equiv mn(\mathbf{r}, t)$$

we obtain

$$\frac{\partial \rho}{\partial t} + \nabla \cdot (\rho\mathbf{u}) = 0 \qquad (5.15)$$

Next we put $\chi = mv_i$, obtaining

$$\frac{\partial}{\partial t}\langle \rho v_i \rangle + \frac{\partial}{\partial x_j}\langle \rho v_i v_j \rangle - \frac{1}{m}\rho F_i = 0 \qquad (5.16)$$

To reduce this further let us write

$$\langle v_i v_j \rangle = \left\langle (v_i - u_i)(v_j - u_j) \right\rangle + \langle v_i \rangle u_j + u_i \langle v_j \rangle - u_i u_j$$

$$= \left\langle (v_i - u_i)(v_j - u_j) \right\rangle + u_i u_j$$

Substituting this into (5.16) we obtain

$$\rho\left(\frac{\partial u_i}{\partial t} + u_j \frac{\partial u_i}{\partial x_j} \right) = \frac{1}{m}\rho F_i - \frac{\partial}{\partial x_j}\left\langle \rho(v_i - u_i)(v_j - u_j) \right\rangle \qquad (5.17)$$

Introducing the abbreviation

$$P_{ij} \equiv \rho\left\langle (v_i - u_i)(v_j - u_j) \right\rangle$$

which is called the *pressure tensor*, we finally have

$$\left(\frac{\partial}{\partial t} + u_j \frac{\partial}{\partial x_j} \right) u_i = \frac{1}{m}F_i - \frac{1}{\rho}\frac{\partial}{\partial x_j}P_{ij} \qquad (5.18)$$

Finally we set $\chi = \frac{1}{2}m|\mathbf{v} - \mathbf{u}|^2$. Then

$$\frac{1}{2}\frac{\partial}{\partial t}\langle\rho|\mathbf{v} - \mathbf{u}|^2\rangle + \frac{1}{2}\frac{\partial}{\partial x_i}\langle\rho v_i|\mathbf{v} - \mathbf{u}|^2\rangle - \frac{1}{2}\rho\langle v_i\frac{\partial}{\partial x_i}|\mathbf{v} - \mathbf{u}|^2\rangle = 0 \quad (5.19)$$

We define the *temperature* by

$$kT \equiv \theta \equiv \frac{1}{3}m\langle|\mathbf{v} - \mathbf{u}|^2\rangle$$

and the *heat flux* by

$$\mathbf{q} = \frac{1}{2}\rho\langle(\mathbf{v} - \mathbf{u})|\mathbf{v} - \mathbf{u}|^2\rangle$$

We then have

$$\frac{1}{2}\rho\langle v_i|\mathbf{v} - \mathbf{u}|^2\rangle = \frac{1}{2}m\rho\langle(v_i - u_i)|\mathbf{v} - \mathbf{u}|^2\rangle + \frac{1}{2}\rho u_i\langle|\mathbf{v} - \mathbf{u}|^2\rangle$$

$$= q_i + \frac{3}{2}\rho\theta u_i$$

and

$$\rho\langle v_i(v_j - u_j)\rangle = \rho\langle(v_i - u_i)(v_j - u_j)\rangle + \rho u_i\langle v_j - u_j\rangle = P_{ij}$$

Thus (5.19) can be written

$$\frac{3}{2}\frac{\partial}{\partial t}(\rho\theta) + \frac{\partial q_i}{\partial x_i} + \frac{3}{2}\frac{\partial}{\partial x_i}(\rho\theta u_i) + mP_{ij}\frac{\partial u_j}{\partial x_i} = 0$$

Since $P_{ij} = P_{ji}$

$$mP_{ij}\frac{\partial u_j}{\partial x_i} = P_{ij}\frac{m}{2}\left(\frac{\partial u_j}{\partial x_i} + \frac{\partial u_i}{\partial x_j}\right) \equiv P_{ij}\Lambda_{ij}$$

The final form is then obtained after a few straightforward steps:

$$\rho\left(\frac{\partial}{\partial t} + u_i\frac{\partial}{\partial x_i}\right)\theta + \frac{2}{3}\frac{\partial}{\partial x_i}q_i = -\frac{2}{3}\Lambda_{ij}P_{ij} \quad (5.20)$$

The three conservation theorems are summarized in (5.21), (5.22), and (5.23).

$$\frac{\partial\rho}{\partial t} + \nabla\cdot(\rho\mathbf{u}) = 0 \qquad \text{(conservation of mass)} \quad (5.21)$$

$$\rho\left(\frac{\partial}{\partial t} + \mathbf{u}\cdot\nabla\right)\mathbf{u} = \frac{\rho}{m}\mathbf{F} - \nabla\cdot\overset{\leftrightarrow}{P} \qquad \text{(conservation of momentum)}$$

$$(5.22)$$

$$\rho\left(\frac{\partial}{\partial t} + \mathbf{u}\cdot\nabla\right)\theta = -\frac{2}{3}\nabla\cdot\mathbf{q} - \frac{2}{3}\overset{\leftrightarrow}{P}\cdot\overset{\leftrightarrow}{\Lambda} \qquad \text{(conservation of energy)} \quad (5.23)$$

where $\overset{\leftrightarrow}{P}$ is a dyadic whose components are P_{ij}, $\nabla\cdot\overset{\leftrightarrow}{P}$ is a vector whose ith component is $\partial P_{ij}/\partial x_j$, and $\overset{\leftrightarrow}{P}\cdot\overset{\leftrightarrow}{\Lambda}$ is a scalar $P_{ij}\Lambda_{ij}$. The auxiliary quantities are defined as follows.

$$\rho(\mathbf{r}, t) \equiv m \int d^3v f(\mathbf{r}, \mathbf{v}, t) \qquad \text{(mass density)} \qquad (5.24)$$

$$\mathbf{u}(\mathbf{r}, t) \equiv \langle \mathbf{v} \rangle \qquad \text{(average velocity)} \qquad (5.25)$$

$$\theta(\mathbf{r}, t) \equiv \tfrac{1}{3}m\langle |\mathbf{v} - \mathbf{u}|^2 \rangle \qquad \text{(temperature)} \qquad (5.26)$$

$$\mathbf{q}(\mathbf{r}, t) \equiv \tfrac{1}{2}m\rho\langle (\mathbf{v} - \mathbf{u})|\mathbf{v} - \mathbf{u}|^2 \rangle \qquad \text{(heat flux vector)} \qquad (5.27)$$

$$P_{ij} \equiv \rho\langle (v_i - u_i)(v_j - u_j) \rangle \qquad \text{(pressure tensor)} \qquad (5.28)$$

$$\Lambda_{ij} \equiv \tfrac{1}{2}m\left(\frac{\partial u_i}{\partial x_j} + \frac{\partial u_j}{\partial x_i} \right) \qquad (5.29)$$

Although the conservation theorems are exact, they have no practical value unless we can actually solve the Boltzmann transport equation and use the distribution function so obtained to evaluate the quantities (5.24)–(5.29). Despite the fact that these quantities have been given rather suggestive names, their physical meaning, if any, can only be ascertained after the distribution function is known. We shall see that when it is known these conservation theorems become the physically meaningful equations of hydrodynamics.

5.4 THE ZERO-ORDER APPROXIMATION

From now on we shall work in the hydrodynamic regime, where the mean free path is small compared to other characteristic lengths. This means that gas molecules make a large number of collisions within a small space. Consequently they come to local equilibrium rapidly. In the lowest-order approximation it is natural to assume that the gas has a local Maxwell-Boltzmann distribution, with slowly varying temperature, density, and average velocity:

$$f(\mathbf{r}, \mathbf{p}, t) \approx f^{(0)}(\mathbf{r}, \mathbf{p}, t) \qquad (5.30)$$

where

$$f^{(0)}(\mathbf{r}, \mathbf{p}, t) = \frac{n}{(2\pi m\theta)^{3/2}} \exp\left[-\frac{m}{2\theta}(\mathbf{v} - \mathbf{u})^2 \right] \qquad (5.31)$$

where $n, \theta, \mathbf{u}$ are all slowly varying functions of $\mathbf{r}$ and t. It is obvious that (5.30) cannot be an exact solution of the Boltzmann transport equation. On the one hand we have

$$\left(\frac{\partial f^{(0)}}{\partial t} \right)_{\text{coll}} = 0 \qquad (5.32)$$

because $n, \theta, \mathbf{u}$ do not depend on $\mathbf{v}$. On the other hand it is clear that in general

$$\left(\frac{\partial}{\partial t} + \mathbf{v} \cdot \nabla_\mathbf{r} + \frac{\mathbf{F}}{m} \cdot \nabla_\mathbf{v} \right) f^{(0)}(\mathbf{r}, \mathbf{p}, t) \neq 0 \qquad (5.33)$$

We postpone the discussion of the accuracy of the approximation (5.30). For the moment let us assume that it is a good approximation and discuss the physical consequences.

If (5.30) is a good approximation, the left side of (5.33) must be approximately equal to zero. This in turn would mean that $n, \theta, \mathbf{u}$ are such that the conservation theorems (5.21)–(5.23) are approximately satisfied. The conservation theorems then become the equations restricting the behaviour of $n, \theta, \mathbf{u}$. To see what they are, we must calculate $\mathbf{q}$ and P_{ij} to the lowest order. The results are denoted respectively by $\mathbf{q}^{(0)}$ and $P_{ij}^{(0)}$. Let $C(\mathbf{r}, t) = n(m/2\pi\theta)^{3/2}$ and $A(\mathbf{r}, t) = m/2\theta$. We easily obtain

$$\mathbf{q}^{(0)} = \frac{1}{2} \frac{\rho}{n} \int d^3v \, (\mathbf{v} - \mathbf{u})|\mathbf{v} - \mathbf{u}|^2 C(\mathbf{r}, t) e^{-A(\mathbf{r},\, t)|\mathbf{v} - \mathbf{u}|^2}$$

$$= \tfrac{1}{2} m^2 C(\mathbf{r}, t) \int d^3U \, \mathbf{U} U^2 e^{-A(\mathbf{r},\, t)U^2} = 0 \tag{5.34}$$

$$P_{ij}^{(0)} = \frac{\rho}{n} C(\mathbf{r}, t) \int d^3v \, (v_i - u_i)(v_j - u_j) e^{-A(\mathbf{r},\, t)|\mathbf{v} - \mathbf{u}|^2}$$

$$= mC(\mathbf{r}, t) \int d^3U \, U_i U_j e^{-A(\mathbf{r},\, t)U^2} = \delta_{ij} P \tag{5.35}$$

where

$$P = \tfrac{1}{3}\rho \left(\frac{m}{2\pi\theta} \right)^{3/2} \int d^3U \, U^2 e^{-A(\mathbf{r},\, t)U^2} = n\theta \tag{5.36}$$

which is the local hydrostatic pressure.

Substituting these into (5.21) and (5.23), and noting that

$$\nabla \cdot \vec{P}^{(0)} = \nabla P$$

$$\vec{P}^{(0)} \cdot \vec{\Lambda} = P \sum_{i=1}^{3} \Lambda_{ii} = mP\nabla \cdot \mathbf{u}$$

We obtain the equations

$$\frac{\partial \rho}{\partial t} + \nabla \cdot (\rho\mathbf{u}) = 0 \qquad \text{(continuity equation)} \tag{5.37}$$

$$\left(\frac{\partial}{\partial t} + \mathbf{u} \cdot \nabla \right)\mathbf{u} + \frac{1}{\rho} \nabla P = \frac{\mathbf{F}}{m} \qquad \text{(Euler's equation)} \tag{5.38}$$

$$\left(\frac{\partial}{\partial t} + \mathbf{u} \cdot \nabla \right)\theta + \frac{1}{c_V}(\nabla \cdot \mathbf{u})\theta = 0 \tag{5.39}$$

where $c_V = \tfrac{3}{2}$. These are the hydrodynamic equations for the nonviscous flow of a gas. They possess solutions describing flow patterns that persist indefinitely. Thus, in this approximation, the local Maxwell-Boltzmann distribution never decays to the true Maxwell-Boltzmann distribution. This is in rough accord with experience, for we know that a hydrodynamic flow, left to itself, takes a long time to die out.

Although derived for dilute gases, (5.37)–(5.39) are also used for liquids because these equations can also be derived through heuristic arguments which indicate that they are of a more general validity.

We shall now briefly point out some of the consequences of (5.37)–(5.39) that are of practical interest.

The quantity $(\partial/\partial t + \mathbf{u} \cdot \nabla) X$ is known as the "material derivative of X," because it is the time rate of change of X to an observer moving with the local average velocity $\mathbf{u}$. Such an observer is said to be moving along a streamline. We now show that in the zero-order approximation a dilute gas undergoes only adiabatic transformations to an observer moving along a streamline. Equations (5.37) and (5.39) may be rewritten as

$$\left(\frac{\partial}{\partial t} + \mathbf{u} \cdot \nabla\right)\rho = -\rho\nabla \cdot \mathbf{u}$$

$$-\frac{3}{2}\frac{\rho}{\theta}\left(\frac{\theta}{\partial t} + \mathbf{u} \cdot \nabla\right)\theta = \rho\nabla \cdot \mathbf{u}$$

Adding these two equations we obtain

$$\left(\frac{\partial}{\partial t} + \mathbf{u} \cdot \nabla\right)\rho - \frac{3}{2}\frac{\rho}{\theta}\left(\frac{\partial}{\partial t} + \mathbf{u} \cdot \nabla\right)\theta = 0$$

or

$$\left(\frac{\partial}{\partial t} + \mathbf{u} \cdot \nabla\right)\left(\rho\theta^{-3/2}\right) = 0 \tag{5.40}$$

Using the equation of state $P = \rho\theta/m$ we can convert (5.40) to the condition

$$P\rho^{-5/3} = \text{constant} \quad \text{(along a streamline)} \tag{5.41}$$

This is the condition for adiabatic transformation for an ideal gas, since $c_P/c_V = \frac{5}{3}$.

Next we derive the linear equation for a sound wave. Let us restrict ourselves to the case in which $\mathbf{u}$ and all the space and time derivatives of $\mathbf{u}$, ρ, and θ are small quantities of the first order. For $\mathbf{F} = 0$, (5.37) and (5.38) may be replaced by

$$\frac{\partial\rho}{\partial t} + \rho\nabla \cdot \mathbf{u} = 0 \tag{5.42}$$

$$\rho\frac{\partial\mathbf{u}}{\partial t} + \nabla P = 0 \tag{5.43}$$

$$\tfrac{3}{2}\rho\frac{\partial\theta}{\partial t} - \theta\frac{\partial\rho}{\partial t} = 0 \tag{5.44}$$

where quantities smaller than first-order ones are neglected. Note that (5.44) is none other than (5.40) or (5.41). Taking the divergence of (5.43) and the time derivative of (5.42), and subtracting one resulting equation from the other, we obtain

$$\nabla^2 P - \frac{\partial^2 \rho}{\partial t^2} = 0 \qquad (5.45)$$

in which higher-order quantities are again neglected. Now P is a function of ρ and θ, but the latter are not independent quantities, being related to each other through the condition of adiabatic transformation (5.44). Hence we may regard P as a function of ρ alone, and write

$$\nabla^2 P = \nabla \cdot \left[\left(\frac{\partial P}{\partial \rho} \right)_S \nabla \rho \right] \approx \left(\frac{\partial P}{\partial \rho} \right)_S \nabla^2 \rho$$

where $(\partial P/\partial \rho)_S$ is the adiabatic derivative, related to the adiabatic compressibility κ_S by

$$\kappa_S = \frac{1}{\rho} \left(\frac{\partial \rho}{\partial P} \right)_S = \frac{3}{5} \frac{m}{\rho \theta} \qquad (5.46)$$

Thus (5.45) can be written in the form

$$\nabla^2 \rho - \rho \kappa_S \frac{\partial^2 \rho}{\partial t^2} = 0 \qquad (5.47)$$

which is a wave equation for ρ, describing a sound wave with a velocity of propagation c given by

$$c = \frac{1}{\sqrt{\rho \kappa_S}} = \sqrt{\frac{5}{3} \frac{\theta}{m}} = \sqrt{\frac{5}{6}} \bar{v} \qquad (5.48)$$

It is hardly surprising that the adiabatic compressibility enters here, because in the present approximation there can be no heat conduction in the gas, as (5.34) indicates.

Finally consider the case of steady flow under the influence of a conservative external force field, i.e., under the conditions

$$\mathbf{F} = -\nabla \phi$$
$$\frac{\partial \mathbf{u}}{\partial t} = 0 \qquad (5.49)$$

Using the vector identity

$$(\mathbf{u} \cdot \nabla)\mathbf{u} = \tfrac{1}{2}\nabla(u^2) - \mathbf{u} \times (\nabla \times \mathbf{u}) \qquad (5.50)$$

we can rewrite (5.38) as follows

$$\nabla \left(\tfrac{1}{2}u^2 + \frac{1}{\rho}P + \frac{1}{m}\phi \right) = \mathbf{u} \times (\nabla \times \mathbf{u}) - \frac{\theta}{m}\frac{\nabla \rho}{\rho} \qquad (5.51)$$

Two further specializations are of interest. First, in the case of uniform density and irrotational flow, namely, $\nabla \rho = 0$ and $\nabla \times \mathbf{u} = 0$, we have

$$\nabla \left(\tfrac{1}{2} u^2 + \frac{1}{\rho} P + \frac{1}{m} \phi \right) = 0 \qquad (5.52)$$

which is *Bernoulli's equation*. Second, in the case of uniform temperature and irrotational flow, namely, $\nabla \theta = 0$ and $\nabla \times \mathbf{u} = 0$, we have

$$\nabla \left(\tfrac{1}{2} u^2 + \frac{1}{m} \phi \right) = -\frac{\theta}{m} \nabla (\log \rho)$$

which may be immediately integrated to yield

$$\rho = \rho_0 \exp \left[-\frac{1}{\theta} \left(\tfrac{1}{2} mu^2 + \phi \right) \right] \qquad (5.53)$$

where ρ_0 is an arbitrary constant.

5.5 THE FIRST-ORDER APPROXIMATION

We now give an estimate of the error incurred in the zero-order approximation (5.30). Let $f(\mathbf{r}, \mathbf{p}, t)$ be the exact distribution function, and let

$$g(\mathbf{r}, \mathbf{p}, t) \equiv f(\mathbf{r}, \mathbf{p}, t) - f^{(0)}(\mathbf{r}, \mathbf{p}, t) \qquad (5.54)$$

We are interested in the magnitude of g as compared to $f^{(0)}$. First let us estimate the order of magnitude of $(\partial f / \partial t)_{\text{coll}}$. We have, by definition,

$$\left(\frac{\partial f}{\partial t} \right)_{\text{coll}} = \int d^3 p_2 \, d^3 p_1' \, d^3 p_2' \, \delta^4 (P_{\text{f}} - P_{\text{i}}) |T_{\text{fi}}|^2 (f_2' f_1' - f_2 f_1)$$

$$\approx \int d^3 p_2 \, d^3 p_1' \, d^3 p_2' \, \delta^4 (P_{\text{f}} - P_{\text{i}}) |T_{\text{fi}}|^2$$

$$\times \left(f_2^{(0)'} g_1' - f_2^{(0)} g_1 + g_2' f_1^{(0)'} - g_2 f_1^{(0)} \right) \qquad (5.55)$$

where we have used (5.54), the fact that $(\partial f^{(0)}/\partial t)_{\text{coll}} = 0$, and the assumption that g is a small quantity whose square can be neglected. An order-of-magnitude estimate of (5.55) may be obtained by calculating the second term of the right side of (5.55), which is

$$-g(\mathbf{r}, \mathbf{p}_1, t) \int d^3 p_2 \, \sigma_{\text{tot}} |\mathbf{v}_2 - \mathbf{v}_1| f_2^{(0)} = -\frac{g(\mathbf{r}, \mathbf{p}_1, t)}{\tau} \qquad (5.56)$$

where τ is a number of the order of magnitude of the collision time. Thus if we put

$$\left(\frac{\partial f}{\partial t} \right)_{\text{coll}} \approx -\frac{f - f^{(0)}}{\tau} \qquad (5.57)$$

we obtain results that are qualitatively correct.* With (5.57) the Boltzmann transport equation becomes

$$\left(\frac{\partial}{\partial t} + \mathbf{v} \cdot \nabla_\mathbf{r} = \frac{\mathbf{F}}{m} \cdot \nabla_\mathbf{v}\right)(f^{(0)} + g) \approx -\frac{g}{\tau} \tag{5.58}$$

Assuming $g \ll f^{(0)}$, we can neglect g on the left side of (5.58). Assume further that $f^{(0)}$ varies by a significant amount (i.e., of the order of itself) only when $|\mathbf{r}|$ varies by a distance L. Then (5.58) furnishes the estimate

$$\bar{v}\frac{f^{(0)}}{L} \approx -\frac{g}{\tau}$$

or $\qquad\qquad\qquad\qquad\qquad\qquad\qquad\qquad\qquad\qquad\qquad\qquad$ (5.59)

$$\frac{g}{f^{(0)}} \approx -\frac{\lambda}{L}$$

where λ is a length of the order of the mean free path. From these considerations we conclude that $f^{(0)}$ is a good approximation if the local density, temperature, and velocity have characteristic wavelengths L much larger than the mean free path λ. The corrections to $f^{(0)}$ would be of the order of λ/L.

A systematic expansion of f in powers of λ/L is furnished by the Chapman-Enskog expansion, which is somewhat complicated. In order not to lose sight of the physical aspects of the problem, we give a qualitative discussion of the first-order approximation based on the approximate equation (5.58). The precise value of τ cannot be ascertained. For the present we have to be content with the knowledge that τ is of the order of the collision time. Thus we put

$$f = f^{(0)} + g \tag{5.60}$$

where, with (5.58), we take

$$g = -\tau\left(\frac{\partial}{\partial t} + \mathbf{v} \cdot \nabla_\mathbf{r} + \frac{\mathbf{F}}{m} \cdot \nabla_\mathbf{v}\right)f^{(0)} \tag{5.61}$$

To calculate g, note that $f^{(0)}$ depends on $\mathbf{r}$ and t only through the functions ρ, θ, and $\mathbf{u}$. Thus we need the derivatives

$$\frac{\partial f^{(0)}}{\partial \rho} = \frac{f^{(0)}}{\rho}$$

$$\frac{\partial f^{(0)}}{\partial \theta} = \frac{1}{\theta}\left(\frac{m}{2\theta}U^2 - \frac{3}{2}\right)f^{(0)}$$

$$\frac{\partial f^{(0)}}{\partial u_i} = \frac{m}{\theta}U_i f^{(0)} \tag{5.62}$$

$$\frac{\partial f^{(0)}}{\partial v_i} = -\frac{m}{\theta}U_i f^{(0)}$$

---

*Techniques useful for solving the Boltzmann transport equation, together with results for a few simple intermolecular potentials, may be found in S. Chapman and T. G. Cowling, *The Mathematical Theory of Non-Uniform Gases*, 2nd ed. (Cambridge University Press, Cambridge, 1952).

where

$$\mathbf{U} \equiv \mathbf{v} - \mathbf{u}(\mathbf{r}, t) \tag{5.63}$$

Hence

$$g = -\tau\left(\frac{\partial}{\partial t} + v_i\frac{\partial}{\partial x_i} + \frac{F_i}{m}\frac{\partial}{\partial v_i}\right)f^{(0)}$$

$$= -\tau f^{(0)}\left[\frac{1}{\rho}D(\rho) + \frac{1}{\theta}\left(\frac{m}{2\theta}U^2 - \frac{3}{2}\right)D(\theta) + \frac{m}{\theta}U_j D(u_j) - \frac{1}{\theta}\mathbf{F}\cdot\mathbf{U}\right] \tag{5.64}$$

where

$$D(X) \equiv \left(\frac{\partial}{\partial t} + v_i\frac{\partial}{\partial x_i}\right)X \tag{5.65}$$

Using the zero-order hydrodynamic equations (5.37)–(5.39), we can show that

$$\begin{aligned}
D(\rho) &= -\rho(\nabla\cdot\mathbf{u}) + \mathbf{U}\cdot\nabla\rho \\
D(\theta) &= -\tfrac{2}{3}\theta\nabla\cdot\mathbf{u} + \mathbf{U}\cdot\nabla\theta \\
D(u_j) &= -\frac{1}{\rho}\frac{\partial P}{\partial x_j} + \frac{F_j}{m} + U_i\frac{\partial u_j}{\partial x_i}
\end{aligned} \tag{5.66}$$

where $P = \rho\theta/m$. Substituting these into (5.64) we obtain

$$g = -\tau f^{(0)}\left[-(\nabla\cdot\mathbf{u}) + \mathbf{U}\cdot\frac{\nabla\rho}{\rho} + \frac{1}{\theta}\left(\frac{m}{2\theta}U^2 - \frac{3}{2}\right)\left(-\frac{2}{3}\theta\nabla\cdot\mathbf{u} + \mathbf{U}\cdot\nabla\theta\right)\right.$$

$$\left. + \frac{m}{\theta}\left(-\mathbf{U}\cdot\frac{\nabla P}{\rho} + \mathbf{U}\cdot\frac{\mathbf{F}}{m} + U_iU_j\frac{\partial u_j}{\partial x_i}\right) - \frac{1}{\theta}\mathbf{F}\cdot\mathbf{U}\right]$$

which, after some rearrangement and cancellation of terms, becomes

$$g = -\tau\left[\frac{1}{\theta}\frac{\partial\theta}{\partial x_i}U_i\left(\frac{m}{2\theta}U^2 - \frac{5}{2}\right) + \frac{1}{\theta}\Lambda_{ij}\left(U_iU_j - \tfrac{1}{3}\delta_{ij}U^2\right)\right]f^{(0)} \tag{5.67}$$

where Λ_{ij} is defined by (5.29).

It is now necessary to calculate $\mathbf{q}$ and P_{ij} with the help of (5.60) to obtain the equations of hydrodynamics to the first order. We have

$$\mathbf{q} = \frac{m\dot\rho}{2n}\int d^3p\,(\mathbf{v} - \mathbf{u})|\mathbf{v} - \mathbf{u}|^2 g$$

Noting that the second term of (5.67) does not contribute to this integral, we obtain

$$\mathbf{q} = -\frac{\tau m^5}{2}\int d^3U\,\mathbf{U}U^2\left(\frac{m}{2\theta}U^2 - \frac{5}{2}\right)\frac{1}{\theta}U_i\frac{\partial\theta}{\partial x_i}f^{(0)}$$

or

$$\mathbf{q} = -K\,\nabla\theta \tag{5.68}$$

where

$$K = \frac{m^5 \tau}{6\theta} \int d^3 U\, U^4 \left(\frac{m}{2\theta} U^2 - \frac{5}{2} \right) f^{(0)} = \tfrac{5}{2}\tau\theta n \tag{5.69}$$

It is clear from (5.68) that K is to be identified as the coefficient of thermal conductivity. It is also clear that $|\mathbf{q}|$ is a small quantity of the first order, being of the order of λ/L.

For the pressure tensor P_{ij}, only the second term of (5.67) contributes:

$$P_{ij} = \frac{\rho}{n} \int d^3 p\, (v_i - u_i)(v_j - u_j)(f^{(0)} + g) = \delta_{ij} P + P'_{ij} \tag{5.70}$$

where $P = \rho\theta/m$ and

$$P'_{ij} = -\frac{\tau\rho m^3}{\theta n} \Lambda_{kl} \int d^3 U\, U_i U_j \left(U_k U_l - \tfrac{1}{3}\delta_{kl} U^2 \right) f^{(0)} \tag{5.71}$$

To evaluate this, note that P'_{ij} is a symmetric tensor of zero trace (i.e., $\sum_{i=1}^{3} P'_{ii} = 0$), and it depends linearly on the symmetric tensor Λ_{ij}. Therefore P'_{ij} must have the form

$$P'_{ij} = -\frac{2\mu}{m} \left(\Lambda_{ij} - \frac{m}{3}\delta_{ij} \nabla \cdot \mathbf{u} \right) \tag{5.72}$$

where $m\nabla \cdot \mathbf{u}$ is none other than the trace of Λ_{ij}:

$$\sum_{i=1}^{3} \Lambda_{ii} = m \sum_{i=1}^{3} \frac{\partial u_i}{\partial x_i} = m\nabla \cdot \mathbf{u} \tag{5.73}$$

and μ is a constant. It remains to calculate μ. For this purpose it suffices to calculate any component of P'_{ij}, e.g., P'_{12}. From (5.71) we have

$$P'_{12} = -\frac{\tau m^4}{\theta} \Lambda_{kl} \int d^3 U\, U_1 U_2 \left(U_k U_l - \tfrac{1}{3}\delta_{kl} U^2 \right) f^{(0)}$$

$$= -2\frac{\tau m^4}{\theta} \Lambda_{12} \int d^3 U\, U_1^2 U_2^2 f^{(0)}$$

Therefore

$$\mu = \frac{\tau m^5}{\theta} \int d^3 U\, U_1^2 U_2^2 f^{(0)} = \tau n\theta \tag{5.74}$$

With this we have

$$P_{ij} = \delta_{ij} P - \frac{2\mu}{m} \left(\Lambda_{ij} - \frac{m}{3}\delta_{ij} \nabla \cdot \mathbf{u} \right) \tag{5.75}$$

The second term is of the order of λ/L. The coefficient μ turns out to be the coefficient of viscosity, as we show shortly.

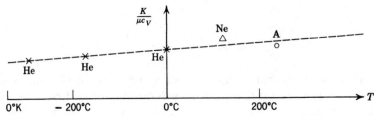

Fig. 5.1 Ratio of thermal conductivity to the product of viscosity and specific heat for different dilute gases.

A comparison of (5.74) with (5.69) shows that

$$\frac{K}{\mu} = \tfrac{5}{2} = \tfrac{5}{3}c_V \tag{5.76}$$

Since the unknown collision time τ drops out in this relation, we might expect (5.76) to be of quantitative significance. A plot of some experimental data for different dilute gases in Fig. 5.1 shows that it is indeed so.

Let us put, with (5.6),

$$\tau \approx \sqrt{\frac{m}{kT}} \frac{1}{na^2} \tag{5.77}$$

where a is the molecular diameter. Then we find that

$$\mu \approx K \approx \frac{\sqrt{mkT}}{a^2} \tag{5.78}$$

5.6 VISCOSITY

To show that (5.74) is the coefficient of viscosity, we independently calculate the coefficient of viscosity using its experimental definition. Consider a gas of uniform and constant density and temperature, with an average velocity given by

$$\begin{aligned} u_x &= A + By \\ u_y &= u_z = 0 \end{aligned} \tag{5.79}$$

where A and B are constants. The gas may be thought of as being composed of different layers sliding over each other, as shown in Fig. 5.2. Draw any plane perpendicular to the y axis, as shown by the dotted line in Fig. 5.2. Let F' be the frictional force experienced by the gas above this plane, per unit area of the plane. Then the coefficient of viscosity μ is experimentally defined by the relation

$$F' = -\mu \frac{\partial u_x}{\partial y} \tag{5.80}$$

The gas above the plane experiences a frictional force by virtue of the fact that it

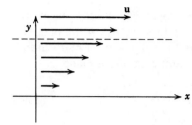

Fig. 5.2 Horizontal flow of a gas with average velocity increasing linearly with height.

suffers a net loss of "x component of momentum" to the gas below. Thus

$$F' \equiv net \text{ amount of "}x \text{ component of momentum"} \\ \text{transported per sec across unit area in the} \\ y \text{ direction} \tag{5.81}$$

The quantity being transported is $m(v_x - u_x)$, whereas the flux effective in the transport is $n(v_y - u_y)$. Hence we have

$$F' = mn\left\langle (v_x - u_x)(v_y - u_y) \right\rangle = m^4 \int d^3v \, (v_x - u_x)(v_y - u_y)(f^{(0)} + g) \tag{5.82}$$

We easily see that the term $f^{(0)}$ does not contribute to the integral in (5.82). The first correction g may be obtained directly from the approximate Boltzmann transport equation

$$\mathbf{v} \cdot \nabla f^{(0)} = -\frac{g}{\tau} \tag{5.83}$$

$$g = -\frac{\tau m}{\theta} v_y (v_x - u_x) B f^{(0)} = -\frac{\tau m}{\theta} U_y U_x \frac{\partial u_x}{\partial y} f^{(0)}$$

where $\mathbf{U} \equiv \mathbf{v} - \mathbf{u}$. Thus

$$F' = -\frac{\partial u_x}{\partial y} \frac{\tau m^5}{\theta} \int d^3U \, U_x^2 U_y^2 f^{(0)} \tag{5.84}$$

A comparison between this and (5.80) yields

$$\mu = \frac{\tau m^5}{\theta} \int d^3U \, U_x^2 U_y^2 f^{(0)} \tag{5.85}$$

which is identical with (5.74).

From the nature of this derivation it is possible to understand physically why μ has the order of magnitude given by (5.78). Across the imaginary plane mentioned previously, a net transport of momentum exists, because molecules constantly cross this plane in both directions. The flux is the same in both directions, being of the order of $n\sqrt{kT/m}$. On the average, however, those that cross from above to below carry more "x component of momentum" than the opposite ones, because the average velocity u_x is greater above than below. Since most molecules that cross the plane from above originated within a mean free path λ above the plane, their u_x is in excess of the local u_x below the plane by the amount $\lambda(\partial u_x/\partial y)$. Hence the net amount of "x component of momentum" transported per second from above to below, per unit area of the plane, is

$$\lambda nm\sqrt{\frac{kT}{m}}\frac{\partial u_x}{\partial y} = \frac{\sqrt{mkT}}{a^2}\frac{\partial u_x}{\partial y} \tag{5.86}$$

Therefore

$$\mu \approx \frac{\sqrt{mkT}}{a^2} \tag{5.87}$$

It is interesting to note that according to (5.87) μ is independent of the density for a given temperature. When Maxwell first derived this fact, he was so surprised that he put it to experimental test by observing the rate of damping of a pendulum suspended in gases of different densities. To his satisfaction, it was verified.

According to (5.87) the coefficient of viscosity increases as the molecular diameter decreases, everything else being constant. This is physically easy to understand because the mean free path λ increases with decreasing molecular diameter. For a given gradient $\partial u_x/\partial y$, the momentum transported across any plane normal to the y axis obviously increases as λ increases. When λ becomes so large that it is comparable to the size of the container of the gas, the whole method adopted here breaks down, and the coefficient of viscosity ceases to be a meaningful concept.

As a topic related to the concept of viscosity we consider the boundary condition for a gas flowing past a wall. A gas, unlike a liquid, does not stick to the wall of its container. Rather, it slips by with an average velocity u_0. To determine u_0, it is necessary to know how the gas molecules interact with the wall. We make the simplifying assumption that a fraction $1 - \alpha$ of the molecules striking the wall is reflected elastically while the remaining fraction α is absorbed by the wall, only to return to the gas later with thermal velocity. The number α is called the *coefficient of accommodation*. Suppose the wall is the xy plane, as shown in Fig. 5.3. Then the downward flux of particles is given by

$$m^3\int_{-\infty}^{\infty}dv_x\int_{-\infty}^{\infty}dv_y\int_{0}^{\infty}dv_z\, nv_z f^{(0)} = n\sqrt{\frac{\theta}{2\pi m}} \tag{5.88}$$

The particles that reach the wall came from a mean free path λ above the wall.

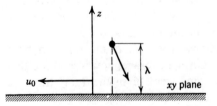

Fig. 5.3 A gas slipping past a wall.

Thus the gas loses to a unit area of the wall an amount of momentum per second equal to

$$F' = -\alpha n m \sqrt{\frac{\theta}{2\pi m}} \left[u_0 + \lambda \left(\frac{\partial u}{\partial z} \right)_0 \right] \tag{5.89}$$

where $(\partial u/\partial z)_0$ is the normal gradient of u at the wall. This is the force of friction per unit area that the wall exerts on the gas, and must equal $-\mu(\partial u/\partial z)_0$. Hence the boundary condition at the wall is

$$\alpha n m \sqrt{\frac{\theta}{2\pi m}} \left[u_0 + \lambda \left(\frac{\partial u}{\partial z} \right)_0 \right] = \mu \left(\frac{\partial u}{\partial z} \right)_0$$

or

$$u_0 = \left(\sqrt{\frac{2\pi}{m\theta}} \frac{\mu}{n\alpha} - \lambda \right) \left(\frac{\partial u}{\partial z} \right)_0 \tag{5.90}$$

Using $\mu = \tau n \theta$ and $\lambda = \beta \tau \sqrt{2\pi\theta/m}$, where β is a constant of the order of unity, we obtain the boundary condition

$$u_0 = s\lambda \left(\frac{\partial u}{\partial z} \right)_0 \tag{5.91}$$

where

$$s = \frac{1 - \alpha\beta}{\alpha\beta}$$

is an empirical constant which may be called the "slipping coefficient." When $s = 0$ there is no slipping at the wall. In general the velocity of slip is equal to the velocity in the gas at a distance of s mean free paths from the wall. Usually $s\lambda$ is a few mean free paths.

5.7 VISCOUS HYDRODYNAMICS

The equations of hydrodynamics in the first-order approximation can be obtained by substituting $\mathbf{q}$ and P_{ij}, given respectively in (5.68) and (5.75), into the conservation theorems (5.21)–(5.23). We first evaluate a few relevant quantities.

$$\nabla \cdot \mathbf{q} = -\nabla(K\nabla\theta) = -K\nabla^2\theta - \nabla K \cdot \nabla\theta \tag{5.92}$$

$$\frac{\partial P_{ij}}{\partial x_j} = \frac{\partial P}{\partial x_i} - \mu \left[\nabla^2 u_i + \frac{1}{3} \frac{\partial}{\partial x_i} (\nabla \cdot \mathbf{u}) \right] - \frac{2}{m} \frac{\partial \mu}{\partial x_j} \left(\Lambda_{ij} - \frac{m}{3} \delta_{ij} \nabla \cdot \mathbf{u} \right) \tag{5.93}$$

$$P_{ij}\Lambda_{ij} = mP(\nabla \cdot \mathbf{u}) - \frac{2\mu}{m} \Lambda_{ij}\Lambda_{ij} + \tfrac{2}{3}\mu m (\nabla \cdot \mathbf{u})^2 \tag{5.94}$$

The quantity $\Lambda_{ij}\Lambda_{ij}$ can be further reduced:

$$\Lambda_{ij}\Lambda_{ij} = \frac{m^2}{4}\left(\frac{\partial u_i}{\partial x_j} + \frac{\partial u_j}{\partial x_i}\right)\left(\frac{\partial u_i}{\partial x_j} + \frac{\partial u_j}{\partial x_i}\right) = \frac{m^2}{2}\frac{\partial u_i}{\partial x_j}\left(\frac{\partial u_i}{\partial x_j} + \frac{\partial u_j}{\partial x_i}\right)$$

Now we reduce the two terms above separately:

$$\frac{\partial u_i}{\partial x_j}\frac{\partial u_i}{\partial x_j} = \frac{\partial}{\partial x_j}\left(u_i\frac{\partial u_i}{\partial x_j}\right) - u_i\frac{\partial^2 u_i}{\partial x_j\partial x_j} = \tfrac{1}{2}\nabla^2(u^2) - \mathbf{u}\cdot\nabla^2\mathbf{u}$$

$$\frac{\partial u_i}{\partial x_j}\frac{\partial u_j}{\partial x_i} = \left(\frac{\partial u_i}{\partial x_j} - \frac{\partial u_j}{\partial x_i}\right)\left(\frac{\partial u_j}{\partial x_i} - \frac{\partial u_i}{\partial x_j}\right) + \frac{\partial u_i}{\partial x_j}\frac{\partial u_i}{\partial x_j} + \frac{\partial u_j}{\partial x_i}\frac{\partial u_j}{\partial x_i} - \frac{\partial u_j}{\partial x_i}\frac{\partial u_i}{\partial x_j}$$

$$= -2(\nabla\times\mathbf{u})^2 + 2\frac{\partial u_i}{\partial x_j}\frac{\partial u_i}{\partial x_j} - \frac{\partial u_j}{\partial x_i}\frac{\partial u_i}{\partial x_j}$$

Hence

$$\frac{\partial u_i}{\partial x_j}\frac{\partial u_j}{\partial x_i} = -(\nabla\times\mathbf{u})^2 + \frac{\partial u_i}{\partial x_j}\frac{\partial u_i}{\partial x_j}$$

and finally

$$\Lambda_{ij}\Lambda_{ij} = \frac{m^2}{2}\left[\nabla^2(u^2) - 2\mathbf{u}\cdot\nabla^2\mathbf{u} - |\nabla\times\mathbf{u}|^2\right] \tag{5.95}$$

Substituting (5.92)–(5.94) into (5.21)–(5.23) we obtain

$$\frac{\partial\rho}{\partial t} + \nabla\cdot(\rho\mathbf{u}) = 0 \tag{5.96}$$

$$\rho\left(\frac{\partial}{\partial t} + \mathbf{u}\cdot\nabla\right)\mathbf{u} = \frac{\mathbf{F}}{m} - \nabla\left(P - \frac{\mu}{3}\nabla\cdot\mathbf{u}\right) + \mu\nabla^2\mathbf{u} + \mathbf{R} \tag{5.97}$$

$$\rho\left(\frac{\partial}{\partial t} + \mathbf{u}\cdot\nabla\right)\theta = \frac{K}{c_V}\nabla^2\theta + \frac{1}{c_V}\nabla K\cdot\nabla\theta - \frac{1}{c_V}$$

$$\times\left[m\rho(\nabla\cdot\mathbf{u}) + \tfrac{2}{3}\mu m(\nabla\cdot\mathbf{u})^2 - \mu m\{\nabla^2(u^2) - 2\mathbf{u}\cdot\nabla^2\mathbf{u} - |\nabla\times\mathbf{u}|^2\}\right] \tag{5.98}$$

where $c_V = \tfrac{3}{2}$ and $\mathbf{R}$ is a vector whose components are given by

$$R_i = \frac{2}{m}\frac{\partial\mu}{\partial x_j}\left(\Lambda_{ij} - \frac{m}{3}\delta_{ij}\nabla\cdot\mathbf{u}\right) \tag{5.99}$$

In these equations the quantities of first-order smallness are μ, K, $\mathbf{u}$, and the derivatives of ρ, θ, and $\mathbf{u}$. Keeping only quantities of first-order smallness, we can neglect all terms involving derivatives of μ and K and the last four terms on the right side of (5.98). We then have the equations of hydrodynamics to the first order:

$$\frac{\partial \rho}{\partial t} + \nabla \cdot (\rho \mathbf{u}) = 0 \qquad \text{(continuity equation)} \tag{5.100}$$

$$\left(\frac{\partial}{\partial t} + \mathbf{u} \cdot \nabla\right)\mathbf{u} = \frac{\mathbf{F}}{m} - \frac{1}{\rho}\nabla\left(P - \frac{\mu}{3}\nabla \cdot \mathbf{u}\right) + \frac{\mu}{\rho}\nabla^2 \mathbf{u}$$

$$\text{(Navier-Stokes equation)} \tag{5.101}$$

$$\left(\frac{\partial}{\partial t} + \mathbf{u} \cdot \nabla\right)\theta = -\frac{1}{c_V}(\nabla \cdot \mathbf{u})\theta + \frac{K}{\rho c_V}\nabla^2 \theta \qquad \text{(heat conduction equation)}$$

$$\tag{5.102}$$

where $c_V = \frac{3}{2}$. The boundary condition to be used when a wall is present is the slip boundary condition (5.91).

If $\mathbf{u} = 0$, (5.102) reduces to

$$\rho c_V \frac{\partial \theta}{\partial t} - K \nabla^2 \theta = 0 \tag{5.103}$$

which is the familiar diffusion equation governing heat conduction. This equation can be derived intuitively from the fact that $\mathbf{q} = -K \nabla \theta$. Although we have proved this fact only for a dilute gas, it is experimentally correct for liquids and solids as well. For this reason (5.103) is often applied to systems other than a dilute gas.

The Navier-Stokes equation can also be derived on an intuitive basis provided we take the meaning of viscosity from experiments. We discuss this derivation in the next section.

5.8 THE NAVIER-STOKES EQUATION

We give a phenomenological derivation of the Navier-Stokes equation to show why it is expected to be valid even for liquids. Some examples of its use are then discussed.

Consider a small element of fluid whose volume is $dx_1 \, dx_2 \, dx_3$ and whose velocity is $\mathbf{u}(\mathbf{r}, t)$. According to Newton's second law the equation of motion of this element of fluid is

$$m \frac{d\mathbf{u}}{dt} = \mathscr{F}$$

where m is the mass of the fluid element and $\mathscr{F}$ is the total force acting on the fluid element. Let the mass density of the fluid be ρ and let there be two forces acting on any element of fluid: A force due to agents external to the fluid, and a force due to neighboring fluid elements. These forces *per unit volume* will be

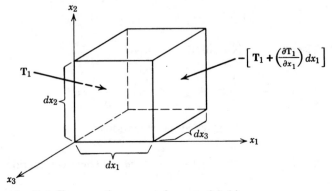

Fig. 5.4 Forces acting on an element of fluid.

respectively denoted by $\mathbf{F}_1$ and $\mathbf{G}$. Thus we can write

$$m = \rho \, dx_1 \, dx_2 \, dx_3$$

$$\mathscr{L} = (\mathbf{F}_1 + \mathbf{G}) \, dx_1 \, dx_2 \, dx_3$$

Therefore Newton's second law for a fluid element takes the form

$$\rho\left(\frac{\partial}{\partial t} + \mathbf{u} \cdot \nabla\right)\mathbf{u} = \mathbf{F}_1 + \mathbf{G} \tag{5.104}$$

Thus the derivation of the Navier-Stokes equation reduces to the derivation of a definite expression for $\mathbf{G}$.

Let us choose a coordinate system such that the fluid element under consideration is a cube with edges along the three coordinate axes, as shown in Fig. 5.4. The six faces of this cube are subjected to forces exerted by neighboring fluid elements. The force on each face is such that its direction is determined by the direction of the normal vector to the face. That is, its direction depends on which side of the face is considered the "outside." This is physically obvious if we remind ourselves that this force arises from hydrostatic pressure and viscous drag. Let $\mathbf{T}_i$ be the force per unit area acting on the face whose normal lies along the x_i axis. Then the forces *per unit area* acting on the two faces normal to the x_i axis are, respectively (see Fig. 5.4),

$$\mathbf{T}_i, \quad -\left(\mathbf{T}_i + \frac{\partial \mathbf{T}_i}{\partial x_i} \, dx_i\right) \qquad (i = 1, 2, 3) \tag{5.105}$$

The total force acting on the cube by neighboring fluid elements is then given by

$$\mathbf{G} \, dx_1 \, dx_2 \, dx_3 = -\left(\frac{\partial \mathbf{T}_1}{\partial x_1} + \frac{\partial \mathbf{T}_2}{\partial x_2} + \frac{\partial \mathbf{T}_3}{\partial x_3}\right) dx_1 \, dx_2 \, dx_3 \tag{5.106}$$

We denote the components of the vectors T_1, T_2, T_3 as follows:

$$T_1 = (P_{11}, P_{12}, P_{13})$$
$$T_2 = (P_{21}, P_{22}, P_{23})$$
$$T_3 = (P_{31}, P_{32}, P_{33})$$
(5.107)

Then

$$G_i = -\frac{\partial P_{ji}}{\partial x_j}$$
(5.108)

or

$$\mathbf{G} = -\nabla \cdot \overset{\leftrightarrow}{P}$$
(5.109)

With this, (5.104) becomes

$$\rho\left(\frac{\partial}{\partial t} + \mathbf{u} \cdot \nabla\right)\mathbf{u} = \mathbf{F}_1 - \nabla \cdot \overset{\leftrightarrow}{P}$$
(5.110)

which is of the same form as (5.22) if we set $\mathbf{F}_1 = \rho\mathbf{F}/m$, where $\mathbf{F}$ is the external force per molecule and m is the mass of a molecule. To derive the Navier-Stokes equation, we only have to deduce a more explicit form for P_{ij}. We postulate that (5.110) is valid, whatever the coordinate system we choose. It follows that P_{ij} is a tensor.

We assume the fluid under consideration to be isotropic, so that there can be no intrinsic distinction among the axes x_1, x_2, x_3. Accordingly we must have

$$P_{11} = P_{22} = P_{33} \equiv P$$
(5.111)

where P is by definition the hydrostatic pressure. Thus P_{ij} can be written in the form

$$P_{ij} = \delta_{ij}P + P'_{ij}$$
(5.112)

where P'_{ij} is a traceless tensor, namely,

$$\sum_{i=1}^{3} P'_{ii} = 0$$
(5.113)

This follows from the fact that (5.113) is true in one coordinate system and that the trace of a tensor is independent of the coordinate system.

Next we make the physically reasonable assumption that the fluid element under consideration, which is really a point in the fluid, has no intrinsic angular momentum. This assumption implies that P_{ij}, and hence P'_{ij}, is a symmetric tensor:

$$P'_{ij} = P'_{ji}$$
(5.114)

To see this we need only remind ourselves of the meaning of, for example, P'_{12}. A glance at Fig. 5.5a makes (5.114) obvious.

Finally we incorporate into P_{ij} the empirical connection between the shear force applied to a fluid element and the rate of deformation of the same fluid element. A shear force F' per unit area acting parallel to a face of a cube of fluid

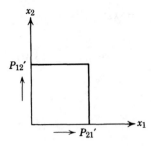

Fig. 5.5a Nonrotation of fluid element implies $P'_{12} = P'_{21}$.

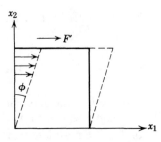

Fig. 5.5b Deformation of fluid element due to shear force.

tends to stretch the cube into a parallelopiped at a rate given by $R' = \mu(d\phi/dt)$, where μ is the coefficient of viscosity and ϕ is the angle shown in Fig. 5.5b.

Consider now the effect of P'_{12} on one fluid element. It can be seen from Fig. 5.5c, where P'_{12} is indicated in its positive sense in accordance with (5.105), that

$$P'_{21} = -\mu\left(\frac{d\phi_1}{dt} + \frac{d\phi_2}{dt}\right) = -\mu\left(\frac{\partial u_2}{\partial x_1} + \frac{\partial u_1}{\partial x_2}\right) \tag{5.115}$$

In general we have

$$P'_{ij} = -\mu\left(\frac{\partial u_i}{\partial x_j} + \frac{\partial u_j}{\partial x_i}\right) \qquad (i \neq j) \tag{5.116}$$

To make P'_{ij} traceless we must take

$$P'_{ij} = -\mu\left[\left(\frac{\partial u_i}{\partial x_j} + \frac{\partial u_j}{\partial x_i}\right) - \frac{2}{3}\delta_{ij}\nabla \cdot \mathbf{u}\right] \tag{5.117}$$

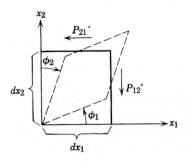

Fig. 5.5c P'_{12} as shear force.

Therefore

$$P_{ij} = \delta_{ij}P - \mu\left[\left(\frac{\partial u_i}{\partial x_j} + \frac{\partial u_j}{\partial x_i}\right) - \frac{2}{3}\delta_{ij}\nabla \cdot \mathbf{u}\right] \tag{5.118}$$

which is identical in form to (5.75). This completes the phenomenological derivation, which makes it plausible that the Navier-Stokes equation is valid for dilute gas and dense liquid alike.

5.9 EXAMPLES IN HYDRODYNAMICS

To illustrate the mathematical techniques of dealing with the equations of hydrodynamics (5.110)–(5.102), we consider two examples of the application of the Navier-Stokes equation to a liquid.

Incompressible Flow

We consider the following problem: A sphere of radius r is moving with instantaneous velocity $\mathbf{u}_0$ in an infinite, nonviscous, incompressible fluid of constant density in the absence of external force. The Navier-Stokes equation reduces to Euler's equation:

$$\rho\left(\frac{\partial}{\partial t} + \mathbf{u} \cdot \nabla\right)\mathbf{u} = -\nabla P \tag{5.119}$$

where $\mathbf{u}$ is the velocity field of the liquid and P the pressure as given by the equation of state of the fluid. Let us choose the center of the sphere to be the origin of the coordinate system and label any point in space by either the rectangular coordinates (x, y, z) or the spherical coordinates (r, θ, ϕ). The boundary conditions shall be such that the normal component of $\mathbf{u}$ vanishes on the surface of the sphere and that the liquid is at rest at infinity:

$$[\mathbf{r} \cdot \mathbf{u(r)}]_{r=a} - (\mathbf{r} \cdot \mathbf{u}_0)_{r=a} = 0$$
$$\mathbf{u(r)} \underset{r=\infty}{\to} 0 \tag{5.120}$$

Note that incompressibility means $\partial V/\partial P = 0$, or that the density is independent of P. Therefore ∇P is arbitrary, and adjusts itself to whatever the boundary condition demands. Since there is no source for the fluid, we must have everywhere

$$\nabla \cdot \mathbf{u} = 0 \tag{5.121}$$

Taking the curl of both sides of (5.119), remembering that ρ is a constant, and neglecting terms of the form $(\partial \mathbf{u}/\partial x_i)(\partial \mathbf{u}/dx_j)$, we find that

$$\left(\frac{\partial}{\partial t} + \mathbf{u} \cdot \nabla\right)(\nabla \times \mathbf{u}) = 0 \tag{5.122}$$

i.e., that $\nabla \times \mathbf{u}$ is constant along a streamline. Note that P drops out because $\nabla \times (\nabla \times P) \equiv 0$. Since very far from the sphere we have $\nabla \times \mathbf{u} = 0$, it follows that everywhere

$$\nabla \times \mathbf{u} = 0 \tag{5.123}$$

This means that $\mathbf{u}$ is the gradient of some function:

$$\mathbf{u} = \nabla \Phi \tag{5.124}$$

where Φ is called the velocity potential. By (5.120) and (5.121) the equation and boundary conditions for Φ are

$$\nabla^2 \Phi(\mathbf{r}) = 0$$

$$\left(\frac{\partial \Phi}{\partial r} \right)_{r=a} = u_0 \cos \theta \tag{5.125}$$

$$\Phi(\mathbf{r}) \underset{r \to \infty}{\to} 0$$

where θ is the angle between u_0 and $\mathbf{r}$, as shown in Fig. 5.6.

The most general solution to $\nabla^2 \Phi = 0$ is a superposition of solid harmonics.* Since the boundary condition involves $\cos \theta$, we try the solution

$$\Phi(\mathbf{r}) = A \frac{\cos \theta}{r^2} \qquad (r \geq a) \tag{5.126}$$

which is a solid harmonic of order 1 and is the potential that would be set up if a dipole source were placed at the center of the sphere. Choosing $A = -\frac{1}{2} u_0 a^3$ satisfies the boundary conditions. Therefore

$$\Phi(r) = -\frac{1}{2} u_0 a^3 \frac{\cos \theta}{r^2} \qquad (r \geq a) \tag{5.127}$$

This is the only solution of (5.125), by the well-known uniqueness theorem of the Laplace equation. The velocity field of the fluid is then given by

$$\mathbf{u}(\mathbf{r}) = -\frac{1}{2} u_0 a^3 \nabla \frac{\cos \theta}{r^2} \qquad (r \geq a) \tag{5.128}$$

The streamlines can be sketched immediately, and they look like the electric field due to a dipole, as shown in Fig. 5.6.

Let us calculate the kinetic energy of the fluid. It is given by the integral

$$
\begin{aligned}
\text{K.E.} &= \int d^3r \, \tfrac{1}{2}\rho |\mathbf{u}|^2 = \frac{\rho}{2} \left(\frac{u_0 a^3}{2} \right)^2 \int_{r \geq a} d^3r \, \nabla \frac{\cos \theta}{r^2} \cdot \nabla \frac{\cos \theta}{r^2} \\
&= \frac{\rho}{2} \left(\frac{u_0 a^3}{2} \right)^2 \int_{r \geq a} d^3r \, \nabla \cdot \left[\frac{\cos \theta}{r^2} \nabla \frac{\cos \theta}{r^2} \right] \\
&= -\frac{\rho}{2} \left(\frac{u_0 a^3}{2} \right)^2 \int_{r=a} d\mathbf{S} \cdot \left[\frac{\cos \theta}{r^2} \nabla \frac{\cos \theta}{r^2} \right] \\
&= -\frac{\rho}{2} \left(\frac{u_0 a^3}{2} \right)^2 a^2 \int_0^{2\pi} d\phi \int_{-1}^{+1} d(\cos \theta) \left(\frac{\cos \theta}{r^2} \frac{\partial}{\partial r} \frac{\cos \theta}{r^2} \right)_{r=a} = \tfrac{1}{2} m' u_0^2
\end{aligned}
\tag{5.129}
$$

---

*A solid harmonic is $r^l Y_{lm}$ or $r^{-l-1} Y_{lm}$, where Y_{lm} is a spherical harmonic.

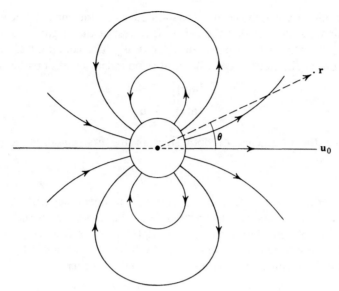

Fig. 5.6 Streamlines in a nonviscous liquid in the presence of a moving sphere.

where m' is half the mass of the displaced fluid:

$$m' = \tfrac{1}{2}\left(\tfrac{4}{3}\pi a^3 \rho\right) \tag{5.130}$$

If the sphere has a mechanical mass m, the total kinetic energy of the system of liquid plus sphere is

$$E = \tfrac{1}{2}(m_0 + m')u_0^2 \tag{5.131}$$

The mass $m_0 + m'$ may be interpreted to be the effective mass of the sphere, since (5.131) is the total energy that has to be supplied for the sphere to move with velocity u_0.

Stokes' Law

We proceed to solve the same problem when the fluid has a nonvanishing coefficient of viscosity μ. The Navier-Stokes equation will be taken to be

$$0 = -\nabla\left(P - \frac{\mu}{3}\nabla \cdot \mathbf{u}\right) + \mu\nabla^2\mathbf{u} \tag{5.132}$$

on the assumption that the material derivative of $\mathbf{u}$, which gives rise to the effective mass, is small compared to the viscous terms. We return to examine the validity of this approximation later. Since there is no source for the fluid, we still require $\nabla \cdot \mathbf{u} = 0$, and (5.132) becomes the simultaneous equations

$$\nabla^2\mathbf{u} = \frac{1}{\mu}\nabla P$$
$$\nabla \cdot \mathbf{u} = 0 \tag{5.133}$$

with the boundary condition that the fluid sticks to the sphere. Let us translate the coordinate system so that the sphere is at rest at the origin while the fluid at infinity flows with uniform constant velocity $\mathbf{u}_0$. The equations (5.133) remain invariant under the translation, whereas the boundary conditions become

$$[\mathbf{u}(\mathbf{r})]_{r=a} = 0$$
$$\mathbf{u}(\mathbf{r}) \underset{r \to \infty}{\to} \mathbf{u}_0 \qquad (5.134)$$

Taking the divergence of both sides of the first equation of (5.133), we obtain

$$\nabla^2 P = 0 \qquad (5.135)$$

Thus the pressure, whatever it is, must be a linear superposition of solid harmonics. A systematic way to proceed would be to write P as the most general superposition of solid harmonics and to determine the coefficient by requiring that (5.133) be satisfied. We take a short cut, however, and *guess* that P is, apart from an additive constant, a pure solid harmonic of order 1:

$$P = P_0 + \mu P_1 \frac{\cos \theta}{r^2} \qquad (5.136)$$

where P_0 and P_1 are constants to be determined later. With this, the problem reduces to solving the inhomogeneous Laplace equation

$$\nabla^2 \mathbf{u} = P_1 \nabla \frac{\cos \theta}{r^2} \qquad (5.137)$$

subject to the conditions

$$\nabla \cdot \mathbf{u} = 0$$
$$[\mathbf{u}(\mathbf{r})]_{r=a} = 0 \qquad (5.138)$$
$$\mathbf{u}(\mathbf{r}) \underset{r \to \infty}{\to} \mathbf{u}_0$$

A particular solution of (5.137) is

$$\mathbf{u}_1 = -\frac{P_1}{6} r^2 \nabla \frac{\cos \theta}{r^2} = -\frac{P_1}{6} \left(\frac{\hat{z}}{r} - 3\mathbf{r} \frac{z}{r^3} \right) \qquad (5.139)$$

where $\hat{z}$ denotes the unit vector along the z axis, which lies along $\mathbf{u}_0$. It is easily verified that (5.139) solves (5.137), if we note that $1/r$ and z/r^3 are both solid harmonics. Thus,

$$\nabla^2 \mathbf{u}_1 = -\frac{P_1}{6} \left[-3\nabla^2 \left(\frac{\mathbf{r}z}{r^3} \right) \right] = P_1 \nabla \left(\frac{z}{r^3} \right) = P_1 \nabla \frac{\cos \theta}{r^2} \qquad (5.140)$$

The complete solution is obtained by adding an appropriate homogeneous solution to (5.139) to satisfy (5.138). By inspection we see that the complete solution is

$$\mathbf{u} = \mathbf{u}_0 \left(1 - \frac{a}{r} \right) + \frac{1}{4} u_0 a (r^2 - a^2) \nabla \frac{\cos \theta}{r^2} \qquad (5.141)$$

where we have set

$$P_1 = -\tfrac{3}{2}u_0 a \qquad (5.142)$$

to have $\nabla \cdot \mathbf{u} = 0$.

We now calculate the force acting on the sphere by the fluid. By definition the force per unit area acting on a surface whose normal point along the x_j axis is $-\mathbf{T}_j$ of (5.107). It follows that the force per unit area acting on a surface element of the sphere is

$$\mathbf{f} = -\left(\frac{x}{r}\mathbf{T}_1 + \frac{y}{r}\mathbf{T}_2 + \frac{z}{r}\mathbf{T}_3\right) = -\hat{r} \cdot \overset{\leftrightarrow}{P} \qquad (5.143)$$

where $\hat{r}$ is the unit vector in the radial direction and $\overset{\leftrightarrow}{P}$ is given by (5.118). The total force experienced by the sphere is

$$\mathbf{F}' = \int dS\, \mathbf{f} \qquad (5.144)$$

where dS is a surface element of the sphere and the integral extends over the entire surface of the sphere. Thus it is sufficient to calculate $\mathbf{f}$ for $r = a$.

The vector $\hat{r} \cdot \overset{\leftrightarrow}{P}$ has the components

$$(\hat{r} \cdot \overset{\leftrightarrow}{P})_i = \frac{1}{r}x_j P_{ji} = \frac{1}{r}x_j \left[\delta_{ji}P - \mu\left(\frac{\partial u_j}{\partial x_i} + \frac{\partial u_i}{\partial x_j}\right)\right]$$

$$= \frac{x_i}{r}P - \frac{\mu}{r}\left[\frac{\partial}{\partial x_i}(x_j u_j) - u_i + x_j\frac{\partial}{\partial x_j}u_i\right]$$

Hence

$$\mathbf{f} = -\hat{r}P + \frac{\mu}{r}\left[\nabla(\mathbf{r} \cdot \mathbf{u}) - \mathbf{u} + (\mathbf{r} \cdot \nabla)\mathbf{u}\right] \qquad (5.145)$$

where P is given by (5.136) and (5.142), and $\mathbf{u}$ is given by (5.141). Since $\mathbf{u} = 0$ when $r = a$, we only need to consider the first and the last terms in the bracket. The first term is zero at $r = a$ by a straightforward calculation. At $r = a$ the second term is found to be

$$\frac{1}{r}[(\mathbf{r} \cdot \nabla)\mathbf{u}]_{r=a} = \left(\frac{\partial \mathbf{u}}{\partial r}\right)_{r=a} = \frac{3}{2}\frac{u_0}{a} - \frac{3}{2}\hat{r}u_0\frac{\cos\theta}{a} \qquad (5.146)$$

When this is substituted into (5.145), the second term exactly cancels the dipole part of $\hat{r}P$, and we obtain

$$(\mathbf{f})_{r=a} = -\hat{r}P_0 + \frac{3}{2}\frac{\mu}{a}\mathbf{u}_0$$

The constant P_0 is unknown, but it does not contribute to the force on the sphere. From (5.144) we obtain

$$\mathbf{F}' = 6\pi\mu a\mathbf{u}_0 \qquad (5.147)$$

which is Stokes' law.

The validity of (5.141) depends on the smallness of the material derivative of $\mathbf{u}$ as compared to $\mu \nabla^2 \mathbf{u}$. Both these quantities can be computed from (5.141). It is then clear that we must require

$$\frac{\rho u_0 a}{\mu} \ll 1 \qquad (5.148)$$

Thus Stokes' law holds only for small velocities and small radii of the sphere. A more elaborate treatment shows that a more accurate formula for $\mathbf{F}'$ is

$$\mathbf{F}' = 6\pi \mu a \mathbf{u}_0 \left(1 + \frac{3}{8} \frac{\rho u_0 a}{\mu} + \cdots \right) \qquad (5.149)$$

The pure number $\rho u_0 a / \mu$ is called the Reynolds number. When the Reynolds number becomes large, turbulence sets in and streamline motion completely breaks down.

PROBLEMS

5.1 Make order-of-magnitude estimates for the mean free path and the collision time for
(a) H_2 molecules in a hydrogen gas in standard condition (diameter of $H_2 = 2.9$ Å);
(b) protons in a plasma (gas of totally ionized H_2) at $T = 3 \times 10^5$ K, $n = 10^{15}$ protons/cm³, $\sigma = \pi r^2$, where $r = e^2 / kT$;
(c) protons in a plasma at the same density as (b) but at $T = 10^7$ K, where thermonuclear reactions occur;
(d) protons in the sun's corona, which is a plasma at $T = 10^6$ K, $n = 10^6$ protons/cc;
(e) slow neutrons of energy 0.5 MeV in ^{238}U ($\sigma \approx \pi r^2$, $r \approx 10^{-13}$ cm).

5.2 A box made of perfectly reflecting walls is divided by a perfectly reflecting partition into compartments 1 and 2. Initially a gas at temperature T_1 was confined in compartment 1, and compartment 2 was empty. A small hole of dimension much less than the mean free path of the gas is opened in the partition for a short time to allow a small fraction of the gas to escape into compartment 2. The hole is then sealed off and the new gas in compartment 2 comes to equilibrium.

(a) During the time when the hole was open, what was the flux dI of molecules crossing into compartment 2 with speed between v and $v + dv$?

(b) During the same time, what was the average energy per particle ϵ of the molecules crossing into compartment 2?

(c) After final equilibrium has been established, what is the temperature T_2 in compartment 2?

   Answer. $T_2 = \frac{4}{3} T_1$.

5.3 (a) Explain why it is meaningless to speak of a sound wave in a gas of strictly noninteracting molecules.

(b) In view of (a), explain the meaning of a sound wave in an ideal gas.

5.4 Show that the velocity of sound in a real substance is to a good approximation given by $c = 1/\sqrt{\rho \kappa_S}$, where ρ is the mass density and κ_S the adiabatic compressibility, by the following steps.

(*a*) Show that in a sound wave the density oscillates adiabatically if

$$K \ll c\lambda\rho c_V$$

where K = coefficient of thermal conductivity

λ = wavelength of sound wave

ρ = mass density

c_V = specific heat

c = velocity of sound

(*b*) Show by numerical examples, that the criterion stated in (*a*) is well satisfied in most practical situations.

5.5 A flat disk of unit area is placed in a dilute gas at rest with initial temperature T. Face A of the disk is at temperature T, and face B is at temperature $T_1 > T$ (see sketch). Molecules striking face A reflect elastically. Molecules striking face B are absorbed by the disk, only to re-emerge from the same face with a Maxwellian distribution of temperature T_1.

(*a*) Assume that the mean free path in the gas is much smaller than the dimension of the disk. Present an argument to show that after a few collision times the gas can be described by the hydrodynamic equations, with face B replaced by a boundary condition for the temperature.

(*b*) Write down the first-order hydrodynamic equations for (*a*), neglecting the flow of the gas. Show that there is no net force acting on the disk.

(*c*) Assume that the mean free path is much larger than the dimensions of the disk. Find the net force acting on the disk.

5.6 A square vane, of area 1 cm², painted white on one side, black on the other, is attached to a vertical axis and can rotate freely about it (see the sketch). Suppose the arrangement is placed in He gas at room temperature and sunlight is allowed to shine on the vane. Explain qualitatively why

(*a*) at high density of the gas the vane does not move;

(*b*) at extremely small densities the vane rotates;

(*c*) at some intermediate density the vane rotates in a sense opposite to that in (*b*). *Estimate* this intermediate density and the corresponding pressure.

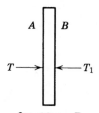

Gas at temp. T **Fig. P5.5**

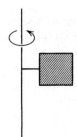

Fig. P5.6

5.7 A dilute gas, infinite in extension and composed of charged molecules, each of charge e and mass m, comes to equilibrium in an infinite lattice of fixed ions. In the absence of an external electric field the equilibrium distribution function is

$$f^{(0)}(\mathbf{p}) = n(2\pi mkT)^{-3/2} e^{-p^2/2mkT}$$

where n and T are constants. A weak uniform electric field E is then turned on, leading to a new equilibrium distribution function. Assume that a collision term of the form

$$\left(\frac{\partial f}{\partial t}\right)_{\text{coll}} = -\frac{f - f^{(0)}}{\tau}$$

where τ is a collision time, adequately takes into account the effect of collisions among molecules and between molecules and lattice. Calculate

(a) the new equilibrium distribution function f, to the first order;

(b) the electrical conductivity σ, defined by the relation

$$ne\langle \mathbf{v} \rangle = \sigma \mathbf{E}$$

B

STATISTICAL MECHANICS

6

CLASSICAL STATISTICAL MECHANICS

6.1 THE POSTULATE OF CLASSICAL STATISTICAL MECHANICS

Statistical mechanics is concerned with the properties of matter in equilibrium in the empirical sense used in thermodynamics.

The aim of statistical mechanics is to derive all the equilibrium properties of a macroscopic molecular system from the laws of molecular dynamics. Thus it aims to derive not only the general laws of thermodynamics but also the specific thermodynamic functions of a given system. Statistical mechanics, however, does not describe how a system approaches equilibrium, nor does it determine whether a system can ever be found to be in equilibrium. It merely states what the equilibrium situation is for a given system.

We recall that in the kinetic theory of gases the process of the approach to equilibrium is rather complicated, but the equilibrium situation, the Maxwell-Boltzmann distribution, is simple. Furthermore, the Maxwell-Boltzmann distribution can be derived in a simple way, independent of the details of molecular interactions. We might suspect that a slight generalization of the method used—the method of the most probable distribution—would enable us to discuss the equilibrium situation of not only a dilute gas but also any macroscopic system. This indeed is true. The generalization is classical statistical mechanics.

We consider a classical system composed of a large number N of molecules occupying a large volume V. Typical magnitudes of N and V are

$$N \approx 10^{23} \qquad \text{molecules}$$
$$V \approx 10^{23} \qquad \text{molecular volumes}$$

Since these are enormous numbers, it is convenient to consider the limiting case

$$N \to \infty$$
$$V \to \infty$$
$$\frac{V}{N} = v \qquad (6.1)$$

where the specific volume v is a given finite number.

The system will be regarded as isolated in the sense that the energy is a constant of the motion. This is clearly an idealization, for we never deal with truly isolated systems in the laboratory. The very fact that measurements can be performed on the system necessitates some interaction between the system and the external world. If the interactions with the external world, however, are sufficiently weak, so that the energy of the system remains approximately constant, we shall consider the system isolated. The walls of the container containing the system (if present) will be idealized as perfectly reflecting walls.

A state of the system is completely and uniquely defined by $3N$ canonical coordinates $q_1, q_2, \ldots, q_{3N}$ and $3N$ canonical momenta $p_1, p_2, \ldots, p_{3N}$. These $6N$ variables are denoted collectively by the abbreviation (p, q). The dynamics of the system is determined by the Hamiltonian $\mathscr{H}(p, q)$, from which we may obtain the canonical equations of motion

$$\frac{\partial \mathscr{H}(p, q)}{\partial p_i} = \dot{q}_i$$

$$\frac{\partial \mathscr{H}(p, q)}{\partial q_i} = -\dot{p}_i$$

$$(6.2)$$

It is convenient to introduce, as we did in Chapter 3, the $6N$-dimensional Γ space, or phase space, of the system, in which each point represents a state of the system, and vice versa. The locus of all points in Γ space satisfying the condition $\mathscr{H}(p, q) = E$ defines a surface called the energy surface of energy E. As the state of the system evolves in time according to (6.2) the representative point traces out a path in Γ space. This path always stays on the same energy surface because by definition energy is conserved.

For a macroscopic system, we have no means, nor desire, to ascertain the state at every instant. We are interested only in a few macroscopic properties of the system. Specifically, we only require that the system has N particles, a volume V, and an energy lying between the values E and $E + \Delta$. An infinite number of states satisfy these conditions. Therefore we think not of a single system, but of an infinite number of mental copies of the same system, existing in all possible states satisfying the given conditions. Any one of these system can be the system we are dealing with. The mental picture of such a collection of systems is the Gibbsian ensemble we introduced in Chapter 3. It is represented by a distribution of points in Γ space characterized by a density function $\rho(p, q, t)$, defined in such a way that

$$\rho(p, q, t) \, d^{3N}p \, d^{3N}q = \text{no. of representative points con-}$$
$$\text{tained in the volume element}$$
$$d^{3N}p \, d^{3N}q \text{ located at } (p, q) \text{ in } \Gamma$$
$$\text{space at the instant } t$$

$$(6.3)$$

We recall Liouville's theorem:

$$\frac{\partial \rho}{\partial t} + \sum_{i=1}^{3N} \left(\frac{\partial \rho}{\partial q_i} \frac{\partial \mathscr{H}}{\partial p_i} - \frac{\partial \mathscr{H}}{\partial q_i} \frac{\partial \rho}{\partial p_i} \right) = 0 \qquad (6.4)$$

In geometrical language it states that the distribution of points in Γ space moves like an incompressible fluid. Since we are interested in the equilibrium situation, we restrict our considerations to ensembles whose density function does not depend explicitly on the time and depends on (p, q) only through the Hamiltonian. That is,

$$\rho(p, q) = \rho'(\mathscr{H}(p, q)) \tag{6.5}$$

where $\rho'(\mathscr{H})$ is a given function of $\mathscr{H}$. It follows immediately that the second term on the left side of (6.4) is identically zero. Therefore

$$\frac{\partial}{\partial t}\rho(p, q) = 0 \tag{6.6}$$

Hence the ensemble described by $\rho(p, q)$ is the same for all times.

Classical statistical mechanics is founded on the following postulate.

Postulate of Equal a Priori Probability When a macroscopic system is in thermodynamic equilibrium, its state is equally likely to be any state satisfying the macroscopic conditions of the system.

This postulate implies that in thermodynamic equilibrium the system under consideration is a member of an ensemble, called the *microcanonical ensemble*, with the density function

$$\rho(p, q) = \begin{cases} \text{Const.} & \text{if } E < \mathscr{H}(p, q) < E + \Delta \\ 0 & \text{otherwise} \end{cases} \tag{6.7}$$

It is understood that all members of the ensemble have the same number of particles N and the same volume V.

Suppose $f(p, q)$ is a measurable property of the system, such as energy or momentum. When the system is in equilibrium, the observed value of $f(p, q)$ must be the result obtained by averaging $f(p, q)$ over the microcanonical ensemble in some manner. If the postulate of equal a priori probability is to be useful, all manners of averaging must yield essentially the same answer.

Two kinds of average values are commonly introduced: the most probable value and the ensemble average. The *most probable value* of $f(p, q)$ is the value of $f(p, q)$ that is possessed by the largest number of systems in the ensemble. The *ensemble average* of $f(p, q)$ is defined by

$$\langle f \rangle \equiv \frac{\int d^{3N}p\, d^{3N}q f(p, q)\rho(p, q)}{\int d^{3N}p\, d^{3N}q \rho(p, q)} \tag{6.8}$$

The ensemble average and the most probable value are nearly equal if the *mean square fluctuation* is small, i.e., if

$$\frac{\langle f^2 \rangle - \langle f \rangle^2}{\langle f \rangle^2} \ll 1 \tag{6.9}$$

If this condition is not satisfied, there is no unique way to determine how the observed value of f may be calculated. When it is not, we should question the validity of statistical mechanics. In all physical cases we shall find that mean square fluctuations are of the order of $1/N$. Thus in the limit as $N \to \infty$ the ensemble average and the most probable value became identical.

Strictly speaking, systems in nature do not obey classical mechanics. They obey quantum mechanics, which contains classical mechanics as a special limiting case. Logically we should start with quantum statistical mechanics and then arrive at classical statistical mechanics as a special case. We do this later. It is only for pedagogical reasons that we begin with classical statistical mechanics.

From a purely logical point of view there is no room for an independent postulate of classical statistical mechanics. It would not be logically satisfactory even if we could show that the postulate introduced here follows from the equations of motion (6.2), for, since the world is quantum mechanical, the foundation of statistical mechanics lies not in classical mechanics but in quantum mechanics. At present we take this postulate to be a working hypothesis whose justification lies in the agreement between results derived from it and experimental facts.

6.2 MICROCANONICAL ENSEMBLE

In the microcanonical ensemble every system has N molecules, a volume V, and an energy between E and $E + \Delta$. It is clear that the average total momentum of the system is zero. We show that it is possible to define quantities that correspond to thermodynamic quantities.

The fundamental quantity that furnishes the connection between the microcanonical ensemble and thermodynamics is the entropy. It is the main task of this section to define the entropy and to show that it possesses all the properties attributed to it in thermodynamics.

Let $\Gamma(E)$ denote the volume in Γ space occupied by the microcanonical ensemble:

$$\Gamma(E) \equiv \int_{E < \mathcal{H}(p,q) < E + \Delta} d^{3N}p \, d^{3N}q \tag{6.10}$$

The dependence of $\Gamma(E)$ on N, V, and Δ is understood. Let $\sum(E)$ denote the volume in Γ space enclosed by the energy surface of energy E:

$$\sum(E) = \int_{\mathcal{H}(p,q) < E} d^{3N}p \, d^{3N}q \tag{6.11}$$

Then

$$\Gamma(E) = \sum(E + \Delta) - \sum(E) \tag{6.12}$$

If Δ is so chosen that $\Delta \ll E$, then

$$\Gamma(E) = \omega(E)\Delta \tag{6.13}$$

where $\omega(E)$ is called the density of states of the system at the energy E and is defined by

$$\omega(E) = \frac{\partial \Sigma(E)}{\partial E} \tag{6.14}$$

The entropy is defined by

$$S(E, V) \equiv k \log \Gamma(E) \tag{6.15}$$

where k is a universal constant eventually shown to be Boltzmann's constant. To justify this definition we show that (6.15) possesses all the properties of the entropy function in thermodynamics, namely,

(a) S is an extensive quantity: If a system is composed of two subsystems whose entropies are, respectively, S_1 and S_2, the entropy of the total system is $S_1 + S_2$, when the subsystems are sufficiently large.

(b) S satisfies the properties of the entropy as required by the second law of the thermodynamics.

To show the extensive property, let the system be divided into two subsystems which have N_1 and N_2 particles and the volumes V_1 and V_2, respectively.* The energy of molecular interaction between the two subsystems is negligible compared to the total energy of each subsystem, if the intermolecular potential has a finite range, and if the surface-to-volume ratio of each subsystem is negligibly small. The total Hamiltonian of the composite system accordingly may be taken to be the sum of the Hamiltonians of the two subsystems:

$$\mathcal{H}(p, q) = \mathcal{H}_1(p_1, q_1) + \mathcal{H}_2(p_2, q_2) \tag{6.16}$$

where (p_1, q_1) and (p_2, q_2) denote, respectively, the coordinates and momenta of the particles contained in the two subsystems.

Let us first imagine that the two subsystems are isolated from each other and consider the microcanonical ensemble for each taken alone. Let the energy of the first subsystem lie between E_1 and $E_1 + \Delta$ and the energy of the second subsystem lie between E_2 and $E_2 + \Delta$. The entropies of the subsystems are, respectively,

$$S_1(E_1, V_1) = k \log \Gamma_1(E_1)$$
$$S_2(E_2, V_2) = k \log \Gamma_2(E_2)$$

where $\Gamma_1(E_1)$ and $\Gamma_2(E_2)$ are the volumes occupied by the two ensembles in their respective Γ spaces. They are schematically represented in Fig. 6.1 by the volumes of the shaded regions, which lie between successive energy surfaces that differ in energy by Δ.

Now consider the microcanonical ensemble of the composite system made up of the two subsystems, and let the total energy lie between E and $E + 2\Delta$.

---

*For simplicity we assume that the same N_1, N_2 particles are always confined respectively to the volumes V_1, V_2. The proof is therefore invalid for a gas, for which S has to be modified (See Section 6.6).

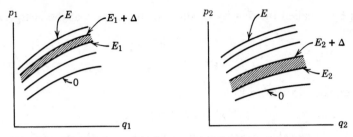

Fig. 6.1 The microcanonical ensemble of the two subsystems.

We choose Δ such that $\Delta \ll E$. This ensemble contains all copies of the composite system for which

(a) the N_1 particles whose momenta and coordinates are (p_1, q_1) are contained in the volume V_1,

(b) the N_2 particles whose momenta and coordinates are (p_2, q_2) are contained in the volume V_2,

(c) the energies E_1, E_2 of the subsystems have values satisfying the condition

$$E < (E_1 + E_2) < E + 2\Delta \tag{6.17}$$

Obviously, the volume of the region of Γ space that corresponds to conditions (a) and (b) with a total energy lying between $E_1 + E_2$ and $E_1 + E_2 + 2\Delta$ is

$$\Gamma_1(E_1)\Gamma_2(E_2)$$

To obtain the total volume of the ensemble specified by (a), (b), and (c), we only have to take the sum of $\Gamma_1(E_1)\Gamma_2(E_2)$ over values of E_1 and E_2 consistent with (c). Since E_1 and E_2 are possible values of the Hamiltonians $\mathscr{H}_1(p_1, q_1)$ and $\mathscr{H}_2(p_2, q_2)$, their spectra of values must be bounded from below; otherwise the subsystems would not be stable. For simplicity we take the lower bounds for both spectra to be 0. If we divide each of the energy spectra E_1 and E_2 into intervals of size Δ, then between 0 and E there are E/Δ intervals in each spectrum. Thus, since $\Delta \ll E$, we can write

$$\Gamma(E) = \sum_{i=1}^{E/\Delta} \Gamma_1(E_i)\Gamma_2(E - E_i) \tag{6.18}$$

where E_i is the energy lying in the center of each energy interval.

The entropy of the composite system of N particles and of volume V, with

$$N = N_1 + N_2$$
$$V = V_1 + V_2$$

is given by

$$S(E, V) = k \log \sum_{i=1}^{E/\Delta} \Gamma_1(E_i)\Gamma_2(E - E_i) \tag{6.19}$$

It will now be shown that when $N_1 \rightarrow \infty$ and $N_2 \rightarrow \infty$ a single term in the sum of (6.18) dominates the sum. The sum in (6.18) is a sum of E/Δ positive terms. Let the largest term in the sum be $\Gamma_1(\bar{E}_1)\Gamma_2(\bar{E}_2)$, where

$$\bar{E}_1 + \bar{E}_2 = E \tag{6.20}$$

Then it is obvious that

$$\Gamma_1(\bar{E}_1)\Gamma_2(\bar{E}_2) \leq \Gamma(E) \leq \frac{E}{\Delta}\Gamma_1(\bar{E}_1)\Gamma_2(\bar{E}_2)$$

or

$$k \log\left[\Gamma_1(\bar{E}_1)\Gamma_2(\bar{E}_2)\right] \leq S(E,V) \leq k \log\left[\Gamma_1(\bar{E}_1)\Gamma_2(\bar{E}_2)\right] + k \log\frac{E}{\Delta} \tag{6.21}$$

If the subsystems are molecular systems with N_1 and N_2 particles, respectively, we expect that as $N_1 \rightarrow \infty$ and $N_2 \rightarrow \infty$,

$$\log \Gamma_1 \propto N_1$$
$$\log \Gamma_2 \propto N_2 \tag{6.22}$$
$$E \propto N_1 + N_2$$

Thus the term $\log(E/\Delta)$ in (6.21) may be neglected because Δ is a constant independent of N. Therefore

$$S(E,V) = S_1(\bar{E}_1,V_1) + S_2(\bar{E}_2,V_2) + O(\log N) \tag{6.23}$$

which proves the extensive property of the entropy. $\blacksquare$

We have actually proved more than the extensive property of the entropy because (6.23) also implies that the energies of subsystems have the definite values $\bar{E}_1$ and $\bar{E}_2$, respectively. They are the values of E_1 and E_2 that maximize the function $\Gamma_1(E_1)\Gamma_2(E_2)$ under the restriction $E_1 + E_2 = E$. That is,

$$\delta\left[\Gamma_1(E_1)\Gamma_2(E_2)\right] = 0, \qquad \delta E_1 + \delta E_2 = 0$$

This leads to the condition

$$\left[\frac{\partial}{\partial E_1}\log \Gamma_1(E_1)\right]_{E_1=\bar{E}_1} = \left[\frac{\partial}{\partial E_2}\log \Gamma_2(E_2)\right]_{E_2=\bar{E}_2}$$

or $\qquad (6.24)$

$$\left[\frac{\partial S_1(E_1)}{\partial E_1}\right]_{E_1=\bar{E}_1} = \left[\frac{\partial S_2(E_2)}{\partial E_2}\right]_{E_2=\bar{E}_2}$$

We define the temperature of any system by

$$\frac{\partial S(E,V)}{\partial E} \equiv \frac{1}{T} \tag{6.25}$$

Then $\bar{E}_1$ and $\bar{E}_2$ are such that the two subsystems have the same temperature:

$$T_1 = T_2 \tag{6.26}$$

The temperature defined by (6.25) is precisely the absolute temperature in thermodynamics. Not only is it a parameter associated with the condition for equilibrium, it is also related to the entropy by (6.25), which is one of the Maxwell relations in thermodynamics. Choosing the standard temperature interval to be the conventional Centigrade degree defines the constant k in (6.15) to be Boltzmann's constant. Thus the proof of the extensive property of the entropy also reveals the meaning of the temperature for an isolated system: *The temperature of an isolated system is the parameter governing the equilibrium between one part of the system and another.*

Although the condition (6.17) allows a range of values of (E_1, E_2) to occur among members of the microcanonical ensemble, the result (6.21) shows that as the number of particles becomes very large almost all members of the ensemble have the values $(\bar{E}_1, \bar{E}_2)$. This fact is fundamental to the success of statistical mechanics as a theory of matter.

A calculation similar to that leading to (6.23) shows that the following definitions of S are equivalent, up to additive constant terms of order $\log N$ or smaller:

$$S = k \log \Gamma(E) \tag{6.27}$$

$$S = k \log \omega(E) \tag{6.28}$$

$$S = k \log \textstyle\sum(E) \tag{6.29}$$

In fact, if these definitions were not equivalent, the validity of statistical mechanics would be in doubt.

To show that S possesses the properties of the entropy as required by the second law of thermodynamics, let us first state the form of the second law that is most convenient for the present purpose. The entropy in thermodynamics, just as S here, is defined only for equilibrium situations. The second law states that if an isolated system undergoes a change of thermodynamic state such that the initial and final states are equilibrium states, the entropy of the final state is not smaller than that of the initial state. For the system we are considering, the only independent macroscopic parameters are N, V, and E. By definition N and E cannot change, for the system is isolated. Thus only V can change. Now V cannot decrease without compressing the system thereby disturbing its isolation. Hence V can only increase. (An example is the free expansion of a gas when one of the containing walls is suddenly removed.) For our purpose the second law states that the entropy is a nondecreasing function of V.

Let us use the definition (6.29):

$$S(E, V) = k \log \textstyle\sum(E)$$

It is obvious that $\sum(E)$ is a nondecreasing function of V, for if $V_1 > V_2$, then the integral (6.11) for $V = V_1$ extends over a domain of integration that includes that for $V = V_2$. This shows that $S(E, V)$ is a nondecreasing function of V.

We conclude that the function $S(E, V)$, as defined by any one of the formulas (6.27)–(6.29), is the entropy of a system of volume V and internal energy E. This conclusion furnishes the connection between the microcanonical ensemble and thermodynamics.

6.3 DERIVATION OF THERMODYNAMICS

We have defined the entropy of a system and have shown that the second law of thermodynamics holds. The complete thermodynamics of a system can now be obtained.

First we discuss the analog of quasistatic thermodynamic transformations. A quasistatic thermodynamic transformation corresponds to a slow variation of E and V, induced by coupling the system to external agents. During such a transformation the ensemble is represented by a collection of representative points uniformly distributed over a slowly changing region in Γ space. The change is so slow that at every instant we have a microcanonical ensemble. Accordingly, the change in the entropy in an infinitesimal transformation is given by

$$dS(E, V) = \left(\frac{\partial S}{\partial E}\right)_V dE + \left(\frac{\partial S}{\partial V}\right)_E dV \qquad (6.30)$$

The coefficient of dE has been defined earlier as the inverse absolute temperature T^{-1}. We now define the pressure of the system to be

$$P \equiv T\left(\frac{\partial S}{\partial V}\right)_E \qquad (6.31)$$

Hence

$$dS = \frac{1}{T}(dE + P\, dV) \qquad (6.32)$$

or

$$dE = T\, dS - P\, dV \qquad (6.33)$$

This is the first law of thermodynamics.

Thus we have succeeded not only in deriving the first and second laws of thermodynamics, but also in finding means to calculate all thermodynamic functions in terms of molecular interactions. The third law of thermodynamics cannot be obtained in classical statistical mechanics, because it is quantum mechanical.

We summarize by giving a practical recipe for finding all the thermodynamic functions of a system.

RECIPE

Consider an isolated system that occupied volume V and has an energy E within a small uncertainty $\Delta \ll E$. The Hamiltonian is presumed known. To find all thermodynamic functions of the system, proceed as follows:

(a) Calculate the density of states $\omega(E)$ of the system from the Hamiltonian.

(b) Find the entropy up to an arbitrary additive constant by the formula

$$S(E,V) = k \log \omega(E)$$

where k is Boltzmann's constant. Alternatively we can use the formula (6.27) or (6.29).

(c) Solve for E in terms of S and V. The resulting function is the thermodynamic internal energy of the system

$$U(S,V) \equiv E(S,V)$$

(d) Find other thermodynamic functions from the following formulas:

$$T = \left(\frac{\partial U}{\partial S} \right)_V \qquad \text{(absolute temperature)}$$

$$P = -\left(\frac{\partial U}{\partial V} \right)_S \qquad \text{(pressure)}*$$

$$A = U - TS \qquad \text{(Helmholtz free energy)}$$

$$G = U + PV - TS \qquad \text{(Gibbs potential)}$$

$$C_V = \left(\frac{\partial U}{\partial T} \right)_V \qquad \text{(heat capacity at constant volume)}$$

(e) To study any equilibrium behavior of the system, use thermodynamics.

6.4 EQUIPARTITION THEOREM

Let x_i be either p_i or q_i ($i = 1, \ldots, 3N$). We calculate the ensemble average of $x_i(\partial \mathscr{H}/\partial x_j)$, where $\mathscr{H}$ is the Hamiltonian. Using the abbreviation $dp\,dq \equiv d^{3N}p\,d^{3N}q$, we can write

$$\left\langle x_i \frac{\partial \mathscr{H}}{\partial x_j} \right\rangle = \frac{1}{\Gamma(E)} \int_{E < \mathscr{H} < E + \Delta} dp\,dq\, x_i \frac{\partial \mathscr{H}}{\partial x_j} = \frac{\Delta}{\Gamma(E)} \frac{\partial}{\partial E} \int_{\mathscr{H} < E} dp\,dq\, x_i \frac{\partial \mathscr{H}}{\partial x_j}$$

Noting that $\partial E/\partial x_j = 0$, we may calculate the last integral in the following

---

*This is equivalent to (6.31) by the chain relation.

manner:

$$\int_{\mathcal{H}<E} dp\, dq\, x_i \frac{\partial \mathcal{H}}{\partial x_j} = \int_{\mathcal{H}<E} dp\, dq\, x_i \frac{\partial}{\partial x_j}(\mathcal{H} - E)$$

$$= \int_{\mathcal{H}<E} dp\, dq\, \frac{\partial}{\partial x_j}[x_i(\mathcal{H} - E)] - \delta_{ij} \int_{\mathcal{H}<E} dp\, dq\,(\mathcal{H} - E)$$

The first integral on the right side vanishes because it reduces to a surface integral over the boundary of the region defined by $\mathcal{H} < E$, and on this boundary $\mathcal{H} - E = 0$. Substituting the latest result into the previous equation, and noting that $\Gamma(E) = \omega(E)\Delta$, we obtain

$$\left\langle x_i \frac{\partial \mathcal{H}}{\partial x_j} \right\rangle = \frac{\delta_{ij}}{\omega(E)} \frac{\partial}{\partial E} \int_{\mathcal{H}<E} dp\, dq\,(E - \mathcal{H})$$

$$= \frac{\delta_{ij}}{\omega(E)} \int_{\mathcal{H}<E} dp\, dq = \frac{\delta_{ij}}{\omega(E)} \Sigma(E)$$

$$= \delta_{ij} \frac{\Sigma(E)}{\partial \Sigma(E)/\partial E} = \delta_{ij} \left[\frac{\partial}{\partial E} \log \Sigma(E) \right]^{-1} = \delta_{ij} \frac{k}{\partial S/\partial E}$$

that is,

$$\left\langle x_i \frac{\partial \mathcal{H}}{\partial x_j} \right\rangle = \delta_{ij} kT \tag{6.34}$$

This is the *generalized equipartition theorem*.

For the special case $i = j$, $x_i = p_i$, we have

$$\left\langle p_i \frac{\partial \mathcal{H}}{\partial p_i} \right\rangle = kT \tag{6.35}$$

For $i = j$ and $x_i = q_i$, we have

$$\left\langle q_i \frac{\partial \mathcal{H}}{\partial q_i} \right\rangle = kT \tag{6.36}$$

According to the canonical equations of motion, $\partial \mathcal{H}/\partial q_i = -\dot{p}_i$. Hence (6.36) leads to the statement

$$\left\langle \sum_{i=1}^{3N} q_i \dot{p}_i \right\rangle = -3NkT \tag{6.37}$$

which is known as the *virial theorem* because $\sum q_i \dot{p}_i$—the sum of the ith coordinate times the ith component of the generalized force—is known in classical mechanics as the virial.

Many physical systems have Hamiltonians that, through a canonical transformation, can be cast in the form

$$\mathscr{H} = \sum_i A_i P_i^2 + \sum_i B_i Q_i^2 \tag{6.38}$$

where P_i, Q_i are canonically conjugate variables and A_i, B_i are constants. For such systems we have

$$\sum_i \left(P_i \frac{\partial \mathscr{H}}{\partial P_i} + Q_i \frac{\partial \mathscr{H}}{\partial Q_i} \right) = 2\mathscr{H} \tag{6.39}$$

Suppose f of the constants A_i and B_i are nonvanishing. Then (6.35) and (6.36) imply that

$$\langle \mathscr{H} \rangle = \tfrac{1}{2} fkT \tag{6.40}$$

That is, each harmonic term in the Hamiltonian contributes $\tfrac{1}{2}kT$ to the average energy of the system. This is known as the *theorem of equipartition of energy*. But (6.40) is the internal energy of the system. Therefore

$$\frac{C_V}{k} = \frac{f}{2} \tag{6.41}$$

Thus the heat capacity is directly related to the number of degrees of freedom of the system.

A paradox arises from the theorem of equipartition of energy. In classical physics every system must in the last analysis have an infinite number of degrees of freedom, for after we have resolved matter into atoms we must continue to resolve an atom into its constituents and the constituents of the constituents, ad infinitum. Therefore the heat capacity of any system is infinite. This is a real paradox in classical physics and is resolved by quantum mechanics. Quantum mechanics possesses the feature that the degree of freedom of a system are manifest only when there is sufficient energy to excite them, and that those degrees of freedom that are not excited can be forgotten. Thus the formula (6.41) is valid only when the temperature is sufficiently high.

6.5 CLASSICAL IDEAL GAS

To illustrate the method of calculation in the microcanonical ensemble we consider the classical ideal gas. This has been considered earlier in our discussion of the kinetic theory of gases. In that discussion we also introduced the microcanonical ensemble, but we obtained all the thermodynamic properties of the ideal gas via the distribution function. For the sake of illustration, we now derive the same results using the recipe given in Section 6.3.

The Hamiltonian is

$$\mathscr{H} = \frac{1}{2m} \sum_{i=1}^{N} p_i^2 \tag{6.42}$$

We first calculate

$$\Sigma(E) = \frac{1}{h^{3N}} \int_{\mathscr{H} < E} d^3p_1 \cdots d^3p_N \, d^3q_1 \cdots d^3q_N \qquad (6.43)$$

where h is a constant of the dimension of momentum × distance, introduced to make $\Sigma(E)$ dimensionless. The integration over q_i can be immediately carried out, giving a factor of V^N. Let

$$R = \sqrt{2mE} \qquad (6.44)$$

Then

$$\Sigma(E) = \left(\frac{V}{h^3}\right)^N \Omega_{3N}(R) \qquad (6.45)$$

where Ω_n is the volume of an n-sphere of radius R:

$$\Omega_n(R) = \int_{x_1^2 + x_2^2 + \cdots + x_n^2 < R^2} dx_1 \, dx_2 \cdots dx_n \qquad (6.46)$$

Clearly,

$$\Omega_n(R) = C_n R^n \qquad (6.47)$$

where C_n is a constant. To find C_n, consider the identity

$$\int_{-\infty}^{+\infty} dx_1 \cdots \int_{-\infty}^{+\infty} dx_n \, e^{-(x_1^2 + \cdots + x_n^2)} = \left(\int_{-\infty}^{+\infty} dx \, e^{-x^2}\right)^n = \pi^{n/2} \qquad (6.48)$$

The left side of (6.48) can be re-expressed as follows. Let $S_n(R) \equiv d\Omega_n(R)/dR$ be the surface area of an n-sphere of radius R. Then

$$\int_{-\infty}^{+\infty} dx_1 \cdots \int_{-\infty}^{+\infty} dx_n \, e^{-(x_1^2 + \cdots + x_n^2)} = \int_0^\infty dR \, S_n(R) \, e^{-R^2}$$

$$= nC_n \int_0^\infty dR \, R^{n-1} e^{-R^2}$$

$$= \tfrac{1}{2} nC_n \int_0^\infty dt \, t^{(n/2)-1} e^{-t} = \tfrac{1}{2} nC_n \Gamma(n/2) \qquad (6.49)$$

where $\Gamma(z)$ is the gamma function. Comparison of (6.49) and (6.48) yields

$$C_n = \frac{\pi^{n/2}}{\Gamma(n/2 + 1)} \qquad (6.50)$$

$$\log C_n \underset{n \to \infty}{\to} \frac{n}{2} \log \pi - \frac{n}{2} \log \frac{n}{2} + \frac{n}{2} \qquad (6.51)$$

Hence

$$\Sigma(E) = C_{3N} \left[\frac{V}{h^3}(2mE)^{3/2}\right]^N \qquad (6.52)$$

The entropy of the ideal gas is

$$S(E, V) = k\left[\log C_{3N} + N \log \frac{V}{h^3} + \tfrac{3}{2}N \log\left(2mE\right)\right] \qquad (6.53)$$

By (6.51), this reduces to

$$S(E, V) = Nk \log\left[V\left(\frac{4\pi m}{3h^2} \frac{E}{N}\right)^{3/2}\right] + \tfrac{3}{2}Nk \qquad (6.54)$$

Solving for E in terms of S and V, and calling the resulting function $U(S, V)$ the internal energy, we obtain

$$U(S, V) = \left(\frac{3}{4\pi} \frac{h^2}{m}\right) \frac{N}{V^{2/3}} \exp\left(\frac{2}{3} \frac{S}{Nk} - 1\right) \qquad (6.55)$$

The temperature is

$$T = \left(\frac{\partial U}{\partial S}\right)_V = \frac{2}{3} \frac{U}{Nk} \qquad (6.56)$$

from which follows

$$C_V = \tfrac{3}{2}Nk \qquad (6.57)$$

Finally the equation of state is

$$P = -\left(\frac{\partial U}{\partial V}\right)_S = \frac{2}{3} \frac{U}{V} = \frac{NkT}{V} \qquad (6.58)$$

This calculation shows that the microcanonical ensemble is clumsy to use. There seems little hope that we can straightforwardly carry out the recipe of the microcanonical ensemble for any system but the ideal gas. We later introduce the canonical ensemble, which gives results equivalent to those of the microcanonical ensemble but which is more convenient for practical calculations.

6.6 GIBBS PARADOX

According to (6.54), the entropy of an ideal gas is

$$S = Nk \log\left(Vu^{3/2}\right) + Ns_0 \qquad (6.59)$$

where

$$u = \tfrac{3}{2}kT$$

$$s_0 = \frac{3k}{2}\left(1 + \log \frac{4\pi m}{3h^2}\right) \qquad (6.60)$$

Consider two ideal gases, with N_1 and N_2 particles, respectively, kept in two separate volumes V_1 and V_2 at the same temperature and the same density. Let us find the change in entropy of the combined system after the gases are allowed to mix in a volume $V = V_1 + V_2$. The temperature will be the same after the mixing process. Hence u remains unchanged. From (6.59) we find that the change in entropy is

$$\frac{\Delta S}{k} = N_1 \log \frac{V}{V_1} + N_2 \log \frac{V}{V_2} > 0 \qquad (6.61)$$

which is the entropy of mixing. If the two gases are different (e.g., argon and neon), this result is experimentally correct.

The Gibbs paradox presents itself if we consider the case in which the two mixing gases are of the same kind. Since the derivation of (6.61) does not depend on the identity of the gases, we would obtain the same increase of entropy (6.61). This is a disastrous result because it implies that the entropy of a gas depends on the history of the gas, and thus cannot be a function of the thermodynamic state alone. Worse, the entropy does not exist, because we can always imagine that the existing state of a gas is arrived at by pulling off any number of partitions that initially divided the gas into any number of compartments. Hence S is larger than any number.

Gibbs resolved the paradox in an empirical fashion by postulating that we have made an error in calculating $\sum(E)$, the number of states of the gas with energy less than E. Gibbs assumed that the correct answer is $N!$ times smaller than we though it was. By this assumption we should subtract from (6.59) the term $\log N! \approx N \log N - N$ and obtain

$$S = Nk \log\left(\frac{V}{N}u^{3/2}\right) + \tfrac{3}{2}Nk\left(\tfrac{5}{3} + \log \frac{4\pi m}{3h^2}\right) \qquad (6.62)$$

This formula does not affect the equation of state and other thermodynamic functions of a system, because the subtracted term is independent of T and V. For the mixing of two different gases (6.62) still predicts (6.61), because N_1 and N_2 are the same constants before and after the mixing. For the mixing of gases that are of the same kind, however, it gives no entropy of mixing because the specific volume V/N is the same before and after mixing.

The formula (6.62) has been experimentally verified as the correct entropy of an ideal gas at high temperatures, if h is numerically set equal to Planck's constant. It is known as the *Sackur-Tetrode equation*.

It is not possible to understand classically why we must divide $\sum(E)$ by $N!$ to obtain the correct counting of states. The reason is inherently quantum mechanical. Quantum mechanically, atoms are inherently indistinguishable in the following sense: A state of the gas is described by an N-particle wave function, which is either symmetric or antisymmetric with respect to the interchange of any two particles. A permutation of the particles can at most change the wave function by a sign, and it does not produce a new state of the system. From this fact it seems reasonable that the Γ-space volume element $dp\,dq$ corresponds to not one but only $dp\,dq/N!$ states of the system. Hence we should divide $\sum(E)$ by $N!$. This rule of counting is known as the "correct Boltzmann counting." It is something that we must append to classical mechanics to get right answers.

The foregoing discussion contains the correct reason for, but is not a derivation of, the "correct Boltzmann counting," because in classical mechanics there is no consistent way in which we can regard the particles as indistinguishable. In all classical considerations other than the counting of states we must continue to regard the particles in a gas as distinguishable.

We may derive the "correct Boltzmann counting" by showing that in the limit of high temperatures quantum statistical mechanics reduces to classical statistical mechanics with "correct Boltzmann counting." This is done in Section 9.2.

PROBLEMS

6.1 Show that the formulas (6.27), (6.28), and (6.29) are equivalent to one another.

6.2 Let the "uniform" ensemble of energy E be defined as the ensemble of all systems of the given type with energy less than E. The equivalence between (6.29) and (6.27) means that we should obtain the same thermodynamic functions from the "uniform" ensemble of energy E as from the microcanonical ensemble of energy E. In particular, the internal energy is E in both ensembles. Explain why this seemingly paradoxical result is true.

6.3 Consider a system of N free particles in which the energy of each particle can assume two and only two distinct values, 0 and E ($E > 0$). Denote by n_0 and n_1 the occupation numbers of the energy level 0 and E, respectively. The total energy of the system is U.

(*a*) Find the entropy of such a system.

(*b*) Find the most probable values of n_0 and n_1, and find the mean square fluctuations of these quantities.

(*c*) Find the temperature as a function of U, and show that it can be negative.

(*d*) What happens when a system of negative temperature is allowed to exchange heat with a system of positive temperature?

  Reference. N. F. Ramsey, *Phys. Rev.* **103**, 20 (1956).

6.4 Using the corrected entropy formula (6.62), work out the entropy of mixing for the case of different gases and for the case of identical gases, thus showing explicitly that there is no Gibbs paradox.

7

CANONICAL ENSEMBLE AND GRAND CANONICAL ENSEMBLE

7.1 CANONICAL ENSEMBLE

We wish to consider the question, "What ensemble is appropriate for the description of a system not in isolation, but in thermal equilibrium with a larger system?" To answer it we must find the probability that the system has energy E, because this probability is proportional to the density in Γ space for the ensemble we want.

We investigated a similar problem in Section 6.2, when we consider the energies of the component parts of a composite system. In the following we discuss the case in which one component part is much smaller than the other.

Consider an isolated composite system made up of two subsystems whose Hamiltonians are, respectively, $\mathcal{H}_1(p_1, q_1)$ and $\mathcal{H}_2(p_2, q_2)$, with number of particles N_1 and N_2, respectively. We assume that $N_2 \gg N_1$ but that both N_1 and N_2 are macroscopically large. We are interested in system 1 only. Consider a microcanonical ensemble of the composite system with total energy between E and $E + 2\Delta$. The energies E_1 and E_2 of the subsystems accordingly can have any values satisfying

$$E < (E_1 + E_2) < E + 2\Delta \tag{7.1}$$

Although this includes a range of values of E_1, E_2, the analysis of Section 6.2 shows that only one set of values, namely $\bar{E}_1, \bar{E}_2$, is important. We assume that $\bar{E}_2 \gg \bar{E}_1$. Let $\Gamma_2(E_2)$ be the volume occupied by system 2 in its own Γ space. The probability of finding system 1 in a state within $dp_1\, dq_1$ of (p_1, q_1), regardless of the state of system 2, is proportional to $dp_1\, dq_1\, \Gamma_2(E_2)$, where $E_2 = E - E_1$. Therefore up to a proportionality constant the density in Γ space for system 1 is

$$\rho(p_1, q_1) \propto \Gamma_2(E - E_1) \tag{7.2}$$

Since only the values near $E_1 = \bar{E}_1$ are expected to be important, and $\bar{E}_1 \ll E$, we may perform the expansion

$$k \log \Gamma_2(E - E_1) = S_2(E - E_1) = S_2(E) - E_1 \left[\frac{\partial S_2(E_2)}{\partial E_2} \right]_{E_2 = E} + \cdots$$

$$\approx S_2(E) - \frac{E_1}{T} \tag{7.3}$$

where T is the temperature of the larger subsystem. Hence

$$\Gamma_2(E - E_1) \approx \exp \left[\frac{1}{k} S_2(E) \right] \exp \left(-\frac{E_1}{kT} \right) \tag{7.4}$$

The first factor is independent of E_1 and is thus a constant as far as the small subsystem is concerned. Owing to (7.2) and the fact that $E_1 = \mathscr{H}_1(p_1, q_1)$, we may take the ensemble density for the small subsystem to be

$$\rho(p, q) = e^{-\mathscr{H}(p, q)/kT} \tag{7.5}$$

where the subscript 1 labeling the subsystem has been omitted, since we may now forget about the larger subsystem, apart from the information that its temperature is T. The larger subsystem in fact behaves like a heat reservoir in thermodynamics. The ensemble defined by (7.5), appropriate for a system whose temperature is determined through contact with a heat reservoir, is called the *canonical ensemble*.

The volume in Γ space occupied by the canonical ensemble is called the *partition function*:

$$Q_N(V, T) \equiv \int \frac{d^{3N}p \, d^{3N}q}{N! h^{3N}} e^{-\beta \mathscr{H}(p, q)} \tag{7.6}$$

where $\beta = 1/kT$, and where we have introduced a constant h, of the dimension of *momentum* $\times$ *distance*, in order to make Q_N dimensionless. The factor $1/N!$ appears, in accordance with the rule of "correct Boltzmann counting." These constants are of no importance for the equation of state.

Strictly speaking we should not integrate over the entire Γ space in (7.6), because (7.2) requires that $\rho(p_1, q_1)$ vanish if $E_1 > E$. The justification for ignoring such a restriction is that in the integral (7.6) only one value of the energy $\mathscr{H}(p, q)$ contributes to the integral and that this value will lie in the range where the approximation (7.4) is valid. We prove this contention in Section 7.2.

The thermodynamics of the system is to be obtained from the formula

$$Q_N(V, T) = e^{-\beta A(V, T)} \tag{7.7}$$

where $A(V, T)$ is the Helmholtz free energy. To justify this identification we show that

(a) A is an extensive quantity,

(b) A is related to the internal energy $U \equiv \langle H \rangle$ and the entropy $S \equiv -(\partial A/\partial T)_V$ by the thermodynamic relation

$$A = U - TS$$

That A is an extensive quantity follows from (7.6), because if the system is made up of two subsystems whose mutual interaction can be neglected, then Q_N is a product of two factors. To prove the relation (b), we first convert (b) into the following differential equation for A:

$$\langle \mathcal{H} \rangle = A - T \left(\frac{\partial A}{\partial T} \right)_V \tag{7.8}$$

To prove (7.8), note the identity

$$\frac{1}{N! h^{3N}} \int dp \, dq \, e^{\beta[A(V,T) - \mathcal{H}(p,q)]} = 1 \tag{7.9}$$

Differentiating with respect to β on both sides, we obtain

$$\frac{1}{N! h^{3N}} \int dp \, dq \, e^{\beta[A(V,T) - \mathcal{H}(p,q)]} \left[A(V,T) - \mathcal{H}(p,q) + \beta \left(\frac{\partial A}{\partial \beta} \right)_V \right] = 0$$

This is the same as

$$A(V,T) - U(V,T) - T \left(\frac{\partial A}{\partial T} \right)_V = 0$$

All other thermodynamic functions may be found from $A(V,T)$ by the Maxwell relations in thermodynamics:

$$P = - \left(\frac{\partial A}{\partial V} \right)_T$$

$$S = - \left(\frac{\partial A}{\partial T} \right)_V$$

$$G = A + PV$$

$$U = \langle H \rangle = A + TS$$

Therefore all calculations in the canonical ensembles begin (and nearly end) with the calculation of the partition function (7.6).

7.2 ENERGY FLUCTUATIONS IN THE CANONICAL ENSEMBLE

We now show that the canonical ensemble is mathematically equivalent to the microcanonical ensemble in the sense that although the canonical ensemble contains systems of all energies the overwhelming majority of them have the same energy. To do this we calculate the mean square fluctuation of energy in the canonical ensemble. The average energy is

$$U = \langle \mathcal{H} \rangle = \frac{\int dp\, dq\, \mathcal{H} e^{-\beta \mathcal{H}}}{\int dp\, dq\, e^{-\beta \mathcal{H}}} \tag{7.10}$$

Hence

$$\int dp\, dq\, [U - \mathcal{H}(p,q)]\, e^{\beta[A(V,T) - \mathcal{H}(p,q)]} = 0 \tag{7.11}$$

Differentiating both sides with respect to β, we obtain

$$\frac{\partial U}{\partial \beta} + \int dp\, dq\, e^{\beta(A - \mathcal{H})} (U - \mathcal{H}) \left(A - \mathcal{H} - T \frac{\partial A}{\partial T} \right) = 0 \tag{7.12}$$

By (7.8) this can be rewritten in the form

$$\frac{\partial U}{\partial \beta} + \left\langle (U - \mathcal{H})^2 \right\rangle = 0 \tag{7.13}$$

Therefore the mean square fluctuation of energy is

$$\langle \mathcal{H}^2 \rangle - \langle \mathcal{H} \rangle^2 = \left\langle (U - \mathcal{H})^2 \right\rangle = - \frac{\partial U}{\partial \beta} = kT^2 \frac{\partial U}{\partial T}$$

or

$$\langle \mathcal{H}^2 \rangle - \langle \mathcal{H} \rangle^2 = kT^2 C_V \tag{7.14}$$

For a macroscopic system $\langle \mathcal{H} \rangle \propto N$ and $C_V \propto N$. Hence (7.14) is a normal fluctuation. As $N \to \infty$, almost all systems in the ensemble have the energy $\langle \mathcal{H} \rangle$, which is the internal energy. Therefore the canonical ensemble is equivalent to the microcanonical ensemble.

It is instructive to calculate the fluctuations in another way. We begin by calculating the partition function in the following manner:

$$\frac{1}{N! h^{3N}} \int dp\, dq\, e^{-\beta \mathcal{H}(p,q)} = \int_0^\infty dE\, \omega(E)\, e^{-\beta E} = \int_0^\infty dE\, e^{-\beta E + \log \omega(E)}$$

$$= \int_0^\infty dE\, e^{\beta[TS(E) - E]} \tag{7.15}$$

where S is the entropy defined in the microcanonical ensemble. Since both S and U are proportional to N, the exponent in the last integrand is enormous. We expect that as $N \to \infty$ the integral receives contribution only from the neighborhood of the maximum of the integrand. The maximum of the integrand occurs at $E = \bar{E}$, where $\bar{E}$ satisfies the conditions

$$T \left(\frac{\partial S}{\partial E} \right)_{E = \bar{E}} = 1 \tag{7.16}$$

$$\left(\frac{\partial^2 S}{\partial E^2} \right)_{E = \bar{E}} < 0 \tag{7.17}$$

The first condition implies $\bar{E} = U$, the internal energy. Next we note that

$$\left(\frac{\partial^2 S}{\partial E^2}\right)_{E=\bar{E}} = \left(\frac{\partial}{\partial E}\frac{1}{T}\right)_{E=\bar{E}} = -\frac{1}{T^2}\left(\frac{\partial T}{\partial E}\right)_{E=\bar{E}} = -\frac{1}{T^2 C_V} \quad (7.18)$$

Thus the condition (7.17) is satisfied if $C_V > 0$, which is true for physical systems. Now let us expand the exponent in (7.15) about $E = \bar{E}$:

$$TS(E) - E = [TS(\bar{E}) - \bar{E}] + \tfrac{1}{2}(E - \bar{E})^2 T\left(\frac{\partial^2 S}{\partial E^2}\right)_{E=\bar{E}} + \cdots$$

$$= [TS(U) - U] - \frac{1}{2TC_V}(E - U)^2 + \cdots \quad (7.19)$$

Hence

$$\frac{1}{N!h^{3N}}\int dp\,dq\, e^{-\beta\mathscr{H}(p,q)} \approx e^{\beta(TS-U)}\int_0^\infty dE\, e^{-(E-U)^2/2kT^2C_V} \quad (7.20)$$

showing that in the canonical ensemble the distribution in *energy* is a Gaussian distribution centered about the value $E = U$ with a width equal to

$$\Delta E = \sqrt{2kT^2 C_V} \quad (7.21)$$

Since $U \propto N$ and $C_V \propto N$, $\Delta E/U$ is negligibly small. As $N \to \infty$ the Gaussian approaches a δ-function. Finally, let us perform the integral in (7.20). It is elementary:

$$\int_0^\infty dE\, e^{-(E-U)^2/2kT^2C_V} = \int_{-U}^\infty dx\, e^{-x^2/2kT^2C_V}$$

$$\approx \int_{-\infty}^{+\infty} dx\, e^{-x^2/2kT^2C_V} = \sqrt{2\pi kT^2 C_V}$$

Therefore

$$\frac{1}{N!h^{3N}}\int dp\,dq\, e^{-\beta H(p,q)} \approx e^{+\beta(TS-U)}\sqrt{2\pi kT^2 C_V} \quad (7.22)$$

$$A \approx (U - TS) - \tfrac{1}{2}kT\log C_V \quad (7.23)$$

This last term is negligible when $N \to \infty$. In that limit we have exactly $A = U - TS$. Statement (7.23) shows that the entropy as defined in the canonical and microcanonical ensemble differs only by terms of the order of $\log N$.

We have shown that almost all systems in the canonical ensemble have the same energy—namely, the energy that is equal to the internal energy of a system at the given temperature T. The reason for this is easy to see, both mathematically and physically.

In the canonical ensemble we distribute systems in Γ space according to the density function $\rho(p,q) = \exp[-\beta\mathscr{H}(p,q)]$, which is represented in Fig. 7.1. The density of points falls off exponentially as we go away from the origin of Γ space. The distribution in energy is obtained by "counting" the number of points

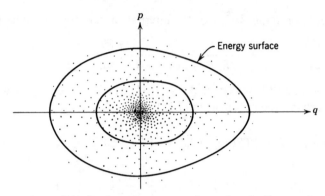

Fig. 7.1 Distribution of representative points in Γ space for the canonical ensemble.

on energy surfaces. As we go away from the origin, the energy increases and the area of the energy surface increases. This is why we get a peak in the distribution in energy. The sharpness of the peak is due to the rapidity with which the area of the energy surface increases as E increases. For an N-body system this area increases like e^E, where $E \propto N$.

From a physical point of view, a microcanonical ensemble must be equivalent to a canonical ensemble, otherwise we would seriously doubt the utility of either. A macroscopic substance has the extensive property, i.e., any part of the substance has the same thermodynamic property as the whole substance. Now consider a piece of substance isolated from everything. Any part of the substance must still be in equilibrium with the rest of the substance, which serves as a heat reservoir that defines a temperature for the part on which we focus our attention. Therefore the whole substance must have a well-defined temperature.

We have seen earlier that in the microcanonical ensemble it matters little whether we take the entropy to be k times the logarithm of the density of states at the energy E, the number of states with energies between E, $E + \Delta$, or all the states with energy below E. In all these cases we arrive at the same thermodynamic behavior. Now we see that it matters little whether we specify the energy of the system or the temperature of the system, for specifying one fixes the other, and we find the same thermodynamic behavior in both cases. All these examples illustrate the insensitivity of thermodynamic results to methods of derivation. The reasons behind this insensitivity are, in all cases, the facts that

(*a*) density of states $\propto e^E$

(*b*) $E \propto N$

(*c*) $N \to \infty$

On these facts depends the validity of statistical mechanics.

7.3 GRAND CANONICAL ENSEMBLE

Although the canonical and the microcanonical ensemble give equivalent results, it may be argued that conceptually the canonical ensemble corresponds more closely to physical situations. In experiments we never deal with a completely isolated system, nor do we ever directly measure the total energy of a macroscopic system. We usually deal with systems with a given temperature—a parameter that we can control in experiments.

By the same token we should not have to specify exactly the number of particles of a macroscopic system, for that is never precisely known. All we can find out from experiments is the average number of particles. This is the motivation for introducing the *grand canonical ensemble*, in which the systems can have any number of particles, with the average number determined by conditions external to the system. This is analogous to the situation in the canonical ensemble, where the average energy of a system is determined by the temperature of the heat reservoir with which it is in contact.

The Γ space for the grand canonical ensemble is spanned by all the canonical momenta and coordinates of systems with $0, 1, 2, \ldots$ number of particles. The density function describing the distribution of representative points in Γ space is denoted by $\rho(p, q, N)$, which gives the density of points representing systems with N particles with the momenta and coordinates (p, q). To find $\rho(p, q, N)$ we consider the canonical ensemble for a system with N particles, volume V, and temperature T, but we focus our attention on a small subvolume V_1 of the system.

Suppose there are N_1 particles in V_1 and $N_2 = N - N_1$ particles in $V_2 = V - V_1$. We assume

$$V_2 \gg V_1$$
$$N_2 \gg N_1$$

and designate the coordinates of the N_1 particle in V_1 by $\{p_1, q_1\}$, and those in V_2 by $\{p_2, q_2\}$. The interactions between particles in V_1 with those in V_2 are surface effects that can be neglected if V_1 is of macroscopic size. Thus the total Hamiltonian can be decomposed in the form

$$\mathcal{H}(p, q, N) = \mathcal{H}(p_1, q_1, N_1) + \mathcal{H}(p_2, q_2, N_2) \tag{7.24}$$

Note that the terms above involve the same function evaluated at different values of its arguments. The partition function of the total system is

$$Q_N(V, T) = \int \frac{dp\, dq}{h^{3N} N!} e^{-\beta \mathcal{H}(p, q, N)}$$

We shall segregate the contributions to the above from different values of N_1. In so doing, we do not care which particles are in V_1 as long as there are N_1 of them, and the coordinates of those that happen to be in V_1 will be designated by $\{p_1, q_1\}$. That is, in carrying out the integration over the phase space of the N-particle system, we always designate by $\{p_1, q_1\}$ the coordinates of those particles that happen to be in V_1, through a change in the variables of integration if necessary. Thus

$$Q_N(V, T) = \frac{1}{h^{3N}N!} \int dp_1 \, dp_2 \sum_{N_1=0}^{N} \frac{N!}{N_1!N_2!}$$

$$\times \int_{v_1} dq_1 \int_{v_2} dq_2 \, e^{-\beta[\mathscr{H}(p_1, q_1, N_1) + \mathscr{H}(p_2, q_2, N_2)]}$$

$$= \sum_{N_1=0}^{N} \frac{1}{h^{3N_1}N_1!} \int dp_1 \int_{v_1} dq_1 \, e^{-\beta\mathscr{H}(p_1, q_1, N_1)} \frac{1}{h^{3N_2}N_2!}$$

$$\times \int dp_2 \int_{v_2} dq_2 \, e^{-\beta\mathscr{H}(p_2, q_2, N_2)} \tag{7.25}$$

The relative probability $\rho(p_1, q_1, N_1)$ that there are N_1 particles in V_1 with coordinates $\{p_1, q_1\}$ is proportional to the summand of $\int dp_1 \, dq_1 \sum_{N_1}$. We choose its normalization such that

$$\rho(p_1, q_1, N_1) = \frac{1}{Q_N(V, T)} \frac{e^{-\beta\mathscr{H}(p_1, q_1, N_1)}}{h^{3N_1}N_1!} \frac{1}{h^{3N_2}N_2!} \int dp_2 \int_{v_2} dq_2 \, e^{-\beta\mathscr{H}(p_2, q_2, N_2)} \tag{7.26}$$

The first factor above is chosen so that

$$\sum_{N_1=0}^{N} \int dp_1 \, dq_1 \, \rho(p_1, q_1, N_1) = 1 \tag{7.27}$$

which is obvious if we rewrite the last expression in (7.25) in terms of ρ as defined in (7.26).

We can rewrite (7.26) in the form

$$\rho(p_1, q_1, N_1) = \frac{Q_{N_2}(V_2, T)}{Q_N(V, T)} \frac{e^{-\beta\mathscr{H}(p_1, q_1, N_1)}}{h^{3N_1}N_1!} \tag{7.28}$$

Using (7.7) we write

$$\frac{Q_{N_2}(V_2, T)}{Q_N(V, T)} = \exp\{-\beta[A(N - N_1, V - V_1, T) - A(N, V, T)]\} \tag{7.29}$$

where $A(N, V, T)$ is the Helmholtz free energy. Since $N \gg N_1$ and $V \gg V_1$, we may use the approximation

$$A(N - N_1, V - V_1, T) - A(N, V, T) \approx -N_1\mu + V_1 P \tag{7.30}$$

where μ and P are, respectively, the chemical potential and the pressure of the part of the system external to the small volume V_1:

$$\mu = \left[\frac{\partial A(N_2, V, T)}{\partial N_2} \right]_{N_2 = N}$$

$$P = - \left[\frac{\partial A(N, V_2, T)}{\partial V_2} \right]_{V_2 = V} \tag{7.31}$$

We now introduce the *fugacity*:

$$z = e^{\beta \mu} \tag{7.32}$$

Substituting (7.32) and (7.30) into (7.29), and then substituting (7.29) into (7.28), we obtain

$$\rho(p, q, N) = \frac{z^N}{N! h^{3N}} e^{-\beta PV - \beta \mathcal{H}(p, q)} \tag{7.33}$$

where the subscript 1 identifying the volume under consideration has been omitted because the system external to the volume can now be forgotten, apart from the information that it has the temperature T, pressure P, and chemical potential μ. We now allow the system external to the volume under consideration to become infinite in size. Then the range of N in (7.33) becomes

$$0 \leq N < \infty$$

The thermodynamic functions for the volume under consideration may be found as follows. First, the internal energy shall be the ensemble average of $\mathcal{H}(p, q)$. Second, the temperature, pressure, and chemical potential shall be respectively equal to T, P, μ. To show that this is a correct recipe, it suffices to remind ourselves that thermodynamics has been derived from the canonical ensemble. It is an elementary thermodynamic exercise to show that if a system is in equilibrium any part of the system must have the same T, P, μ as any other part; but this is the desired result.

To obtain a convenient formal recipe for finding all the thermodynamic functions we define the *grand partition function* as follows:

$$\mathcal{Q}(z, V, T) \equiv \sum_{N=0}^{\infty} z^N Q_N(V, T) \tag{7.34}$$

which in principle can be calculated from a knowledge of the Hamiltonian. Integrating both sides of (7.33) over all (p, q) for a given N, and then summing N from 0 to ∞, we find that

$$\frac{PV}{kT} = \log \mathcal{Q}(z, V, T) \tag{7.35}$$

Thus the grand partition function directly gives the pressure as a function of z, V, and T. The average number $\overline{N}$ of particles in the volume V is by definition the ensemble average

$$\bar{N} \equiv \langle N \rangle = \frac{\sum\limits_{N=0}^{\infty} Nz^{N}Q_{N}(V,T)}{\sum\limits_{N=0}^{\infty} z^{N}Q_{N}(V,T)} = z\frac{\partial}{\partial z}\log \mathscr{Q}(z,V,T) \qquad (7.36)$$

The equation of state, which is the equation expressing P as a function of N, V, and T, is obtained by eliminating z between (7.35) and (7.36).

All other thermodynamic functions may be obtained from the internal energy:

$$U = -\frac{\partial}{\partial \beta}\log \mathscr{Q}(z,V,T) \qquad (7.37)$$

After eliminating z with the help of (7.36), U becomes a function of N, V, and T. We can then use the formulas

$$C_{V} = \left(\frac{\partial U}{\partial T}\right)_{V}$$

$$S = \int_{0}^{T} dT\, \frac{C_{V}}{T}$$

$$A = U - TS$$

7.4 DENSITY FLUCTUATIONS IN THE GRAND CANONICAL ENSEMBLE

We now calculate the density fluctuations in the grand canonical ensemble. By differentiating (7.36) with respect to z, one can easily show

$$\langle N^{2} \rangle - \langle N \rangle^{2} = z\frac{\partial}{\partial z}z\frac{\partial}{\partial z}\log \mathscr{Q}(z,V,T) = kTV\frac{\partial^{2}P}{\partial \mu^{2}} \qquad (7.38)$$

where the last equality is obtained through the use of (7.34) and (7.36). To express the above in terms of conveniently measurable quantities, assume that the Helmholtz free energy of the system, being an extensive quantity, can be written in the form

$$A(N,V,T) = Na(v), \qquad v \equiv V/N \qquad (7.39)$$

where the temperature dependence of $a(v)$ has been suppressed for brevity. Then the two equations in (7.31) can be rewritten as

$$\mu = a(v) - v\frac{\partial a(v)}{\partial v}$$

$$P = -\frac{\partial a(v)}{\partial v} \qquad (7.40)$$

Regarding both μ and P as functions of v and T, we obtain from the above

$$\frac{\partial \mu}{\partial v} = -v \frac{\partial^2 a(v)}{\partial v^2}$$

$$\frac{\partial P}{\partial \mu} = \frac{\partial P/\partial v}{\partial \mu/\partial v} = \frac{1}{v}$$

(7.41)

Hence

$$\frac{\partial^2 P}{\partial \mu^2} = -\frac{1}{v^2} \frac{\partial v}{\partial \mu} = \frac{1}{v^3 \partial^2 a/\partial v^2} = -\frac{1}{v^3 \partial P/\partial v}$$

(7.42)

Substituting this relation into (7.38), we finally obtain, after some minor rewriting,

$$\langle N^2 \rangle - \langle N \rangle^2 = \overline{N} k T \kappa_T / v, \qquad \kappa_T = \frac{1}{v(-\partial P/\partial v)}$$

(7.43)

This shows that the density fluctuations are vanishing small in the thermodynamic limit, provided the isothermal compressibility κ_T is finite (i.e., not infinite). This is true except in the transition region of a first-order phase transition, including the critical point.

The relation (7.43) is similar to (7.14), whereby a fluctuation is related to an appropriate "susceptibility." In the case of energy fluctuations the relevant susceptibility is the specific heat at constant volume, and in the case of density fluctuations it is the isothermal compressibility. These are special cases of a more general rule known as the *fluctuation-dissipation theorem*, the historically earliest form of which is the Einstein relation (2.61) pertaining to Brownian motion.

The probability that a system in the grand canonical ensemble has N particles is proportional to

$$W(N) \equiv z^N Q_N(V, T) = \exp \beta [\mu N - A(N, V, T)]$$

(7.44)

where A is the Helmholtz free energy calculated from the canonical ensemble with N particles. When the density fluctuations are small, $W(N)$ is strongly peaked about $N = \overline{N}$, with a width of the order of $\sqrt{\overline{N}}$, and we may obtain the Helmholtz free energy directly from the grand partition function through the formula

$$A(\overline{N}, V, T) = kT \overline{N} \log z - kT \log \mathcal{Q}(z, V, T)$$

(7.45)

where z is to be eliminated through (7.36).

When $\partial P/\partial v = 0$, as happens at the critical point, the density fluctuations become very large, as is borne out experimentally by the phenomenon of critical opalescence. However, even in this case, (7.45) is still valid. To show this requires a more detailed analysis, which we postpone until the end of this chapter.

7.5 THE CHEMICAL POTENTIAL

Thermodynamics

The chemical potential μ is defined such that the Helmholtz free energy A changes by $\mu\, dN$, when the number of particles change by dN, at constant T and V. Hence

$$dA = -P\, dV - S\, dT + \mu\, dN \qquad (7.46)$$

from which we can deduce a more general form of the first law of thermodynamics:

$$dU = -P\, dV + T\, dS + \mu\, dN \qquad (7.47)$$

When μ is positive, it tends to drive N to smaller values, in order to lower the energy. Hence the name chemical potential.*

From (7.46) we can also deduce the change of the Gibbs free energy:

$$dG = -V\, dP - S\, dT + \mu\, dN \qquad (7.48)$$

Thus we have the equivalent Maxwell relations

$$\mu = \left(\frac{\partial A}{\partial N}\right)_{V,T} = \left(\frac{\partial G}{\partial N}\right)_{P,T}$$

A useful result is the chemical potential of an ideal gas, which can be easily calculated from the partition function for an ideal gas:

$$Q_N = \frac{1}{h^{3N}N!}\int dp\, dq \exp\left[\beta \sum_{i=1}^{N} p_i^2/2m\right] = \frac{1}{N!}\left(\frac{V}{\lambda^3}\right)^N \qquad (7.49)$$

$$\lambda = \sqrt{2\pi\hbar^2/mkT}$$

Hence

$$A = kT \log Q_N = -kTN\left[\log\left(\frac{V}{N\lambda^3}\right) + 1\right] \qquad (7.50)$$

$$\mu = \partial A/\partial N = kT \log(\lambda^3 n)$$

where n is the density.

Conservation of Particle Number

For ordinary matter, it makes sense to speak of a system of N atoms, because N is an effectively conserved quantity. The chemical potential may be viewed as the Lagrange multiplier to take that into account. The conservation law has its origin in the more fundamental law of baryon conservation, which states that the number of baryons (such as protons or neutrons) minus the number of anti-baryons is conserved. This means, for example, that a proton can be created or

---

*The name fugacity for $\exp(\beta\mu)$ has a dictionary meaning of "the tendency to flee," or "volatility." *The fugacity of pleasure, the fragility of beauty* (Samuel Johnson).

annihilated only in conjunction with an antiproton. At low temperatures the thermal energy is not sufficient to create pairs, nor are there antiprotons present. Thus the number of protons (and neutrons) is effectively conserved.

The same thing can be said about electrons, whose number appears to be conserved at low temperatures only because there is insufficient energy to create electron-positron pairs, and there are usually no positrons present. The truly conserved quantity is the number of electrons minus the number of positrons.

A correct description of matter at high temperatures must take into account the possibility of pair creation. There will be an average number of particles and antiparticles present in equilibrium, there will also be fluctuations about the average values. It is the difference between particle and antiparticle number that remains strictly constant and is determined by the initial conditions. For example, the reaction $e^+ + e^- \rightleftarrows \gamma$ can occur in the interior of stars, and establishes the equilibrium density of electrons, positrons, and radiation.

The detailed mechanism for pair creation and annihilation is not relevant for the equilibrium situation, and affects only the relaxation time for the establishment of equilibrium. Thus, to treat the equilibrium situation we may describe the system in the grand canonical ensemble, using the effective Hamiltonian

$$\mathcal{H} = \mathcal{H}_1 + \mathcal{H}_2 - \mu(N_1 - N_2) \tag{7.51}$$

where the subscripts 1 and 2 refer to particle and antiparticle, and μ is the Lagrange multiplier introduced to enable us to treat N_1 and N_2 as unconstrained variables. The grand partition function is

$$\begin{aligned}
\mathcal{Q} &= \sum_{N_1=0}^{\infty} \sum_{N_2=0}^{\infty} Q_{N_1} Q_{N_2}\, e^{\beta\mu(N_1 - N_2)} \\
&= \sum_{N_1=0}^{\infty} \sum_{N_2=0}^{\infty} \exp - \beta\left[A_{N_1} + A_{N_2} - \mu(N_1 - N_2)\right]
\end{aligned} \tag{7.52}$$

where Q_N is a partition function, and A_N the corresponding Helmholtz free energy. In the thermodynamic limit we keep only the largest term in the summand, with N_1 and N_2 determined by the conditions

$$\begin{aligned}
\partial A_{N_1}/\partial N_1 &= \mu \\
\partial A_{N_2}/\partial N_2 &= -\mu
\end{aligned} \tag{7.53}$$

We should calculate the Helmholtz free energies using relativistic kinematics. For the purpose of illustration, however, we shall pretend that the energy of a particle is $E = mc^2 + p^2/2m$. The inclusion of the rest energy is important, for we are concerned with processes that can convert it into other forms of energy. Thus we take the chemical potential to be (7.50) plus the rest energy. The conditions for equilibrium then become

$$\begin{aligned}
kT \log\left(\lambda^3 n_1\right) + mc^2 &= \mu \\
kT \log\left(\lambda^3 n_2\right) + mc^2 &= -\mu
\end{aligned} \tag{7.54}$$

where n_1 and n_2 are the densities of particles and antiparticles. Adding the two equations gives

$$\lambda^6 n_1 n_2 = e^{-2mc^2/kT} \tag{7.55}$$

The rest energy of an electron corresponds to a temperature of 6×10^9 K. That for a proton is 2000 times higher. Thus the right side of (7.55) is essentially zero for ordinary temperatures. This means that, if either n_1 or n_2 is not essentially zero, the other must be. This shows why we can completely ignore antiparticles when $kT \ll mc^2$.

Chemical Equilibrium

Suppose we have a reaction such as

$$2H_2 + O_2 \rightleftarrows 2H_2O$$

What are the fractions of each species of molecules present in an equilibrium mixture? More generally we consider a reaction in which a group of particles $X_1, X_2, \ldots$ participate in a reaction to yield a group of particles $Y_1, Y_2, \ldots$ or vice versa:

$$\nu_1 X_1 + \nu_2 X_2 + \cdots \rightleftarrows \nu_1' Y_1 + \nu_2' Y_2 + \cdots \tag{7.56}$$

The numbers ν_i are called *stoichiometric coefficients*. The process is a generalization of our previous discussion of particle-antiparticle reaction, for which $\nu_1 = \nu_2 = 1$. For notational convenience, rewrite (7.56) as

$$\sum_{i=1}^{K} \nu_i X_i = 0 \tag{7.57}$$

where both X's and Y's are denoted by X, and $\nu_i' = -\nu_i$. The conservation law in this case is

$$\frac{\delta N_1}{\nu_1} = \frac{\delta N_2}{\nu_2} = \cdots = \frac{\delta N_K}{\nu_K} \tag{7.58}$$

where δN_i is the increase in the number of particles of the ith type. This means $\delta N_i / \nu_i$ is independent of i:

$$\delta N_i = \nu_i \, \delta N \qquad i = 1, \ldots, K \tag{7.59}$$

To find the equilibrium condition consider the reaction proceeding at constant V and T. In equilibrium the Helmholtz energy A is at a minimum. Hence any variation of the number N_i from their equilibrium value will not change A to first order. Assume A is the sum of the component free energies. Then

$$0 = \delta A = \sum_{i=1}^{K} \delta A_i = \sum_{i=1}^{K} \frac{\partial A_i}{\partial N_i} \delta N_i = \sum_{i=1}^{K} \mu_i \nu_i \, \delta N \tag{7.60}$$

Since δN is arbitrary, we obtain as a condition for equilibrium

$$\sum_{i=1}^{K} \mu_i \nu_i = 0 \tag{7.61}$$

where μ_i is the chemical potential of X_i. If the reaction proceeds at constant P and T, one considers the Gibbs free energy instead, and arrives at the same condition.

7.6 EQUIVALENCE OF THE CANONICAL ENSEMBLE AND THE GRAND CANONICAL ENSEMBLE

We have seen that if $\partial P / \partial v < 0$ then almost all systems in the grand canonical ensemble have the same number of particles N. Then the grand canonical ensemble is trivially equivalent to the canonical ensemble for N particles.

To complete our investigation of the equivalence between the canonical and the grand canonical ensemble it is necessary to consider values of v for which $\partial P / \partial v = 0$. It will be shown that in such cases the function $W(N)$ given in (7.44) will no longer have a sharp maximum; the equation of state as given by the recipe in the grand canonical ensemble nevertheless still agrees with that given by the recipe in the canonical ensemble. In this sense the two ensembles are always equivalent.

Physically the values of v for which $\partial P / \partial v = 0$ correspond to the transition region of a first-order phase transition. In this region, (7.43) leads us to expect that the fluctuations of density in a given volume of the system will be large. This is also expected physically, for in such a region the system is composed of two or more phases of *different* densities. Therefore the number of particles in any given volume can have a whole range of values, depending on the amounts of each phase present. At the critical point of a gas-liquid system fluctuations in density are also expected to be large, because throughout the system molecules are spontaneously forming large clusters and breaking up. It is clear that under these conditions the grand canonical ensemble must continue to yield thermodynamic predictions that are in agreement with those obtained by the canonical ensemble. Otherwise the validity of either as a description of matter would be in doubt, for it is a basic experimental fact that we can obtain the same thermodynamic information whether we look at the whole system or at only a subvolume of the system.

The mathematical questions that we try to answer are as follows. Suppose $Q_N(V, T)$ is given, and we wish to calculate

$$\mathcal{Q}(z, V, T) \equiv \sum_{N=0}^{\infty} z^N Q_N(V, T) \tag{7.62}$$

for given values of z, V, and T.

(a) For a given value of z is the following true for some N?

$$\mathcal{Q}(z, V, T) \approx z^N Q_N(V, T) \tag{7.63}$$

(b) Does there always exist a value of z for which N has any given positive value?

The answers are obviously no, if $Q_N(V, T)$ is *any* function of N, V, T. We are only interested, however, in the answers when $Q_N(V, T)$ is the partition function of a physical system. Thus we must first make some assumptions about $Q_N(V, T)$.

To incorporate the salient features of a physical system into our considerations, and yet keep the mathematics simple, we assume that we are dealing with a system

(*a*) whose molecules interact through an intermolecular potential that contains a hard-sphere repulsion of finite diameter plus a finite potential of finite range, and

(*b*) whose Helmholtz free energy has the form

$$A(N, V) \equiv -\frac{1}{\beta} \log Q_N(V) = -\frac{V}{\beta} f(v) \tag{7.64}$$

where $v \equiv V/N$, $\beta = 1/kT$, and $f(v)$ is finite. The temperature will be fixed throughout our discussions and will not be displayed unless necessary. The function $f(v)$ is related to the pressure $P(v)$ of the canonical ensemble by

$$f(v) = \frac{1}{v} \int_{v_0}^{v} dv' \, \beta P(v') \tag{7.65}$$

where the integration is carried out along an isotherm and v_0 is an arbitrary constant corresponding to an arbitrary additive constant in the Helmholtz free energy.

(*c*) We further assume that $f(v)$ is such that

$$\frac{\partial P}{\partial v} \leq 0 \tag{7.66}$$

This immediately implies that

$$\frac{\partial^2 f(v)}{\partial (1/v)^2} \leq 0 \tag{7.67}$$

With these assumptions the grand partition function may be written in the form

$$\mathscr{Q}(z, V) = \sum_{N=0}^{\infty} e^{V\phi(V/N, z)} \tag{7.68}$$

where z is an arbitrary fixed number and

$$\phi(v, z) \equiv f(v) + \frac{1}{v} \log z \tag{7.69}$$

Using (7.65) we obtain

$$\phi(v, z) = \frac{1}{v} \log z + \frac{1}{v} \int_{v_0}^{v} dv' \, \beta P(v') \tag{7.70}$$

By (7.67), we have $\partial^2 \phi / \partial (1/v)^2 \le 0$, or

$$\frac{\partial^2 \phi}{\partial v^2} + \frac{2}{v} \frac{\partial \phi}{\partial v} \le 0 \tag{7.71}$$

We now calculate the grand partition function. For a fixed volume V the partition function $Q_N(V)$ vanishes whenever

$$N > N_0(V)$$

where $N_0(V)$ is the maximum number of particles that can be accommodated in the volume V, such that no two particles are separated by a distance less than the diameter of the hard sphere in the interparticle potential. Therefore $\mathscr{2}(z, V)$ is a polynomial of degree $N_0(V)$. For large V it is clear that

$$N_0(V) = aV \tag{7.72}$$

where a is a constant. Let the largest value among the terms in this polynomial be $\exp[V\phi_0(z)]$, where

$$\phi_0(z) = \max \left[\phi\left(\frac{V}{N}, z \right) \right] \quad (N = 0, 1, 2, \dots) \tag{7.73}$$

Then the following inequality holds:

$$e^{V\phi_0(z)} \le \mathscr{2}(z, v) \le N_0(V) \, e^{V\phi_0(z)}$$

Using (7.72) we obtain

$$e^{V\phi_0(z)} \le \mathscr{2}(z, V) \le aV e^{V\phi_0(z)}$$

or

$$\phi_0(z) \le \frac{1}{V} \log \mathscr{2}(z, V) \le \phi_0(z) + \frac{\log(aV)}{V} \tag{7.74}$$

Therefore

$$\lim_{V \to \infty} \frac{1}{V} \log \mathscr{2}(z, V) = \phi_0(z) \tag{7.75}$$

Let $\bar{v}$ be a value of v at which $\phi(v, z)$ assumes its largest possible value. Since $\phi(v, z)$ is differentiable, $\bar{v}$ is determined by the conditions

$$\left(\frac{\partial \phi}{\partial v} \right)_{v = \bar{v}} = 0 \tag{7.76}$$

$$\left(\frac{\partial^2 \phi}{\partial v^2} \right)_{v = \bar{v}} \le 0 \tag{7.77}$$

By virtue of (7.71) the first condition implies the second. Therefore $\bar{v}$ is de-

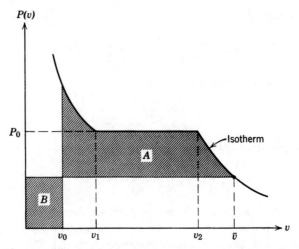

Fig. 7.2 Typical isotherm of a substance in the transition region of a first-order phase transition.

termined by (7.76) alone. By (7.69) and (7.65) we may rewrite it in the form

$$\int_{v_0}^{\bar{v}} dv' \, P(v') - \bar{v} P(\bar{v}) = -kT \log z$$

or

$$\left[\int_{v_0}^{\bar{v}} dv' \, P(v') - (\bar{v} - v_0) P(\bar{v}) \right] - v_0 P(\bar{v}) = -kT \log z \qquad (7.78)$$

A geometrical representation of this condition is shown in Fig. 7.2. The value of $\bar{v}$ is such that the difference between the area of the region A and that of the region B is numerically equal to $-kT \log z$. The result is shown in Fig. 7.3. It is seen that to every value of $\bar{v}$ greater than the close-packing volume there corresponds a value of z. This answers question (b) in the affirmative.

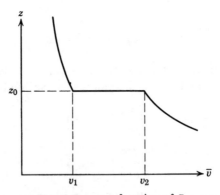

Fig. 7.3 z as a function of $\bar{v}$.

There is a value of z that corresponds to all the values of $\bar{v}$ lying in the interval $v_1 \leq \bar{v} \leq v_2$. This value, denoted by z_0, is given by

$$\log z_0 = \beta v_1 P(v_1) - \int_{v_0}^{v_1} dv' \, \beta P(v') \tag{7.79}$$

7.7 BEHAVIOR OF W(N)

In (7.44) we introduced the quantity $W(N)$, which is the (unnormalized) probability that a system in the grand canonical ensemble has N particles. Comparing (7.44) to (7.68) we see that

$$W(N) = \exp\left[V\phi\left(\frac{V}{N}, z \right) \right] \tag{7.80}$$

Hence it is of some interest to examine the function $\phi(v, z)$ in more detail. Suppose $P(v)$ has the form shown in the $P - v$ diagram of Fig. 7.2. For values of v lying in the range $v_1 \leq v \leq v_2$, P has the constant value P_0. For this range of v we have

$$\phi(v, z) = \frac{1}{v}\left[\log z + \int_{v_0}^{v_1} dv' \, \beta P(v') - \beta P_0 v_1 \right] + \beta P_0$$

which is the same as

$$\phi(v, z) = \frac{1}{v} \log\left(\frac{z}{z_0} \right) + \beta P_0 \qquad (v_1 \leq v \leq v_2) \tag{7.81}$$

where z_0 is defined by (7.79). Hence we can immediately make a qualitative sketch of a family of curves, one for each z, for the function $\phi(v, z)$ in the interval $v_1 \leq v \leq v_2$. The result is shown in Fig. 7.4.

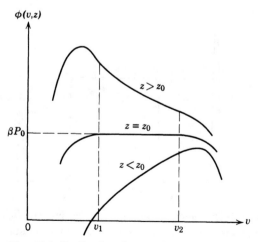

Fig. 7.4 Qualitative form of $\phi(v, z)$ for a physical substance.

To deduce the behavior of $\phi(v, z)$ outside the interval just discussed we use the following facts:

(a) $\partial\phi/\partial v$ is everywhere continuous. This is implied by (7.70).

(b) $\partial\phi/\partial v = 0$ implies $\partial^2\phi/\partial v^2 \leq 0$. That is, as a function of v, ϕ cannot have a minimum. This follows from (7.71).

(c) For $z \neq z_0$, ϕ has one and only one maximum. This follows from (b).

Guided by these facts we obtain the curves shown in Fig. 7.4.

The behavior of $W(N)$ can be immediately obtained from that of $\phi(v, z)$. It is summarized by the series of graphs in Fig. 7.5. For $z \neq z_0$, $W(N)$ has a single sharp peak at some value of N. This peak becomes infinitely sharp as $V \rightarrow \infty$. For $z = z_0$, all values of N in the interval

$$v_1 \leq \frac{V}{N} \leq v_2 \tag{7.82}$$

are equally probable. The number of N values corresponding to (7.82) is

$$\left(\frac{1}{v_1} - \frac{1}{v_2}\right)V \tag{7.83}$$

This situation corresponds to the large fluctuation of density in the transition

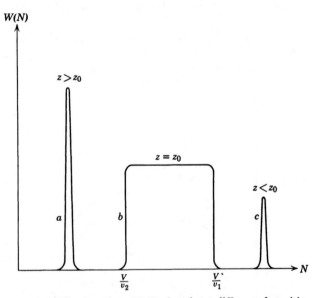

Fig. 7.5 The function $W(N)$ for three different fugacities (hence three different densities). For curves a and c the system is in a single pure phase. For curve b the system is undergoing a first-order phase transition.

region and may be stated in more physical terms as follows: The pressure is unchanged if we take any number of particles from one phase and deliver them to the other. Each time we do this, however, the total number of particles in a given volume changes, because the densities of the two phases are generally different. Let us start with the system in one pure phase and then transfer the particles one by one to the other phase, until the system exists purely in the other phase. The number of transfers we can make is proportional to V. Each transfer corresponds to a term in the grand partition function, and all these terms have the same value.

7.8 THE MEANING OF THE MAXWELL CONSTRUCTION

It has been shown that if the pressure P calculated in the canonical ensemble satisfies the condition $\partial P/\partial v \leq 0$, the pressure calculated in the grand canonical ensemble is also P. We show that the converse is also true. We then have the statement

(a) The pressure P calculated in the canonical ensemble agrees with that calculated in the grand canonical ensemble if and only if $\partial P/\partial v \leq 0$. It will further be shown that

(b) If $\partial P/\partial v > 0$ for some v, the pressure in the grand canonical ensemble is obtainable from P by making the Maxwell construction.

Suppose the pressure calculated in the canonical ensemble is given and is denoted by $P_{can}(v)$. At a certain temperature we assume $P_{can}(v)$ to have the qualitative form shown in the $P - v$ diagram of Fig. 7.6.

The partition function of the system under consideration is

$$Q_N(V) = e^{VF(v)} \tag{7.84}$$

where

$$F(v) = \frac{1}{v} \int_{v_0}^{v} dv' \, \beta P_{can}(v') \tag{7.85}$$

It is easily seen that

$$\beta P_{can}(v) = F(v) + v \frac{\partial F(v)}{\partial v} \tag{7.86}$$

Let

$$\Phi(v, z) \equiv F(v) + \frac{1}{v} \log z \tag{7.87}$$

It is easily verified that

$$\frac{\partial^2 \Phi}{\partial v^2} + \frac{2}{v} \frac{\partial \Phi}{\partial v} = \frac{\beta}{v} \frac{\partial P_{can}}{\partial v} \begin{cases} > 0 & (a < v < b) \\ \leq 0 & (\text{otherwise}) \end{cases} \tag{7.88}$$

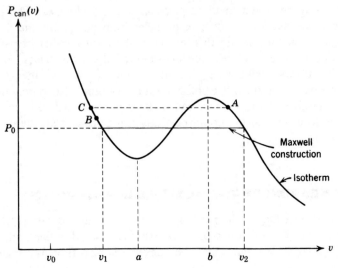

Fig. 7.6 Isotherm with $\partial P_{\mathrm{can}}/\partial v > 0$ for v lying in the range $a < v < b$.

To calculate the grand partition function we recall that the derivation of (7.75) is independent of the sign of $\partial P/\partial v$. Hence, in analogy with (7.75), we have in the present case

$$\lim_{V \to \infty} \frac{1}{V} \log \mathscr{2}(z, V) = \Phi(\bar{v}, z) \tag{7.89}$$

where

$$\Phi(\bar{v}, z) = \max \left[\Phi(v, z) \right] \tag{7.90}$$

This determines $\bar{v}$ in terms of z, or vice versa. The pressure in the grand canonical ensemble, denoted by $P_{\mathrm{gr}}(\bar{v})$, is given by

$$\beta P_{\mathrm{gr}}(\bar{v}) = \Phi(\bar{v}, z) \tag{7.91}$$

From (7.87) and (7.85) we see that both Φ and $\partial \Phi/\partial v$ are continuous functions of v. Hence (7.90) is equivalent to the conditions

$$\begin{aligned} \left(\frac{\partial \Phi}{\partial v} \right)_{v = \bar{v}} &= 0 \\ \left(\frac{\partial^2 \Phi}{\partial v^2} \right)_{v = \bar{v}} &\leq 0 \end{aligned} \tag{7.92}$$

with the following additional rule: If (7.92) determines more than one value of $\bar{v}$, we must take only the value that gives the largest $\Phi(\bar{v}, z)$.

The first condition of (7.92) is the same as

$$\left(v^2 \frac{\partial F}{\partial v} \right)_{v = \bar{v}} = \log z \tag{7.93}$$

Substituting this into (7.86) we obtain

$$\beta P_{\text{can}}(\bar{v}) = F(\bar{v}) + \frac{1}{\bar{v}} \log z = \Phi(\bar{v}, z) \tag{7.94}$$

Comparing this with (7.91) we obtain

$$P_{\text{can}}(\bar{v}) = P_{\text{gr}}(\bar{v}) \tag{7.95}$$

That is, if there is a value $\bar{v}$ that satisfies (7.92), then at this value of the specific volume the pressure is the same in the canonical and grand canonical ensemble. Therefore it only remains to investigate the possible values of $\bar{v}$.

It is obvious that $\bar{v}$ can never lie between the values a and b shown in Fig. 7.6, because, as we can see from (7.88), in that region $\partial \Phi / \partial v = 0$ implies $\partial^2 \Phi / \partial v^2 > 0$, in contradiction to (7.92). On the other hand, outside this region, $\partial \Phi / \partial v = 0$ implies $\partial^2 \Phi / \partial v^2 \leq 0$. Hence the first condition of (7.92) alone determines $\bar{v}$. Using (7.85) we can write this condition in the form

$$\int_{v_0}^{\bar{v}} dv' \, P_{\text{can}}(v') - \bar{v} P_{\text{can}}(\bar{v}) = -kT \log z \tag{7.96}$$

There is a value of z, denoted by z_0, at which (7.96) has two roots v_1 and v_2 for which $\Phi(v_1, z) = \Phi(v_2, z)$. The conditions for this to be so are that

$$-kT \log z_0 = \int_{v_0}^{v_1} dv' \, P_{\text{can}}(v') - v_1 P_{\text{can}}(v_1) = \int_{v_0}^{v_2} dv' \, P_{\text{can}}(v') - v_2 P_{\text{can}}(v_2)$$

$$\Phi(v_1, z_0) = \Phi(v_2, z_0) \tag{7.97}$$

The second condition is equivalent to $P_{\text{can}}(v_1) = P_{\text{can}}(v_2)$, by virtue of (7.94). Combining these conditions, we obtain

$$\int_{v_1}^{v_2} dv' \, P_{\text{can}}(v') = (v_2 - v_1) P_{\text{can}}(v_1) \tag{7.98}$$

which means that v_1 and v_2 are the end points of a Maxwell construction on P_{can}, as shown in Fig. 7.6.

In general we can find z as a function of $\bar{v}$ by solving (7.96) graphically, in a manner similar to that used in the last section for (7.78). The result is qualitatively sketched in Fig. 7.7. As explained before, the interval $a < \bar{v} < b$ must be excluded. By definition of the Maxwell construction, the portions of the curves outside the interval $v_1 \leq \bar{v} \leq v_2$, shown in solid lines in Fig. 7.7, coincide with the corresponding portions in Fig. 7.3. We need to discuss further only the dashed portions of the curves.

Consider the points A and B in Fig. 7.7. Let their volumes be, respectively, v_A and v_B and let their common z value be z'. The fact that they are both solutions of (7.96) means that the function $\Phi(v, z')$ has two maxima, located respectively at $v = v_A$ and $v = v_B$. These maxima cannot be of the same height, because that would mean that v_A and v_B are, respectively, v_2 and v_1, which they are not. To determine which maximum is higher we note that by (7.85), (7.94),

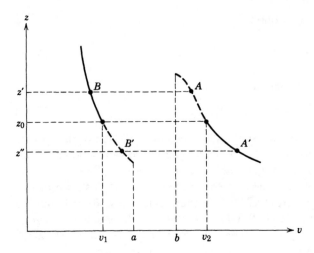

Fig. 7.7 z as a function of $\bar{v}$.

and the fact that z' is common to both,

$$\int_{v_B}^{v_A} dv' \, P_{\text{can}}(v') = v_A P_{\text{can}}(v_A) - v_B P_{\text{can}}(v_B) \qquad (7.99)$$

Suppose $P_{\text{can}}(v_B) < P_{\text{can}}(v_A)$. Consider the point C indicated in Fig. 7.6. By inspection of Fig. 7.6 we see that

$$\int_{v_C}^{v_A} dv' \, P_{\text{can}}(v') < (v_A - v_C) P_{\text{can}}(v_A)$$

Subtracting (7.99) from this inequality, we obtain

$$\int_{v_C}^{v_B} dv' \, P_{\text{can}}(v') < v_B P_{\text{can}}(v_B) - v_C P_{\text{can}}(v_A)$$

which, by the original assumption, implies

$$\int_{v_C}^{v_B} dv' \, P_{\text{can}}(v') < (v_B - v_C) P_{\text{can}}(v_B)$$

By inspection of Fig. 7.6 we see that this is impossible. Therefore we must have $P_{\text{can}}(v_B) > P_{\text{can}}(v_A)$. By (7.94), this means that

$$\Phi(v_B, z') > \Phi(v_A, z')$$

In a similar fashion we can prove that, for the points A' and B' in Fig. 7.7,

$$\Phi(v_{A'}, z'') > \Phi(v_{B'}, z'')$$

Therefore the dashed portions of the curves in Fig. 7.7 must be discarded.

In Fig. 7.8, $P_{\text{gr}}(\bar{v})$ is shown as the solid curve. It is the same as $P_{\text{can}}(\bar{v})$ except that the portion between v_1 and v_2 is missing because there is no z that will give a $\bar{v}$ lying in that interval. In other words, in the grand canonical ensemble the system cannot have a volume in that interval. We can, however, fill in a horizontal line at P_0 by the usual arguments, namely, that since the systems

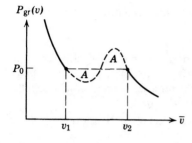

Fig. 7.8 The pressure in the grand canonical ensemble (solid lines).

at v_1 and v_2 have the same temperature, pressure, and chemical potential, a system at v_1 can coexist with a system at v_2 with any relative amount of each present.

It is an experimental fact that $\partial P/\partial v \le 0$. It could not be otherwise, for then the system would be in the highly unstable situation in which releasing the pressure on it leads to a shrinkage. The quantity P_{can} is the result of a (generally approximate) calculation, and may or may not have this desirable property. However, the corresponding pressure in the grand canonical ensemble always satisfies the stability condition because the ensemble explicitly includes all possible density fluctuations of the system.

PROBLEMS

7.1 (*a*) Obtain the pressure of a classical ideal gas as a function of N, V, and T, by calculating the partition function.

(*b*) Obtain the same by calculating the grand partition function.

7.2 Consider a classical system of N noninteracting diatomic molecules enclosed in a box of volume V at temperature T. The Hamiltonian for a *single* molecule is taken to be

$$\mathscr{H}(\mathbf{p}_1, \mathbf{p}_2, \mathbf{r}_1, \mathbf{r}_2) = \frac{1}{2m}\left(p_1^2 + p_2^2 \right) + \tfrac{1}{2}K|\mathbf{r}_1 - \mathbf{r}_2|^2$$

where $\mathbf{p}_1, \mathbf{p}_2, \mathbf{r}_1, \mathbf{r}_2$, are the momenta and coordinates of the two atoms in a molecule. Find

(*a*) the Helmholtz free energy of the system;

(*b*) the specific heat at constant volume;

(*c*) the mean square molecule diameter $\langle |\mathbf{r}_1 - \mathbf{r}_2|^2 \rangle$.

7.3 Repeat the last problem, using the Hamiltonian

$$\mathscr{H}(\mathbf{p}_1, \mathbf{p}_2, \mathbf{r}_1, \mathbf{r}_2) = \frac{1}{2m}\left(p_1^2 + p_2^2 \right) + \epsilon|r_{12} - r_0|$$

where ϵ and r_0 are given positive constants and $r_{12} \equiv |\mathbf{r}_1 - \mathbf{r}_2|$.

Answer.

$$\frac{C_V}{Nk} = 6 - \frac{x^2\left[2(x^2 - 2) + (x + 2)^2\, e^{-x} \right]}{(x^2 + 2 - e^{-x})^2} \qquad (x \equiv \epsilon r_0/kT)$$

7.4 Prove *Van Leeuwen's Theorem*: The phenomenon of diamagnetism does not exist in classical physics.

The following hints may be helpful:

(a) If $\mathcal{H}(\mathbf{p}_1,\ldots,\mathbf{p}_N; \mathbf{q}_1,\ldots,\mathbf{q}_N)$ is the Hamiltonian of a system of charged particles in the absence of an external magnetic field, then $\mathcal{H}[\mathbf{p}_1 - (e/c)\mathbf{A}_1,\ldots,\mathbf{p}_N - (e/c)\mathbf{A}_N; \mathbf{q}_1,\ldots,\mathbf{q}_N]$ is the Hamiltonian of the same system in the presence of an external magnetic field $\mathbf{H} = \nabla \times \mathbf{A}$, where $\mathbf{A}_i$ is the value of $\mathbf{A}$ at the position $\mathbf{q}_i$.

(b) The induced magnetization of the system along the direction of H is given by

$$M = \left\langle -\frac{\partial \mathcal{H}}{\partial H} \right\rangle = kT\frac{\partial}{\partial H} \log Q_N$$

where $\mathcal{H}$ is the Hamiltonian in the presence of $\mathbf{H}$, $H = |\mathbf{H}|$, and Q_N is the partition function of the system in the presence of $\mathbf{H}$.

7.5 Langevin's Theory of Paramagnetism. Consider a system of N atoms, each of which has an intrinsic magnetic moment of magnitude μ. The Hamiltonian in the presence of an external magnetic field $\mathbf{H}$ is

$$\mathcal{H}(p,q) - \mu H \sum_{i=1}^{N} \cos \alpha_i$$

where $\mathcal{H}(p, q)$ is the Hamiltonian of the system in the absence of an external magnetic field, and α_i is the angle between $\mathbf{H}$ and the magnetic moment of the ith atom. Show that

(a) The induced magnetic moment is

$$M = N\mu\left(\coth \theta - \frac{1}{\theta}\right) \qquad (\theta \equiv \mu H/kT)$$

(b) The magnetic susceptibility per atom is

$$\chi = \frac{\mu^2}{kT}\left(\frac{1}{\theta^2} - \operatorname{csch}^2 \theta\right)$$

(c) At high temperatures χ satisfies Curie's law, namely $\chi \propto T^{-1}$. Find the proportionality constant, which is called Curie's constant.

7.6 Imperfect Gas. Consider a system of N molecules ($N \to \infty$) contained in a box of volume V ($V \to \infty$). The Hamiltonian of the system is

$$\mathcal{H} = \sum_{i=1}^{N} \frac{p_i^2}{2m} + \sum_{i<j} v_{ij}$$

$$v_{ij} = v(|\mathbf{r}_i - \mathbf{r}_j|)$$

where $\mathbf{p}_i$ and $\mathbf{r}_i$ are, respectively, the momentum and the position of the ith molecule. The intermolecular potential $v(r)$ has the qualitative form shown in the accompanying figure. Let

$$f_{ij} \equiv f(|\mathbf{r}_i - \mathbf{r}_j|)$$

$$f(r) \equiv e^{-\beta v(r)} - 1$$

A sketch of $f(r)$ is also shown in the same figure.

(a) Show that the equation of state of the system is

$$\frac{Pv}{kT} = 1 + v\frac{\partial Z(v,T)}{\partial v}$$

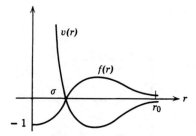

where $v \equiv V/N$ and

$$Z(v,T) \equiv \frac{1}{N} \log \left[\frac{1}{V^N} \int d^3 r_1 \cdots d^3 r_N \prod_{i<j} (1 + f_{ij}) \right]$$

(b) By expanding the product $\prod (1 + f_{ij})$, show that

$$Z(v,T) = \frac{1}{N} \log \left[\frac{1}{V^N} \int d^3 r_1 \cdots d^3 r_N \left(1 + \sum_{i<j} f_{ij} + \cdots \right) \right]$$

$$= \log \left[1 + \frac{N}{2V} \int d^3 r f(r) + \cdots \right]^{1/N}$$

(c) Show that at low densities, i.e.,

$$r_0^3 / v \ll 1$$

it is a good approximation to retain only the first two terms in the series appearing in the expression $Z(v, T)$. Hence the equation of state is approximately given by

$$\frac{Pv}{kT} \approx 1 - \frac{1}{2v} \int_0^\infty dr\, 4\pi r^2 f(r)$$

The coefficient of $1/v$ is called the *second virial coefficient*.

Note. (i) Retaining the first two terms in the series appearing in $Z(v, T)$ is a good approximation because $Z(v, T)$ is the logarithm of the Nth root of the series. The approximation is certainly invalid for the series itself

(ii) If all terms in the expansion of $\prod (1 + f_{ij})$ were kept, we would have obtained a systematic expansion of Pv/kT in powers of $1/v$. Such an expansion is known as the *virial expansion*.

(iii) The complete virial expansion is difficult to obtain by the method described in this problem. It is obtained in Chapter 10 via the grand canonical ensemble. See (10.27) and (10.30).

7.7 Van der Waals Equation of State

(a) Show that for low densities the Van der Waals equation of state (2.28) reduces to

$$\frac{Pv}{kT} \approx 1 + \frac{1}{v} \left(b' - \frac{a'}{kT} \right)$$

(b) Show that the imperfect gas of Problem 7.6 has an equation of state of the same form as shown in (a), with

$$b' = \frac{2\pi}{3} \sigma^3$$

$$a' = -2\pi kT \int_\sigma^\infty dr\, r^2 \left(1 - e^{-\beta v(r)} \right)$$

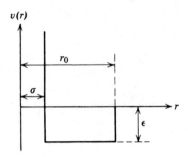

7.8 The equation of state for an N_2 gas can be written in the form

$$PV/NkT = 1 + a_2(T)(N/V)$$

for low densities. The second virial coefficient $a_2(T)$ has been measured as a function of temperature and is given in the accompanying table. Assume that the intermolecular potential $v(r)$ between N_2 molecules has the form shown in the accompanying sketch. From the data given, determine what you consider to be the best choice for the constants σ, r_0, and ϵ.

Temperature, K	$a_2(T)$, K/atm
100	-1.80
200	-4.26×10^{-1}
300	-5.49×10^{-2}
400	1.12×10^{-1}
500	2.05×10^{-1}

7.9 A dilute mixture of H_2 and O_2 gases is kept at constant temperature T. Initially the density of H_2 was n_0, the density of O_2 was $n_0/2$, and there was no H_2O present. After a certain time, the mixture becomes an equilibrium mixture of H_2, O_2, and H_2O. Find the equilibrium densities of the three components n_1, n_2, n_3, as a function of T and n_0.

8

QUANTUM STATISTICAL MECHANICS

8.1 THE POSTULATES OF QUANTUM STATISTICAL MECHANICS

All systems in nature obey quantum mechanics. In quantum mechanics an observable of a system is associated with a Hermitian operator, which operates on a Hilbert space. A state of the system is a vector $|\Psi\rangle$ in the same Hilbert space. If $|q\rangle$ is an eigenvector of the position operators of all the particles in the system, then $\langle q|\Psi\rangle \equiv \Psi(q)$ is the wave function of the system in the state $|\Psi\rangle$. The wave function furnishes a complete description of the state.

At any instant of time the wave function Ψ of a truly isolated system may be expressed as a linear superposition of a complete orthonormal set of stationary wave functions $\{\Phi_n\}$:

$$\Psi = \sum_n c_n \Phi_n \tag{8.1}$$

where c_n is a complex number and is a function of time. The index n stands for a set of quantum numbers, which are eigenvalues of certain chosen dynamical operators of the system. The square modulus $|c_n|^2$ is the probability that a measurement performed on the system will find it to have the quantum numbers n.

In statistical mechanics we always deal with systems that interact with the external world. Here we can regard the system plus the external world as a truly isolated system. The wave function Ψ for this whole system will depend on both the coordinates of the system under consideration and the coordinates of the external world. If $\{\Phi_n\}$ denotes a complete set of orthonormal stationary wave functions of the system, then Ψ is still formally given by (8.1), but c_n is to be interpreted as a wave function of the external world. It depends on the coordinates of the external world as well as on the time.

Suppose $\mathcal{O}$ is an operator corresponding to an observable of the system. According to the rules of quantum mechanics, the average result of a large number of measurements of this observable is instantaneously given by the expectation value

$$\frac{(\Psi, \mathcal{O}\Psi)}{(\Psi, \Psi)} = \frac{\sum_n \sum_m (c_n, c_m)(\Phi_n, \mathcal{O}\Phi_m)}{\sum_n (c_n, c_n)} \tag{8.2}$$

where (c_n, c_m), the scalar product of the nth and the mth wave function of the external world, is a function of time. The denominator of (8.2), being identical with (Ψ, Ψ), is independent of time, because the Hamiltonian of the system plus external world is Hermitian. When we actually measure an observable in the laboratory, we measure not its instantaneous value but a time average. Thus the directly measurable quantity is not (8.2) but the following quantity:

$$\langle \mathcal{O} \rangle \equiv \frac{\overline{(\Psi, \mathcal{O}\Psi)}}{(\Psi, \Psi)} = \frac{\sum_n \sum_m \overline{(c_n, c_m)}(\Phi_n, \mathcal{O}\Phi_m)}{\sum_n \overline{(c_n, c_n)}} \tag{8.3}$$

where $\overline{(c_n, c_m)}$ is the average of (c_n, c_m) over a time interval that is short compared to the resolving time of the measuring apparatus but long compared to molecular times (e.g., collision times or periods of molecular motion). We note that $\sum_n \overline{(c_n, c_n)}$ is identical with $\sum_n (c_n, c_n)$, because the latter is independent of time.

The postulates of quantum statistical mechanics are postulates concerning the coefficients $\overline{(c_n, c_m)}$, when (8.3) refers to a macroscopic observable of a macroscopic system in thermodynamic equilibrium.

For definiteness, we consider a macroscopic system which, although not truly isolated, interacts so weakly with the external world that its energy is approximately constant. Let the number of particles in the system be N and the volume of the system be V, and let its energy lie between E and $E + \Delta (\Delta \ll E)$. Let $\mathcal{H}$ be the Hamiltonian of the system. For such a system it is convenient (but not necessary) to choose a standard set of complete orthonormal wave functions $\{\Phi_n\}$ such that every Φ_n is a wave function for N particles contained in the volume V and is an eigenfunction of $\mathcal{H}$ with the eigenvalue E_n:

$$\mathcal{H}\Phi_n = E_n\Phi_n \tag{8.4}$$

The postulates of quantum statistical mechanics are the following:

Postulate of Equal a Priori Probability

$$\overline{(c_n, c_n)} = \begin{cases} 1 & (E < E_n < E + \Delta) \\ 0 & (\text{otherwise}) \end{cases} \tag{8.5}$$

Postulate of Random Phases

$$\overline{(c_n, c_m)} = 0 \qquad (n \neq m) \tag{8.6}$$

As a consequence of these postulates we may *effectively* regard the wave function of the system as given by

$$\Psi = \sum_n b_n \Phi_n \tag{8.7}$$

where

$$|b_n|^2 = \begin{cases} 1 & (E < E_n < E + \Delta) \\ 0 & (\text{otherwise}) \end{cases} \tag{8.8}$$

and where the phases of the complex numbers $\{b_n\}$ are random numbers. In this manner the effect of the external world is taken into account in an average way. The observed value of an observable associated with the operator $\mathcal{O}$ is then given by

$$\langle \mathcal{O} \rangle = \frac{\sum_n |b_n|^2 (\Phi_n, \mathcal{O}\Phi_n)}{\sum_n |b_n|^2} \tag{8.9}$$

It should be emphasized that for (8.7) and (8.8) to be valid the system must interact with the external world. Otherwise the postulate of random phases is false. By the randomness of the phases we mean no more and no less than the absence of interference of probability amplitudes, as expressed by (8.9). For a truly isolated system such a circumstance may be true at an instant, but it cannot be true for all times.

The postulate of random phases implies that the state of a system in equilibrium may be regarded as an *incoherent* superposition of eigenstates. It is possible to think of the system as one member of an infinite collection of systems, each of which is in an eigenstate whose wave function is Φ_n. Since these systems do not interfere with one another, it is possible to form a mental picture of each system *one at a time*. This mental picture is the quantum mechanical generalization of the Gibbsian ensemble. The ensemble defined by the previous postulates is the *microcanonical ensemble*.

The postulates of quantum statistical mechanics are to be regarded as working hypotheses whose justification lies in the fact that they lead to results in agreement with experiments. Such a point of view is not entirely satisfactory, because these postulates cannot be independent of, and should be derivable from, the quantum mechanics of molecular systems. A rigorous derivation is at present lacking. We return to this subject very briefly at the end of this chapter.

We should recognize that the postulates of quantum statistical mechanics, even if regarded as phenomenological statements, are more fundamental than the laws of thermodynamics. The reason is twofold. First, the postulates of quantum statistical mechanics not only imply the laws of thermodynamics, they also lead to definite formulas for all the thermodynamic functions of a given substance.

Second, they are more directly related to molecular dynamics than are the laws of thermodynamics.

The concept of an ensemble is a familiar one in quantum mechanics. A trivial example is the description of an incident beam of particles in the theory of scattering. The incident beam of particles in a scattering experiment is composed of many particles, but in the theory of scattering we consider the particles one at a time. That is, we calculate the scattering cross section for a single incident particle and then add the cross sections for all the particles to obtain the physical cross section. Inherent in this method is the assumption that the wave functions of the particles in the incident beam bear no definite phase with respect to one another. What we have described is in fact an ensemble of particles.

A less trivial example is the description of a beam of incident electrons whose spin can be polarized. If an electron has the wave function

$$\left[A \binom{1}{0} + B \binom{0}{1} \right] e^{i \mathbf{k} \cdot \mathbf{r}}$$

where A and B are definite complex numbers, the electron has a spin pointing in some definite direction. This corresponds to an incident beam of completely polarized electrons. In the cross section calculated with this wave function there will appear interference terms proportional to $A^*B + AB^*$. If we have an incident beam that is partially polarized, we first calculate the cross section with a wave function proportional to $\binom{1}{0}$ and then do the same thing for $\binom{0}{1}$, adding the two cross sections with appropriate weighting factors. This is equivalent to describing the incident beam by an ensemble of electrons in which the states $\binom{1}{0}$ and $\binom{0}{1}$ occur with certain relative probabilities.

8.2 DENSITY MATRIX

An ensemble is an incoherent superposition of states. Its relevance to physics has been postulated in the previous section. We note that only the square moduli $|b_n|^2$ appear in (8.9). Hence it should be possible to describe an ensemble in such a way that the random phases of the states never need to be mentioned. Such a goal is achieved by introducing the density matrix.

Before we define the density matrix let us note that an operator is defined when all its matrix elements with respect to a complete set of states are defined. Its matrix elements with respect to any other complete set of states can be found by the well-known rules of transformation theory in quantum mechanics. Therefore, if all the matrix elements of an operator are defined in one representation, the operator is thereby defined in any representation.

We define the density matrix ρ_{mn} corresponding to a given ensemble by

$$\rho_{mn} \equiv (\Phi_n, \rho\Phi_m) \equiv \delta_{mn}|b_n|^2 \tag{8.10}$$

where Φ_n and b_n have the same meaning as in (8.7). In this particular representation ρ_{mn} is a diagonal matrix, but in some other representation it need not be. Equation (8.10) also defines the density operator ρ whose matrix elements are ρ_{mn}. The operator ρ operates on state vectors in the Hilbert space of the system under consideration.

In terms of the density matrix, (8.9) can be rewritten in the form

$$\langle \mathcal{O} \rangle = \frac{\sum\limits_n (\Phi_n, \mathcal{O}\rho\Phi_n)}{\sum\limits_n (\Phi_n, \rho\Phi_n)} = \frac{\mathrm{Tr}\,(\mathcal{O}\rho)}{\mathrm{Tr}\,\rho} \tag{8.11}$$

The notation $\mathrm{Tr}\,A$ denotes the trace of the operator A and is the sum of all the diagonal matrix elements of A in *any* representation. An elementary property of the trace is that

$$\mathrm{Tr}\,(AB) = \mathrm{Tr}\,(BA)$$

It follows immediately that $\mathrm{Tr}\,A$ is independent of the representation; if $\mathrm{Tr}\,A$ is calculated in one representation, its value in another representation is

$$\mathrm{Tr}\,(SAS^{-1}) = \mathrm{Tr}\,(S^{-1}SA) = \mathrm{Tr}\,A$$

The introduction of the density matrix merely introduces a notation. It does not introduce new physical content. The usefulness of the density matrix lies solely in the fact that with its help (8.11) is presented in a form that is manifestly independent of the choice of the basis $\{\Phi_n\}$, although this independence is a property that this expectation value always possesses.

The density operator ρ defined by (8.10) contains all the information about an ensemble. It is independent of time if it commutes with the Hamiltonian of the system and if the Hamiltonian is independent of time. This statement is an immediate consequence of the equation of motion of ρ:

$$i\hbar \frac{\partial \rho}{\partial t} = [\mathcal{H}, \rho] \tag{8.12}$$

which is the quantum mechanical version of Liouville's theorem.

Formally we can represent the density operator ρ as

$$\rho = \sum_n |\Phi_n\rangle |b_n|^2 \langle \Phi_n| \tag{8.13}$$

where $|\Phi_n\rangle$ is the state vector whose wave function is Φ_n. To prove (8.13), we verify that it has the matrix elements (8.10):

$$\rho_{mn} \equiv (\Phi_m, \rho\Phi_n) \equiv \langle \Phi_m|\rho|\Phi_n\rangle = \sum_k \langle \Phi_m|\Phi_k\rangle |b_k|^2 \langle \Phi_k|\Phi_n\rangle = \delta_{mn}|b_n|^2 \quad \blacksquare$$

The time-averaging process through which we averaged out the effect of the external world on the system under consideration may be reformulated in terms of the density matrix.

Formula (8.2) is a general formula for the expectation value of any operator $\mathcal{O}$ with respect to an arbitrary wave function Ψ. It may be trivially rewritten in the form

$$\frac{(\Psi, \mathcal{O}, \Psi)}{(\Psi, \Psi)} = \frac{\sum\limits_{n}\sum\limits_{m} R_{mn}\mathcal{O}_{nm}}{\sum\limits_{n} R_{nn}} = \frac{\mathrm{Tr}\,(R\mathcal{O})}{\mathrm{Tr}\,R}$$

where $R_{nm} \equiv (c_m, c_n) \equiv (\Phi_n, R\Phi_m)$, the last identity being a definition of the operator R, and $\mathcal{O}_{nm} \equiv (\Phi_n, \mathcal{O}\Phi_m)$. Although R may depend on the time, $\mathrm{Tr}\,R$ is independent of time. The density operator is the time average of R:

$$\rho \equiv \overline{R}$$

8.3 ENSEMBLES IN QUANTUM STATISTICAL MECHANICS

Microcanonical Ensemble

The density matrix for the microcanonical ensemble in the representation in which the Hamiltonian is diagonal is

$$\rho_{mn} = \delta_{mn}|b_n|^2 \tag{8.14}$$

where

$$|b_n|^2 = \begin{cases} \text{Const.} & (E < E_n < E + \Delta) \\ 0 & (\text{otherwise}) \end{cases} \tag{8.15}$$

where $\{E_n\}$ are the eigenvalues of the Hamiltonian. The density operator is

$$\rho = \sum_{E < E_n < E+\Delta} |\Phi_n\rangle\langle\Phi_n| \tag{8.16}$$

The trace of ρ is equal to the number of states whose energy lies between E and $E + \Delta$:

$$\mathrm{Tr}\,\rho = \sum_n \rho_{nn} \equiv \Gamma(E) \tag{8.17}$$

For macroscopic systems the spectrum $\{E_n\}$ almost forms a continuum. For $\Delta \ll E$, we may take

$$\Gamma(E) = \omega(E)\Delta \tag{8.18}$$

where $\omega(E)$ is the density of states at energy E. The connection between the microcanonical ensemble and thermodynamics is established by identifying the entropy as

$$S(E, V) = k \log \Gamma(E) \tag{8.19}$$

where k is Boltzmann's constant. This definition is the same as in classical statistical mechanics, except that $\Gamma(E)$ must be calculated in quantum mechanics. From this point on all further developments become exactly the same as in classical statistical mechanics and so they need not be repeated. No Gibbs paradox will result from (8.19) because the correct counting of states is automatically implied by the definition of $\Gamma(E)$ in (8.17).

The only new result following from (8.19) that is not obtainable in classical statistical mechanics is the third law of thermodynamics, which we discuss separately in Section 8.4.

Canonical Ensemble

The derivation of the canonical ensemble from the microcanonical ensemble given in Chapter 8 did not make essential use of classical mechanics. That derivation continues to be valid in quantum statistical mechanics, with the trivial change that the integration over Γ space is replaced by a sum over all the states of the system:

$$\frac{1}{N!h^{3N}} \int dp \, dq \rightarrow \sum_n \tag{8.20}$$

Thus the canonical ensemble is defined by the density matrix

$$\rho_{mn} = \delta_{mn} e^{-\beta E_n} \tag{8.21}$$

where $\beta = 1/kT$. This result states that at the temperature T the relative probability for the system to have the energy eigenvalue E_n is $e^{-\beta E_n}$, which is called the *Boltzmann factor*. The partition function is given by

$$Q_N(V, T) = \text{Tr}\,\rho = \sum_n e^{-\beta E_n} \tag{8.22}$$

where it must be emphasized that *the sum on the right side is a sum over states and not over energy eigenvalues*. The connection with thermodynamics is the same as in classical statistical mechanics.

The density operator ρ is

$$\rho = \sum_n |\Phi_n\rangle e^{-\beta E_n} \langle \Phi_n| = e^{-\beta \mathcal{H}} \sum_n |\Phi_n\rangle\langle\Phi_n|$$

where $\mathcal{H}$ is the Hamiltonian operator. Now the operator $\sum_n |\Phi_n\rangle\langle\Phi_n|$ is the identity operator, by the completeness property of eigenstates. Therefore

$$\rho = e^{-\beta \mathcal{H}} \tag{8.23}$$

The partition function can be written in the form

$$Q_N(V, T) = \text{Tr}\, e^{-\beta \mathcal{H}} \tag{8.24}$$

where the trace is to be taken over all states of the system that has N particles in the volume V. This form, which is explicitly independent of the representation, is sometimes convenient for calculations. The ensemble average of $\mathcal{O}$ in the canonical ensemble is

$$\langle \mathcal{O} \rangle = \frac{\text{Tr}\,(\mathcal{O}\, e^{-\beta \mathcal{H}})}{Q_N} \tag{8.25}$$

Grand Canonical Ensemble

For the grand canonical ensemble the density operator ρ operates on a Hilbert space with an indefinite number of particles. We do not display it because we do not need it. It is sufficient to state that the grand partition function is

$$\mathscr{2}(z, V, T) = \sum_{N=0}^{\infty} z^N Q_N(V, T) \tag{8.26}$$

where Q_N is the partition function for N particles. The connection between $\log \mathscr{2}$ and thermodynamics is the same as in classical statistical mechanics. The ensemble average of $\mathscr{O}$ in the grand canonical ensemble is

$$\langle \mathscr{O} \rangle = \frac{1}{\mathscr{2}} \sum_{N=0}^{\infty} z^N \langle \mathscr{O} \rangle_N \tag{8.27}$$

where $\langle \mathscr{O} \rangle_N$ is the ensemble average (8.25) in the canonical ensemble for N particles. These equations can be written more generally in the forms

$$\mathscr{2}(z, V, T) = \mathrm{Tr}\, e^{-\beta(\mathscr{H} - \mu N)}$$

$$\langle \mathscr{O} \rangle = \frac{1}{\mathscr{2}} \mathrm{Tr} \left[\mathscr{O}\, e^{-\beta(\mathscr{H} - \mu N)} \right] \tag{8.28}$$

where N is an operator representing a conserved quantity (i.e., one that commutes with the Hamiltonian), and the trace is taken over all states without restriction on the eigenvalues of N. The only restrictions on the trace are boundary conditions, which specify the volume containing the system, and the symmetry property of the states under the interchange of identical particles.

8.4 THIRD LAW OF THERMODYNAMICS

The definition of entropy is given by (8.19). At the absolute zero of temperature a system is in its ground state, i.e., a state of lowest energy. For a system whose energy eigenvalues are discrete, (8.19) implies that at absolute zero $S = k \log G$, where G is the degeneracy of the ground state. If the ground state is unique, then $S = 0$ at absolute zero. If the ground state is not unique, but $G \leq N$, where N is the total number of molecules in the system, then at absolute zero $S \leq k \log N$. In both of these cases the third law of thermodynamics holds, because the entropy per molecule at absolute zero is of order $(\log N)/N$.

The energy eigenvalues for most macroscopic systems, however, essentially form a continuous spectrum. For these systems the previous argument only shows that the entropy per molecule approaches zero when the temperature T is so low that

$$kT \ll \Delta E$$

where ΔE is the energy difference between the first excited state and the ground state. As an estimate let us put

$$\Delta E \approx \frac{\hbar^2}{mV^{2/3}}$$

where m is the mass of a nucleon, $V = 1$ cm^3. Then we find that $T \approx 5 \times 10^{-15}$ K. Clearly this phenomenon has nothing to do with the third law of thermodynamics, which is a phenomenological statement based on experiments performed above 1 K.

To verify the third law of thermodynamics for systems having an almost continuous energy spectrum we must study the behavior of the density of states $\omega(E)$ near $E = 0$. Most of the substances known to us become crystalline solids near absolute zero. For these substances all thermodynamic functions near absolute zero may be obtained through Debye's theory, which is discussed in Section 12.2. It is shown there that the third law of thermodynamics is fulfilled.

The only known substance that remains a liquid at absolute zero is helium, which is discussed in Chapter 13. There it is shown that near absolute zero the density of states for liquid helium is qualitatively the same as that for a crystalline solid. Therefore the third law of thermodynamics is also fulfilled for liquid helium.

Apart from these specific examples, which include all known substances, we cannot give a more universal proof of the third law of thermodynamics. But this is perhaps sufficient; after all, the third law of thermodynamics is a summary of empirical data gathered from known substances.

8.5 THE IDEAL GASES: MICROCANONICAL ENSEMBLE

The simplest system of N identical particles is that composed of N noninteracting members. The Hamiltonian is

$$\mathcal{H} = \sum_{i=1}^{N} \frac{p_i^2}{2m} \tag{8.29}$$

where $p_i^2 = \mathbf{p}_i \cdot \mathbf{p}_i$, and $\mathbf{p}_i$ is the momentum operator of the ith particle. The Hamiltonian is independent of the positions of the particles or any other coordinates, e.g., spin, if any.

In nature a system of N identical particles is one of two types: A *Bose system* or a *Fermi system*.* A complete set of eigenfunctions for a Bose system is the collection of those eigenfunctions of $\mathcal{H}$ that are symmetric under an interchange of any pair of particle coordinates. A complete set of eigenfunctions for a Fermi system is the collection of those eigenfunctions of $\mathcal{H}$ that are antisymmetric under an interchange of any pair of particle coordinates. Particles forming a Bose system are called *bosons*, and particles forming a Fermi system are called *fermions*.

*See the Appendix, Section A.1.

In addition to these two types of systems we define, for mathematical comparison, what is called a Boltzmann system. It is defined as a system of particles whose eigenfunctions are *all* the eigenfunctions of $\mathcal{H}$; but the rule for counting these eigenfunctions shall be the "correct Boltzmann counting." The set of eigenfunctions for a Boltzmann system includes those for a Bose system, those for a Fermi system, and more. There is no known system of this type in nature. It is a useful model, however, because at high temperatures the thermodynamic behavior of both the Bose system and the Fermi system approaches that of the Boltzmann system.

For noninteracting identical particles we have three cases: The ideal Bose gas, the ideal Fermi gas, and the ideal Boltzmann gas. We first work out the thermodynamics of these ideal gases in the formalism of the microcanonical ensemble. For this purpose it is necessary to find out, for each of the three cases, the number of states $\Gamma(E)$ of the system having an energy eigenvalue that lies between E and $E + \Delta$. That is, we must learn how to count.

To avoid all unnecessary complications we confine our discussion to spinless particles. Any energy eigenvalue of an ideal system is a sum of single-particle energies, called *levels*. These are given by

$$\epsilon_{\mathbf{p}} = \frac{p^2}{2m} \tag{8.30}$$

where $p \equiv |\mathbf{p}|$ and $\mathbf{p}$ is the momentum eigenvalue of the single particle:

$$\mathbf{p} = \frac{2\pi\hbar}{L}\mathbf{n} \tag{8.31}$$

in which $\mathbf{n}$ is a vector whose components are 0 or $\pm$ integers and L is the cube root of the volume of the system:

$$L \equiv V^{1/3}$$

In the limit as $V \to \infty$ the possible values of $\mathbf{p}$ form a continuum. Then a sum over $\mathbf{p}$ can sometimes be replaced by an integration

$$\sum_{\mathbf{p}} \to \frac{V}{h^3}\int d^3p \tag{8.32}$$

where $h = 2\pi\hbar$ is Planck's constant.*

A state of an ideal system can be specified by specifying a set of occupation numbers $\{n_{\mathbf{p}}\}$ so defined that there are $n_{\mathbf{p}}$ particles having the momentum $\mathbf{p}$ in the state under consideration. Obviously the total energy E and the total number of particles N of the state are given by

$$E = \sum_{\mathbf{p}}\epsilon_{\mathbf{p}}n_{\mathbf{p}}$$

$$N = \sum_{\mathbf{p}}n_{\mathbf{p}} \tag{8.33}$$

---

*For an explanation of (8.31) and (8.32), see the Appendix, Section A.2.

For spinless bosons and fermions $\{n_{\mathbf{p}}\}$ uniquely defines a state of the system. The allowed values for any $n_{\mathbf{p}}$ are

$$n_{p} = \begin{cases} 0,1,2,\ldots & \text{(for bosons)} \\ 0,1 & \text{(for fermions)} \end{cases} \tag{8.34}$$

For a Boltzmann gas $n_{\mathbf{p}} = 0,1,2,\ldots$, but $\{n_{\mathbf{p}}\}$ specifies $N!/\prod_{\mathbf{p}}(n_{\mathbf{p}}!)$ states of the N-particle system. This is because an interchange of the momenta of two particles in the system in general leads to a new state but leaves $\{n_{\mathbf{p}}\}$ unchanged.

The total energy is a given number E to within a small uncertainty Δ, whose value is unimportant. Hence $\Gamma(E)$ may be found as follows. As $V \to \infty$, the levels (8.30) form a continuum. Let us divide the spectrum of (8.30) into groups of levels containing respectively $g_1, g_2,\ldots$ levels, as shown in Fig. 8.1. Each group is called a cell and has an average energy ϵ_i. The occupation number of the ith cell, denoted by n_i, is the sum of $n_{\mathbf{p}}$ over all the levels in the ith cell. Each g_i is assumed to be very large, but its exact value is unimportant. Let

$$W\{n_i\} \equiv \begin{array}{l} \text{no. of states of the system corresponding to the set of} \\ \text{occupation numbers } \{n_i\} \end{array} \tag{8.35}$$

Then

$$\Gamma(E) = \sum_{\{n_i\}} W\{n_i\} \tag{8.36}$$

where the sum extends over all sets of integers $\{n_i\}$ satisfying the conditions

$$E = \sum_i \epsilon_i n_i \tag{8.37}$$

$$N = \sum_i n_i \tag{8.38}$$

To find $W\{n_i\}$ for a Bose gas and a Fermi gas it is sufficient to find w_i, the number of ways in which n_i particles can be assigned to the ith cell (which

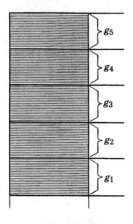

Fig. 8.1 Division of the single-particle energy spectrum into cells.

contains g_i levels). Since interchanging particles in different cells does not lead to a new state of the system, we have $W\{n_i\} = \prod_j w_j$. For a Boltzmann gas interchanging particles in different cells leads to a new state of the system, and we consider all N particles together. The three cases are worked out as follows.

Bose Gas. Each level can be occupied by any number of particles. Picture the ith cell to have g_i subcells, with $g_i - 1$ partitions, as follows:

$$\cdot \cdot | \cdot | \cdot \cdot | \cdot | \quad | \cdot \cdot | \qquad | \quad \cdot$$

$$\text{subcell 1} \quad 2 \quad 3 \qquad\qquad g_i - 1 \quad g_i$$

The number w_i is the number of permutations of the n_i particles plus the $g_i - 1$ partitions that give rise to distinct arrangements:

$$w_i = \frac{(n_i + g_i - 1)!}{n_i!(g_i - 1)!}$$

Hence

$$W\{n_i\} = \prod_i w_i = \prod_i \frac{(n_i + g_i - 1)!}{n_i!(g_i - 1)!} \qquad \text{(Bose)} \qquad (8.39)$$

Fermi Gas. The number of particles in each of the g_i subcells of the ith cell is either 0 or 1. Therefore w_i is equal to the number of ways in which n_i things can be chosen from g_i things:

$$w_i = \binom{g_i}{n_i} = \frac{g_i!}{n_i!(g_i - n_i)!} \qquad (8.40)$$

Hence

$$W\{n_i\} = \prod_i w_i = \prod_i \frac{g_i!}{n_i!(g_i - n_i)!} \qquad \text{(Fermi)} \qquad (8.41)$$

Boltzmann Gas. The N particles are first placed into cells, the ith cell having n_i particles. There are $N!/\prod_i(n_i!)$ ways to do this. Within the ith cell there are g_i levels. Among the n_i particles in the ith cells, the first one can occupy these levels g_i ways. The second and all subsequent ones also can occupy the levels g_i ways. Therefore there are $(g_i)^{n_i}$ ways in which n_i particles can occupy the g_i levels. The total number of ways to obtain $\{n_i\}$ is therefore

$$N!\prod_i \frac{g_i^{n_i}}{n_i!}$$

However, $W\{n_i\}$ is defined to be $1/N!$ of the last quantity:

$$W\{n_i\} = \prod_i \frac{g_i^{n_i}}{n_i!} \qquad \text{(Boltzmann)} \qquad (8.42)$$

This definition corresponds to the rule of "correct Boltzmann counting" and does not correspond to any physical property of the particles in the system. It is just a rule that defines the mathematical model.

The fact that the rule for the counting of states is different for the three cases gives rise to the terminology *Bose statistics*, *Fermi statistics*, and *Boltzmann statistics*, which refer to the three rules of counting, respectively.

To obtain the entropy $S = k \log \Gamma(E)$ we need to sum $W\{n_i\}$ over $\{n_i\}$ in accordance with (8.35). This is a formidable task. For the Boltzmann gas it was explicitly done in Section 6.5. As we might correctly guess, however, $\Gamma(E)$ is quite well approximated by $W\{\bar{n}_i\}$, where $\{\bar{n}_i\}$ is the set of occupation numbers that maximizes $W\{\bar{n}_i\}$ subject to (8.37) and (8.38). We adopt this approximation and verify its correctness by showing that the fluctuations are small. Accordingly the entropy is taken to be

$$S = k \log W\{\bar{n}_i\} \tag{8.43}$$

To find $\{\bar{n}_i\}$ we maximize $W\{n_i\}$ by varying the n_i subject to (8.37) and (8.38). The details of this calculation are similar to that in Section 4.3 and will not be reproduced. We merely give the answers:

$$\bar{n}_i = \begin{cases} \dfrac{g_i}{z^{-1} e^{\beta \epsilon_i} \mp 1} & \text{(Bose and Fermi)} \\[2mm] g_i z e^{-\beta \epsilon_i} & \text{(Boltzmann)} \end{cases} \tag{8.44}$$

We deduce from this that

$$\bar{n}_{\mathbf{p}} = \begin{cases} \dfrac{1}{z^{-1} e^{\beta \epsilon_{\mathbf{p}}} \mp 1} & \text{(Bose and Fermi)} \\[2mm] z e^{-\beta \epsilon_{\mathbf{p}}} & \text{(Boltzmann)} \end{cases} \tag{8.45}$$

The parameters z and β are two Lagrange multipliers to be determined from the conditions

$$\sum_{\mathbf{p}} \epsilon_{\mathbf{p}} \bar{n}_{\mathbf{p}} = E$$
$$\sum_{\mathbf{p}} \bar{n}_{\mathbf{p}} = N \tag{8.46}$$

The first of these leads to the identification $\beta = 1/kT$, and the second identifies z as the fugacity.

Using Stirling's approximation and neglecting 1 compared to g_i we have from (8.43) and (8.44),

$$\frac{S}{k} = \log W\{\bar{n}_i\} = \begin{cases} \displaystyle\sum_i \left[\bar{n}_i \log\left(1 + \frac{g_i}{\bar{n}_i}\right) + g_i \log\left(1 + \frac{\bar{n}_i}{g_i}\right) \right] & \text{(Bose)} \\[4mm] \displaystyle\sum_i \left[\bar{n}_i \log\left(\frac{g_i}{\bar{n}_i} - 1\right) - g_i \log\left(1 - \frac{\bar{n}_i}{g_i}\right) \right] & \text{(Fermi)} \\[4mm] \displaystyle\sum_i \bar{n}_i \log(g_i/\bar{n}_i) & \text{(Boltzmann)} \end{cases}$$

$$\tag{8.47}$$

More explicitly,

$$
\frac{S}{k} = \begin{cases}
\sum_i g_i \left[\dfrac{\beta \epsilon_i - \log z}{z^{-1} e^{\beta \epsilon_i} - 1} - \log \left(1 - z e^{-\beta \epsilon_i} \right) \right] & \text{(Bose)} \\[2ex]
\sum_i g_i \left[\dfrac{\beta \epsilon_i - \log z}{z^{-1} e^{\beta \epsilon_i} + 1} + \log \left(1 + z e^{-\beta \epsilon_i} \right) \right] & \text{(Fermi)} \\[2ex]
z \sum_i g_i e^{-\beta \epsilon_i} \left(\beta \epsilon_i - \log z \right) & \text{(Boltzmann)}
\end{cases}
\qquad (8.48)
$$

The validity of these equations depends on the assumption that

$$
\overline{n_i^2} - \bar{n}_i^2 \ll \bar{n}_i^2 \qquad (8.49)
$$

This is best discussed in the grand canonical ensemble (see Problem 8.4). From (8.48) all other thermodynamic functions can be determined after z is determined in terms of N from (8.46).

The Boltzmann gas will be worked out explicitly. From (8.38) and (8.44) we have

$$
N = z \sum_i g_i e^{-\beta \epsilon_i} = z \sum_{\mathbf{p}} e^{-\beta \epsilon_{\mathbf{p}}} = \frac{zV}{h^3} \int_0^\infty dp \, 4\pi p^2 \, e^{-\beta p^2 / 2m} = \frac{zV}{\lambda^3} \qquad (8.50)
$$

where

$$
\lambda = \sqrt{\frac{2\pi \hbar^2}{mkT}} \qquad (8.51)
$$

This quantity is called the *thermal wavelength* because it is of the order of the de Broglie wavelength of a particle of mass m with the energy kT. Writing $v = V/N$ we obtain

$$
z = \frac{\lambda^3}{v} \qquad (8.52)
$$

The condition $E = \sum n_i \epsilon_i$ requires that

$$
E = z \sum_i g_i \epsilon_i \, e^{-\beta \epsilon_i} = z \sum_{\mathbf{p}} \epsilon_{\mathbf{p}} e^{-\beta \epsilon_{\mathbf{p}}} = \frac{zV}{h^3} \int_0^\infty dp \, 4\pi p^2 \left(\frac{p^2}{2m} \right) e^{-\beta p^2 / 2m} = \tfrac{3}{2} NkT \qquad (8.53)
$$

Therefore T is the absolute temperature. The entropy is, by (8.48) and (8.46),

$$
\frac{S}{k} = z \sum_{\mathbf{p}} e^{-\beta \epsilon_{\mathbf{p}}} \left(\beta \epsilon_{\mathbf{p}} - \log z \right) = \beta E - N \log z
$$

$$
= \tfrac{3}{2} N - N \log \left[\frac{N}{V} \left(\frac{2\pi \hbar^2}{mkT} \right)^{3/2} \right] \qquad (8.54)
$$

This is the Sackur-Tetrode equation. The fact that the constant $h = 2\pi \hbar$ is

Planck's constant follows from (8.31), where $\hbar$ first makes its appearance. The equation of state is deduced from the function $U(S, V)$, which is E expressed in terms of S and V. We straightforwardly find $PV = NkT$. It is to be noted that (8.54) does not satisfy the third law of thermodynamics. This should cause no concern, because a Boltzmann gas is not a physical system. It is only a model toward which the Bose and Fermi gas converge at high temperatures. This shows, however, that the third law of thermodynamics is not an automatic consequence of the general principles of quantum mechanics, but depends on the nature of the density of states near the ground state.

The Bose and Fermi gases can be worked out along similar lines. They are more conveniently discussed, however, in the grand canonical ensemble, which we consider in the next section.

8.6 THE IDEAL GASES: GRAND CANONICAL ENSEMBLE

The partition functions for the ideal gases are

$$Q_N(V, T) = \sum_{\{n_\mathbf{p}\}} g\{n_\mathbf{p}\} \, e^{-\beta E\{n_\mathbf{p}\}} \tag{8.55}$$

where

$$E\{n_\mathbf{p}\} = \sum_\mathbf{p} \epsilon_\mathbf{p} n_\mathbf{p} \tag{8.56}$$

and the occupation numbers are subject to the condition

$$\sum_\mathbf{p} n_\mathbf{p} = N \tag{8.57}$$

For a Bose gas and a Boltzmann gas $n_\mathbf{p} = 0, 1, 2, \ldots$. For a Fermi gas $n_\mathbf{p} = 0, 1$. The number of states corresponding to $\{n_\mathbf{p}\}$ is

$$g\{n_\mathbf{p}\} = \begin{cases} 1 & \text{(Bose and Fermi)} \\ \dfrac{1}{N!}\left(\dfrac{N!}{\prod_\mathbf{p} n_\mathbf{p}!}\right) & \text{(Boltzmann)} \end{cases} \tag{8.58}$$

We first work out the Boltzmann gas:

$$Q_N = \sum_{\substack{n_0, n_1, \ldots \\ \Sigma n_i = N}} \left(\frac{e^{-\beta n_0 \epsilon_0}}{n_0!} \frac{e^{-\beta n_1 \epsilon_1}}{n_1!} \cdots \right) = \frac{1}{N!} \left(e^{-\beta \epsilon_0} + e^{-\beta \epsilon_1} + \cdots \right)^N$$

This equality is the multinomial theorem. In the limit as $V \to \infty$ we can write

$$\sum_\mathbf{p} e^{-\beta \epsilon_\mathbf{p}} = \frac{V}{h^3} \int_0^\infty dp \, 4\pi p^2 \, e^{-\beta p^2/2m} = V\left(\frac{mkT}{2\pi\hbar^2}\right)^{3/2} \tag{8.59}$$

Therefore

$$\frac{1}{N} \log Q_N = \log \left[\frac{V}{N} \left(\frac{mkT}{2\pi\hbar^2} \right)^{3/2} \right] \tag{8.60}$$

from which easily follows the Sackur-Tetrode equation for the entropy and the equation of state $PV = NkT$. The grand partition function is trivial and will not be considered.

For the Bose gas and the Fermi gas the partition function cannot be evaluated easily because of the condition (8.57). Instead of the partition function we consider the grand partition function

$$\mathcal{Q}(z, V, T) = \sum_{N=0}^{\infty} z^N Q_N(V, T) = \sum_{N=0}^{\infty} \sum_{\substack{\{n_p\} \\ \Sigma n_p = N}} z^N e^{-\beta \Sigma \epsilon_p n_p}$$

$$= \sum_{N+0}^{\infty} \sum_{\substack{\{n_p\} \\ \Sigma n_p = N}} \prod_p (z e^{-\beta \epsilon_p})^{n_p} \tag{8.61}$$

Now it is to be noted that the double summation just given is equivalent to summing each n_p independently. To prove this we must show that every term in one case appears once and only once in the other, and vice versa. This is easily done mentally. Therefore

$$\mathcal{Q}(z, V, T) = \sum_{n_0} \sum_{n_1} \cdots \left[(z e^{-\beta \epsilon_0})^{n_0} (z e^{-\beta \epsilon_1})^{n_1} \cdots \right]$$

$$= \left[\sum_{n_0} (z e^{-\beta \epsilon_0})^{n_0} \right] \left[\sum_{n_1} (z e^{-\beta \epsilon_1})^{n_1} \right] \cdots$$

$$= \prod_p \left[\sum_n (z e^{-\beta \epsilon_p})^n \right]$$

where the sum $\sum_n$ extends over the values $n = 0, 1, 2, \ldots$ for the Bose gas and the values $n = 0, 1$ for the Fermi gas. The results are

$$\mathcal{Q}(z, V, T) = \begin{cases} \displaystyle\prod_p \frac{1}{1 - z e^{-\beta \epsilon_p}} & \text{(Bose)} \\[2ex] \displaystyle\prod_p (1 + z e^{-\beta \epsilon_p}) & \text{(Fermi)} \end{cases} \tag{8.62}$$

The equations of state are

$$\frac{PV}{kT} = \log \mathcal{Q}(z, V, T) = \begin{cases} \displaystyle -\sum_p \log (1 - z e^{-\beta \epsilon_p}) & \text{(Bose)} \\[2ex] \displaystyle \sum_p \log (1 + z e^{-\beta \epsilon_p}) & \text{(Fermi)} \end{cases} \tag{8.63}$$

from which z is to be eliminated with the help of the equations

$$N = z \frac{\partial}{\partial z} \log \mathscr{Q}(z, V, T) = \begin{cases} \sum_{\mathbf{p}} \dfrac{z e^{-\beta \epsilon_{\mathbf{p}}}}{1 - z e^{-\beta \epsilon_{\mathbf{p}}}} & \text{(Bose)} \\[4mm] \sum_{\mathbf{p}} \dfrac{z e^{-\beta \epsilon_{\mathbf{p}}}}{1 + z e^{-\beta \epsilon_{\mathbf{p}}}} & \text{(Fermi)} \end{cases}$$

(8.64)

The average occupation numbers $\langle n_{\mathbf{p}} \rangle$ are given by

$$\langle n_{\mathbf{p}} \rangle \equiv \frac{1}{\mathscr{Q}} \sum_{N=0}^{\infty} z^N \sum_{\substack{\{n_{\mathbf{p}}\} \\ \Sigma n_{\mathbf{p}} = N}} n_{\mathbf{p}} e^{-\beta \Sigma \epsilon_{\mathbf{p}} n_{\mathbf{p}}} = -\frac{1}{\beta} \frac{\partial}{\partial \epsilon_{\mathbf{p}}} \log \mathscr{Q}$$

$$= \frac{z e^{-\beta \epsilon_{\mathbf{p}}}}{1 \mp z e^{-\beta \epsilon_{\mathbf{p}}}} \qquad \text{(Bose and Fermi)}$$

(8.65)

which are the same as (8.45). The equations (8.64) are non other than the statement

$$N = \sum_{\mathbf{p}} \langle n_{\mathbf{p}} \rangle$$

(8.66)

The results here are completely equivalent to those in the microcanonical ensemble, as they should be.

Now we let $V \to \infty$, and replace sums over $\mathbf{p}$ by integrals over $\mathbf{p}$ in the manner indicated in (8.32), whenever possible. Such a replacement is clearly valid if the summand in question is finite for all $\mathbf{p}$. In (8.63) and (8.64), the fugacity z is nonnegative for both the ideal Fermi gas and the ideal Bose gas because, if z were negative, then (8.64) cannot be satisfied for positive N. We see immediately that for the ideal Fermi gas it is permissible to replace the sums in (8.63) and (8.64) by integrals over $\mathbf{p}$. We then obtain the following equation of state.

Ideal Fermi Gas

$$\begin{cases} \dfrac{P}{kT} = \dfrac{4\pi}{h^3} \displaystyle\int_0^\infty dp\, p^2 \log\left(1 + z e^{-\beta p^2/2m}\right) \\[4mm] \dfrac{1}{v} = \dfrac{4\pi}{h^3} \displaystyle\int_0^\infty dp\, p^2 \dfrac{1}{z^{-1} e^{\beta p^2/2m} + 1} \end{cases}$$

(8.67)

where $v = V/N$. It can be verified in a straightforward fashion that (8.67) can also be written in the form

$$\begin{cases} \dfrac{P}{kT} = \dfrac{1}{\lambda^3} f_{5/2}(z) \\[4mm] \dfrac{1}{v} = \dfrac{1}{\lambda^3} f_{3/2}(z) \end{cases}$$

(8.68)

where $\lambda = \sqrt{2\pi\hbar^2/mkT}$ and

$$f_{5/2}(z) \equiv \frac{4}{\sqrt{\pi}} \int_0^\infty dx\, x^2 \log\left(1 + z\, e^{-x^2}\right) = \sum_{l=1}^\infty \frac{(-1)^{l+1} z^l}{l^{5/2}} \qquad (8.69)$$

$$f_{3/2}(z) \equiv z\frac{\partial}{\partial z} f_{5/2}(z) = \sum_{l=1}^\infty \frac{(-1)^{l+1} z^l}{l^{3/2}} \qquad (8.70)$$

For the ideal Bose gas the summands appearing in (8.63) and (8.64) diverge as $z \to 1$, because the single term corresponding to $\mathbf{p} = 0$ diverges. Thus the single term $\mathbf{p} = 0$ may be as important as the entire sum.* We split off the terms in (8.63) and (8.64) corresponding to $\mathbf{p} = 0$ and replace the rest of the sums by integrals. We then obtain the following equation of state.

Ideal Bose Gas

$$\begin{cases} \dfrac{P}{kT} = -\dfrac{4\pi}{h^3} \int_0^\infty dp\, p^2 \log\left(1 - z\, e^{-\beta p^2/2m}\right) - \dfrac{1}{V} \log\left(1 - z\right) \\[2mm] \dfrac{1}{v} = \dfrac{4\pi}{h^3} \int_0^\infty dp\, p^2 \dfrac{1}{z^{-1} e^{\beta p^2/2m} - 1} + \dfrac{1}{V}\dfrac{z}{1 - z} \end{cases} \qquad (8.71)$$

where $v = V/N$. It can be verified in a straightforward fashion that (8.71) can also be written in the form

$$\begin{cases} \dfrac{P}{kT} = \dfrac{1}{\lambda^3} g_{5/2}(z) - \dfrac{1}{V} \log\left(1 - z\right) \\[2mm] \dfrac{1}{v} = \dfrac{1}{\lambda^3} g_{3/2}(z) + \dfrac{1}{V}\dfrac{z}{1 - z} \end{cases} \qquad (8.72)$$

where $\lambda = \sqrt{2\pi\hbar^2/mkT}$, and

$$g_{5/2}(z) \equiv -\frac{4}{\sqrt{\pi}} \int_0^\infty dx\, x^2 \log\left(1 - z\, e^{-x^2}\right) = \sum_{l=1}^\infty \frac{z^l}{l^{5/2}} \qquad (8.73)$$

$$g_{3/2}(z) \equiv z\frac{\partial}{\partial z} g_{5/2}(z) = \sum_{l=1}^\infty \frac{z^l}{l^{3/2}} \qquad (8.74)$$

As (8.65) implies, the quantity $z/(1 - z)$ is the average occupation number $\langle n_0 \rangle$ for the single-particle level with $\mathbf{p} = 0$:

$$\frac{z}{1 - z} = \langle n_0 \rangle \qquad (8.75)$$

This term contributes significantly to (8.72) if $\langle n_0 \rangle/V$ is a finite number, i.e., if a

---

*That this is in fact the case is shown in Section 12.3 in connection with the Bose-Einstein condensation.

finite fraction of all the particles in the system occupy the single level with $\mathbf{p} = 0$. We shall see in Section 11.3 that such a circumstance gives rise to the phenomenon of Bose-Einstein condensation.

The internal energy for both the Fermi and the Bose gases may be found from the formula

$$U(z, V, T) = \frac{1}{\mathscr{Q}} \sum_{N=0}^{\infty} z^N \sum_{\substack{\{n_\mathbf{p}\} \\ \Sigma n_\mathbf{p} = N}} \left[e^{-\beta \Sigma \epsilon_\mathbf{p} n_\mathbf{p}} \sum_\mathbf{p} \epsilon_\mathbf{p} n_\mathbf{p} \right] = -\frac{\partial}{\partial \beta} [\log \mathscr{Q}(z, V, T)]$$

(8.76)

Since $\log \mathscr{Q} = PV/kT$, we obtain from (8.68) and (8.72) the results

$$\frac{1}{V} U(z, V, T) = \begin{cases} \dfrac{3}{2} \dfrac{kT}{\lambda^3} f_{5/2}(z) & \text{(Fermi)} \\ \dfrac{3}{2} \dfrac{kT}{\lambda^3} g_{5/2}(z) & \text{(Bose)} \end{cases}$$

(8.77)

To express U in terms of N, V, and T, we must eliminate z. The result would be a very complicated function. A comparison between (8.77), (8.68), and (8.72), however, shows that U is directly related to the pressure by*

$$U = \tfrac{3}{2} PV \qquad \text{(Bose and Fermi)}$$

(8.78)

This relation also holds for the ideal Boltzmann gas.

The detailed study of the ideal gases together with their applications is taken up in Chapters 11 and 12.

8.7 FOUNDATIONS OF STATISTICAL MECHANICS

The present section contains no derivations. It merely furnishes an orientation on the subject of the derivation of statistical mechanics from molecular dynamics.[†]

It is recalled that a special case of statistical mechanics, the classical kinetic theory of gases, can be derived almost rigorously from molecular dynamics. The only ad hoc assumption in that derivation is the assumption of molecular chaos, which, however, plays a well-understood role, namely, the reduction of *reversible microscopic phenomena* to *irreversible macroscopic phenomena*. Since irreversibility is a necessary result of any successful derivation, an assumption of this kind is not only unavoidable but also desirable, because it serves to mark clearly the point at which irreversibility enters. An improvement on the existing derivation consists of replacing this assumption by one less ad hoc, but not of doing away with it altogether.

---

*It is assumed that the term $V^{-1} \log(1 - z)$ in (8.72) can be neglected. This is justified in Section 12.3.

[†] For a source of literature see *Fundamental Problems in Statistical Mechanics*, edited by E. G. D. Cohen (North-Holland, Amsterdam, 1962).

The derivation of the classical kinetic theory of gases may be considered largely satisfactory. When we consider the more general problem of the derivation of statistical mechanics, we may well keep this theory in mind as a model example. From this example, we learn that a satisfactory derivation of statistical mechanics must simultaneously fulfill two requirements:

(*a*) It must clearly display the connection between microscopic reversibility and macroscopic irreversibility.

(*b*) It must provide a detailed description of the approach to equilibrium.

Thus a satisfactory derivation of statistical mechanics must satisfy not only the philosophical desire of the physicist to base all natural phenomena on molecular dynamics, but also the practical desire of the physicist to calculate numbers with which to compare with experiments.

Logically speaking, it suffices to derive quantum statistical mechanics, of which classical statistical mechanics is a special case. If we want to understand nonequilibrium phenomenon in the classical domain, however, it is expedient to use classical mechanics as a starting point. For this reason attempts to derive classical statistical mechanics from classical mechanics can be of great practical value.

Attempts to derive statistical mechanics have so far been one of two types: Some appeal to the ergodic theorem, while others aim at establishing the "master equation." Only the latter seems capable of fulfilling both the requirements set forth previously.

The master equation is an equation governing the time development of the quantity $P_n(t)$, which is the probability that at the instant t the system is in the state n. If the word "state" is appropriately interpreted, $P_n(t)$ can be defined either in classical or quantum mechanics. To justify statistical mechanics, we have to show that $P_n(t)$ approaches the quantity $\overline{(c_n, c_n)}$ of (8.5) when t is much longer than a characteristic time of the system called the relaxation time, e.g., molecular collision time.

The master equation is

$$\frac{dP_n(t)}{dt} = \sum_m \left[W_{nm} P_m(t) - W_{mn} P_n(t) \right] \tag{8.79}$$

where W_{mn} is the transition probability per second from the state n to the state m. It was first derived by Pauli under the assumption that n refers to a single quantum state of the system and that the coefficients in the expansion (8.1) have random phases at all times. All subsequent work after Pauli's has been concerned with the improvement of these assumptions and with the solution of the master equation itself.

It can be shown that solutions to the master equation approach the desired limit as $t \rightarrow \infty$. Hence the task of deriving statistical mechanics reduces to the justification of the master equation and the calculation of the relaxation time.

The similarity between the master equation and the Boltzmann transport equation may be noted, although we should remember that the latter refers to μ space whereas the former refers to Γ space. The random-phase assumption here is similar to the assumption of molecular chaos in the Boltzmann transport equation. In both cases the solution for $t \to \infty$ is relatively easy to obtain, but the relaxation time is difficult to calculate.

The approach involving the master equation seems to hold greater promise for a satisfactory derivation of statistical mechanics and the concomitant understanding of general nonequilibrium phenomena. Further discussion of the master equation, however, is beyond the scope of this book.*

PROBLEMS

8.1 Find the density matrix for a partially polarized incident beam of electrons in a scattering experiment, in which a fraction f of the electrons are polarized along the direction of the beam and a fraction $1 - f$ is polarized opposite to the direction of the beam.

8.2 Derive the equations of state (8.67) and (8.71), using the microcanonical ensemble.

8.3 Prove (7.14) in quantum statistical mechanics.

8.4 Verify (8.49) for Fermi and Bose statistics, i.e., the fluctuations of cell occupations are small.

Solution. By (8.65),

$$\langle n_k \rangle = -\frac{1}{\beta} \frac{\partial}{\partial \epsilon_k} \log \mathscr{Q}$$

Differentiating this again with respect to ϵ_k leads to

$$\langle n_k^2 \rangle - \langle n_k \rangle^2 = -\frac{1}{\beta} \frac{\partial}{\partial \epsilon_k} \langle n_k \rangle$$

from which we can deduce

$$\langle n_k^2 \rangle - \langle n_k \rangle^2 = \langle n_k \rangle \pm \langle n_k \rangle^2 \tag{A}$$

with the plus sign for Bose statistics, and the minus sign for Fermi statistics. (For Fermi statistics the results is obvious because $n_k^2 = n_k$.) The fluctuations are not necessarily small. Note, however, that (A) refers to the fluctuations of the occupation of individual states, and not the cell occupations.

As a calculation useful for later purposes, we note

$$\langle n_k n_p \rangle - \langle n_k \rangle \langle n_p \rangle = -\frac{1}{\beta} \frac{\partial}{\partial \epsilon_k} \langle n_p \rangle, \qquad (p \neq k)$$

The right side is zero because $\langle n_p \rangle$ depends only on ϵ_p. Thus we have

$$\langle n_k n_p \rangle = \langle n_k \rangle \langle n_p \rangle, \qquad (p \neq k) \tag{B}$$

---

*For a general discussion of the master equation, see N. G. Van Kampen, in Cohen, *op. cit.* An improvement on the random phase approximation is described by L. Van Hove, in Cohen, *op. cit.*

In the infinite-volume limit the spectrum of states becomes a continuum. The physically interesting question concerns the fluctuations in the occupation of a group of states, or a cell. Let

$$n_i = \sum_k n_k$$

where the sum extends over a group of states in cell i. We are interested in

$$\langle n_i^2 \rangle - \langle n_i \rangle^2 = \left\langle \left(\sum_k n_k \right)^2 \right\rangle - \left\langle \sum_k n_k \right\rangle^2$$

By using (B), it is easily shown that the right side is equal to

$$\sum_k \left(\langle n_k^2 \rangle - \langle n_k \rangle^2 \right)$$

Hence using (A) we obtain

$$\langle n_i^2 \rangle - \langle n_i \rangle^2 = \langle n_i \rangle \pm \sum_k \langle n_k \rangle^2$$

where the plus sign holds for Bose statistics, and the minus sign for Fermi statistics. In the infinite-volume limit, the k sum is replaced by an integral over a region in k space. No matter how small this region is, the integral is proportional to the volume V of the system. (This is equivalent to the statement that a finite fraction of the particles occupies a cell.) Thus the left side is of order V^2, but the right side is only of order V.

8.5 Calculate the grand partition function for a system of N noninteracting quantum mechanical harmonic oscillators, all of which have the same natural frequency ω_0. Do this for the following two cases:

(a) Boltzmann statistics

(b) Bose statistics.

Suggestions. Write down the energy levels of the N-oscillator system and determine the degeneracies of the energy levels for the two cases mentioned.

8.6 What is the equilibrium ratio of ortho- to parahydrogen at a temperature of 300 K? What is the ratio in the limit of high temperatures? Assume that the distance between the protons in the molecule is 0.74 Å.

The following hints may be helpful.

(a) Boltzmann statistics is valid for H_2 molecules at the temperatures considered.

(b) The energy of a single H_2 molecule is a sum of terms corresponding to contributions from rotational motion, vibrational motion, translational motion, and excitation of the electronic cloud. Only the rotational energy need be taken into account.

(c) The rotational energies are

$$E_{para} = \frac{\hbar^2}{2I} l(l+1) \qquad (l = 0, 2, 4, \ldots)$$

$$E_{ortho} = \frac{\hbar^2}{2I} l(l+1) \qquad (l = 1, 3, 5, \ldots)$$

where I is the moment of inertia of the H_2 molecule.

Answer. Let T = absolute temperature and $\beta = 1/kT$. Then

$$\frac{N_{ortho}}{N_{para}} = \frac{3 \sum_{l \text{ odd}} (2l+1) \, e^{-(\beta \hbar^2 / 2I) l(l+1)}}{\sum_{l \text{ even}} (2l+1) \, e^{-(\beta \hbar^2 / 2I) l(l+1)}}$$

GENERAL PROPERTIES OF THE PARTITION FUNCTION

9.1 THE DARWIN-FOWLER METHOD

Although the canonical ensemble may be derived from the microcanonical ensemble, as we have shown in Section 7.1, it may also be derived directly. Indeed, if we are not too concerned with rigor, the derivation is very simple. Consider an ensemble of M systems such that the energy averaged over all the systems is a given number U. We wish to find the most probable distribution of energies among these M systems in the limit as $M \to \infty$. By definition of an ensemble, the systems do not interact with one another; they may be considered one at a time, and they are consequently distinguishable from one another. Therefore our problem is mathematically identical with the problem of the most probable distribution for a classical ideal gas of particles. The answer as we know is the Maxwell-Boltzmann distribution, i.e., the energy value E_n occurs among the systems with relative probability $e^{-\beta E_n}$, where β is determined by the average energy U. This ensemble is the canonical ensemble. It is obvious that this derivation holds equally well in quantum and in classical statistical mechanics.

We want to present here a more rigorous derivation that avoids the use of Stirling's approximation, which is necessary in the usual derivation of the Maxwell-Boltzmann distribution. The purpose of this presentation is not only to derive the canonical ensemble directly but also to introduce the method of saddle point integration, which is a useful mathematical tool in statistical mechanics. The considerations that follow hold equally well for quantum and for classical statistical mechanics.

The method we shall describe is due to Darwin and Fowler. Assume that a system in the ensemble may have any one of the energy values E_k ($k = 0, 1, 2, \dots$). By choosing the unit of energy to be sufficiently small, we can regard E_k as an integer. Among the systems in the ensemble let

$$m_0 \quad \text{systems have energy } E_0$$
$$m_1 \quad \text{systems have energy } E_1$$
$$\vdots$$
$$m_k \quad \text{systems have energy } E_k$$
$$\vdots$$

(9.1)

The set of integers $\{m_k\}$ describes an arbitrary distribution of energy among the systems. It must satisfy the conditions

$$\sum_{k=0}^{\infty} m_k = M$$

$$\sum_{k=0}^{\infty} E_k m_k = MU$$

(9.2)

where both M and U are integers. Our purpose is to find the most probable set $\{\overline{m}_k\}$.

Given an arbitrary set $\{m_k\}$ satisfying (9.2) there are generally more ways than one to construct an ensemble corresponding to (9.1), because the interchange of any two systems (which are distinguishable) leaves $\{m_k\}$ unchanged. Let $W\{m_k\}$ be the number of distinct ways in which we can assign energy values to systems so as to satisfy (9.1). Obviously

$$W\{m_k\} = \frac{M!}{m_0! m_1! m_2! \cdots}$$

(9.3)

For the present case the postulate of equal a priori probability states that all distributions in energy among the systems are equally probable, subject to the conditions (9.2). Thus $\{\overline{m}_k\}$ is the set that maximizes (9.3). In anticipation of the fact that in the limit as $M \to \infty$ almost all possible sets $\{m_k\}$ are identical with $\{\overline{m}_k\}$, we can also find $\{\overline{m}_k\}$ by calculating the value of m_k averaged over all possible distributions in energy:

$$\langle m_k \rangle \equiv \frac{\sum_{\{m_i\}}' m_k W\{m_i\}}{\sum_{\{m_i\}}' W\{m_i\}}$$

(9.4)

where a prime over the sums indicate that they are sums over all sets $\{m_k\}$ subject to (9.2). We must also calculate the mean square fluctuation $\langle m_k^2 \rangle - \langle m_k \rangle^2$. If this vanishes as $M \to \infty$, then in that limit $\langle m_k \rangle \to \overline{m}_k$.

For convenience we modify the definition of $W\{m_k\}$ to

$$W\{m_k\} = \frac{M! g_0^{m_0} g_1^{m_1} \cdots}{m_0! m_1! \cdots}$$

(9.5)

where g_k is a number which at the end of the calculation will be set equal to unity. Let

$$\Gamma(M, U) \equiv \sum_{\{m_i\}}' W\{m_i\} \tag{9.6}$$

Then

$$\langle m_k \rangle = g_k \frac{\partial}{\partial g_k} \log \Gamma \tag{9.7}$$

The mean square fluctuation can be obtained as follows:

$$\langle m_k^2 \rangle = \frac{1}{\Gamma} \sum_{\{m_i\}}' m_k^2 W\{m_i\} = \frac{1}{\Gamma} g_k \frac{\partial}{\partial g_k} \left(g_k \frac{\partial \Gamma}{\partial g_k} \right)$$

$$= g_k \frac{\partial}{\partial g_k} \left(\frac{1}{\Gamma} g_k \frac{\partial \Gamma}{\partial g_k} \right) - \left(\frac{\partial}{\partial g_k} \frac{1}{\Gamma} \right) g_k^2 \frac{\partial \Gamma}{\partial g_k}$$

$$= g_k \frac{\partial}{\partial g_k} \left(g_k \frac{\partial}{\partial g_k} \log \Gamma \right) + \left(g_k \frac{\partial}{\partial g_k} \log \Gamma \right)^2$$

Therefore

$$\langle m_k^2 \rangle - \langle m_k \rangle^2 = g_k \frac{\partial}{\partial g_k} \left(g_k \frac{\partial}{\partial g_k} \log \Gamma \right) \tag{9.8}$$

Thus it is sufficient to calculate $\log \Gamma$.

By (9.6) and (9.5)

$$\Gamma = M! \sum_{m_0, m_1, \ldots}' \left(\frac{g_0^{m_0}}{m_0!} \cdot \frac{g_1^{m_1}}{m_1!} \cdots \right) \tag{9.9}$$

This cannot be explicitly evaluated because of the restriction (9.2). We are, however, only interested in this quantity in the limit as $M \to \infty$. To proceed, we define a generating function for Γ in the following manner. For any complex number z, let

$$G(M, z) \equiv \sum_{U=0}^{\infty} z^{MU} \Gamma(M, U) \tag{9.10}$$

Using (9.9) and (9.2) we obtain

$$G(M, z) = M! \sum_{U=0}^{\infty} \sum_{m_0, m_1, \ldots}' \left[\frac{\left(g_0 z^{E_0} \right)^{m_0}}{m_0!} \cdot \frac{\left(g_0 z^{E_1} \right)^{m_1}}{m_1!} \cdots \right] \tag{9.11}$$

It is easily seen that the double sum in (9.11) is equivalent to summing over all sets $\{m_k\}$ subject only to the condition $\sum m_k = M$. To show this we need only verify that every term in one sum appears once in the other and vice versa. Hence

$$G(M, z) = \sum_{\substack{m_0, m_1, \ldots \\ \sum m_k = M}} \frac{M!}{m_0! m_1! \cdots} \left[\left(g_0 z^{E_0} \right)^{m_0} \left(g_1 z^{E_1} \right)^{m_1} \cdots \right]$$

$$= \left(g_0 z^{E_0} + g_1 z^{E_1} + \cdots \right)^M \tag{9.12}$$

The last step follows by use of the multinomial theorem. Let

$$f(z) \equiv \sum_{k=0}^{\infty} g_k z^{E_k} \tag{9.13}$$

Then

$$G(M, z) = [f(z)]^M \tag{9.14}$$

To obtain $\Gamma(M, U)$ from $G(M, z)$ we note that by definition $\Gamma(M, U)$ is the coefficient of z^{MU} in the expansion of $G(M, z)$ in powers of z. Therefore

$$\Gamma(M, U) = \frac{1}{2\pi i} \oint dz \, \frac{[f(z)]^M}{z^{MU+1}} \tag{9.15}$$

where the contour of integration is a closed path in the complex z plane about $z = 0$.

We may assume without loss of generality that the sequence $E_0, E_1, \ldots$ is a sequence of nondecreasing integers with no common divisor, because any common division τ can be removed by choosing the unit of energy τ times larger. Furthermore, we can set $E_0 = 0$, since this would only change the zero point of the energy. In so doing U would be changed to $U - E_0$, which we can again call U. The numbers $g_0, g_1, \ldots$ are as close to unity as we wish. For the immediate calculations we omit them temporarily. Hence

$$f(z) = 1 + z^{E_1} + z^{E_2} + \cdots \qquad (E_1 \leq E_2 \leq E_3 \leq \cdots) \tag{9.16}$$

where $E_1, E_2, \ldots$ are integers with no common divisor. When z is a real positive number x, $f(x)$ is a monotonically increasing function of x with a radius of convergence at, say, $x = R$. The same is true for $[f(x)]^M$, as illustrated in Fig. 9.1. The function $1/z^{MU+1}$, on the other hand, is a monotonically decreasing function along the real positive axis. The product of these two functions has a minimum at x_0 between 0 and R, as shown in Fig. 9.1. Now $f(z)$ is an analytic function for $|z| < R$, and z^{-MU-1} is analytic everywhere except at $z = 0$. Therefore the integrand of (9.18)

$$I(z) = \frac{[f(z)]^M}{z^{MU+1}} \tag{9.17}$$

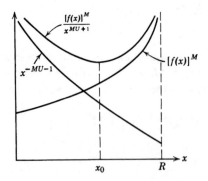

Fig. 9.1 The function $[f(x)]^M/x^{MU+1}$ for real positive x.

is analytic at $z = x_0$. An analytic function has a unique derivative at a given point. Furthermore it satisfies the Cauchy-Riemann equation

$$\left(\frac{\partial^2}{\partial x^2} + \frac{\partial^2}{\partial y^2} \right) I(z) = 0 \qquad (z \equiv x + iy) \qquad (9.18)$$

Hence

$$\left(\frac{\partial I}{\partial z} \right)_{z=x_0} = 0, \qquad \left(\frac{\partial^2 I}{\partial x^2} \right)_{z=x_0} > 0, \qquad \left(\frac{\partial^2 I}{\partial y^2} \right)_{z=x_0} < 0 \qquad (9.19)$$

That is, in the complex z plane, $I(z)$ has a minimum at $z = x_0$ along a path on the real axis but has a maximum at $z = x_0$ along a path parallel to the imaginary axis passing through x_0. The point x_0 is a saddle point, as illustrated in Fig. 9.2. Let $g(z)$ be defined by

$$\begin{aligned} I(z) &\equiv e^{Mg(z)} \\ g(z) &= \log f(z) - U \log z \end{aligned} \qquad (9.20)$$

where we have neglected $1/M$ as compared to U. Then x_0 is the root of the equation

$$g'(x_0) = 0$$

or

$$\frac{\sum_k E_k x_0^{E_k}}{\sum_k x_0^{E_k}} = U \qquad (9.21)$$

Furthermore

$$\left(\frac{\partial^2 I}{\partial x^2} \right)_{z=x_0} = Mg''(x_0) \exp \left[Mg(x_0) \right] \underset{M \to \infty}{\to} \infty \qquad (9.22)$$

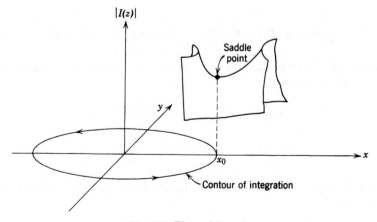

Fig. 9.2 The saddle point.

Hence the saddle point touches an infinitely sharp peak and an infinitely steep valley in the limits as $M \to \infty$. If we choose the contour of integration to be a circle about $z = 0$ with radius x_0, the contour will pass through x_0 in the imaginary direction. Thus along the contour the integrand has an extremely sharp maximum at the point $z = x_0$. If elsewhere along the contour there is no maximum comparable in height to this one, the contribution to the integral comes solely from the neighborhood of x_0. This is in fact true because for $z = x_0 e^{i\theta}$

$$|I(z)| = \frac{1}{x_0^{MU}} \left|1 + \left(x_0 e^{i\theta}\right)^{E_1} + \left(x_0 e^{i\theta}\right)^{E_2} + \cdots \right|^M \qquad (9.23)$$

The series (9.23) is maximum when all terms are real. This happens when and only when $\theta E_k = 2\pi n_k$, where n_k is 0 or an integer. If $\theta \neq 0$, then $2\pi/\theta$ must be a rational number, and this would mean that $E_k = (2\pi/\theta)n_k$, which is impossible unless $\theta = 2\pi$, because the E_k have no common divisor. Hence we conclude that the largest value of $I(z)$ occurs at $z = x_0$.

To do the integral (9.15) we expand the integrand about $z = x_0$:

$$g(z) = g(x_0) + \tfrac{1}{2}(z - x_0)^2 g''(x_0) + \cdots$$

Hence

$$\Gamma(M, U) = \frac{1}{2\pi i} \oint dz\, e^{Mg(z)} \approx e^{Mg(x_0)} \frac{1}{2\pi i} \oint dz\, e^{1/2 Mg''(x_0)(z-x_0)^2}$$

Putting $(z - x_0) = iy$, we obtained

$$\Gamma(M, U) \approx e^{Mg(x_0)} \frac{1}{2\pi} \int_{-\infty}^{+\infty} dy\, e^{-1/2 Mg''(x_0)y^2} = \frac{e^{Mg(x_0)}}{\sqrt{2\pi Mg''(x_0)}} \qquad (9.24)$$

Hence

$$\frac{1}{M} \log \Gamma(M, U) \approx g(x_0) - \frac{1}{2M} \log\left[2\pi Mg''(x_0)\right] \qquad (9.25)$$

As $M \to \infty$ the first term gives the exact result. To evaluate $g(x_0)$ we first obtain from (9.20) the formulas

$$g(x_0) = \log f(x_0) - U \log x_0$$

$$g''(x_0) = \frac{f''(x_0)}{f(x_0)} - \frac{U(U-1)}{x_0^2}$$

Using $f(x_0)$ from (9.13) (restoring now the numbers g_k) and defining a parameter β by

$$x_0 \equiv e^{-\beta} \qquad (9.26)$$

we obtain

$$g(x_0) = \log\left(\sum_{k=0}^{\infty} g_k e^{-\beta E_k}\right) + \beta U$$

$$g''(x_0) = \frac{e^{2\beta} \sum_{k=0}^{\infty} g_k (E_k^2 - U^2) e^{-\beta E_k}}{\sum_{k=0}^{\infty} g_k e^{-\beta E_k}} \equiv e^{2\beta} \left\langle (E - U)^2 \right\rangle \tag{9.27}$$

Hence

$$\frac{1}{M} \log \Gamma(M, U) = \log\left(\sum_{k=0}^{\infty} g_k e^{-\beta E_k}\right) + \beta U - \frac{1}{2M} \log\left[2\pi M g''(x_0)\right] \tag{9.28}$$

from which we find, using (9.7) and (9.8),

$$\frac{\langle m_k \rangle}{M} = \frac{e^{-\beta E_k}}{\sum_{k=0}^{\infty} e^{-\beta E_k}} \tag{9.29}$$

$$\frac{\langle m_k^2 \rangle - \langle m_k \rangle^2}{M^2} = \frac{1}{M} \frac{\langle m_k \rangle}{M} \left[1 - \frac{\langle m_k \rangle}{M} - \frac{\langle m_k \rangle}{M} \frac{(E_k - U)^2}{\langle (E - U)^2 \rangle} \right] \tag{9.30}$$

This is an exact formula in the limit as $M \to \infty$. We see that the fluctuations vanish in this limit. Therefore $\langle m_k \rangle = \overline{m}_k$. The parameter β is determined by (9.21) and (9.26):

$$U = \frac{\sum_{k=0}^{\infty} E_k e^{-\beta E_k}}{\sum_{k=0}^{\infty} e^{-\beta E_k}} = \langle E \rangle \tag{9.31}$$

Hence β can be identified as $1/kT$, where T is the absolute temperature.

In the most probable distribution the probability of finding a system in the ensemble having the energy E_k is (9.29). The ensemble with such an energy distribution is the canonical ensemble.

9.2 CLASSICAL LIMIT OF THE PARTITION FUNCTION

let $\mathscr{H}$ be the Hamiltonian operator of a system of N identical spinless particles.* Let $\mathscr{H}$ be the sum of two operators, the kinetic energy operator K, and the potential energy operator Ω:

$$\mathscr{H} = K + \Omega \tag{9.32}$$

---

*It is straightforward to generalize the following considerations to the case of particles with spin and to the case of a mixed system of two or more more different kinds of particles.

If $\Psi(\mathbf{r}_1, \ldots, \mathbf{r}_N)$ is a wave function of the system, then

$$K\Psi(\mathbf{r}_1, \ldots, \mathbf{r}_N) = -\frac{\hbar^2}{2m} \sum_{i=1}^{N} \nabla_i^2 \Psi(\mathbf{r}_1, \ldots, \mathbf{r}_N) \tag{9.33}$$

$$\Omega\Psi(\mathbf{r}_1, \ldots, \mathbf{r}_N) = \Omega(\mathbf{r}_1, \ldots, \mathbf{r}_N)\Psi(\mathbf{r}_1, \ldots, \mathbf{r}_N) \tag{9.34}$$

where m is the mass of a particle, ∇_i^2 is the Laplacian operator with respect to $\mathbf{r}_i$, and $\Omega(\mathbf{r}_1, \ldots, \mathbf{r}_N)$ is a sum of two-body potentials:

$$\Omega(\mathbf{r}_1, \ldots, \mathbf{r}_N) = \sum_{i<j} v_{ij} \tag{9.35}$$

where

$$v_{ij} \equiv v(|\mathbf{r}_i - \mathbf{r}_j|) \tag{9.36}$$

Whenever convenient, we use the abbreviation $(1, \ldots, N)$ for $(\mathbf{r}_1, \ldots, \mathbf{r}_N)$.

The partition function of the system is

$$Q_N(V, T) = \mathrm{Tr}\, e^{-\beta\mathscr{H}} = \sum_n \left(\Phi_n, e^{-\beta\mathscr{H}}\Phi_n\right) \tag{9.37}$$

where $\Phi_n(1, \ldots, N)$ is a member of any complete orthonormal set of wave functions of the system and $\Phi_n^*(1, \ldots, N)$ is its complex conjugate. For any operator $\mathcal{O}$,

$$\left(\Phi_n, \mathcal{O}\Phi_n\right) \equiv \int d^{3N}\mathbf{r}\, \Phi_n^*(1, \ldots, N)\mathcal{O}\Phi_n(1, \ldots, N) \tag{9.38}$$

Each Φ_n satisfies the boundary conditions imposed on the system and is normalized in the box of volume V containing the system. It is a symmetric (antisymmetric) function of $\mathbf{r}_1, \ldots, \mathbf{r}_N$, if the system is a system of bosons (fermions).

It will be shown that when the temperature is sufficiently high we can make the approximation

$$Q_N(V, T) \approx \frac{1}{N!h^{3N}} \int d^{3N}p\, d^{3N}r\, e^{-\beta\mathscr{H}(p, r)} \tag{9.39}$$

where h is Planck's constant and $\mathscr{H}(p, r)$ is the classical Hamiltonian

$$\mathscr{H}(p, r) \equiv \sum_{i=1}^{N} \frac{p_i^2}{2m} + \Omega(\mathbf{r}_1, \ldots, \mathbf{r}_N) \tag{9.40}$$

This will prove that at sufficiently high temperatures the partition function approaches the classical partition function with "correct Boltzmann counting." In the course of proving (9.39) we obtain the criterion for a "sufficiently high temperature."

Free Particles

Let us first consider an ideal gas, for which $\Omega(1, \ldots, N) \equiv 0$. The eigenfunctions of the Hamiltonian are the free-particle wave functions $\Phi_p(1, \ldots, N)$ described in the Appendix, Section A.2. They are labeled by a set of N momenta

$$p \equiv \{\mathbf{p}_1, \ldots, \mathbf{p}_N\} \tag{9.41}$$

and satisfy the eigenvalue equation

$$K\Phi_p(1, \ldots, N) = K_p\Phi_p(1, \ldots, N) \tag{9.42}$$

where

$$K_p \equiv \frac{1}{2m}\left(p_1^2 + \cdots + p_N^2\right) \tag{9.43}$$

For convenience we impose periodic boundary conditions with respect to the volume V. It follows that each $\mathbf{p}_i$ has the allowed values

$$\mathbf{p}_i = \frac{2\pi\hbar\mathbf{n}}{V^{1/3}} \tag{9.44}$$

where $\mathbf{n}$ is a vector whose components may be $0, \pm 1, \pm 2, \ldots$. More explicitly, $\Phi_p(1, \ldots, N)$ is given by

$$\Phi_p(1, \ldots, N) = \frac{1}{\sqrt{N!}} \sum_P \delta_P\left[u_{\mathbf{p}_1}(P1) \cdots u_{\mathbf{p}_N}(PN)\right]$$

$$= \frac{1}{\sqrt{N!}} \sum_P \delta_P\left[u_{P\mathbf{p}_1}(1) \cdots u_{P\mathbf{p}_N}(N)\right] \tag{9.45}$$

where

$$u_{\mathbf{p}}(\mathbf{r}) = \frac{1}{\sqrt{V}} e^{i\mathbf{p}\cdot\mathbf{r}/\hbar} \tag{9.46}$$

The notation is explained in the Appendix, Section A.2. A permutation of the momenta $\mathbf{p}_1, \ldots, \mathbf{p}_N$ does not produce a new state, for Φ_p is either invariant or changes sign. Therefore a sum over states is $1/N!$ times a sum over all the momenta independently. In the limit $V \to \infty$ a momentum sum may be replaced by an integral:

$$\sum_P \to \frac{V}{h^3} \int d^3p \tag{9.47}$$

Therefore

$$\operatorname{Tr} e^{-\beta K} = \frac{V^N}{N!h^{3N}} \int d^{3N}p\, d^{3N}r\, |\Phi_p(1, \ldots, N)|^2 e^{-\beta K_p} \tag{9.48}$$

Using (9.45) we can write

$$|\Phi_p(1, \ldots, N)|^2 = \frac{1}{N!} \sum_P \sum_{P'} \delta_P\delta_{P'}\left[u_{\mathbf{p}_1}^*(P1)u_{P'_{\mathbf{p}_1}}(1)\right] \cdots \left[u_{\mathbf{p}_N}^*(PN)u_{P'_{\mathbf{p}_N}}(N)\right] \tag{9.49}$$

Now every term in the P' sum will give the same contribution to the integral in

(9.48). Thus we may replace the above by $N!$ times any one term in the P' sum:

$$|\Phi(1,\ldots,N)|^2 \to \sum_P \delta_P \left[u_{\mathbf{p}_1}^*(P1)u_{\mathbf{p}_1}(1)\right] \cdots \left[u_{\mathbf{p}_N}^*(PN)u_{\mathbf{p}_N}(N)\right]$$

$$= \frac{1}{V^N} \sum_P \delta_P \exp\frac{i}{\hbar}\left[\mathbf{p}_1 \cdot (\mathbf{r}_1 - P\mathbf{r}_1) + \cdots + \mathbf{p}_N \cdot (\mathbf{r}_1 - P\mathbf{r}_N)\right]$$

$$(9.50)$$

When this is substituted into (9.48), each momentum integral can be expressed in term of the function

$$f(\mathbf{r}) \equiv \frac{\int d^3p\, e^{-\beta(p^2/2m)+i\mathbf{p}\cdot\mathbf{r}/\hbar}}{\int d^3p\, e^{-\beta p^2/2m}} = e^{-\pi r^2/\lambda^2} \qquad (9.51)$$

where $r \equiv |\mathbf{r}|$ and $\lambda = \sqrt{2\pi\hbar^2/mkT}$, the thermal wave length. The result is

$$\mathrm{Tr}\, e^{-\beta K} = \frac{1}{N!h^{3N}} \int d^{3N}p\, d^{3N}r\, e^{-\beta(p_1^2+\cdots+p_N^2)/2m}$$

$$\times \sum_P \delta_P[f(\mathbf{r}_1 - P\mathbf{r}_1) \cdots f(\mathbf{r}_N - P\mathbf{r}_N)] \qquad (9.52)$$

This is an exact identity. For high temperatures the integrand may be approximated as follows. The sum $\sum_P$ contains $N!$ terms. The term corresponding to the unit permutation $P = 1$ is $[f(0)]^N = 1$. The term corresponding to a permutation which only interchanges $\mathbf{r}_i$ and $\mathbf{r}_j$ is $[f(\mathbf{r}_i - \mathbf{r}_j)]^2$. Thus by enumerating the permutations in increasing order of the number of coordinates interchanged we arrive at the expansion

$$\sum_P \delta_P[f(\mathbf{r}_1 - P\mathbf{r}_1) \cdots f(\mathbf{r}_N - P\mathbf{r}_N)] = 1 \pm \sum_{i<j} f_{ij}^2 + \sum_{i,j,k} f_{ij}f_{ik}f_{kj} \pm \cdots$$

$$(9.53)$$

where $f_{ij} \equiv f(\mathbf{r}_i - \mathbf{r}_j)$ and where the plus sign applies to bosons and the minus sign to fermions. According to (9.51), f_{ij} vanishes extremely rapidly if $|\mathbf{r}_i - \mathbf{r}_j| \gg \lambda$. Therefore, when the temperature is so high that

$$\text{(thermal wavelength)} \ll \text{(average interparticle distance)} \qquad (9.54)$$

we have

$$\mathrm{Tr}\, e^{-\beta K} \approx \frac{1}{N!h^{3N}} \int d^{3N}p\, d^{3N}r\, e^{-\beta(p_1^2+\cdots+p_N^2)/2m} \qquad (9.55)$$

which proves (9.39) for an ideal gas.

It is of some interest to examine the first quantum correction to the classical partition function of an ideal gas. If $|\mathbf{r}_i - \mathbf{r}_j| \gg \lambda$, we may approximate the

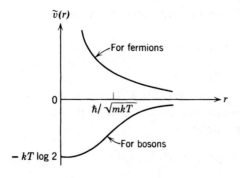

Fig. 9.3 The "statistical potential" between particles in an ideal gas arising from the symmetry properties of the N-particle wave function.

right side of (9.53) by $1 \pm \sum f_{ij}^2$. To the same order of approximation, however, we can also write

$$1 \pm \sum_{i<j} f_{ij}^2 \approx \prod_{i<j}\left(1 \pm f_{ij}^2\right) = \exp\left(-\beta \sum_{j<i} \tilde{v}_{ij}\right) \tag{9.56}$$

where

$$\tilde{v}_{ij} \equiv -kT \log\left(1 \pm f_{ij}^2\right) = -kT \log\left[1 \pm \exp\left(-\frac{2\pi|\mathbf{r}_i - \mathbf{r}_j|^2}{\lambda^2}\right)\right] \tag{9.57}$$

with the plus sign for bosons and the minus sign for fermions. Therefore an improvement over (9.55) is the formula

$$\mathrm{Tr}\, e^{-\beta K} \approx \frac{1}{N! h^{3N}} \int d^{3N}p\, d^{3N}r \exp\left[-\beta\left(\sum_i \frac{p_i^2}{2m} + \sum_{i<j} \tilde{v}_{ij}\right)\right] \tag{9.58}$$

This shows that the first quantum correction to the partition function of an ideal gas has the same effect as that of endowing the particles with an interparticle potential* $\tilde{v}(r)$ and treating the problems classically. The potential $\tilde{v}(r)$ is attractive for bosons and repulsive for fermions, as illustrated in Fig. 9.3. In this sense we sometimes speak of the "statistical attraction" between bosons and the "statistical repulsion" between fermions. It must be emphasized, however, that $\tilde{v}(r)$ originates solely from the symmetry properties of the wave function. Furthermore, it depends on the temperature and thus cannot be regarded as a true interparticle potential.

Interacting Particles

We now turn our attention to the more general case in which the particles of the system interact with one another. For the calculation of traces we may continue

*First discussed by G. E. Uhlenbeck and L. Gropper, *Phys. Rev.* **41**, 79 (1932).

to use the free-particle wave functions Φ_p, because any complete orthonormal set of wave functions will do.

First it is to be noted that in general K does not commute with Ω. Hence

$$e^{-\beta\mathcal{H}} = e^{-\beta(K+\Omega)} \neq e^{-\beta K} e^{-\beta\Omega} \tag{9.59}$$

because the left side is invariant under the exchange of K and Ω whereas the right side is not. To find a suitable approximation for $e^{-\beta\mathcal{H}}$ when $\beta \to 0$, let us assume that the following expansion is possible:

$$e^{-\beta(K+\Omega)} = e^{-\beta K} e^{-\beta\Omega} e^{C_0} e^{\beta C_1} e^{\beta^2 C_2} \cdots \tag{9.60}$$

where $C_0, C_1, C_2, \ldots$ are operators to be determined by taking the nth derivatives of both sides of (9.60) with respect to β and then setting $\beta = 0$. Letting $n = 0, 1, 2, \ldots$, we successively find that*

$$C_0 = 0$$
$$C_1 = 0$$
$$C_2 = -\tfrac{1}{2}[K, \Omega] \tag{9.61}$$
$$\vdots$$

If $[K, \Omega]$ commutes with both K and Ω, we would find that $C_n = 0$ $(n > 2)$. In our case this is untrue but we shall assume that for β sufficiently small the following is a good approximation:

$$e^{-\beta(K+\Omega)} \approx e^{-\beta K} e^{-\beta\Omega} e^{-1/2\beta^2[K, \Omega]} \tag{9.62}$$

Consequently

$$Q_N(V, T) \approx \mathrm{Tr}\left(e^{-\beta K} e^{-\beta\Omega} e^{-1/2\beta^2[K, \Omega]}\right) \tag{9.63}$$

From (9.33) and (9.34) it can be easily verified that

$$[K, \Omega] = -\frac{\hbar^2}{2m} \sum_{i=1}^{N} \nabla_j^2 \Omega + 2(\nabla_i\Omega) \cdot \nabla_i$$

$$= -\frac{\hbar^2}{2m} \sum_{i \neq j} \left(w_{ij} - 2\mathbf{F}_{ij} \cdot \nabla_i\right) \tag{9.64}$$

where

$$w_{ij} = w_{ji} = w\left(|\mathbf{r}_i - \mathbf{r}_j|\right), \qquad w(r) \equiv \nabla^2 v(r)$$
$$\mathbf{F}_{ij} = -\mathbf{F}_{ji} \equiv \nabla_i v\left(|\mathbf{r}_i - \mathbf{r}_j|\right) \tag{9.65}$$

When we exponentiate (9.64), we may act as if the two terms in it commute with

---

*The exact expansion is known as the Baker-Campbell-Hausdorf theorem. For an elementary derivation see R. M. Wilcox, *J. Math. Phys.* **8**, 962 (1967).

each other because the correction belongs to a higher order in β than we are considering. With this in mind, we substitute (9.64) into (9.63), and again use free-particle states to calculate the trace. The operator ∇_i may then be replaced by $i\mathbf{p}_i/\hbar$. Thus we have

$$Q_N(V, T) \approx \frac{V^N}{N! h^{3N}} \int d^{3N}p \, d^{3N}r \, |\Phi_p(1, \ldots, N)|^2$$

$$\times \exp\left[-\beta \sum_{j=i}^{N} \left(\frac{p_j^2}{2m} + \frac{i\beta\hbar}{2m}\mathbf{G}_j \cdot \mathbf{p}_j\right) - \beta \sum_{i<j}\left(v_{ij} - \frac{\beta\hbar^2}{2m}w_{ij}\right)\right]$$

(9.66)

where

$$\mathbf{G}_i = \sum_{\substack{j=1 \\ (j\neq i)}}^{N} \mathbf{F}_{ji}$$

(9.67)

The momentum integrations can be done, again in term of the function $f(\mathbf{r})$ of (9.51). We then obtain

$$Q_N(V, T) \approx \frac{1}{N! h^{3N}} \int d^{3N}p \, d^{3N}r \, e^{-\beta \mathcal{H}(p, r) + \beta^2 \Omega'} X$$

$$X \equiv \sum_P \delta_P[f(\mathbf{r}_1 - P\mathbf{r}_1 + \mathbf{G}_1') \cdots f(\mathbf{r}_N - P\mathbf{r}_N + \mathbf{G}_N')]$$

(9.68)

where

$$\Omega' = \frac{\beta\hbar^2}{2m}\sum_{i<j} w_{ij}$$

$$\mathbf{G}_i' = \frac{\beta\hbar^2}{2m}\mathbf{G}_i$$

(9.69)

Making an expansion similar to (9.53), and an approximation similar to (9.56), we write

$$X \approx \prod_{i=1}^{N} f(\mathbf{G}_i')\prod_{i<j}\left[1 \pm \frac{f(\mathbf{r}_i - \mathbf{r}_j + \mathbf{G}_i')f(\mathbf{r}_j - \mathbf{r}_i + \mathbf{G}_j')}{f(\mathbf{G}_i')f(\mathbf{G}_j')}\right]$$

(9.70)

The first product can be rewritten as

$$\exp\frac{\lambda^2}{16\pi}\sum_{i, j, k}\mathbf{F}_{ji} \cdot \mathbf{F}_{ik}$$

which gives rise to an effective three-body force among the particles. The second product is generally complicated, involving many-body forces. We shall assume that the range and depth of the intermolecular potential is such that we can neglect the terms G_i' in this product. Then we can state

$$Q_N(V, T) \approx \frac{1}{N! h^{3N}} \int d^{3N}p \, d^{3N}r \, e^{-\beta \mathcal{H}_{\text{eff}}} \tag{9.71}$$

$$\mathcal{H}_{\text{eff}} = \mathcal{H} + \Omega_1 + \Omega_2 + \Omega_3 \tag{9.72}$$

$$\Omega_1 = \sum_{i<j} \tilde{v}_{ij} \tag{9.73}$$

$$\Omega_2 = \frac{\lambda^2}{4\pi} \sum_{i<j} w_{ij} \tag{9.74}$$

$$\Omega_3 = \frac{\lambda^2}{16\pi} \sum_{i,j,k} \mathbf{F}_j \cdot \mathbf{F}_{ik} \tag{9.75}$$

9.3 SINGULARITIES AND PHASE TRANSITIONS

Phase transitions are manifested in experiments by the occurrence of singularities in thermodynamic functions, such as the pressure in a liquid–gas system, or the magnetization in a ferromagnet. How is it possible that such singularities arise from the partition function, which seems to be an analytic function of its arguments? The answer lies in the fact that a macroscopic body is close to the idealized thermodynamic limit—the limit of infinite volume with particle density held fixed. As we approach this limit, the partition function can develop singularities, because the limit function of a sequence of analytic functions need not be analytic.

Yang and Lee propose a definite scenario for the occurrence of singularities in the thermodynamic limit, which we shall now describe. It is formal in character, and belongs to a field sometimes known as "rigorous statistical mechanics."[*]

As a concrete model consider a classical system consisting of N molecules in volume V, interacting with one another through a pairwise potential as depicted in Fig. 9.4. Each molecule is taken to be a hard sphere surrounded by an attractive potential of finite range. Thus, a finite volume V can accommodate at most a finite number of molecules $M(V)$. For $N > M(V)$ the partition function vanishes because at least two molecules must "touch," rendering the energy infinite:

$$Q_N(V) = 0, \quad \text{for } N > M(V) \tag{9.76}$$

where we have suppressed the temperature to simplify the notation. The grand

---

[*]C. N. Yang and T. D. Lee, *Phys. Rev.* **87**, 404 (1952); T. D. Lee and C. N. Yang, *Phys. Rev.* **87**, 410 (1952). For rigorous stuff see also D. Ruelle, *Statistical Mechanics* (Benjamin, New York, 1969), Chapters 3 and 5; and J. Glimm and A. Jaffe, *Quantum Physics* (Springer-Verlag, New York, 1981), Chapter 2.

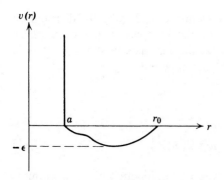

Fig. 9.4 Idealized interparticle potential.

partition function is a polynomial of degree $M(V)$ in the fugacity z:

$$\mathscr{Q}(z, V) = 1 + zQ_1(V) + z^2Q_2(V) + \cdots + z^M Q_M \qquad (9.77)$$

Since all the coefficients $Q_N(V)$ are positive, the polynomial has no real positive roots. The parametric form of the equation of state is

$$\frac{P}{kT} = V^{-1} \log \mathscr{Q}(z, V)$$

$$\frac{1}{v} = V^{-1} z \frac{\partial}{\partial z} \log \mathscr{Q}(z, V) \qquad (9.78)$$

For any finite value of V, however large, both P and v are analytic functions of z in a region of the complex z plane that includes the entire real axis. Therefore P is an analytic function of v in a region of the complex v plane that include all physical values of v, i.e., the real axis. Hence all thermodynamic functions must be free of singularities. From (9.78) and (9.77) we see that $P > 0$, and

$$\frac{\partial P}{\partial v} = \frac{\partial P}{\partial z} \frac{\partial z}{\partial v} = \frac{1}{vz(\partial v/\partial z)} = \frac{kT}{v^4 [\langle n^2 \rangle - \langle n \rangle^2]} \le 0 \qquad (9.79)$$

where n is the density. To have the possibility of singularities, we must go to the limit $V \to \infty$ at fixed v—the thermodynamic limit:

$$\frac{P}{kT} = \lim_{V \to \infty} V^{-1} \log \mathscr{Q}(z, V)$$

$$\frac{1}{v} = \lim_{V \to \infty} V^{-1} z \frac{\partial}{\partial z} \log \mathscr{Q}(z, V) \qquad (9.80)$$

Note that, in the second equation, the order of the operations $\lim$ and $z(\partial/\partial z)$ can be interchanged only if the limits above are approached uniformly. (For an example see Problem 9.5.)

The above conclusions are valid also for a quantum mechanical system since they depend only on the assumption that a finite volume can accommodate at most a finite number of molecules, which is true for molecules with a hard core, even in quantum mechanics.

Yang and Lee show that phase transitions are controlled by the distribution of roots of the grand partition function in the complex z plane. A phase transition occurs whenever a root approaches the real axis in the limit $V \to 0$. The precise results are stated in the form of two theorems.

THEOREM 1
The limit

$$F_\infty(z) \equiv \lim_{V \to \infty} \frac{1}{V} \log \mathscr{Q}(z, V) \tag{9.81}$$

exists for all $z > 0$, and is a continuous nondecreasing function of z. This limit is also independent of the shape of V, if the surface area of V increases no faster than $V^{2/3}$.

THEOREM 2
Suppose R is a region in the complex z plane that includes a segment of the positive real axis, and contains no roots of the grand partition function. Then in this region the quantity $V^{-1} \log \mathscr{Q}$ converges uniformly to its limit as $V \to \infty$. The limit is analytic for all z in R.

We refer the reader to the original literature for the proofs, and merely discuss their consequences here.

A thermodynamic phase is defined by those values of z contained in any single region R of theorem 2. Since in any region R the convergence to the limit $F_\infty(z)$ is uniform, we can interchange the order of lim and $z(\partial/\partial z)$ in (9.80). In any single phase, therefore, the equation of state is given in parametric form by

$$\begin{aligned} \frac{P(z)}{kT} &= F_\infty(z) \\ \frac{1}{v(z)} &= z\frac{\partial}{\partial z}F_\infty(z) \end{aligned} \tag{9.82}$$

The properties $P > 0$ and $\partial P/\partial v < 0$ are maintained in the thermodynamic limit. We illustrate some possible behaviors of the equation of state.

Suppose the region R includes the entire positive z axis. Then the system always exists in a single phase. The equation state may be obtained graphically by eliminating z. The situation is illustrated in Fig. 9.5.

If on the other hand a zero of the grand partition function approaches a point z_0 on the real positive z axis, then there will be two distinct regions R_1 and R_2 in which theorem 2 holds separately. At $z = z_0$, $P(z)$ must be continuous, as required by theorem 1. However, its derivative may be discontinuous. The system then possesses two phases, corresponding respectively to the regions $z < z_0$ and $z > z_0$. Now $1/v(z)$ is a nondecreasing function of z:

$$z\frac{\partial}{\partial z}\frac{1}{v(z)} = z\frac{\partial}{\partial z}z\frac{\partial}{\partial z}V^{-1} \log \mathscr{Q}(z, V) = \langle n^2 \rangle - \langle n \rangle^2 \geq 0 \tag{9.83}$$

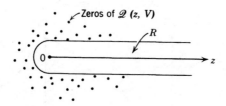

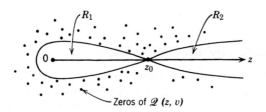

Fig. 9.5 Region R that is free of zeros of $\mathcal{Q}(z, V)$, leading to an equation of state that exhibits only a single phase.

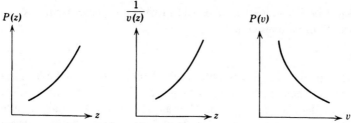

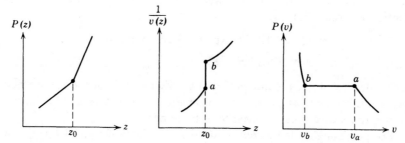

Fig. 9.6 Two regions R_1, R_2 each free of zeros of $\mathcal{Q}(z, V)$, corresponding to an equation of state with two phases connected by a first-order phase transition.

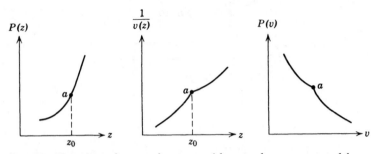

Fig. 9.7 Equation of state of system with two phases connected by a second-order transition.

Hence, if $\partial P/\partial z$ is discontinuous, $1/v(z)$ must make a discontinuous upward jump when z increases through z_0. The result is a first-order phase transition, as depicted in Fig. 9.6. The fact that $1/v(z)$ actually assumes all the values between the points a and b follows from the fact that the curve of $1/v(z)$ is the limiting curve of a sequence of continuous curves.

If in the same example $\partial P/\partial z$ is continuous at $z = z_0$, but $\partial^2 P/\partial z^2$ is discontinuous, then we would have a second-order phase transition, as illustrated in Fig. 9.7.

9.4 THE LEE-YANG CIRCLE THEOREM

The scenario for the occurrence of a phase transition proposed above can be explicitly demonstrated in the case of a lattice gas with attractive interactions. The system consists of point atoms located on the sites of a discrete lattice, with the condition that no two atoms occupy the same site, and that atoms on different sites have attractive pairwise interactions. That is, the interatomic potential u is such that

$$u = \infty, \quad \text{(if the two atoms are on the same site)}$$

$$u < 0, \quad \text{(otherwise)}$$

The detailed nature of the interaction (e.g., whether it is of the nearest-neighbor type) is unspecified, as is the dimensionality or structure of the lattice. For example, the lattice does not even have to be periodic.

The Lee-Yang circle theorem states that, for the lattice gas defined above, all roots of the equation

$$\mathscr{Q}(z, V) = 0$$

lie on the unit circle in the complex z plane.

For finite V the roots occur in complex-conjugate pairs, and none can touch the real axis. Only in the limit $V \to \infty$ is it possible for any of them to approach the real axis at $z = 1$. Thus, there can be at most one transition point.

Denoting the phase angles of the roots by θ_k, we have

$$\mathscr{Q}(z, V) = \mathscr{C} \prod_k (z - e^{i\theta_k}), \qquad \mathscr{C} = \prod_k (-e^{i\theta_k})^{-1} \tag{9.84}$$

In the thermodynamic limit the roots become continuously distributed on the unit circle. We can define a distribution function $g(\theta)$ by writing

$$V^{-1} \sum_k \rightarrow \int_0^{2\pi} d\theta \, g(\theta) \tag{9.85}$$

The fact that the roots occur in complex conjugate pairs means that $g(\theta) = g(-\theta)$. Using this fact, we can deduce the equation of state in the form

$$\frac{P}{kT} = \int_0^\pi d\theta \, g(\theta) \log(1 - 2z \cos \theta + z^2)$$

$$\frac{1}{v} = 2z \int_0^\pi d\theta \, g(\theta) \frac{z - \cos \theta}{1 - 2z \cos \theta + z^2} \tag{9.86}$$

We can see from this general form that, as z varies along the real axis, the only point where singularities can occur is $z = 1$. A singularity at this point will occur only if the integrals diverge at $\theta = 0$. Therefore, no phase transitions occur if $g(0) = 0$. On the other hand, if $g(0) \neq 0$, then $z = 1$ will be a singular point, and (9.86) will give different functions for $z > 1$ and for $z < 1$, which cannot be analytically continued into each other.

PROBLEMS

9.1 Derive with the help of the saddle point integration method a formula for the partition function for an ideal Bose gas of N particles.

9.2 (*a*) Find the equations of state for an ideal Bose gas and an ideal Fermi gas in the limit of high temperatures. Include the first correction due to quantum effects. (Consultation with Problem 7.6 may be helpful.)

(*b*) Estimate, for each of the following ideal gases, the temperature below which quantum effects would become important: H_2, He, N_2.

9.3 Pair Correlation Function. The pair correlation function $D(r_1, r_2)$ of a system of particles is defined as follows:

$D(r_1, r_2) \, d^3r_1 \, d^3r_2 \equiv$ probability of simultaneously finding a particle in the volume element d^3r_1 about r_1 and a particle in the volume element d^3r_2 and r_2.

Calculate $D(r_1, r_2)$ for an ideal Bose gas and an ideal Fermi gas in the limit of high temperatures. Include quantum corrections only to the lowest approximation.

Solution. Classically we have

$$D(r_1, r_2) = \frac{N(N-1) \int d^{3N}p \, d^3r_3 \cdots d^3r_N \, e^{-\beta \mathscr{H}(p, r)}}{\int d^{3N}p \, d^{3N}r \, e^{-\beta \mathscr{H}(p, r)}}$$

For our problem we use this formula with

$$\mathcal{H}(p, r) = \sum_{i=j}^{N} \frac{p_i^2}{2m} + \sum_{i<j} \bar{v}_{ij}$$

To avoid complications assume that the density of the gas is almost zero. The limit $N \to \infty$, $V \to \infty$ should be so taken that $N/V \to 0$. Then

$$D(\mathbf{r}_1, \mathbf{r}_2) \approx \frac{N(N-1)V^{N-2}\left[1 \pm f_{12}^2 \pm \dfrac{N(N-1)}{2V}\int d^3r f^2(r)\right]}{1 \pm \dfrac{N(N-1)}{2V}\int d^3r f^2(r)}$$

$$\approx \frac{1}{v^2}\left[1 \pm \exp\left(-\frac{2\pi}{\lambda^2}|\mathbf{r}_1 - \mathbf{r}_2|^2\right)\right]$$

This result continues to hold for finite v with $\lambda^3/v \ll 1$, although our derivation did not justify such a conclusion.

9.4 Show that the equation of state (9.86) of the Lee-Yang lattice gas has the following electrostatic analog:

(a) Consider a circular cylinder of unit radius perpendicular to the complex z plane, cutting it at the unit circle. Suppose the cylinder is charged with a surface charge density that depends only on the angle θ around the unit circle (with $\theta = 0$ corresponding to $z = 1$). The charge density (per unit area) is equal to $g(\theta)$, with $g(\theta) = g(-\theta)$. Let $\phi(z)$ and $E(z)$ be, respectively, the electrostatic potential and the electric field at a point z on the real axis. Then

$$P/kT = -\tfrac{1}{2}\phi(z)$$
$$n = \tfrac{1}{2}zE(z)$$

where $n = 1/v$ is the density.

(b) Assume $g(0) \neq 0$. Show by electrostatic argument that P is continuous at $z = 1$, but n jumps discontinuously. This shows that there is a first-order phase transition. Using Gauss' theorem in electrostatics, show the discontinuity in density is given by $\Delta n = 2\pi g(0)$.

9.5 Consider the grand partition function

$$\mathcal{Q}(z, V) = (1 + z)^V (1 + z^{\alpha V})$$

where α is a positive constant.

(a) Write down the equation of state in a parametric form, eliminate z graphically, and show that there is a first-order phase transition. Find the specific volumes of the two phases.

(b) Find the roots of $\mathcal{Q}(z, V) = 0$ in the complex z plane, at fixed V. Show that as $V \to \infty$ the roots converge toward the real axis at $z = 1$.

(c) Find the equation of state in the "gas" phase. Show that a continuation of this equation beyond the phase-transition density fails to show any sign of the transition. This will demonstrate that the order of the operations $z(\partial/\partial z)$ and $V \to \infty$ can be interchanged only within a single-phase region.

10

APPROXIMATE METHODS

10.1 CLASSICAL CLUSTER EXPANSION

Many systems of physical interest can be treated classically. A large class of such systems is described by a classical Hamiltonian for N particles of the form

$$\mathscr{H} = \sum_{i=1}^{N} \frac{p_i^2}{2m} + \sum_{i<j} v_{ij} \tag{10.1}$$

where $\mathbf{p}_i$ is the momentum of the ith particle and $v_{ij} = v(|\mathbf{r}_i - \mathbf{r}_j|)$ is the potential energy of interaction between the ith and the jth particle. If the system occupies a volume V, the partition function is

$$Q_N(V, T) = \frac{1}{N! h^{3N}} \int d^{3N}p \, d^{3N}r \exp\left(-\beta \sum_i \frac{p_i^2}{2m} - \beta \sum_{i<j} v_{ij}\right) \tag{10.2}$$

where each coordinate $\mathbf{r}_i$ is integrated over the volume V. The integrations over momenta can be immediately effected, leading to

$$Q_N(V, T) = \frac{1}{\lambda^{3N} N!} \int d^{3N}r \exp\left(-\beta \sum_{i<j} v_{ij}\right) \tag{10.3}$$

where $\lambda = \sqrt{2\pi\hbar^2/mkT}$ is the thermal wavelength. The integral in (10.3) is called the *configuration integral*. For potentials v_{ij} of the usual type between molecules, a systematic method for the calculation of the configuration integral consists of expanding the integrand in powers of $\exp(-\beta v_{ij}) - 1$. This leads to the cluster expansion of Ursell and Mayer.* As we shall see, this expansion is of practical use if the system is a dilute gas.

---

*For original literature, see J. E. Mayer and M. G. Mayer, *Statistical Mechanics* (Wiley, New York, 1940), Chapter 13.

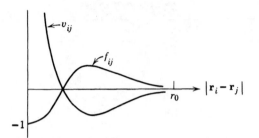

Fig. 10.1 Intermolecular potential v_{ij} and the function f_{ij}.

Let the configuration integral be denoted by $Z_N(V, T)$:

$$Z_N(V, T) \equiv \int d^3r_1 \cdots d^3r_N \exp\left(-\beta \sum_{i<j} v_{ij}\right) \tag{10.4}$$

in terms of which the partition function may be written as

$$Q_N(V, T) = \frac{1}{N!\lambda^{3N}} Z_N(V, T) \tag{10.5}$$

and the grand partition function as

$$\mathscr{Q}(z, V, T) = \sum_{N=0}^{\infty} \left(\frac{z}{\lambda^3}\right)^N \frac{Z_N(V, T)}{N!} \tag{10.6}$$

Let f_{ij} be defined by

$$e^{-\beta v_{ij}} \equiv 1 + f_{ij} \tag{10.7}$$

For the usual type of intermolecular potentials, v_{ij} and f_{ij} have the qualitative forms shown in Fig. 10.1. Thus f_{ij} is everywhere bounded and is negligibly small when $|\mathbf{r}_i - \mathbf{r}_j|$ is larger than the range of the intermolecular potential. In terms of f_{ij} the configuration integral may be represented by

$$Z_N(V, T) = \int d^3r_1 \cdots d^3r_N \prod_{i<j}(1 + f_{ij}) \tag{10.8}$$

in which the integrand is a product of $\frac{1}{2}N(N - 1)$ terms, one for each distinct pair of particles. Expanding this product we obtain

$$Z_N(V, T) = \int d^3r_1 \cdots d^3r_N \left[1 + (f_{12} + f_{13} + \cdots)\right.$$
$$\left. + (f_{12}f_{13} + f_{12}f_{14} + \cdots) + \cdots \right] \tag{10.9}$$

A convenient way to enumerate all the terms in the expansion (10.9) is to associate each term with a graph, defined as follows:

An N-particle graph is a collection of N distinct circles numbered $1, 2, \ldots, N$, with any number of lines joining the same number of distinct pairs of circles. If the *distinct pairs joined by lines are the pairs* $\alpha, \beta, \ldots, \lambda$, *then the graph represents the* *term*

$$\int d^3r_1 \cdots d^3r_N f_\alpha f_\beta \cdots f_\gamma \qquad (10.10)$$

appearing in the expansion (10.9).

If the set of distinct pairs $\{\alpha, \beta, \ldots, \gamma\}$ is joined by lines in a given graph, replacing this set by a set $\{\alpha', \beta', \ldots, \gamma'\}$ that is not identical with $\{\alpha, \beta, \ldots, \gamma\}$ gives rise to a graph that is counted as distinct from the original one (although the integrals represented by the respective graphs have the same numerical value). For example, for $N = 3$, the following graphs are distinct:

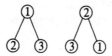

but the following graphs are identical:

①—② ③   ②—① ③

We may regard a graph as a picturesque way of writing the integral (10.10). For example, we may write, for $N = 10$,

$$\begin{bmatrix} \text{①} & \text{③—⑨} & \text{⑥—⑧} \\ \text{|} & & \text{⨉} \\ \text{②} & \text{④} \quad \text{⑤} & \text{⑦} \quad \text{⑩} \end{bmatrix} = \int d^3 r_1 \cdots d^3 r_{10}\, f_{12} f_{39} f_{67} f_{68} f_{8,10} f_{6,10} f_{78} \qquad (10.11)$$

With such a convention, we can state that

$$Z_N = (\text{sum of all distinct } N\text{-particle graphs}) \qquad (10.12)$$

The proof is obvious.

Any graph can in general be decomposed into smaller units. For example, the graph (14.11) is a product of five factors, namely

$$\begin{bmatrix} \text{①} & \text{③—⑨} & \text{⑥—⑧} \\ \text{|} & & \text{⨉} \\ \text{②} & \text{④} \quad \text{⑤} & \text{⑦} \quad \text{⑩} \end{bmatrix}$$

$$= [\text{④}] \cdot [\text{⑤}] \cdot [\text{①—②}] \cdot [\text{③—⑨}] \cdot \begin{bmatrix} \text{⑥—⑧} \\ \text{⨉} \\ \text{⑦} \quad \text{⑩} \end{bmatrix}$$

Each factor corresponds to a *connected graph*, in which every circle is attached to at least one line, and every circle is joined directly or indirectly to all other circles in the graph.

It would facilitate the analysis of Z_N if we first defined the basic units out of which an arbitrary graph can be composed. Accordingly we define an *l*-cluster to be an l particle connected graph. For example, the following is a 6-cluster:

$$= \int d^3r_1 \cdots d^3r_6 \, f_{12}f_{23}f_{14}f_{46}f_{56} \tag{10.13}$$

We define a *cluster integral* $b_l(V, T)$ by

$$b_l(V, T) \equiv \frac{1}{l!\lambda^{3l-3}V} \text{ (sum of all possible l-clusters)} \tag{10.14}$$

The normalization factor is so chosen that

    (a) $b_l(V, T)$ is dimensionless;
    (b) $\bar{b}_l(T) \equiv \lim_{V \to \infty} b_l(V, T)$ is a finite number.

The property (b) follows from the fact that f_{ij} has a finite range, so that in an l-cluster the only integration that gives rise to a factor V is the integration over the "center of gravity" of the l particles. Some of the cluster integrals are

$$b_1 = \frac{1}{V}[①] = \frac{1}{V} \int d^3r_1 = 1 \tag{10.15}$$

$$b_2 = \frac{1}{2!\lambda^3 V}[①-②] = \frac{1}{\lambda^3 2V} \int d^3r_1 \, d^3r_2 \, f_{12} = \frac{1}{2\lambda^3} \int d^3r_{12} \, f_{12} \tag{10.16}$$

$$b_3 = \frac{1}{3!\lambda^6 V}\left[\vcenter{\hbox{}} + \vcenter{\hbox{}} + \vcenter{\hbox{}} + \vcenter{\hbox{}} \right] \tag{10.17}$$

Any N-particle graph is a product of a number of clusters, of which m_l are l-clusters, with

$$\sum_{l=1}^{N} lm_l = N \tag{10.18}$$

A given set of integers $\{m_l\}$ satisfying (10.18), however, does not uniquely specify a graph, because

    (a) there are in general many ways to form an l-cluster, e.g.,

    (b) there are in general many ways to assign which particle belongs to which cluster, e.g.,

Thus a set of integers $\{m_l\}$ specifies a collection of graphs. Let the sum of all the graphs corresponding to $\{m_l\}$ be denoted by $S\{m_l\}$. Then

$$Z_N = \sum_{\{m_l\}} S\{m_l\} \qquad (10.19)$$

where the summation extends over all sets $\{m_l\}$ satisfying (10.18).

By definition, $S\{m_l\}$ can be obtained as follows. First write down an arbitrary N-particle graph that contains m_1 1-clusters, m_2 2-clusters, etc.; e.g.,

$$\times \left\{ \begin{bmatrix} \\ \end{bmatrix}\begin{bmatrix} \\ \end{bmatrix}\begin{bmatrix} \\ \end{bmatrix} \cdots \begin{bmatrix} \\ \end{bmatrix} \right\} \cdots \qquad (10.20)$$

m_3 factors

There are exactly N circles appearing in (10.20), and these N circles are to be filled in by the numbers $1, 2, \ldots, N$ in an arbitrary but definite order. We can write down many more examples like (10.20); e.g., we may change the choice of some of the 3-clusters (there being four distinct topological shapes for a 3-cluster). Again we may permute the numbering of all the N circles in (10.20), and that would lead to a distinct graph. If we add up all these possibilities, we obtain $S\{m_l\}$. Thus we may write

$$S\{m_l\} = \sum_P [\bigcirc]^{m_1} [\bigcirc\!-\!\bigcirc]^{m_2}$$

The meaning of this formula is as follows. Each bracket contains the sum over all l-clusters. If all the brackets $[\cdots]^{m_l}$ are expanded in multinomial expansions, the summand of $\sum_P$ will itself be a sum of a large number of terms in which every term contains exactly N circles. The sum $\sum_P$ extends over all distinct ways of numbering these circles from 1 to N.

Now each graph is an integral whose value is independent of the way its circles are numbered. Therefore $S\{m_l\}$ is equal to the number of terms in the sum $\sum_P$ times the value of any term in the sum. The number of terms in the sum $\sum_P$ can be found by observing that

(a) there are m_l l-clusters, and a permutation of these m_l things does not lead to a new graph;

(b) in the sum over all l-clusters, such as (10.17), a permutation of the l particles within it does not lead to a new graph. Hence the number of terms in the sum $\sum\limits_{P}$ is*

$$\frac{N!}{[(1!)^{m_1}(2!)^{m_2}\cdots][m_1!m_2!\cdots]} \tag{10.22}$$

and the value of any term is

$$(1!Vb_1)^{m_1}(2!\lambda^3 Vb_2)^{m_2}(3!\lambda^6 Vb_3)^{m_3}\cdots \tag{10.23}$$

Therefore

$$S\{m_l\} = N!\prod_{l=1}^{N}\frac{(V\lambda^{3l-3}b_l)^{m_l}}{m_l!} = N!\lambda^{3N}\prod_{l=1}^{N}\frac{1}{m_l!}\left(\frac{V}{\lambda^3}b_l\right)^{m_l} \tag{10.24}$$

From (10.5), (10.9), and (10.24) we obtain

$$Q_N(V,T) = \sum_{\{m_l\}}\prod_{l=1}^{N}\frac{1}{m_l!}\left(\frac{V}{\lambda^3}b_l\right)^{m_l} \tag{10.25}$$

This formula is complicated by the restriction (10.18). The grand partition function is simpler in appearance:

$$\mathscr{Q}(z,V,T) = \sum_{m_1=0}^{\infty}\sum_{m_2=0}^{\infty}\cdots\left[\frac{1}{m_1!}\left(\frac{V}{\lambda^3}zb_1\right)^{m_1}\frac{1}{m_2!}\left(\frac{V}{\lambda^3}z^2b_2\right)^{m_2}\cdots\right]$$

or

$$\frac{1}{V}\log\mathscr{Q}(z,V,T) = \frac{1}{\lambda^3}\sum_{l=1}^{\infty}b_l z^l \tag{10.26}$$

from which we obtain the equation of state in parametric form:

$$\begin{cases} \dfrac{P}{kT} = \dfrac{1}{\lambda^3}\displaystyle\sum_{l=1}^{\infty}b_l z^l \\[3mm] \dfrac{1}{v} = \dfrac{1}{\lambda^3}\displaystyle\sum_{l=1}^{\infty}lb_l z^l \end{cases} \tag{10.27}$$

This is known as the cluster expansion for the equation of state.[†]

What we have described is historically the first graphical representation of a perturbation series. Graphs have become indispensable tools in the many-body problem and in quantum field theory, in which the analog of (10.26), known generally by the name of the *linked cluster theorem*, plays an important role. Generally it states that the sum of all graphs is the exponential of the sum of all connected graphs.

---

*To understand the method of counting the reader is advised to work out some simple examples.
[†]Compare this derivation with that outlined in Problem 7.6.

If the system under consideration is a dilute gas, we may expand the pressure in powers of $1/v$ and obtain the virial expansion. For this purpose we may take the equation of state to be

$$
\begin{cases}
\dfrac{P}{kT} = \dfrac{1}{\lambda^3} \sum_{l=1}^{\infty} \mathring{b}_l z^l \\[4mm]
\dfrac{1}{v} = \dfrac{1}{\lambda^3} \sum_{l=1}^{\infty} l \mathring{b}_l z^l
\end{cases}
\tag{10.28}
$$

where

$$
\mathring{b}_l(T) \equiv \lim_{V \to \infty} b_l(V, T)
\tag{10.29}
$$

The virial expansion of the equation of state is defined to be

$$
\frac{Pv}{kT} = \sum_{l=1}^{\infty} a_l(T) \left(\frac{\lambda^3}{v} \right)^{l-1}
\tag{10.30}
$$

where $a_l(T)$ is called the lth virial coefficient. We can find the relationship between the virial coefficients a_l and the cluster integrals $\mathring{b}_l$ by substituting (10.30) into (10.28) and requiring that the resulting equation be satisfied for every z:

$$
\sum_{l=1}^{\infty} a_l \left(\sum_{n=1}^{\infty} n \mathring{b}_n z^n \right)^{l-1} = \frac{\displaystyle\sum_{l=1}^{\infty} \mathring{b}_l z^l}{\displaystyle\sum_{l=1}^{\infty} l \mathring{b}_l z^l}
\tag{10.31}
$$

This is equivalent to the condition

$$
\left(\mathring{b}_1 z + 2\mathring{b}_2 z^2 + 3\mathring{b}_3 z^3 + \cdots \right) \left[a_1 + a_2 \left(\sum_{n=1}^{\infty} n \mathring{b}_n z^n \right) + a_3 \left(\sum_{n=1}^{\infty} n \mathring{b}_n z^n \right)^2 + \cdots \right]
$$
$$
= \mathring{b}_1 z + \mathring{b}_2 z^2 + \mathring{b}_3 z^3 + \cdots
\tag{10.32}
$$

By equating the coefficient of each power of z we obtain

$$
\begin{aligned}
a_1 &= \mathring{b}_1 = 1 \\
a_2 &= -\mathring{b}_2 \\
a_3 &= 4\mathring{b}_2^2 - 2\mathring{b}_3 \\
a_4 &= -20\mathring{b}_2^3 + 18\mathring{b}_2\mathring{b}_3 - 3\mathring{b}_4
\end{aligned}
\tag{10.33}
$$
$$
\cdots
$$

Each virial coefficient therefore involves only a straightforward computation of a number of integrals.

Note that (10.28) differs from (10.27) in that the limit $V \to \infty$ is taken term by term in (10.28). In so doing we have lost all information about possible phase transitions, as we have remarked earlier in Section 9.3. The equation of state

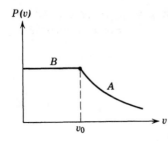

Fig. 10.2 Equation of state obtained by taking the virial expansion to be exact.

(10.30) of the gas phase cannot tell us if and when a phase transition will occur. Mayer* has demonstrated that the equation of state (10.30) has the general form shown in Fig. 10.2. The portion of the isotherm marked A is valid for $v > v_1$, but the value of v_1 is unrelated to v_0, and cannot be obtained from (10.30). The portion marked B is purely mathematical, and unrelated to how the isotherm actually behaves in that region.

10.2 QUANTUM CLUSTER EXPANSION

Kahn and Uhlenbeck[†] develop a cluster expansion in quantum statistical mechanics. The method they introduce applies equally well to classical statistical mechanics.

Consider N identical particles enclosed in a volume V. Let the Hamiltonian $\mathscr{H}$ of the system have the same form as (10.1) but be an operator instead of a number. In the coordinate representation, $\mathbf{p}_j = -i\hbar\nabla_j$, and v_{ij} is the same function of the number $|\mathbf{r}_i - \mathbf{r}_j|$ as that shown in Fig. 10.1. The partition function is

$$Q_N(V, T) = \mathrm{Tr}\, e^{-\beta H} = \int d^{3N}r \sum_\alpha \Psi_\alpha^*(1,\ldots,N)\, e^{-\beta\mathscr{H}}\Psi_\alpha(1,\ldots,N) \quad (10.34)$$

where $\{\Psi_\alpha\}$ is a complete set of orthonormal wave functions appropriate to the system considered, and the set of coordinates $\{\mathbf{r}_1,\ldots,\mathbf{r}_N\}$ is denoted in abbreviation by $\{1,\ldots,N\}$. It is important to use symmetric or antisymmetric wave functions, as required by the statistics of the particles (see Problem 10.4). Let us define

$$W_N(1,\ldots,N) \equiv N!\lambda^{3N}\sum_\alpha \Psi_\alpha^*(1,\ldots,N)\, e^{-\beta\mathscr{H}}\Psi_\alpha(1,\ldots,N) \quad (10.35)$$

The partition function can be written in the form

$$Q_N(V, T) = \frac{1}{N!\lambda^{3N}} \int d^{3N}r\, W_N(1,\ldots,N) \quad (10.36)$$

*See Mayer and Mayer, *loc. cit.*
[†] B. Kahn and G. E. Uhlenbeck, *Physica* **5**, 399 (1938).

The integral appearing in (10.36) approaches the classical configuration integral in the limit of high temperatures. Some properties of the function $W_N(1,\ldots,N)$ are

(a) $W_1(1) = 1$

Proof

$$W_1(1) = W_1(\mathbf{r}_1) = \frac{\lambda^3}{V} \sum_{\mathbf{p}} e^{-i\mathbf{p}\cdot\mathbf{r}_1/\hbar} e^{(\beta\hbar^2/2m)\nabla^2} e^{i\mathbf{p}\cdot\mathbf{r}_1/\hbar}$$

$$= \left(\frac{\lambda}{h}\right)^3 \int d^3p \, e^{-\beta p^2/2m} = 1 \qquad \blacksquare$$

(b) $W_N(1,\ldots,N)$ is a symmetric function of its arguments.

(c) $W_N(1,\ldots,N)$ is invariant under a unitary transformation of the complete set of wave functions $\{\Psi_\alpha\}$ appearing in (10.35).

Proof

Suppose $\Psi_\alpha = \sum_\lambda S_{\alpha\lambda}\Phi_\lambda$, where $S_{\alpha\lambda}$ is a unitary matrix:

$$\sum_\alpha S_{\alpha\lambda}^* S_{\alpha\gamma} = \delta_{\lambda\gamma}$$

Then

$$\sum_\alpha (\Psi_\alpha, e^{-\beta\mathscr{H}}\Psi_\alpha) = \sum_{\alpha,\lambda} S_{\alpha\lambda}^* S_{\alpha\gamma} (\Phi_\lambda, e^{-\beta\mathscr{H}}\Phi_\gamma) = \sum_\lambda (\Phi_\lambda, e^{-\beta\mathscr{H}}\Phi_\lambda) \qquad \blacksquare$$

The following property appears to be intuitively obvious, but it is difficult to establish quantitatively. Suppose the coordinates $\mathbf{r}_1,\ldots,\mathbf{r}_N$ have such values that they can be divided into two groups containing respectively A and B coordinates, with the property that any two coordinates $\mathbf{r}_i$ and $\mathbf{r}_j$ belonging to different groups must satisfy the condition

$$\begin{aligned} |\mathbf{r}_i - \mathbf{r}_j| &\gg r_0 \\ |\mathbf{r}_i - \mathbf{r}_j| &\gg \lambda \end{aligned} \qquad (10.37)$$

Then

$$W_N(\mathbf{r}_1,\ldots,\mathbf{r}_N) \approx W_A(r_A)W_B(r_B) \qquad (10.38)$$

where r_A and r_B denote collectively the respective coordinates in the two groups.

Consider first the case $N = 2$. According to (10.38) we should expect that as $|\mathbf{r}_1 - \mathbf{r}_2| \to \infty$,

$$W_2(1,2) \to W_1(1)W_2(2) \qquad (10.39)$$

If we define a function $U_2(1,2)$ by $W_2(1,2) = W_1(1)W_2(2) + U_2(1,2)$, we should expect that, as $|\mathbf{r}_1 - \mathbf{r}_2| \to \infty$,

$$U_2(1,2) \to 0 \qquad (10.40)$$

Hence the integral of $U_2(1,2)$ over $\mathbf{r}_1$ and $\mathbf{r}_2$ should be the analog of the 2-cluster in classical statistical mechanics.

We proceed systematically in the following manner. Let a sequence of cluster functions $U_l(1, \ldots, l)$ be successively defined by the following scheme, in which the lth equation is a definition of $U_l(1, \ldots, l)$:

$$W_1(1) = U_1(1) = 1 \tag{10.41}$$

$$W_2(1,2) = U_1(1)U_1(2) + U_2(1,2) \tag{10.42}$$

$$W_3(1,2,3) = U_1(1)U_1(2)U_1(3) + U_1(1)U_2(2,3)$$
$$+ U_1(2)U_2(3,1) + U_1(3)U_2(1,2) + U_3(1,2,3) \tag{10.43}$$

$$\vdots$$

The last equation in this scheme, defining $U_N(1, \ldots, N)$, is

$$W_N(1, \ldots, N)$$
$$= \sum_{\{m_l\}} \sum_P \underbrace{[U_1(\ ) \cdots U_1(\ )]}_{m_1 \text{ factors}} \underbrace{[U_2(,) \cdots U_2(,)]}_{m_2 \text{ factors}} \cdots \underbrace{[U_N(,,\cdots,)]}_{m_N \text{ factors}} \tag{10.44}$$

where m_l is zero or a positive integer and the set of integers $\{m_l\}$ satisfies the condition

$$\sum_{l=1}^{N} lm_l = N \tag{10.45}$$

The sum over $\{m_l\}$ in (10.44) extends over all sets $\{m_l\}$ satisfying (10.45). The arguments of the U_l are left blank in (10.44). There are exactly N such blanks, and they are to be filled by the N coordinates $r_1, \ldots, r_N$ in any order. The sum $\sum_P$ is a sum over all *distinct* ways of filling these blanks.

We can solve the equations (10.41)–(10.44) successively for U_1, U_2, etc., and obtain

$$U_1(1) = W_1(1) = 1 \tag{10.46}$$

$$U_2(1,2) = W_2(1,2) - W_1(1)W_1(2) \tag{10.47}$$

$$U_3(1,2,3) = W_3(1,2,3) - W_2(1,2)W_1(3) - W_2(2,3)W_1(1)$$
$$- W_2(3,1)W_1(2) + 2W_1(1)W_1(2)W_1(3) \tag{10.48}$$

$$\vdots$$

We see that $U_l(1, \ldots, l)$ is a symmetric function if its arguments and is determined by all the $W_{N'}$ with $N' \le l$. By the property (10.38) we expect that $U_l \to 0$ as $|r_i - r_j| \to \infty$, where r_i and r_j are any two of the arguments of U_l.

The l cluster integral $b_l(V, T)$ is defined by

$$b_l(V, T) \equiv \frac{1}{l! \lambda^{3l-3} V} \int d^3 r_1 \cdots d^3 r_l U_l(1, \ldots, l) \tag{10.49}$$

It is clear that b_l is dimensionless. If U_l vanishes sufficiently rapidly whenever any two of its arguments are far apart from each other, the integral appearing in (10.49) is proportional to V as $V \to \infty$, and the limit $b_l(\infty, T)$ may be expected to exist. Whether this is true depends on the nature of the interparticle potential. We assume that it is.

We now show that the partition function is expressible directly in terms of the cluster integrals. According to (10.36) we need to integrate W_N over all the coordinates. Let us make use of the formula (10.44). An integration over all the coordinates will yield the same result for every term in the sum $\sum_P$. Thus the result of the integration is the number of terms in the sum $\sum_P$ times the integral of any term in the sum $\sum_P$. The number of terms in the sum $\sum_P$ is given by (10.22). Hence

$$\int d^{3N}r \, W(1,\ldots,N)$$

$$= \sum_{\{m_l\}} \frac{N!}{[(1!)^{m_1}(2!)^{m_2}\cdots](m_1!m_2!\cdots)}$$

$$\times \int d^{3N}r \left[(U_1 \cdots U_1)(U_2 \cdots U_2)\cdots\right]$$

$$= N! \sum_{\{m_l\}} \frac{1}{m_1!}\left[\frac{1}{1!}\int d^3r_1 \, U_1(1)\right]^{m_1} \frac{1}{m_2!}\left[\frac{1}{2!}\int d^3r_1 \, d^3r_2 \, U_2(1,2)\right]^{m_2} \cdots$$

$$= N! \sum_{\{m_l\}} \prod_{l=1}^{N} \frac{(V\lambda^{3l-3}b_l)^{m_l}}{m_l!} = N!\lambda^{3N} \sum_{\{m_l\}} \prod_{l=1}^{N} \frac{1}{m_l!}\left(\frac{V}{\lambda^3}b_l\right)^{m_l} \qquad (10.50)$$

Therefore the partition function is given by

$$Q_N(V,T) = \sum_{\{m_l\}} \prod_{l=1}^{N} \frac{1}{m_l!}\left(\frac{V}{\lambda^3}b_l\right)^{m_l} \qquad (10.51)$$

This is of precisely the same form as (10.25) for the classical partition function. The discussion following (10.25) therefore applies equally well to the present case and will not be repeated. We point out only the main differences between the quantum cluster integrals and the classical ones.

For an ideal gas we have seen in earlier chapters that

$$b_l^{(0)} = \begin{cases} l^{-5/2} & \text{(ideal Bose gas)} \\ (-1)^{l+1}l^{-5/2} & \text{(ideal Fermi gas)} \end{cases} \qquad (10.52)$$

Thus for a Bose and a Fermi gas b_l does not vanish for $l > 1$, even in the absence of interparticle interactions, in contradistinction to the classical ideal gas.

The calculation of b_l in the classical case only involves the calculation of a number of integrals—a finite task. In the quantum case, however, the calculation of b_l necessitates a knowledge of U_l, which in turn necessitates a knowledge of $W_{N'}$ for $N' \leq l$. Thus to find b_l for $l > 1$ we would have to solve an l-body problem. There is no finite prescription for doing this except for the case $l = 2$, which is the subject of the next section.

10.3 THE SECOND VIRIAL COEFFICIENT

To calculate the second virial coefficient a_2 for any system it is sufficient to calculate b_2, since $a_2 = -b_2$. A general formula for b_2 (in fact, for all b_l) has already been given for the classical case. Only the quantum case is considered here.*

To find b_2 we need to know $W_2(1, 2)$, which is a property of the two-body system. Let the Hamiltonian for the two-body system in question be

$$\mathscr{H} = -\frac{\hbar^2}{2m}(\nabla_1^2 + \nabla_2^2) + v(|\mathbf{r}_1 - \mathbf{r}_2|) \tag{10.53}$$

and let its normalized eigenfunctions be $\Psi_\alpha(1, 2)$, with eigenvalues E_α:

$$\mathscr{H}\Psi_\alpha(1, 2) = E_\alpha\Psi_\alpha(1, 2) \tag{10.54}$$

Let

$$\begin{align}
\mathbf{R} &= \tfrac{1}{2}(\mathbf{r}_1 + \mathbf{r}_2) \\
\mathbf{r} &= \mathbf{r}_2 - \mathbf{r}_1
\end{align} \tag{10.55}$$

Then

$$\begin{align}
\Psi_\alpha(1, 2) &= \frac{1}{\sqrt{V}} e^{i\mathbf{P}\cdot\mathbf{R}} \psi_n(\mathbf{r}) \\
E_\alpha &= \frac{P^2}{4m} + \epsilon_n
\end{align} \tag{10.56}$$

where the quantum number α refers to the set of quantum numbers $(\mathbf{P}, n)$. The relative wave function $\psi_n(\mathbf{r})$ satisfies the eigenvalue equation

$$\left[-\frac{\hbar^2}{m}\nabla^2 + v(r)\right]\psi_n(\mathbf{r}) = \epsilon_n\psi_n(\mathbf{r}) \tag{10.57}$$

with the normalization condition

$$\int d^3r\, |\psi_n(\mathbf{r})|^2 = 1 \tag{10.58}$$

Using (10.56) to be the wave functions for the calculation of $W_2(1, 2)$, we find from (10.35) that

$$W_2(1, 2) = 2\lambda^6 \sum_\alpha |\Psi_\alpha(1, 2)|^2 e^{-\beta E_\alpha} = \frac{2\lambda^6}{V}\sum_{\mathbf{P}}\sum_n |\psi_n(\mathbf{r})|^2 e^{-\beta P^2/4m} e^{-\beta\epsilon_n} \tag{10.59}$$

In the limit as $V \to \infty$ the sum over $\mathbf{P}$ can be effected immediately:

$$\frac{1}{V}\sum_{\mathbf{P}} e^{-\beta P^2/4m} = \frac{4\pi}{h^3}\int_0^\infty dP\, P^2 e^{-\beta P^2/4m} = \frac{2^{3/2}}{\lambda^3} \tag{10.60}$$

*The following development is due to E. Beth and G. E. Uhlenbeck, *Physica* 4, 915 (1937).

where $\lambda = \sqrt{2\pi\hbar^2/mkT}$, the thermal wavelength. Therefore

$$W_2(1,2) = 2^{5/2}\lambda^3 \sum_n |\psi_n(\mathbf{r})|^2 e^{-\beta\epsilon_n} \tag{10.61}$$

If we repeat all the calculations so far for a two-body system of noninteracting particles, we obtain

$$W_2^{(0)}(1,2) = 2^{5/2}\lambda^3 \sum_n |\psi_n^{(0)}(\mathbf{r})|^2 e^{-\beta\epsilon_n^{(0)}} \tag{10.62}$$

where the superscript $^{(0)}$ refers to quantities of the noninteracting system. From (10.49) and (10.47) we have

$$b_2 = \frac{1}{2\lambda^3 V}\int d^3r_1\, d^3r_2\, U_2(1,2) = \frac{1}{2\lambda^3 V}\int d^3R\, d^3r\, [W_2(1,2) - 1]$$

Hence

$$b_2 - b_2^{(0)} = \frac{1}{2\lambda^3 V}\int d^3R\, d^3r\, [W_2(1,2) - W_2^{(0)}(1,2)]$$

$$= 2\sqrt{2}\int d^3r \sum_n [|\psi_n(\mathbf{r})|^2 e^{-\beta\epsilon_n} - |\psi_n^{(0)}(\mathbf{r})|^2 e^{-\beta\epsilon_n^{(0)}}]$$

$$= 2\sqrt{2}\sum_n (e^{-\beta\epsilon_n} - e^{-\beta\epsilon_n^{(0)}}) \tag{10.63}$$

where

$$b_2^{(0)} = \begin{cases} 2^{-5/2} & \text{(ideal Bose gas)} \\ -2^{-5/2} & \text{(ideal Fermi gas)} \end{cases} \tag{10.64}$$

To analyze (10.63) further we must study the energy spectra $\epsilon_n^{(0)}$ and ϵ_n. For the noninteracting system, $\epsilon_n^{(0)}$ forms a continuum. We write

$$\epsilon_n^{(0)} = \frac{\hbar^2 k^2}{m} \tag{10.65}$$

which defines the relative wave number k. For the interacting system the spectrum of ϵ_n in general contains a discrete set of values ϵ_B, corresponding to two-body bound states, and a continuum. In the continuum, we define the wave number k for the interacting system by putting

$$\epsilon_n = \frac{\hbar^2 k^2}{m} \tag{10.66}$$

Let $g(k)\, dk$ be the number of states with wave number lying between k and $k + dk$, and let $g^{(0)}(k)\, dk$ denote the corresponding quantity for the noninteracting system. Then (10.63) can be written in the form

$$b_2 - b_2^{(0)} = 2^{3/2}\left\{ \sum_B e^{-\beta\epsilon_B} + \int_0^\infty dk\, [g(k) - g^{(0)}(k)]\, e^{-\beta\hbar^2 k^2/m} \right\} \tag{10.67}$$

where ϵ_B denotes the energy of a bound state of the interacting two-body system.

We remark in passing that the factor $2^{3/2}$ in front of (10.67) is the ratio $(\lambda/\lambda_{cm})^{3/2}$, where λ is the thermal wavelength, and λ_{cm} is the thermal wavelength of the center-of-mass motion of the two-body system.

Let $\eta_l(k)$ be the scattering phase shift of the potential $v(r)$ for the lth partial wave of wave number k. It will be shown that

$$g(k) - g^{(0)}(k) = \frac{1}{\pi} \sum_l{}' (2l+1) \frac{\partial \eta_l(k)}{\partial k} \tag{10.68}$$

where the sum $\sum'$ extends over the values

$$l = \begin{cases} 0, 2, 4, 6, \ldots & \text{(bosons)} \\ 1, 3, 5, 7, \ldots & \text{(fermions)} \end{cases} \tag{10.69}$$

Therefore

$$\mathit{b}_2 - \mathit{b}_2^{(0)} = 2^{3/2} \left\{ \sum_B e^{-\beta \epsilon_B} = \frac{1}{\pi} \int_0^\infty dk \sum_l{}' (2l+1) \frac{\partial \eta_l(k)}{\partial k} e^{-\beta \hbar^2 k^2/m} \right\} \tag{10.70}$$

A partial integration leads finally to the formula

$$\mathit{b}_2 - \mathit{b}_2^{(0)} = 2^{3/2} \left\{ \sum_B e^{-\beta \epsilon_B} + \frac{\lambda^2}{\pi^2} \sum_l{}' (2l+1) \int_0^\infty dk\, k \eta_l(k)\, e^{-\beta \hbar^2 k^2/m} \right\} \tag{10.71}$$

It remains to prove (10.68). We may choose both $\psi_n(\mathbf{r})$ and $\psi_n^{(0)}(\mathbf{r})$ to be pure spherical harmonics, because $v(r)$ does not depend on the angles of $\mathbf{r}$ with respect to any fixed axis. Thus we write

$$\psi_{klm}(\mathbf{r}) = A_{klm} Y_l^m(\theta, \phi) \frac{u_{kl}(r)}{r}$$

$$\psi_{klm}^{(0)}(\mathbf{r}) = A_{klm}^{(0)} Y_l^m(\theta, \phi) \frac{u_{kl}^{(0)}(r)}{r} \tag{10.72}$$

For bosons $\psi(\mathbf{r}) = \psi(-\mathbf{r})$, and for fermions $\psi(\mathbf{r}) = -\psi(-\mathbf{r})$. Therefore

$$l = \begin{cases} 0, 2, 4, 6, \ldots & \text{(bosons)} \\ 1, 3, 5, 7, \ldots & \text{(fermions)} \end{cases} \tag{10.73}$$

Let the boundary conditions be

$$u_{kl}(R) = u_{kl}^{(0)}(R) = 0 \tag{10.74}$$

where R is a very large radius which approaches infinity at the end of the calculation. The asymptotic forms of u_{kl} and $u_{kl}^{(0)}$ are

$$u_{kl}(r) \xrightarrow[r \to \infty]{} \sin\left[kr + \frac{l\pi}{2} + \eta_l(k) \right]$$

$$u_{kl}^{(0)}(r) \xrightarrow[r \to \infty]{} \sin\left(kr + \frac{l\pi}{2} \right) \tag{10.75}$$

This defines $\eta_l(k)$. The eigenvalues k are determined by the boundary conditions (10.74):

$$kR + \frac{l\pi}{2} + \eta_l(k) = \pi n \quad \text{(interacting system)}$$

$$kR + \frac{l\pi}{2} = \pi n \quad \text{(noninteracting system)}$$

(10.76)

where $n = 0, 1, 2, \ldots$. It is seen that the eigenvalues k depends on n and l but not on m. Since there are $2l + 1$ spherical harmonics Y_l^m for a given l, each eigenvalue k is $(2l + 1)$-fold degenerate.

For a given l, changing n by one unit causes k to change by the respective amounts $\Delta k, \Delta k^{(0)}$:

$$\Delta k = \frac{\pi}{R + [\partial \eta_l(k)/\partial k]}$$

$$\Delta k^{(0)} = \frac{\pi}{R}$$

(10.77)

These are the spacings of eigenvalues for a given l. Let the number of states of a given l with wave number lying between k and $k + dk$ be denoted by $g_l(k)\, dk$ and $g_l^{(0)}(k)\, dk$ for the two cases. We must have

$$\frac{g_l(k)\,\Delta k}{2l + 1} = 1$$

$$\frac{g_l^{(0)}(k)\,\Delta k^{(0)}}{2l + 1} = 1$$

(10.78)

or

$$g_l(k) = \frac{2l + 1}{\pi}\left[R + \frac{\partial \eta_l(k)}{\partial k} \right]$$

$$g_l^{(0)}(k) = \frac{2l + 1}{\pi} R$$

(10.79)

Therefore

$$g_l(k) - g_l^{(0)}(k) = \frac{2l + 1}{\pi} \frac{\partial \eta_l(k)}{\partial k}$$

(10.80)

Summing (10.80) over all l consistent with (10.73) we obtain (10.68).

For $l > 2$ there is no known formula for b_l comparable in simplicity to (14.71), because there is no known treatment of the l-body problem for $l > 2$ comparable to the phase shift analysis of the two-body problem.*

Lastly, we remark that there is no essential difference between a sharp scattering resonance and a bound state, as far as the second virial coefficient is concerned. In the neighborhood of a sharp resonance, the scattering phase shift

*An analysis of b_3 is given by A. Pais and G. E. Uhlenbeck, *Phys. Rev.* **116**, 250 (1959).

increases by π over a small energy interval. In the idealized limit of an infinitely sharp resonance, we can represent the phase shift by

$$\frac{\partial \eta(k)}{\partial k} = \pi\delta(k - k_0) \qquad (10.81)$$

where k_0 marks the position of the resonance. From (10.70), it is clear that each sharp resonance contributes to the second virial coefficient a term of the same form as that from a bound state. This supports what we expect, namely, that a sharp resonance can be treated as a particle.

10.4 VARIATIONAL PRINCIPLES

In quantum mechanics we are familiar with the variational principle, which states that the lowest energy eigenvalue of the system is the minimum of the expectation value of the Hamiltonian, taken with respect to a wave function that is completely arbitrary, except for normalization and the imposed boundary conditions of the problem. By using a trial wave function with adjustible parameters, one can use the variational principle to obtain an upper bound for the ground state energy, and improve on the bound by giving the trial wave function more freedom to vary. We shall describe here similar variational principles for the partition function.

Gibbs Variational Principle

Let ρ denote a normalized density function for an ensemble, classical or quantum mechanical. That is, it is a real positive quantity in the classical case, and a Hermitian operator with positive eigenvalues in the quantum case, and that

$$\text{Tr}\,\rho = 1 \qquad (10.82)$$

In the classical case the operation Tr means $\int dp\,dq$. Now define

$$\psi(\rho) \equiv \text{Tr}\,(\mathcal{H}\rho) + \beta^{-1}\,\text{Tr}\,(\rho \log \rho) \qquad (10.83)$$

where $\mathcal{H}$ is the Hamiltonian of the system under consideration, and β is a constant. The Gibbs variational principle states the following:

(a) Minimize $\psi\{\rho\}$ by varying ρ, subject only to the condition that it be a legitimate normalized density function. The function $\bar{\rho}$ that minimizes $\psi(\rho)$ is the density function of the canonical ensemble with $kT = \beta^{-1}$.

(b) The Helmholtz free energy is given by $A = \psi(\bar{\rho})$.

To prove this, first calculate the variations of ψ when ρ changes by $\delta\rho$:

$$\delta\psi = \text{Tr}\,\{[\mathcal{H} + \beta^{-1}(1 + \log \rho)]\,\delta\rho\}$$
$$\delta^2\psi = \text{Tr}\,[(\beta\rho)^{-1}(\delta\rho)^2] \qquad (10.84)$$

The second of these shows that ψ is a convex function, so that $\delta\psi = 0$ gives a minimum. We now vary ρ, taking the normalization constraint into account through a Lagrange multiplier λ:

$$0 = \delta\psi + \lambda\delta(\mathrm{Tr}\,\rho) = \mathrm{Tr}\left\{\left[\mathscr{H} + \beta^{-1}(1 + \log\rho) + \lambda\right]\delta\rho\right\} \quad (10.85)$$

Solving for ρ, and determining λ by (10.82), we obtain

$$\bar{\rho} = e^{-\beta\mathscr{H}}/\mathrm{Tr}\,e^{-\beta\mathscr{H}} \quad (10.86)$$

which proves (a). Substituting this into (10.83) gives

$$\psi(\bar{\rho}) = -\beta^{-1}\log\mathrm{Tr}\,e^{-\beta\mathscr{H}} \quad (10.87)$$

which proves (b). ∎

Peierls Variational Principle

Consider a quantum mechanical system with Hamiltonian $\mathscr{H}$, and partition function $Q = \mathrm{Tr}\exp(-\beta\mathscr{H})$. The Peierls variational principle* states that

$$Q \geq \sum_n e^{-\beta(\Phi_n,\,\mathscr{H}\Phi_n)} \quad (10.88)$$

where $\{\Phi_n\}$ is an arbitrary set of wave functions of the system. Obviously, the equality holds when $\{\Phi_n\}$ is the set of eigenfunctions of $\mathscr{H}$. Note that the set $\{\Phi_n\}$ does not have to be complete, for the inequality holds a fortiori for an incomplete set since the terms on the right side of (10.88) are all positive. Thus it suffices to prove (10.88) under the assumption that $\{\Phi_n\}$ is a complete set of wave functions.

The Peierls variational principle is a special case of a more general theorem on convex functions. Suppose $f(x)$ is a real convex function of a real variable x, (i.e., $f''(x) \geq 0$.) Let us denote by $\bar{f}$ the average of $f(x)$ over a selected set of x's, with specified weights:

$$\overline{f(x)} \equiv \sum_n c_n f(x_n) \quad (10.89)$$

where $\{x_n\}$ is a arbitrary set of real numbers, and $\{c_n\}$ is a set of real numbers such that

$$c_n \geq 0, \qquad \sum_n c_n = 1 \quad (10.90)$$

By the mean-value theorem,

$$f(x) = f(\bar{x}) + \tfrac{1}{2}(x - \bar{x})^2 f''(x_1) \quad (10.91)$$

for some x_1. Now average both sides:

$$\overline{f(x)} = f(\bar{x}) + \tfrac{1}{2}\overline{(x - \bar{x})^2 f''(x_1)} \quad (10.92)$$

*R. E. Peierls, *Phys. Rev.* **54**, 918 (1938).

Since $f''(x_1) \geq 0$, we have the following theorem:

$$\overline{f(x)} \geq f(\bar{x}) \tag{10.93}$$

Now let $\{\Phi_n\}$ be a complete set of wave functions, and S the unitary matrix which relates it to the eigenfunctions $\{\Psi_n\}$ of $\mathscr{H}$:

$$\Phi_n = \sum_m S_{nm}\Psi_m$$
$$\sum_n |S_{nm}|^2 = 1 \tag{10.94}$$

from which we can see that

$$(\Phi_n, \mathscr{H}\Phi_n) = \sum_m |S_{nm}|^2 E_m \tag{10.95}$$

where $\{E_m\}$ are the eigenvalues of $\mathscr{H}$. We can write the partition function in the form

$$Q = \sum_n e^{-\beta E_n} = \sum_n \sum_m |S_{nm}|^2 e^{-\beta E_m} \tag{10.96}$$

and define

$$q = \sum_n e^{-\beta(\Phi_n, \mathscr{H}\Phi_n)} = \sum_n \exp\left[-\beta \sum_m |S_{nm}|^2 E_m\right] \tag{10.97}$$

Let $f(x) = \exp(-\beta x)$. For each n, the following definitions fulfill the requirements of the previous theorem:

$$\bar{E}_n = \sum_m |S_{nm}|^2 E_m$$
$$\overline{f(E)}_n = \sum_m |S_{nm}|^2 f(E_m) \tag{10.98}$$

Thus we can write

$$Q - q = \sum_n \left[\overline{f(E)}_n - f(\bar{E}_n)\right] \tag{10.99}$$

According to the theorem, $Q - q \geq 0$ term by term. ∎

10.5 IMPERFECT GASES AT LOW TEMPERATURES

An imperfect gas is an extremely dilute system of particles that interact among themselves through an interparticle potential of finite range and of such a nature that there exists no two-particle bound state. The diluteness of the gas enables us to treat the interparticle interaction as a small perturbation on the ideal gas. An imperfect gas, therefore, is the first improvement on the ideal gas as a model for a physical gas. We shall consider an imperfect gas at extremely low temperatures.

For such a system there are two important parameters of the dimension of length: the thermal wavelength λ and the average interparticle separation $v^{1/3}$. These two lengths may be of comparable magnitude, but they must be much larger than the range of the interparticle potential, or any other length in the problem, except that size of the container.

In quantum mechanics a particle cannot be localized within its de Broglie wavelength, which in the present case may be replaced by the thermal wavelength. Thus in the present case a particle "spreads" over a distance much larger than the range of the interaction potential. Within the range of interaction of any given particle, the probability of finding another particle is small. Therefore

- (a) the effective interaction experienced by a particle is small, even though the interparticle potential may have large values;
- (b) the details of the interparticle potential are unimportant, because a particle that is spread out in space sees only an averaged effect of the potential.

In the quantum theory of scattering it is known that at low energies the scattering of a particle by a potential does not depend on the shape of the potential, but depends only on a single parameter obtainable from the potential — the scattering length a. The total scattering cross section at low energies is $4\pi a^2$. Hence roughly speaking a is the effective diameter of the potential. We may also say that at low energies the scattering from a potential looks like that from a hard sphere of diameter a. This makes it plausible that at extremely low temperatures it is possible to describe an imperfect gas solely in terms of the three parameters λ, $v^{1/3}$, and a. Our problem is to formulate a method by which all the thermodynamic functions of the imperfect gas can be obtained to lowest order in the small parameters a/λ and $a/v^{1/3}$.

We first show that, for the purpose of calculating the low-lying energy levels of an imperfect gas, the Hamiltonian of the system may be replaced by an effective Hamiltonian in which only scattering parameters, such as the scattering length, appear explicitly. The partition function of the imperfect gas can then be calculated with the help of the effective Hamiltonian. This method, first introduced by Fermi,* is known as the *method of pseudopotentials.*

Consider first a system of two particles interacting through a finite-ranged potential which has no bound state. The object of the method of pseudopotentials is to obtain all the energy levels of the system in terms of the scattering phase shifts of the potential. For the sake of concreteness we first assume that the potential is the hard-sphere potential with diameter a. The wave function for the two particles may be written in the form

$$\Psi(\mathbf{r}_1, \mathbf{r}_2) = e^{i\mathbf{P}\cdot\mathbf{R}} \, \psi(\mathbf{r}) \tag{10.100}$$

---

*E. Fermi, *Ricerca Sci.* **7**, 13 (1936). Our presentation follows that of K. Huang and C. N. Yang, *Phys. Rev.* **105**, 767 (1957).

where

$$\mathbf{R} = \tfrac{1}{2}(\mathbf{r}_1 + \mathbf{r})$$
$$\mathbf{r} = \mathbf{r}_2 - \mathbf{r}_1 \tag{10.101}$$

and $\mathbf{P}$ is the total momentum vector. The Schrödinger equation in the center-of-mass system is

$$(\nabla^2 + k^2)\psi(\mathbf{r}) = 0 \qquad (r > a)$$
$$\psi(\mathbf{r}) = 0 \qquad (r \leq a) \tag{10.102}$$

The hard-sphere potential is no more than a boundary condition for the relative wave function $\psi(\mathbf{r})$. It is understood that some boundary condition for $r \to \infty$ is specified, but what it is is irrelevant to our considerations. The number k is the relative wave number, and (10.102) presents an eigenvalue problem for k. When the allowed values of k are known, the energy eigenvalues of the system are given by

$$E(\mathbf{P}, \mathbf{k}) = \frac{P^2}{2M} + \frac{\hbar^2 k^2}{2\mu}$$

where M is the total mass and μ the reduced mass of the system.

The aim of the method of pseudopotentials is to replace the hard-sphere boundary condition by an inhomogeneous term for the wave equation. Such an idea is familiar in electrostatics, where to find the electrostatic potential in the presence of a metallic sphere (with some given boundary condition at infinity) we may replace the sphere by a distribution of charges on the surface of the sphere and find the potential set up by the fictitious charges. We can further replace the surface charges by a collection of multipoles at the center of the sphere with appropriate strengths. If we solve the Poisson equation with these multipole sources, we obtain the exact electrostatic potential *outside the sphere*. In an analogous way, the method of pseudopotentials replaces the boundary condition on $\psi(\mathbf{r})$ by a collection of sources at the point $\mathbf{r} = 0$. Instead of producing electrostatic multipole potentials, however, these sources will produce scattered S waves, P waves, D waves, etc.

Let us first consider spherically symmetric (S wave) solutions of (10.102) at very low energies ($k \to 0$). The equations (10.102) become

$$\frac{1}{r^2}\frac{d}{dr}\left(r^2 \frac{d\psi}{dr}\right) = 0 \qquad (r > a)$$
$$\psi(r) = 0 \qquad (r \leq a) \tag{10.103}$$

The solution is obviously

$$\psi(r) = \begin{cases} \text{const.}\left(1 - \dfrac{a}{r}\right) & (r > a) \\ 0 & (r \leq a) \end{cases} \tag{10.104}$$

Now define an extended wave function $\psi_{ex}(r)$ such that

$$(\nabla^2 + k^2)\psi_{ex}(r) = 0 \qquad \text{(everywhere except at } r = 0) \qquad (10.105)$$

with the boundary condition

$$\psi_{ex}(a) = 0 \qquad (10.106)$$

For $k \to 0$ we have

$$\psi_{ex}(r) \xrightarrow[r \to 0]{} \left(1 - \frac{a}{r}\right)\chi \qquad (10.107)$$

where χ is a constant that depends on the boundary condition at $r = \infty$. We can avoid explicit use of this boundary condition by writing

$$\chi = \left[\frac{\partial}{\partial r}(r\psi_{ex})\right]_{r=0} \qquad (10.108)$$

which is an immediate consequence of (10.107). To eliminate the explicit requirement (10.106), we generalize the equation (10.106) to include the point $r = 0$. This can be easily done by finding the behavior of $(\nabla^2 + k^2)\psi_{ex}$ near $r = 0$, as required by (10.107). Since $k \to 0$, it is sufficient to note that according to (10.107)

$$\nabla^2\psi_{ex}(r) \xrightarrow[r \to 0]{} 4\pi a\,\delta(\mathbf{r})\chi = 4\pi a\,\delta(\mathbf{r})\frac{\partial}{\partial r}(r\psi_{ex}) \qquad (10.109)$$

Therefore as $k \to 0$ the function $\psi_{ex}(r)$ everywhere satisfies the equation

$$(\nabla^2 + k^2)\psi_{ex}(r) = 4\pi a\,\delta(\mathbf{r})\frac{\partial}{\partial r}(r\psi_{ex}) \qquad (10.110)$$

The operator $\delta(\mathbf{r})(\partial/\partial r)r$ is the pseudopotential.* For small k and for $r \geq a$, $\psi_{ex}(r)$ satisfies the same equation and the same boundary condition as $\psi(r)$. Therefore $\psi_{ex}(r) = \psi(r)$ for $r \geq a$, and the eigenvalues of k are the same in both cases.

The equation (10.110) is not the exact equation we desire, because only the S-wave solutions with small k coincide with the actual solutions of the physical problem. To obtain an equation for an extended wave function that rigorously coincides with $\psi(r)$ for $r \geq a$ it is necessary to generalize (10.110) to arbitrary values of k and to nonspherically symmetric solutions.† It suffices for the present to state that the result of the generalization consists of the following modifications of (10.110):

(a) The exact S-wave pseudopotential is

$$-\frac{4\pi}{k \cot \eta_0}\delta(\mathbf{r})\frac{\partial}{\partial r}r \qquad (10.111)$$

---

*The foregoing derivation is due to J. M. Blatt and V. F. Weisskopf, *Theoretical Nuclear Physics* (Wiley, New York, 1952), p. 74.

†See Huang and Yang, *op. cit.*

where η_0 is the S-wave phase shift for the hard-sphere potential:

$$-\frac{1}{k\cot\eta_0} = \frac{\tan ka}{k} = a\left[1 + \tfrac{1}{3}(ka)^2 + \cdots\right] \qquad (10.112)$$

(b) An infinite series of pseudopotentials is added to the right side of (10.110), representing the effects of P-wave scattering, D-wave scattering, etc. The lth-wave pseudopotential is proportional to a^{2l+1}.

From these results it is seen that (10.110) is correct up to the order a^2. That is, if the wave function $\psi(r)$ and the eigenvalue k are expanded in a power series in a, then (10.110) correctly gives the coefficients of a and a^2.

The differential operator $(\partial/\partial r)r$ in the pseudopotential (10.110) may be replaced by unity if $\psi_{ex}(\mathbf{r})$ is well behaved at $r = 0$, for then

$$\left[\frac{\partial}{\partial r}(r\psi_{ex})\right]_{r=0} = \psi_{ex}(0) + \left[r\frac{\partial}{\partial r}\psi_{ex}\right]_{r=0} = \psi_{ex}(0) \qquad (10.113)$$

If $\psi_{ex}(\mathbf{r})\xrightarrow[r\to 0]{}Ar^{-1} + B$, however, then

$$\left[\frac{\partial}{\partial r}(r\psi_{ex})\right]_{r=0} = B \qquad (10.114)$$

An illustration of the effect of $(\partial/\partial r)r$ is given in Problem 10.7.

We now turn to the method of pseudopotentials for the case of two particles interacting through a general finite-ranged potential which has no bound state. Here (10.102) is replaced by the equation

$$\frac{\hbar^2}{2\mu}(\nabla^2 + k^2)\psi(\mathbf{r}) = v(r)\psi(\mathbf{r}) \qquad (10.115)$$

with some given boundary condition for $r \to \infty$. At low energies only S-wave scattering is important. Therefore let us consider only spherically symmetric solutions. Then (10.115) reduces to

$$u''(r) + k^2 u(r) = \frac{\mu}{\hbar^2}v(r)u(r) \qquad (10.116)$$

where

$$u(r) \equiv r\psi(r) \qquad (10.117)$$

By assumption $v(r)$ is finite-ranged and has no bound state. Therefore, as $r \to \infty$, $u(r)$ approaches a sinusoidal function:

$$u(r)\xrightarrow[r\to\infty]{}u_\infty(r) \qquad (10.118)$$

where

$$u_\infty(r) \equiv r\psi_\infty(r) = \text{const.}\,(\sin kr + \tan\eta_0\cos kr) \qquad (10.119)$$

where η_0 is by definition the S-wave phase shift. For $k \to 0$,

$$\psi_\infty(r)\xrightarrow[r\to 0]{}\text{const.}\left(1 + \frac{\tan\eta_0}{kr}\right) \qquad (10.120)$$

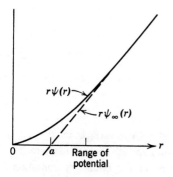

Fig. 10.3 Wave function in a repulsive potential with positive scattering length.

In general η_0 is a function of k. For small k there is a well-known expansion analogous to (10.112), known as the effective range expansion:

$$k \cot \eta_0 = -\frac{1}{a} + \frac{1}{2} k^2 r_0 + \cdots \qquad (10.121)$$

where a is called the scattering length and r_0 the effective range. The meaning of the scattering length can be seen by substituting (10.121) into (10.119). For $k \to 0$ we obtain (10.117). As illustrated in Figs. 10.3 and 10.4, the scattering length is the intercept of the asymptotic wave function $r\psi_\infty(r)$ with the coordinate axis. For the hard-sphere potential the scattering length is the hard-sphere diameter. In general a may be either positive or negative. It is positive for a predominantly repulsive potential (Fig. 10.3) and negative for a predominantly attractive potential (Fig. 10.4).

At low energies we may neglect all terms in (10.121) except $-1/a$ and obtain

$$-\frac{1}{k \cot \eta_0} \approx a$$

This approximation, known as the "shape-independent approximation," states that at low energies the potential acts as if it were a hard-sphere potential of

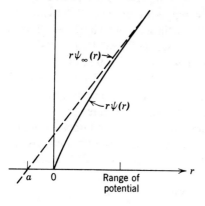

Fig. 10.4 Wave function in an attractive potential with negative scattering length.

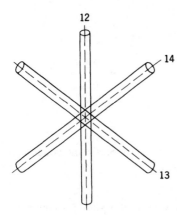

Fig. 10.5 The three-like hypersurface in the $3N$-dimensional configuration space. The hard-sphere interactions are equivalent to the boundary condition that the wave function vanishes on the surface of the "tree."

diameter a. Therefore (10.110) can be taken over.* In general (10.110) is certainly valid for the calculation of the energy to the lowest order in the scattering length a. Whether it is still meaningful to use (10.110) for higher orders in a depends on the potential.

Having introduced the pseudopotentials in the two-body problem we are now in a position to discuss the generalization to the N-body problem. The considerations that follow are independent of statistics.

Let us first consider the n-body problem with hard-sphere interactions. The Schrödinger equation for the system is

$$-\frac{\hbar^2}{2m}\left(\nabla_1^2 + \cdots + \nabla_N^2\right)\Psi = E\Psi \qquad \left(|\mathbf{r}_i - \mathbf{r}_j| > a, \quad \text{all } i \neq j\right)$$

$$\Psi = 0 \qquad \text{(otherwise)}$$

(10.122)

We also require that Ψ satisfies some boundary condition on the surface of a large cube, e.g., that Ψ satisfies periodic boundary conditions. The hard-sphere interactions are equivalent to a boundary condition that requires Ψ to vanish whenever $|\mathbf{r}_i - \mathbf{r}_j| = a$, for all $i \neq j$. In the $3N$-dimensional configuration space the collection of all points for which $|\mathbf{r}_i - \mathbf{r}_j| = a$ represents a tree-like hypersurface, a portion of which we schematically represent by Fig. 10.5. Thus we draw a cylinder, labeled 12, to represent the surface in which $|\mathbf{r}_1 - \mathbf{r}_2| = a$, whereas $\mathbf{r}_3, \ldots, \mathbf{r}_N$ may have arbitrary values. The whole "tree" is the totality of all such cylinders, $\frac{1}{2}N(N-1)$ in number, which mutually intersect in a complicated way. If the hard-sphere diameter a is small, these cylinders have a small radius. To find the wave function outside the "tree," it is natural to replace the "tree'" by a series of "multipoles" at the "axes," i.e., at the lines $|\mathbf{r}_i - \mathbf{r}_j| = 0$.

It can be easily shown that replacing the effect of each cylinder by multipoles along its axis amount to introducing the two-body pseudopotentials described in the previous section. Our extended wave function would then satisfy a

*The derivation of (10.110) remains valid if a is negative.

Schrödinger equation containing the sum of $\frac{1}{2}N(N-1)$ two-body pseudopotentials. These two-body pseudopotentials, however, do not exactly replace the effect of the "tree." Although they correctly give the behavior of Ψ near a cylinder and far away from any intersection of cylinders, they do not necessarily give the correct behavior of Ψ near an intersection of two or more cylinders. For example, the intersection corresponding to $|\mathbf{r}_1 - \mathbf{r}_2| = a$ and $|\mathbf{r}_1 - \mathbf{r}_3| = a$ represents a configuration in which particles 1, 2, and 3 collide simultaneously—an intrinsically three-body effect which has not been taken into account in the two-body pseudopotentials. The sum of two-body pseudopotentials accounts only for the effects of binary collisions.

Using our geometrical picture, we see that in addition to the two-body pseudopotentials it may be necessary to place additional multipoles (pseudopotentials) at each intersection of two or more cylinders. To find the exact magnitude of these three- and more-body pseudopotentials we would have to solve three- and more-body problems. Their dependence on the hard-sphere diameter a, however, can be found by a dimensional argument.

As an example, the three-body pseudopotential needed at the intersection of the lines $|\mathbf{r}_1 - \mathbf{r}_2| = 0$ and $|\mathbf{r}_1 - \mathbf{r}_3| = 0$ must appear in the three-body Schrödinger equation in the form

$$\left(\nabla_1^2 + \nabla_2^2 + \nabla_3^2 + k^2\right)\Psi$$
$$= (\text{sum of two-body pseudopotentials}) + \delta(\mathbf{r}_1 - \mathbf{r}_2)\,\delta(\mathbf{r}_1 - \mathbf{r}_3)K\Psi$$

The quantity K must be of the dimension $(\text{length})^4$. At low energies ($k \to 0$) the only length in the problem is a. Therefore K must be of the order a^4. In a similar way we deduce that four-body pseudopotentials are of the order a^7, and so forth. These pseudopotentials may be ignored, if we are only interested in an accuracy up to the order a^2. The necessity for such n-body pseudopotentials shows that the pseudopotentials are not additive. This is analogous to the well-known situation in electrostatics that image charges are not additive. For example, the images of a point charge in front of two mutually orthogonal plane conductors are not simply the two images produced by each plane conductor taken separately.

If the interparticle potential is not the hard-sphere potential but a finite-ranged potential that has no bound state, the considerations just given can be taken over. The effective Hamiltonian for an imperfect gas of N identical particles of mass m may be taken to be

$$\mathscr{H} = -\frac{\hbar^2}{2m}\left(\nabla_1^2 + \cdots + \nabla_N^2\right) + \frac{4\pi a\hbar^2}{m}\sum_{i<j}\delta(\mathbf{r}_i - \mathbf{r}_j)\frac{\partial}{\partial r_{ij}}r_{ij} \quad (10.123)$$

where a is the scattering length. This is valid for both fermions and bosons. The eigenvalues of this Hamiltonian will be the correct eigenvalues for an imperfect hard-sphere gas up to order a^2. For a general imperfect gas they will be correct to the lowest order in a.

We note that (10.123) is not a Hermitian operator because $(\partial/\partial r)r$ is not a Hermitian operator. This need not cause concern because, by its derivation, (10.123) has been shown to have real eigenvalues that are the approximate eigenvalues of the real problem. The non-Hermiticity reflects the fact that the eigenfunctions of (10.123) do not everywhere coincide with the eigenfunctions of the real problem, but do so only in the asymptotic region. The fact, however, that (10.123) is not Hermitian means that we cannot find its eigenvalues by variational methods.

If the pseudopotentials in (10.123) are regarded as small perturbations to be treated only to the first order in perturbation theory, then the operators $(\partial/\partial r)r$ will always act on unperturbed free-particle wave functions, which are well-behaved. Hence the operators $(\partial/\partial r)r$ can be set equal to unity, and we can work with the Hamiltonian

$$\mathcal{H}' = -\frac{\hbar^2}{2m}\left(\nabla_1^2 + \cdots + \nabla_N^2\right) + \frac{4\pi a\hbar^2}{m}\sum_{i<j}\delta(\mathbf{r}_i - \mathbf{r}_j) \qquad (10.124)$$

It is to be emphasized that *this Hamiltonian is valid only for the purpose of applying first-order perturbation theory*. We must not diagonalize (10.124) exactly, because the exact eigenvalues are the same as those for a free-particle system—it being well known that a three-dimensional δ-function potential produces no scattering.

The first-order energy levels of $\mathcal{H}'$ are calculated in the Appendix a. The result for bosons is given in (A.36); that for fermions in (A.42).

PROBLEMS

10.1 (*a*) Calculate b_2 and b_3 for a classical hard-sphere gas with hard-sphere diameter a.

(*b*) Express the equation of state of a classical hard-sphere gas in the form of a virial expansion. Include terms up to the third virial coefficient.

10.2 Find b_2 for an ideal Bose gas and compare it with b_2. Is the difference significant? (See (8.72).)

10.3 Calculate the second virial coefficients for a spinless hard-sphere Bose gas and a spinless hard-sphere Fermi gas to the *two* lowest nonvanishing orders in a/λ, where a is the hard sphere diameter and λ is the thermal wavelength.

Answers.

$$b_2 = 2^{-5/2} - \frac{2a}{\lambda} - \frac{10\pi^2}{3}\left(\frac{a}{\lambda}\right)^5 + \cdots \qquad \text{(Bose)}$$

$$b_2 = -2^{-5/2} - 6\pi\left(\frac{a}{\lambda}\right)^3 + 18\pi^2\left(\frac{a}{\lambda}\right)^5 + \cdots \qquad \text{(Fermi)}$$

10.4 In calculating W_N defined in (10.35), the symmetry or antisymmetry of the wave function makes the calculation complicated. The following is a method to deal with this problem.

Let the free-particle wave functions for a system of distinguishable particles be

$$\chi_p(1,\ldots,N) \equiv \frac{1}{V^{N/2}} e^{i(\mathbf{p}_1 \cdot \mathbf{r}_1 + \cdots + \mathbf{p}_N \cdot \mathbf{r}_N)}$$

Let

$$\langle 1,\ldots,N| e^{-\beta\mathcal{H}} |1',\ldots,N'\rangle$$
$$= \sum_{\mathbf{p}_1} \cdots \sum_{\mathbf{p}_N} \chi_p^*(1,\ldots,N)\, e^{-\beta\mathcal{H}} \chi_p (1',\ldots,N')$$

The symbol $|1,\ldots,N\rangle$ may be regarded as an eigenvector of the position operators of N distinguishable particles. Show that with the help of this quantity (10.35) may be expressed in the form

$$\frac{1}{\lambda^{3N}} W_N(1,\ldots,N) = \sum_P \delta_P \langle 1,\ldots,N| e^{-\beta H} |P1,\ldots,PN\rangle$$

10.5 Models for Ferromagnetism. Consider a lattice of N fixed atoms of spin $\frac{1}{2}$. The quantum mechanical spin operators of the ith atom are the Pauli spin matrices σ_i. Assuming that only nearest neighbors interact via a spin-spin interaction, we obtain the *Heisenberg model of ferromagnetism*. The Hamiltonian is

$$\mathcal{H}_{\text{Heisenberg}} = -\epsilon \sum_{\langle ij \rangle} \sigma_i \cdot \sigma_j - \mu \sum_{i=1}^{N} \sigma_i \cdot \mathbf{H}$$

where $\langle ij \rangle$ denotes a nearest-neighbor pair, $\mathcal{H}$ is a uniform external magnetic field, and ϵ and μ are positive constants.

Another model, the *Ising model*, is constructed by associating with the ith atom a *number* s_i that is either $+1$ or -1 and taking the Hamiltonian to be

$$\mathcal{H}_{\text{Ising}} = -\epsilon \sum_{\langle ij \rangle} s_i s_j - \mu \sum_{i=1}^{N} s_i H$$

where H is the z component of $\mathbf{H}$.

Using the Peierls variational principle prove that, for the same temperature, the Helmholtz free energy of the Heisenberg model is not greater than that of the Ising model.

10.6 Mean-Field Approximation. Consider the Ising model, whose Hamiltonian is given in the last problem. In the mean-field approximation one assumes that each spin sees a mean field due to all its neighbors. Determine this mean field with the help of the Gibbs variational principle, as follows:

(*a*) Assume a product form for the trial density function

$$\rho(s_1,\ldots,s_N) = g(s_1) \cdots g(s_N)$$
$$g(s) = C e^{Bs}$$

Find C by normalizing $g(s)$. The mean field B is to be determined.

(*b*) Instead of B, use as variational parameter the magnetization per spin

$$m = \sum_s s g(s)$$

Show that $B = \tanh^{-1} m$.

(c) Show that the Gibbs function $\psi(\rho)$, as defined in (10.83) is given by

$$\psi = N\left[-\tfrac{1}{2}\epsilon\gamma m^2 - \mu Hm + kT(Bm + \log C)\right]$$

where γ is the number of nearest neighbors.

(d) Show that ψ is minimized by $\overline{m}$, which satisfies

$$\overline{m} = \tanh\left[(\epsilon\gamma\overline{m} + \mu H)/kT\right]$$

(e) Show that the Helmholtz free energy per spin is given by

$$\frac{A}{N} = -\tfrac{1}{2}\epsilon\gamma\overline{m}^2 + \mu H\overline{m} + kT\left[\frac{1+\overline{m}}{2}\log\frac{1+\overline{m}}{2} + \frac{1-\overline{m}}{2}\log\frac{1-\overline{m}}{2}\right]$$

These results are the same as those of the Bragg-Williams approximation, which we shall derive and analyze in Chapter 14.

10.7 (a) Find all spherically symmetric solutions and corresponding eigenvalues of the equation

$$\left(\nabla^2 + k_n^2\right)\psi_n(r) = 0$$

in the region between two concentric spheres of radii R and a $(R > a)$, with the boundary conditions

$$\psi(R) = \psi(a) = 0$$

(b) Expand the eigenvalues k_n^2 in powers of a, keeping terms up to order a^2.

(c) Using the method of pseudopotentials, calculate the eigenvalue k_n^2 up to order a^2 and show that it agrees with the answer to (b).

Reference. K. Huang and C. N. Yang, *Phys. Rev.* **105**, 767 (1957), §2(b).

11

FERMI SYSTEMS

In this chapter we study various examples of systems of fermions. The dominant common characteristic is the existence of the Fermi surface, which is a direct consequence of the Pauli exclusion principle.

11.1 THE EQUATION OF STATE OF AN IDEAL FERMI GAS

The equation of state of a spinless ideal Fermi gas is obtained by eliminating z from Eqs. (8.67). We first study the behavior of z as determined by the second equation of (8.67), namely

$$\frac{\lambda^3}{v} = f_{3/2}(z) \tag{11.1}$$

where $v = V/N$, $\lambda = \sqrt{2\pi\hbar^2/mkT}$ is the thermal wavelength, and

$$f_{3/2}(z) = \frac{4}{\sqrt{\pi}} \int_0^\infty dx \, \frac{x^2}{z^{-1}e^{x^2} + 1} \tag{11.2}$$

is a monotonically increasing function of z. For small z we have the power series expansion

$$f_{3/2}(z) = z - \frac{z^2}{2^{3/2}} + \frac{z^3}{3^{3/2}} - \frac{z^4}{4^{3/2}} + \cdots \tag{11.3}$$

For large z an asymptotic expansion may be obtained through a method due to Sommerfeld, as follows. For convenience put $z = e^\nu$, so that ν is related to the chemical potential μ by

$$\nu = \log z = \mu/kT \tag{11.4}$$

Then

$$f_{3/2}(z) = \frac{4}{\sqrt{\pi}} \int_0^\infty dx \, \frac{x^2}{e^{x^2-\nu}+1} = \frac{2}{\sqrt{\pi}} \int_0^\infty dy \, \frac{\sqrt{y}}{e^{y-\nu}+1}$$

$$= \frac{4}{3\sqrt{\pi}} \int_0^\infty dy \, \frac{y^{3/2}e^{y-\nu}}{(e^{y-\nu}+1)^2} \tag{11.5}$$

The last step is obtained through a partial integration. Expanding $y^{3/2}$ in a Taylor series about ν, we obtain

$$f_{3/2}(z) = \frac{4}{3\sqrt{\pi}} \int_0^\infty dy \, \frac{e^{y-\nu}}{(e^{y-\nu}+1)^2}$$

$$\times \left[\nu^{3/2} + \tfrac{3}{2}\nu^{1/2}(y-\nu) + \tfrac{3}{8}\nu^{-1/2}(y-\nu)^2 + \cdots \right]$$

$$= \frac{4}{3\sqrt{\pi}} \int_{-\nu}^\infty dt \, \frac{e^t}{(e^t+1)^2} \left(\nu^{3/2} + \tfrac{3}{2}\nu^{1/2}t + \tfrac{3}{8}\nu^{-1/2}t^2 + \cdots \right) \tag{11.6}$$

Now we write

$$\int_{-\nu}^\infty = \int_{-\infty}^{+\infty} - \int_{-\infty}^{-\nu}$$

The second integral is of order $e^{-\nu}$. Therefore

$$f_{3/2}(z) = \frac{4}{3\sqrt{\pi}} \int_{-\infty}^{+\infty} dt \, \frac{e^t}{(e^t+1)^2} \left(\nu^{3/2} + \tfrac{3}{2}\nu^{1/2}t + \tfrac{3}{8}\nu^{-1/2}t^2 + \cdots \right) + O(e^{-\nu})$$

$$= \frac{4}{3\sqrt{\pi}} \left(I_0\nu^{3/2} + \tfrac{3}{2}I_1\nu^{1/2} + \tfrac{3}{8}I_2\nu^{-1/2} + \cdots \right) + O(e^{-\nu}) \tag{11.7}$$

where

$$I_n \equiv \int_{-\infty}^{+\infty} dt \, \frac{t^n e^t}{(e^t+1)^2} \tag{11.8}$$

Apart from the factor t^n, the integrand is an even function of t. Hence $I_n = 0$ for odd n. For $n = 0$ we have

$$I_0 = -2\int_0^\infty dt \, \frac{d}{dt} \frac{1}{(e^t+1)} = 1 \tag{11.9}$$

and for even $n > 0$,

$$I_n = -2\left[\frac{\partial}{\partial \lambda} \int_0^\infty dt \, \frac{t^{n-1}}{e^{\lambda t}+1} \right]_{\lambda=1} = 2n\int_0^\infty du \, \frac{u^{n-1}}{e^u+1}$$

$$= (n-1)!(2n)(1-2^{1-n})\zeta(n) \tag{11.10}$$

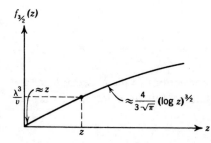

Fig. 11.1 The function $f_{3/2}(z)$.

where $\zeta(n)$ is the Riemann zeta function,* some special value of which are

$$\zeta(2) = \frac{\pi^2}{6}, \qquad \zeta(4) = \frac{\pi^4}{90}, \qquad \zeta(6) = \frac{\pi^6}{945}$$

Hence

$$f_{3/2}(z) = \frac{4}{3\sqrt{\pi}} \left[(\log z)^{3/2} + \frac{\pi^2}{8}(\log z)^{-1/2} + \cdots \right] + O(z^{-1}) \quad (11.11)$$

A graph of $f_{3/2}(z)$ is shown in Fig. 11.1. For any given positive value of λ^3/v, the value of z determined by (11.1) can be read off such a graph. It is seen that z increases monotonically as λ^3/v increases. For fixed v, z increases monotonically as the temperature decreases.

High Temperatures and Low Densities ($\lambda^3/v \ll 1$)

For $\lambda^3/v \ll 1$ the average interparticle separation $v^{1/3}$ is much larger than the thermal wavelength λ. We expect quantum effects to be negligible. From (11.1) and (11.3),

$$\frac{\lambda^3}{v} = z - \frac{z^2}{2^{3/2}} + \cdots$$

which may be solved to give

$$z = \frac{\lambda^3}{v} + \frac{1}{2^{3/2}}\left(\frac{\lambda^3}{v}\right)^2 + \cdots \quad (11.12)$$

Thus z reduces to that of the Boltzmann gas (Eq. (8.52)) when $\lambda^3 \to 0$ ($T \to \infty$). The average occupation number (8.65) reduces to Maxwell-Boltzmann form:

$$\langle n_\mathbf{p} \rangle \approx \frac{\lambda^3}{v} e^{-\beta \epsilon_\mathbf{p}} \quad (11.13)$$

---

*cf. *Handbook of Mathematical Functions*, M. Abramovitz and I. A. Stegun, eds., (National Bureau of Standards, Washington, D.C., 1964), Chapter 23.

The equation of state (8.67) then becomes

$$\frac{Pv}{kT} = \frac{v}{\lambda^3}\left(z - \frac{z^2}{2^{5/2}} + \cdots\right) = 1 + \frac{1}{2^{5/2}}\frac{\lambda^3}{v} + \cdots \tag{11.14}$$

This is in the form of a virial expansion. The corrections to the classical ideal gas law, however, are not due to molecular interactions, but to quantum effects. The second virial coefficient in this case is

$$\frac{\lambda^3}{2^{5/2}} = \frac{1}{2}\left(\frac{\pi\hbar^2}{mkT}\right)^{3/2} \tag{11.15}$$

All other thermodynamic functions reduce to those for a classical ideal gas plus small corrections.

Low Temperatures and High Densities $(\lambda^3/v \gg 1)$

For $\lambda^3/v \gg 1$ the average de Broglie wavelength of a particle is much greater than the average interparticle separation. Thus quantum effects, in particular the effects of the Pauli exclusion principle, become all important.

In the neighborhood of absolute zero we have, from (11.1) and (11.11),

$$\frac{1}{v}\left(\frac{2\pi\hbar^2}{mkT}\right)^{3/2} \approx \frac{4}{3\sqrt{\pi}}(\log z)^{3/2} \tag{11.16}$$

Hence

$$z \approx e^{\beta\epsilon_F} \tag{11.17}$$

where ϵ_F, the chemical potential at absolute zero, is called the *Fermi energy*:

$$\epsilon_F \equiv \frac{\hbar^2}{2m}\left(\frac{6\pi^2}{v}\right)^{2/3} \tag{11.18}$$

To study its physical significance, let us examine $\langle n_{\mathbf{p}}\rangle$ near absolute zero:

$$\langle n_{\mathbf{p}}\rangle \approx \frac{1}{e^{\beta(\epsilon_{\mathbf{p}}-\epsilon_F)} + 1} \tag{11.19}$$

If $\epsilon_{\mathbf{p}} < \epsilon_F$, then the exponential in the denominator vanishes as $T \to 0$ $(\beta \to \infty)$. Hence $\langle n_{\mathbf{p}}\rangle = 1$. Otherwise, $\langle n_{\mathbf{p}}\rangle = 0$. Thus

$$\langle n_{\mathbf{p}}\rangle_{T=0} = \begin{cases} 1 & (\epsilon_{\mathbf{p}} < \epsilon_F) \\ 0 & (\epsilon_{\mathbf{p}} > \epsilon_F) \end{cases} \tag{11.20}$$

The physical meaning of this formula is clear. Because of the Pauli exclusion principle no two particles can be in the same state. Therefore, in the ground state of the system, the particles occupy the lowest possible levels and fill the levels up to the finite energy level ϵ_F. Thus ϵ_F is simply the single-particle energy level below which there are exactly N states. In momentum space the particles fill a sphere of radius p_F, the surface of which is called the *Fermi surface*.

With this interpretation, let us now calculate the Fermi energy independently, under more general conditions. Suppose all single-particle energy levels are g-fold degenerate. For example, $g = 2s + 1$ for a particle of spin s. The condition determining ϵ_F is then

$$g\sum_p \langle n_p \rangle_{T=0} = N \tag{11.21}$$

In view of (11.20), this states that there are N states with energy below the Fermi energy. Putting $\epsilon_F = p_F^2/2m$, we find

$$\frac{g}{(2\pi\hbar)^3}\frac{4\pi}{3}p_F^3 = \frac{N}{V} \tag{11.22}$$

Hence

$$\epsilon_F = \frac{\hbar^2}{2m}\left(\frac{6\pi^2}{gv}\right)^{2/3} \tag{11.23}$$

which reduces to (11.18) when $g = 1$. We can also interpret (11.21) as follows. Particles with different quantum numbers are not constrained by any symmetry requirement with respect to the interchange of their *positions*. Thus we may consider a system of N fermions, each with degeneracy g, to be made up of g independent Fermi gases each with N/g particles whose energies are nondegenerate.

To obtain the thermodynamic functions for low temperatures and high densities we first obtain the expansion for the chemical potential from (11.1) and (11.11):

$$kT\nu = kT \log z = \epsilon_F\left[1 - \frac{\pi^2}{12}\left(\frac{kT}{\epsilon_F}\right)^2 + \cdots\right] \tag{11.24}$$

The expansion parameter is kT/ϵ_F. If we define the *Fermi temperature* T_F, which is a function of density, by

$$kT_F \equiv \epsilon_F \tag{11.25}$$

then low temperature and high density means $T \ll T_F$. In this domain the gas is said to be *degenerate* because the particles tend to go to the lowest energy levels possible. For this reason T_F is also called the *degeneracy temperature*.

The average occupation number is

$$\langle n_p \rangle = \frac{1}{e^{\beta\epsilon_p - \nu} + 1} \tag{11.26}$$

where ν is given by (11.24). Since $\epsilon_p = p^2/2m$, n_p depends on $\mathbf{p}$ only through p^2. A sketch of n_p is shown in Fig. 11.2.

The internal energy is

$$U = \sum_p \epsilon_p \langle n_p \rangle = \frac{V}{h^3}\frac{4\pi}{2m}\int_0^\infty dp\, p^4 \langle n_p \rangle$$

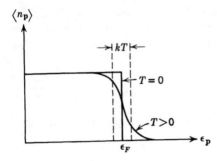

Fig. 11.2 Average occupation number in an ideal Fermi gas.

After a partial integration we obtain

$$U = \frac{V}{4\pi^2 m\hbar^3} \int_0^\infty dp \, \frac{p^5}{5} \left(-\frac{\partial}{\partial p} \langle n_\mathbf{p} \rangle \right) = \frac{\beta V}{20\pi^2 m^2 \hbar^3} \int_0^\infty dp \, \frac{p^6 e^{\beta \epsilon_\mathbf{p} - \nu}}{(e^{\beta \epsilon_\mathbf{p} - \nu} + 1)^2}$$

(11.27)

It is apparent from Fig. 11.2 that $\partial \langle n_\mathbf{p} \rangle / \partial p$ is sharply peaked at $p = p_F$. In fact, at absolute zero it is a δ function at $p = p_F$. Therefore the integral in (11.27) can be evaluated by expanding the factor p^6 about $p = p_F$. The procedure is similar to that used in obtaining (11.11). After inserting ν from (11.24) we obtain the asymptotic expansion

$$U = \tfrac{3}{5} N \epsilon_F \left[1 + \tfrac{5}{12}\pi^2 \left(\frac{kT}{\epsilon_F} \right)^2 + \cdots \right]$$

(11.28)

The first term is the ground state energy of the Fermi gas at the given density, as we can verify by showing the following:

$$\sum_{|\mathbf{p}| < p_F} \frac{p^2}{2m} = \tfrac{3}{5} N \epsilon_F$$

(11.29)

The specific heat at constant volume can be immediately obtained from (11.28)

$$\frac{C_V}{Nk} \approx \frac{\pi^2}{2} \frac{kT}{\epsilon_F}$$

(11.30)

It vanishes linearly as $T \to 0$, thus verifying the third law of thermodynamics. We know that C_V/Nk approaches $\tfrac{3}{2}$ as $T \to \infty$. Thus a rough sketch of C_V/Nk can be made, as shown in Fig. 11.3. The fact that it is proportional to T at these low temperatures can be understood as follows. At a temperature $T > 0$, $\langle n_\mathbf{p} \rangle$ differs from that at $T = 0$ because a certain number of particles are excited to energy levels $\epsilon_\mathbf{p} > \epsilon_F$. Roughly speaking, particles with energies of order kT below ϵ_F are excited to energies of order kT above ϵ_F (see Fig. 11.2). The number of particles excited is therefore of the order of $(kT/\epsilon_F)N$. Therefore the total excitation energy above the ground state is $\Delta U \approx (kT/\epsilon_F)NkT$, from which follows $C_V \approx (kT/\epsilon_F)Nk$.

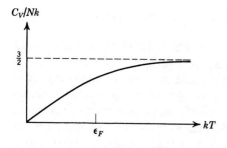

Fig. 11.3 Specific heat of an ideal Fermi gas.

From (8.78) and (11.28) follows the equation of state

$$P = \frac{2}{3}\frac{U}{V} = \frac{2}{5}\frac{\epsilon_F}{v}\left[1 + \frac{5\pi^2}{12}\left(\frac{kT}{\epsilon_F}\right)^2 + \cdots\right] \qquad (11.31)$$

This shows that even at absolute zero it is necessary to contain the ideal Fermi gas with externally fixed walls because the pressure does not vanish. This is a manifestation of the Pauli exclusion principle, which allows only one particle to have zero momentum. All other particles must have finite momentum and give rise to the zero-point pressure.

To obtain the thermodynamic function for arbitrary values of λ^3/v numerical methods must be employed to calculate the functions $f_{3/2}(z)$ and $f_{5/2}(z)$.

11.2 THE THEORY OF WHITE DWARF STARS

It is an empirical rule that the brightness of a star is proportional to its color (i.e., predominant wavelength emitted). The proportionality constant is roughly the same for all stars. Thus if we make a plot of brightness against color, we obtain what is known as the Hertzsprung-Russell diagram, in which most stars fall within a linear strip called the main sequence, as shown in Fig. 11.4. There are, however, stars that are exceptions to this rule. There are the red giant stars, huge stars which are abnormally bright for their red color; and there are the white dwarf stars, small stars which are abnormally faint for their white color. The white

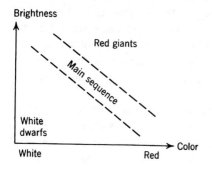

Fig. 11.4 Russell-Hertzprung diagram.

dwarf star makes an interesting subject for our study, because to a good approximation it is a degenerate Fermi gas.

A detailed study of the constitution of white dwarf stars leads to the conclusion that they lack brightness because the hydrogen supply, which is the main energy source of stars, has been used up, and they are composed mainly of helium. What little brightness they have is derived from the gravitational energy released through a slow contraction of the star. Probably these stars have reached the end point of stellar evolution. One of the nearest stars to the solar system, the companion of Sirius, 8 light years from us, is a white dwarf. So faint that it escapes the naked eye, it was first predicted by the calculations of Bessel, who tried to explain why Sirius apparently moves about a point in empty space.

An idealized model of a white dwarf may be constructed from some typical data for such a star:

Content: mostly helium

Density $\approx 10^7$ g/cm^3 $\approx 10^7\,\rho_\odot$

Mass $\approx 10^{33}$ g $\approx M_\odot$

Central temperature $\approx 10^7$ K $\approx T_\odot$

where the subscript $\odot$ denotes quantities referring to the sun. Thus a white dwarf star is a mass of helium at an extremely high temperature and under extreme compression. The temperature 10^7 K corresponds to a thermal energy of 1000 eV. Hence the helium atoms are expected to be completely ionized, and the star may be regarded as a gas composed of helium nuclei and electrons. We regard the gas of electrons as an ideal Fermi gas, with a density of approximately 10^{30} electrons/cm^3. This corresponds to a Fermi energy of

$$\epsilon_F \approx \frac{\hbar^2}{2m}\frac{1}{v^{2/3}} \approx 20 \text{ MeV}$$

and a Fermi temperature of

$$T_F \approx 10^{11} \text{ K}$$

Since the Fermi temperature is much greater than the temperature of the star, the electron gas is a highly degenerate Fermi gas, which behaves no differently from an electron gas at absolute zero. In fact we regard the electron gas to be an ideal Fermi gas in its ground state. The enormous zero-point pressure exerted by the electron gas is counteracted by the gravitational attraction that binds the star. This gravitational binding is due almost entirely to the helium nuclei in the star. The pressure due to kinetic motion of the helium nuclei, and to any radiation that may be present, will be neglected.

Thus we arrive at the following idealized model: A white dwarf is taken to be a system of N electrons in its ground state, at such a density that the electrons must be treated by relativistic dynamics. The electrons move in a background of $N/2$ motionless helium nuclei which provide the gravitational attraction to hold the entire system together.* This model must then exhibit properties that are the combined effects of the Pauli principle, relativistic dynamics, and the gravitational law.

First let us work out the pressure exerted by a Fermi gas of relativistic electrons in the ground state. The states for a single electron are specified by the momentum $\mathbf{p}$ and the spin quantum number $s = \pm \frac{1}{2}$. The single-particle energy levels are independent of s:

$$\epsilon_{\mathbf{p}s} = \sqrt{(pc)^2 + (m_ec^2)^2}$$

where m_e is the mass of an electron. The ground state energy of the Fermi gas is

$$E_0 = 2 \sum_{|\mathbf{p}| < p_F} \sqrt{(pc)^2 + (m_ec^2)^2} = \frac{2V}{h^3} \int_0^{p_F} dp\, 4\pi p^2 \sqrt{(pc)^2 + (m_ec^2)^2} \quad (11.32)$$

where p_F, the Fermi momentum, is defined by

$$\frac{V}{h^3}\left(\frac{4}{3}\pi p_F^3\right) = \frac{N}{2}$$

or

$$p_F = \hbar\left(\frac{3\pi^2}{v}\right)^{1/3} \quad (11.33)$$

Changing the variable of integration in (11.32) to $x = p/m_e c$ we obtain

$$\frac{E_0}{N} = \frac{m_e^4 c^5}{\pi^2 \hbar^3} v f(x_F) \quad (11.34)$$

where

$$f(x_F) = \int_0^{x_F} dx\, x^2 \sqrt{1 + x^2} = \begin{cases} \frac{1}{3}x_F^3\left(1 + \frac{3}{10}x_F^2 + \cdots\right) & (x_F \ll 1) \\[2mm] \frac{1}{4}x_F^4\left(1 + \frac{1}{x_F^2} + \cdots\right) & (x_F \gg 1) \end{cases} \quad (11.35)$$

and

$$x_F \equiv \frac{p_F}{m_e c} = \frac{\hbar}{m_e c}\left(\frac{3\pi^2}{v}\right)^{1/3} \quad (11.36)$$

---

*The temperature in an actual white dwarf star is so high that electron-positron pairs can be created in electron-electron collisions. These pairs in turn annihilate into radiation. Therefore in equilibrium there should be a certain number of electron-positron pairs and a certain amount of radiation present. We neglect the effects of these. It has been speculated that neutrinos can also be created in electron-electron, electron-positron, and photon-photon collisions with appreciable probability. This leads to some interesting phenomena, for neutrinos interact so weakly with matter that they do not come to thermal equilibrium with the rest of the system. They simply leave the star and cause a constant drain of energy. (H. Y. Chiu and P. Morrison, *Phys. Rev. Lett.* **5**, 573 (1960).) Our model is based on the neglect of these effects.

If the total mass of the star is M and the radius of the star is R, then

$$M = (m_e + 2m_p)N \approx 2m_p N$$

$$R = \left(\frac{3V}{4\pi}\right)^{1/3} \tag{11.37}$$

where m_p is the mass of a proton. In terms of M and R we have

$$v = \frac{8\pi}{3} \frac{m_p R^3}{M} \tag{11.38}$$

and

$$x_F = \frac{\hbar}{m_e c} \frac{1}{R} \left(\frac{9\pi}{8} \frac{M}{m_p}\right)^{1/3} \equiv \frac{\bar{M}^{1/3}}{\bar{R}} \tag{11.39}$$

where

$$\bar{M} = \frac{9\pi}{8} \frac{M}{m_p}$$

$$\bar{R} = \frac{R}{(\hbar/m_e c)} \tag{11.40}$$

The pressure exerted by the Fermi gas is

$$
P_0 = -\frac{\partial E_0}{\partial V} = \frac{m_e^4 c^5}{\pi^2 \hbar^3}\left[-f(x_F) - \frac{\partial f(x_F)}{\partial x_F} v \frac{\partial x_F}{\partial v}\right]
$$

$$
= \frac{m_e^4 c^5}{\pi^2 \hbar^3}\left[\frac{1}{3} x_F^3 \sqrt{1 + x_F^2} - f(x_F)\right] \tag{11.41}
$$

The nonrelativistic and extreme relativistic limits of P_0 are given by

$$P_0 \approx \left(\frac{m_e^4 c^5}{15\pi^2 \hbar^3}\right) x_F^5 = \tfrac{4}{5} K \frac{\bar{M}^{5/3}}{\bar{R}^5} \qquad \text{(nonrel.: } x_F \ll 1\text{)} \tag{11.42}$$

$$P_0 \approx \left(\frac{m_e^4 c^5}{12\pi^2 \hbar^3}\right)(x_F^4 - x_F^2) = K\left(\frac{\bar{M}^{4/3}}{\bar{R}^4} - \frac{\bar{M}^{2/3}}{\bar{R}^2}\right) \qquad \text{(extreme rel.: } x_F \gg 1\text{)} \tag{11.43}$$

where

$$K = \frac{m_e c^2}{12\pi^2}\left(\frac{m_e c}{\hbar}\right)^3 \tag{11.44}$$

A qualitative plot of P_0 against R for fixed M is shown in Fig. 11.5. It is seen that, for small R, P_0 becomes smaller than what is expected on the basis of nonrelativistic dynamics.

The condition for equilibrium of the star may be obtained through the following argument. Let us first imagine that there is no gravitational interaction. Then the density of the system will be uniform, and external walls will be needed to keep the Fermi gas at a given density. The amount of work that an external

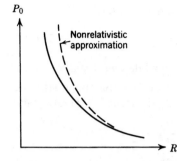

P_0

Nonrelativistic approximation

Fig. 11.5 Pressure of an ideal Fermi gas at absolute zero.

R

agent has to do to compress the star of given mass from a state of infinite diluteness to a state of finite density would be given by

$$-\int_{\infty}^{R} P_0 4\pi r^2 \, dr \tag{11.45}$$

where P_0 is the pressure of a uniform Fermi gas and R is the radius of the star. Now imagine that the gravitational interaction is "switched on." Different parts of the star will now attract one another, resulting in a decrease of the energy of the star by an amount that is called the gravitational self-energy. On dimensional grounds the gravitational self-energy must have the form

$$-\frac{\alpha\gamma M^2}{R} \tag{11.46}$$

where γ is the gravitational constant and α is a pure number of the order of unity. The exact value of α depends on the functional form of the density as a function of spatial distance and cannot be determined by our argument. If R is the equilibrium radius of the star, the gravitational self-energy must exactly compensate the work done in bringing the star together. Hence

$$\int_{\infty}^{R} P_0 4\pi r^2 \, dr = -\frac{\alpha\gamma M^2}{R} \tag{11.47}$$

Differentiating (11.47) with respect to R we obtain the condition for equilibrium:

$$P_0 = \frac{\alpha}{4\pi}\frac{\gamma M^2}{R^4} = \frac{\alpha}{4\pi}\gamma\left(\frac{8m_p}{9\pi}\right)^2\left(\frac{m_e c}{\hbar}\right)^4\frac{\overline{M}^2}{\overline{R}^4} \tag{11.48}$$

Strictly speaking, (11.47) merely defines α. Its physical content is furnished by the assumption that α is of the order of unity. We now determine the relation between M and R by inserting an appropriate expression for P_0 into (11.48). This will be done for the following three different cases:

(*a*) Suppose the temperature of the electron gas is much higher than the Fermi temperature. Then the electron gas may be considered as an ideal Boltzmann gas, with

$$P_0 = \frac{kT}{v} = \frac{3kT}{8\pi m_p}\frac{M}{R^3}$$

Substitution of this into (11.48) yields the linear relation

$$R = \tfrac{2}{3}\alpha M \frac{m_p \gamma}{kT} \tag{11.49}$$

This case, however, is never applicable for a white dwarf star.

(b) Suppose the electron gas is at such a low density that nonrelativistic dynamics may be used $(x_F \ll 1)$. Then P_0 is given by (11.42), and (11.48) leads to the equilibrium condition

$$\tfrac{4}{5}K\frac{\overline{M}^{5/3}}{\overline{R}^5} = K'\frac{\overline{M}^2}{\overline{R}^4}$$

where

$$K' = \frac{\alpha}{4\pi}\gamma\left(\frac{8m_p}{9\pi}\right)^2\left(\frac{m_e c}{\hbar}\right)^4 \tag{11.50}$$

Thus the radius of the star decreases as the mass of the star increases:

$$\overline{M}^{1/3}\overline{R} = \frac{4}{5}\frac{K}{K'} \tag{11.51}$$

This condition is valid when the density is low. Hence it is valid for small M and large R.

(c) Suppose the electron gas is at such a high density that relativistic effects are important $(x_F \gg 1)$. Then P_0 is given by (11.43). The equilibrium condition becomes

$$K\left(\frac{\overline{M}^{4/3}}{\overline{R}^4} - \frac{\overline{M}^{2/3}}{\overline{R}^2}\right) = K'\frac{\overline{M}^2}{\overline{R}^4} \tag{11.52}$$

or

$$\overline{R} = \overline{M}^{2/3}\sqrt{1 - (\overline{M}/\overline{M}_0)^{2/3}} \tag{11.53}$$

where

$$\overline{M}_0 = \left(\frac{K}{K'}\right)^{3/2} = \left(\frac{27\pi}{64\alpha}\right)^{3/2}\left(\frac{\hbar c}{\gamma m_p^2}\right)^{3/2} \tag{11.54}$$

Numerically,

$$\frac{\hbar c}{\gamma m_p^2} \approx 10^{39} \tag{11.55}$$

This interesting pure number is the rest energy of X divided by the gravitational attraction of two protons separated by the Compton wavelength of X, where X is anything. The mass M_0 corresponding to the reduced quantity $\overline{M}_0$ is (taking $\alpha \approx 1$):

$$M_0 = \frac{8}{9\pi}m_p\overline{M}_0 \approx 10^{33}\,\text{g} \approx M_\odot \tag{11.56}$$

the mass of the sun. The formula (11.53) is valid for high densities or for $R \to 0$.

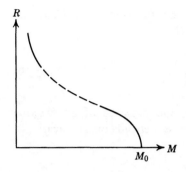

Fig. 11.6 Radius-mass relationship of a white dwarf star.

Hence it is valid for M near M_0. Our model yields the remarkable prediction that no white dwarf star can have a mass larger than M_0, because otherwise (11.53) would give an imaginary radius. The physical reason underlying this result is that if the mass is greater than a certain amount, the pressure coming from the Pauli exclusion principle is not sufficient to support the gas against gravitational collapse.

The radius-mass relationship of a white dwarf star, according to our model, has the form shown in Fig. 11.6, where the solid lines indicate the regions covered by formulas (11.51) and (11.53). We have not been able to calculate α, so that an exact value of M_0 cannot be obtained. More refined considerations* give the result

$$M_0 = 1.4 M_\odot \tag{11.57}$$

This mass is known as the Chandrasekhar limit. Thus according to our model no star can become a white dwarf unless its mass is less than $1.4 M_\odot$. This conclusion has so far been verified by astronomical observations. If the mass of a star is greater than the Chandrasekhar limit then it will eventually collapse under its own gravitational attraction. When the density becomes so high that new interactions, dormant thus far, are awakened, a new regime takes over. For example, the star could explode as a supernova.

11.3 LANDAU DIAMAGNETISM

Van Leeuwen's theorem[†] states that the phenomenon of diamagnetism is absent in classical statistical mechanics. Landau[‡] first showed how diamagnetism arises from the quantization of the orbits of charged particles in a magnetic field.

The magnetic susceptibility per unit volume of a system is defined to be

$$\chi \equiv \frac{\partial \mathcal{M}}{\partial H} \tag{11.58}$$

---

*S. Chandrasekhar, *Stellar Structure* (Dover, New York, 1957), Chapter XI.
[†]See Problem 8.7.
[‡]L. Landau, *Z. Phys.* **64**, 629 (1930).

where $\mathscr{M}$ is the average induced magnetic moment per unit volume of the system along the direction of an external magnetic field H:

$$\mathscr{M} \equiv \frac{1}{V} \left\langle -\frac{\partial \mathscr{H}}{\partial H} \right\rangle \tag{11.59}$$

where $\mathscr{H}$ is the Hamiltonian of the system in the presence of an external magnetic field H. For weak fields the Hamiltonian $\mathscr{H}$ depends on H linearly. In the canonical ensemble we have

$$\mathscr{M} = kT\frac{\partial}{\partial H} \frac{\log Q_N}{V} \tag{11.60}$$

and in the grand canonical ensemble we have

$$\mathscr{M} = kT\frac{\partial}{\partial H} \left(\frac{\log \mathscr{Q}}{V} \right)_{T,V,z} \tag{11.61}$$

where z is to be eliminated in terms of N by the usual procedure.

A system is said to be diamagnetic if $\chi < 0$; paramagnetic if $\chi > 0$. To understand diamagnetism in the simplest possible terms, we construct an idealized model of a physical substance that exhibits diamagnetism. The magnetic properties of a physical substance are mainly due to the electrons in the substance. These electrons are either bound to atoms or nearly free. In the presence of an external magnetic field two effects are important for the magnetic properties of the substance: (a) The electrons, free or bound, move in quantized orbits in the magnetic field. (b) The spins of the electrons tend to be aligned parallel to the magnetic field. The atomic nuclei contribute little to the magnetic properties except through their influence on the wave functions of the electrons. They arc too massive to have significant orbital magnetic moments, and their intrinsic magnetic moments are about 10^{-3} times smaller than the electron's. The alignment of the electron spin with the external magnetic field gives rise to paramagnetism, whereas the orbital motions of the electrons give rise to diamagnetism. In a physical substance these two effects compete. We completely ignore paramagnetism for the present, however. The effect of atomic binding on the electrons is also ignored. Thus we consider the idealized problem of a free spinless electron gas in an external magnetic field.

Landau Levels

The Hamiltonian of a nonrelativistic electron in an external magnetic field is

$$\mathscr{H} = \frac{1}{2m} \left(\mathbf{p} + \frac{e}{c}\mathbf{A} \right)^2 \tag{11.62}$$

where e is positive, (i.e., the charge of the electron is $-e$). The Schrödinger equation $\mathscr{H}\psi = \epsilon\psi$ is invariant under the gauge transformation

$$\mathbf{A}(\mathbf{r}) \to \mathbf{A}(\mathbf{r}) - \nabla\omega(\mathbf{r})$$

$$\psi(\mathbf{r}) \to \exp\left[-\frac{ie}{\hbar c}\omega(\mathbf{r})\right]\psi(\mathbf{r}) \tag{11.63}$$

where $\omega(\mathbf{r})$ is an arbitrary continuous function. We consider a uniform external magnetic field H pointing along the z axis, and choose the vector potential, via a gauge transformation if necessary, such that

$$A_x = -Hy, \qquad A_y = A_z = 0 \tag{11.64}$$

This is called "choosing the gauge." The Hamiltonian then reads

$$\mathscr{H} = \frac{1}{2m}\left\{\left[p_x - (eH/c)y\right]^2 + p_y^2 + p_z^2\right\} \tag{11.65}$$

We solve the Schrödinger equation by assuming a wave function of the form

$$\psi(x, y, z) = e^{i(k_x x + k_z z)}f(y) \tag{11.66}$$

Then $f(y)$ satisfies the equation for a harmonic oscillator:

$$\left[\frac{1}{2m}p_y^2 + \tfrac{1}{2}m\omega_0^2(y - y_0)^2\right]f(y) = \epsilon'f(y)$$

$$\omega_0 = eH/mc, \qquad y_0 = (\hbar c/eH)k_x \tag{11.67}$$

where $\epsilon' = \epsilon - \hbar^2 k_z^2/2m$. The natural frequency of the harmonic oscillator ω_0 is the "cyclotron frequency," that of a classical charge moving in a circular orbit normal to a uniform magnetic field. The energy eigenvalues are thus

$$\epsilon(p_z, j) = \frac{p_z^2}{2m} + \hbar\omega_0(j + \tfrac{1}{2}), \qquad (j = 0, 1, 2, \dots) \tag{11.68}$$

where $p_z = \hbar k_2$. These are the Landau energy levels. Since they are independent of k_x, they have a degeneracy equal to the number of allowed values of k_x, such that y_0 lies within the container of the system.

Let us put the system in a large cube of size L, and impose periodic boundary conditions. The allowed values of k_x are of the form $2\pi n_x/L$, where $n_x = 0, \pm 1, \pm 2, \dots$. For y_0 to lie between 0 and L, the values of n_x must be positive and bounded by

$$g = (eH/hc)L^2 \tag{11.69}$$

which is the degeneracy of a Landau level. The proportionality to L^2 reflects the fact that the projection of the electron orbit onto the xy plane can be centered anywhere in the plane without changing the energy. Thus, when the external field is turned on, the energy spectrum associated with the motion in the xy plane changes from a continuous spectrum to a discrete one, and the level spacing and degeneracy increases with the external field. This is illustrated in Fig. 11.7.

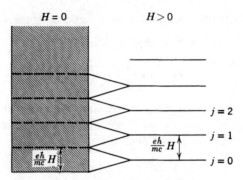

Fig. 11.7 Comparison of the energy spectra of a charged particle with and without magnetic field.

Flux Quantization

The Landau levels and the degeneracies derived above are all we need to calculate the partition function. However, we take the opportunity to discuss flux quantization briefly, to help us better understand the wave functions.

Consider a plane with a hole in it, which contains a certain amount of magnetic flux Φ, as shown in Fig. 11.8. Suppose there is no magnetic field anywhere else. Then the vector potential in the plane must be "pure gauge," i.e., of the form

$$\mathbf{A} = \nabla \omega$$

We cannot transform this to zero through any continuous gauge transformation, because necessarily

$$\int_C d\mathbf{S} \cdot \mathbf{A} = \Phi \tag{11.70}$$

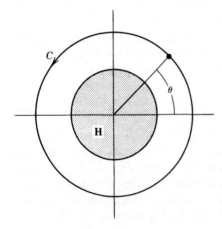

Fig. 11.8 Charged particle moving in plane with a hole containing magnetic flux. The particle will not notice the flux if either (a) it is in a localized state, or (b) the flux is quantized in units of hc/e.

where the closed path C encloses the hole, as indicated in Fig. 11.7. A solution to (11.70) is

$$\omega = \Phi\theta/2\pi \qquad (11.71)$$

where θ is the angle around the hole, measured from some arbitrary axis.

Now consider an electron moving in the plane, with the boundary condition that its wave function vanish in the hole. In general it is affected by the flux, because the Schrödinger equation involves $\mathbf{A}$, which is nonzero where the electron moves. But since $\mathbf{A}$ is pure gauge, we are tempted to try to remove it from the Schrödinger equation through the gauge transformation

$$\mathbf{A} \to \mathbf{A} - \nabla\omega$$

In so doing, the wave function of the electron acquires the phase factor

$$\exp\left(-\frac{ie\omega}{\hbar c}\right) = \exp\left[-i\theta\left(\frac{e\Phi}{hc}\right)\right] \qquad (11.72)$$

which is generally unacceptable because it will render the wave function discontinuous in space (for θ increases by 2π each time we go around the hole). The objection is circumvented under either of the following circumstances:

(*a*) The electron is "localized," i.e., its wave function is nonvanishing only in the neighborhood of some point. In this case, where (11.72) might lead to a discontinuity, the wave function vanishes anyway. This is not relevant to free electrons, but may be relevant, for example, for an electron trapped by an impurity in a metal.

(*b*) The electron is "extended," with a wave function that is phase-coherent around a closed path about the hole, but the flux is quantized in integer multiples of the flux quantum

$$\Phi_0 = hc/e \qquad (11.73)$$

In this case, (11.72) becomes a periodic function of θ and represents a legitimate gauge transformation. Thus, the vector potential can be transformed away, and the electron does not "know" there is flux through the hole. This is the basis of the Aharanov-Bohm effect,[*] and the flux quantization in superconductivity,[†] which have been experimentally verified.

The relevance of flux quantization to the Landau levels lies in the fact that the degeneracy (11.69) is just the total magnetic flux measured in units of the flux quantum:

$$g = \Phi/\Phi_0 \qquad (11.74)$$

---

[*]Y. Aharanov and D. Bohm, *Phys. Rev.* **115**, 465 (1959).

[†]See P. G. De Gennes, *Superconductivity of Metal and Alloys* (Benjamin, New York, 1966), p. 149.

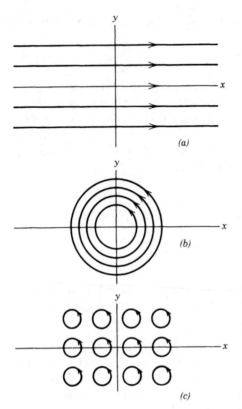

Fig. 11.9 Different bases for electron states in a Landau level, which is highly degenerate: (a) Member wave functions are peaked at different elevations (y direction), and are eigenstates of momentum in the x direction. (b) Member wave functions are eigenstates of orbital angular momentum. They are peaked at concentric circles, with equal areas between successive circles. Each ring between circles supports one magnetic flux quantum. (c) Member wave functions are "vortices" of flux quanta, forming a lattice.

We have been working in a gauge in which the wave functions have definite linear momentum in the x direction. Thus the probability densities are independent of x, and peaked about parallel ridges at $y = y_0$. The spacing between successive values of y_0 is hc/eHL, and hence the area of the strip in the xy plane between two neighboring ridges is hc/eH. Thus, exactly one flux quantum hc/e goes through the strip.

Since the energy levels are highly degenerate, we can make linear transformations on the wave functions belonging to the same Landau level to obtain equivalent sets. Such transformations are equivalent to gauge transformations. We can make them eigenfunctions of angular momentum about the z axis, in which case the probability distributions will be peaked about circles in the xy plane about the z axis, with exactly one quantum of flux going through the annular ring between two successive circles. We can also make them into individual orbitals, whose centers form a regular lattice in the xy plane. In this case each orbital will link exactly one flux quantum (a "vortex.") These different bases are illustrated in Fig. 11.9.

The qualitative fact relevant to our immediate purpose is that the flux quantum sets a finite minimum size of an orbit, and thereby provides the escape from Van Leeuwen's theorem.

Magnetic Susceptibility

The grand partition function is

$$\mathcal{Q} = \prod_{\lambda}(1 + ze^{-\beta\epsilon_\lambda}) \tag{11.75}$$

where λ denotes the set of quantum numbers $\{p_z, j, \alpha\}$, with $\alpha = 1, \ldots, g$. Thus

$$\log \mathcal{Q} = \sum_{\alpha=1}^{g} \sum_{j=0}^{\infty} \sum_{p_z} \log[1 + ze^{-\beta\epsilon(p_z, j)}]$$

$$= \frac{2gL}{h} \sum_{j=0}^{\infty} \int_0^{\infty} dp \log[1 + ze^{-\beta\epsilon(p, j)}] \tag{11.76}$$

The average number of electrons is

$$N = \frac{2gL}{h} \sum_{j=0}^{\infty} \int_0^{\infty} dp \frac{1}{z^{-1}e^{\beta\epsilon(p, j)} + 1} \tag{11.77}$$

To calculate the magnetization in the classical domain we take the high-temperature limit. The condition (11.76) requires that $z \to 0$ to keep N finite. Thus we expand the above equations in powers of z, and retain only the first-order term:

$$\log \mathcal{Q} \approx \frac{2zgL}{h} \sum_{j=0}^{\infty} \int_0^{\infty} dp\, e^{-\beta[p^2/2m + \hbar\omega_0(j+1/2)]}$$

$$= \frac{zgL}{\lambda} \frac{e^{-x}}{1 - e^{-2x}} \tag{11.78}$$

where $\lambda = \sqrt{2\pi\hbar^2/mkT}$ and $x = \hbar\omega_0/2kT$. We keep only the lowest-order contribution in x:

$$\log \mathcal{Q} \approx \frac{zgL}{\lambda} \frac{1}{2x}\left(1 - \frac{x^2}{6}\right) = \frac{zV}{\lambda^3}\left[1 - \frac{1}{24}\left(\frac{\hbar\omega_0}{kT}\right)^2\right] \tag{11.79}$$

from which follows

$$\chi \approx -\frac{z}{3kT\lambda^3}\left(\frac{e\hbar}{2mc}\right)^2 \tag{11.80}$$

To eliminate z, we note from (11.77) that to first order in z, N is the same as $\log \mathcal{Q}$. Hence

$$\frac{N}{V} \approx \frac{z}{\lambda^3} \tag{11.81}$$

Solving for z and substituting the result into (11.80), we obtain the final answer

$$\chi \approx -\frac{1}{3kTv}\left(\frac{e\hbar}{2mc}\right)^2 \tag{11.82}$$

which conforms to Curie's $1/T$ law. Since the lowest energy of an electron is

$$\hbar\omega_0/2 = (e\hbar/2mc)H$$

we see that the magnetic moment of the minimal orbit is just the Bohr magneton $e\hbar/2mc$.

11.4 THE DE HAAS-VAN ALPHEN EFFECT

We now turn to the low-temperature limit of electrons in Landau levels. The electrons will tend to occupy the lowest available levels. As the magnetic field is decreased, each Landau level can accomodate fewer electrons because the degeneracy is decreased. Consequently, some electrons will be forced to jump up to a higher level. This causes the de Haas-Van Alphen effect, the oscillation of the low-temperature magnetic susceptibility as the magnetic field is decreased. To study this effect in a simple context we shall assume $kT \ll \hbar\omega_0$, so that we can set $T = 0$. We shall also ignore the motion in the z direction.*

Our problem is to calculate the ground state energy of a two-dimensional electron system of total area L^2 in a uniform magnetic field H. We write the Landau levels ϵ'_j and their degeneracy g in the following notation:

$$\epsilon_j = 2\mu_0 H\left(j + \tfrac{1}{2}\right), \qquad \mu_0 = e\hbar/2mc$$
$$g = NH/H_0, \qquad H_0 = nhc/e \qquad (11.83)$$

where $n = N/L^2$ is the number of electrons per unit area. The field H_0 is the value of H above which the Landau level can hold all the N particles.

If $H/H_0 > 1$, then all particles can be accommodated in the lowest Landau level, and the ground state energy per particle is

$$E_0/N = \mu_0 H, \qquad (H/H_0 > 1) \qquad (11.84)$$

If $H < H_0$, then some particles will have to occupy higher levels. Suppose H is such that the j lowest Landau levels are completely filled, the $(j + 1)$th level is partially filled, and all higher levels are empty. The condition for H is

$$(j + 1)g < N < (j + 2)g$$

or

$$\frac{1}{j + 2} < \frac{H}{H_0} < \frac{1}{j + 1} \qquad (11.85)$$

For H in this interval,

$$E_0/N = g \sum_{i=0}^{j} \epsilon_i + [N - (j + 1)g]\epsilon_{j+1}$$
$$= \mu_0(H/H_0)[2j + 3 - (j + 1)(j + 2)(H/H_0)] \qquad (11.86)$$

---

*The experimental effect was discovered by W. J. De Haas and P. M. Van Alphen, *Leiden Commun.*, 212 (1931). Our simplified model is that of R. E. Peierls, *Z. Phys.* **81**, 186 (1933). For a more realistic treatment see J. M. Luttinger, *Phys. Rev.* **121**, 1251 (1961).

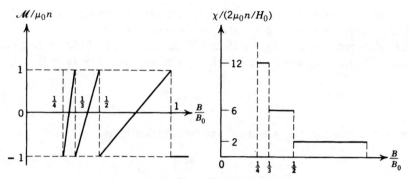

Fig. 11.10 De Haas-Van Alphen effect.

Introducing the parameter

$$x = H/H_0 \tag{11.87}$$

we can summarize the results as follows:

$$\frac{1}{N} E_0(B) = \begin{cases} \mu_0 H_0 x & (x > 1) \\ \mu_0 H_0 x [(2j + 3) - (j + 1)(j + 2)x] \\ \quad \left(\dfrac{1}{j + 2} < x < \dfrac{1}{j + 1}, \; j = 0, 1, 2, \ldots \right) \end{cases} \tag{11.88}$$

The magnetization per unit volume and the magnetic susceptibility per unit volume are respectively given by

$$\mathscr{M} = \begin{cases} -\mu_0 n & (x > 1) \\ \mu_0 n [2(j + 1)(j + 2)x - (2j + 3)] \\ \quad \left(\dfrac{1}{j + 2} < x < \dfrac{1}{j + 1}, \; j = 0, 1, 2, \ldots \right) \end{cases} \tag{11.89}$$

$$\chi = \begin{cases} 0 & (x > 1) \\ \dfrac{2\mu_0 n}{H_0} (j + 1)(j + 2) & \left(\dfrac{1}{j + 2} < x < \dfrac{1}{j + 1}, \; j = 0, 1, 2, \ldots \right) \end{cases} \tag{11.90}$$

These are shown in Fig. 11.10

11.5 THE QUANTIZED HALL EFFECT

The Hall effect was discovered in the nineteenth century: When crossed magnetic and electric fields are applied to a metal, a voltage is induced in a direction orthogonal to the crossed fields, as evidenced by an induced current flowing in that direction—the Hall current. This effect is easy to understand on the basis of the free electron theory of a metal, as follows. Crossed magnetic and electric fields, denoted, respectively, by **H** and **E**, act as velocity filters to free charges, letting through only whose those velocity v is such that $\mathbf{E} + (\mathbf{v}/c)\mathbf{B} = 0$, or

$$\frac{v}{c} = \frac{E}{B} \tag{11.91}$$

For free charge carriers in a metal, the current density is

$$j = qnv \tag{11.92}$$

where q is the charge, and n the density. The Hall resistivity ρ_{xy} is defined as the ratio of the electric field (in the y direction) to the Hall current density (in the x direction):

$$j = \frac{E}{\rho_{xy}} \tag{11.93}$$

Substituting this into (11.92) and then into (11.91), we obtain

$$\rho_{xy} = \frac{H}{qnc} \tag{11.94}$$

Measurements of the Hall resistivity in various metals has yielded charge carrier densities and provided the first demonstrations that there are not only negative charge carriers (electrons), but also positive ones (holes).

The two-dimensional electron system used as a model in the last section can now be created in the laboratory, thanks to developments in the transistor technology. It can be made by injecting electrons into the interface of an alloy sandwich, which confines the electrons in a thin film about 500 Å thick. The Hall experiment has been performed on such two-dimensional electron systems at very low temperatures, and the direct resistivities ρ_{xx} and the Hall resistivities ρ_{xy} have been measured, as indicated in Fig. 11.11.

The experimental results are quite dramatic, as shown in Fig. 11.12. As the magnetic field H increases the degeneracy of the Landau levels increases. Since the electron density does not depend on the field the filling fraction ν of the lowest Landau level decreases:

$$\nu \equiv \frac{hcn}{eH} \tag{11.95}$$

The Hall resistivity exhibits plateaus at $\nu = 1, \frac{2}{3}, \frac{1}{3}$, with values equal to $1/\nu$, in units of h/e^2. At the same time, the conventional resistivity ρ_{xx} drops to very low values. This indicates that in the neighborhood of these special filling fractions the two-dimensional electron fluid flows with almost no resistance. The value at $\nu = 1$, called the integer quantized Hall effect, was first observed in a MOSFIT (metal-oxide semiconductor field-effect transistor) at $T = 1.5$ K. The Hall resis-

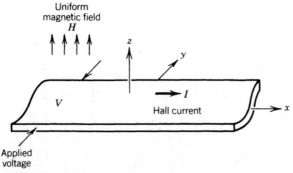

Fig. 11.11 The Hall effect. A current I flows in a direction orthogonal to crossed electric and magnetic fields. The Hall resistivity is defined as $\rho_{xy} = V/I$. The conventional resistivity ρ_{xx} can be obtained by measuring the voltage drop along the direction of the current.

tivity was found to be quantized with a precision of one part in 10^5.* The fractional values were found soon after.[†]

The integer effect is easy to understand on a naive basis. Since at $\nu = 1$ the lowest Landau level is completely filled, there is an energy gap above the Fermi level. Low-energy excitations are therefore impossible, and so the centers of the electron orbits flow like a free gas. Using (11.94) with $n = eH/hc$, the Landau degeneracy per unit area, we immediately obtain the desired result.

$$\rho_{xy} = \frac{h}{e^2} \tag{11.96}$$

But this does not explain why the Hall resistivity continues to be quantized even when the field is changed somewhat, so that there is a plateau of the quantized value, as seen in the data.

Laughlin[‡] offers the important insight that the integer effect is due to the phase coherence of the electronic wave function over the entire sample, and that the effect of impurities are important in producing the observed plateau. Consider a sample in the form of a ribbon forming a closed loop, as shown in Fig. 11.13. A magnetic field H pierces the ribbon everywhere normal to its surface, and a voltage V is applied across the edges of the ribbon. Our object is to deduce the relation between the Hall current I and V.

The Hall current produces a magnetic moment $\mu = IA/c$, where A is the area enclosed by the ribbon loop. Imagine that a small amount of magnetic flux $\delta\Phi$ is introduced through the loop, corresponding to an increase in the magnetic

*K. V. Klitzing, G. Dorda, and M. Pepper, *Phys. Rev. Lett.* **45**, 494 (1980).

†D. C. Tsui, H. L. Stormer, and A. C. Gossard, *Phys. Rev. Lett.* **48**, 1559 (1982).

‡R. B. Laughlin, *Phys. Rev. B* **23**, 5632 (1981). See also B. I. Halperin, *Phys. Rev. B* **25**, 2185 (1982).

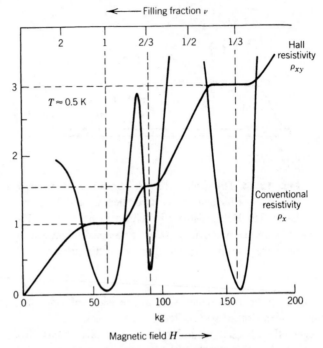

Fig. 11.12 Quantized Hall effect: Schematic representation of experimental data. The filling fraction ν is the fraction of degenerate states in the lowest Landau levels occupied by electrons. The Hall resistivity exhibits plateau of value $1/\nu$, at $\nu = 1, \frac{2}{3}, \frac{1}{3}$ (in units of h/e^2.) The conventional resistivity becomes very small at these values. The quantization is accurate to at least one part in 10^4.

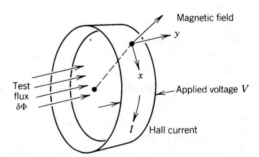

Fig. 11.13 Hall effect in idealized geometry.

field $\delta H = \delta\Phi/A$ normal to the plane of the loop. The energy of the system increases by $\delta E = \mu\,\delta H = (IA/c)(\delta\Phi/A)$. Hence we can find the current from the formula

$$I/c = \frac{\delta E}{\delta\Phi} \tag{11.97}$$

We recall from our discussion of flux quantization in Section 11.3 that the "localized" electrons will not respond to the flux, but the "extended" ones may. Electrons in Landau levels do have extended wave functions, and thus will respond to the flux and contribute to the Hall current. Across the ribbon, in the y direction, the wave function of an electron is peaked about some value of y, say y_0. The allowed values of y_0 extends from one edge of the ribbon to the other. (We are using here the "strip" representation of the wave functions, as shown in Fig. 11.9a.)

Consider now a completely filled Landau level. The electron density across the ribbon may be represented schematically as in Fig. 11.14. The electrons lying closer to the right edge have a higher electrostic energy because of the applied voltage. Now imagine that the flux through the loop is increased slowly from zero. The electrons will respond to the change until the flux reaches the quantum value hc/e, at which point they cannot feel the flux. During the slow increase, the energy of the electrons must rise by the transfer of electrons from one edge of the ribbon to the other. When the flux reaches one quantum, the electron distribution must look exactly the same as before. Overall, therefore, the electrons play musical chairs, moving up one position per quantum of flux penetration, as indicated in Fig. 11.14. Since the gain in energy is $\delta E = eV$, and the change in flux is $\delta\Phi = hc/e$, we have from (11.97) $I = (e^2/h)V$, whence

$$\rho_{xy} = \frac{h}{e^2} \tag{11.98}$$

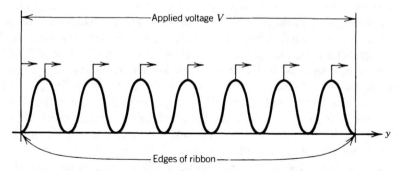

Fig. 11.14 Schematic representation of electron density across the ribbon in Fig. 11.13, when the lowest Landau level is completely filled. The electrons move to the right by one "musical chair," when one unit of test flux pushes through the loop in Fig. 11.13.

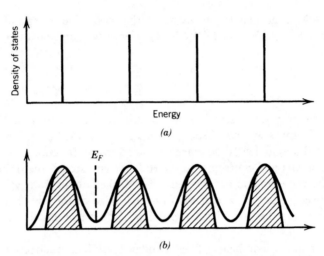

Fig. 11.15 Density of state of an electron in a magnetic field. (*a*) In a pure metal, a series of delta functions mark the positions of the Landau levels. (*b*) In the presence of impurities, the Landau levels broaden to bands (shaded region). Localized electrons states fill the gap between Landau bands. E_F denotes the Fermi level.

If the total number of electrons is fixed, then changing the magnetic field will change the filling fraction. A filled level will either become underfilled, or it will spill electrons over to a higher level. In either case, the previous analysis breaks down. However, in a physical sample there are always impurities that trap electrons into localized states. It is the presence of impurities that give rise to the stability of the effect, as shown by the plateau in the data.

In Fig. 11.15 we show the density of states in a pure sample as compared with one with impurities. In a pure sample we have a series of δ functions at the Landau levels, while in the presence of impurities each Landau level is broadened to a band, shown shaded in Fig. 11.15. At low impurity densities these bands do not overlap. The important point is that between the Landau bands the density of states is not zero, as would be the case for a pure sample, but is filled by contributions from localized states. The Fermi level can lie in a continuum between Landau bands, and it can shift in response to a change in the occupancy of the filled Landau band, so that the band beneath it remains filled. Thus, for a certain range of the external magnetic field the lowest Landau band remains completely filled, and our argument applies.

The fractional effects are more intriguing. What accounts for the stability of the electron film when the Landau level is only one-third full? The answer must lie in the Coulomb interaction among the electrons, but so far we only have preliminary guesses.*

*R. B. Laughlin, *Phys. Rev. Lett.* **50**, 1395 (1983).

11.6 PAULI PARAMAGNETISM

The Hamiltonian of a nonrelativistic free electron in an external magnetic field B is given by

$$\frac{1}{2m}\left(\mathbf{p} + \frac{e}{c}\mathbf{A}\right)^2 - \mu_0\boldsymbol{\sigma}\cdot\mathbf{H} \tag{11.99}$$

where $\mu_0 = eh/2mc$, and $\boldsymbol{\sigma}$ are the Pauli spin matrices. The first term gives rise to diamagnetism, as we have studied. The second term gives rise to paramagnetism. We now consider its effect alone, and take the single-particle Hamiltonian to be*

$$\mathscr{H} = \frac{p^2}{2m} - \mu_0\boldsymbol{\sigma}\cdot\mathbf{H} \tag{11.100}$$

The eigenvalues of $\boldsymbol{\sigma}\cdot\mathbf{H}$ are sH, where $s = \pm 1$. Hence the single-particle energy levels are

$$\epsilon_{\mathbf{p},s} = \frac{p^2}{2m} - s\mu_0 H \tag{11.101}$$

An energy eigenvalue of the N-particle system may be labeled by the occupation numbers $n_{\mathbf{p},s}$ of the single-particle levels $\epsilon_{\mathbf{p},s}$:

$$E_n = \sum_{\mathbf{p}}\sum_s \epsilon_{\mathbf{p},s} n_{\mathbf{p},s} = \sum_{\mathbf{p}}\left[\left(\frac{p^2}{2m} - \mu_0 H\right)n_{\mathbf{p},+1} + \left(\frac{p^2}{2m} + \mu_0 H\right)n_{\mathbf{p},-1}\right] \tag{11.102}$$

where

$$n_{\mathbf{p},s} = 0, 1$$

$$\sum_s\sum_{\mathbf{p}} n_{\mathbf{p},s} = N \tag{11.103}$$

Let

$$n_{\mathbf{p},+1} \equiv n_{\mathbf{p}}^+$$

$$n_{\mathbf{p},-1} \equiv n_{\mathbf{p}}^-$$

$$\sum_{\mathbf{p}} n_{\mathbf{p},+1} \equiv N_+ \tag{11.104}$$

$$\sum_{\mathbf{p}} n_{\mathbf{p},-1} \equiv N_- = N - N_+$$

Then an energy eigenvalue of the system can also be written in the form

$$E_n = \sum_{\mathbf{p}}\left(n_{\mathbf{p}}^+ + n_{\mathbf{p}}^-\right)\frac{p^2}{2m} - \mu_0 H(N_+ - N_-) \tag{11.105}$$

*Following W. Pauli, Z. *Phys.* **41**, 81 (1927).

The partition function is

$$Q_N = \underset{\{n_{\mathbf{p}}^+\},\{n_{\mathbf{p}}^-\}}{\sum}{}' \exp\left[-\beta\sum_{\mathbf{p}}(n_{\mathbf{p}}^+ + n_{\mathbf{p}}^-)\frac{p^2}{2m} + \beta\mu_0 H(N_+ - N_-)\right] \quad (11.106)$$

where the prime over the sum denotes the restrictions (11.103). The sum can be evaluated as follows. First we choose an arbitrary integer N_+ and sum over all sets $\{n_{\mathbf{p}}^+\}, \{n_{\mathbf{p}}^-\}$ such that $\sum_{\mathbf{p}} n_{\mathbf{p}}^+ = N_+$, and $\sum_{\mathbf{p}} n_{\mathbf{p}}^- = N - N_+$. Then we sum over all integers N_+ from 0 to N. In this manner we arrive at the formula

$$Q_n = \sum_{N_+=0}^{N} e^{\beta\mu_0 H(2N_+ - N)} \underset{\{n_{\mathbf{p}}^+\}}{\sum}{}'' \exp\left(-\beta\sum_{\mathbf{p}}\frac{p^2}{2m}n_{\mathbf{p}}^+\right) \underset{\{n_{\mathbf{p}}^-\}}{\sum}{}''' \exp\left(-\beta\sum_{\mathbf{p}}\frac{p^2}{2m}n_{\mathbf{p}}^-\right)$$

$$(11.107)$$

where $\sum''$ is subject to the restriction $\sum_{\mathbf{p}} n_{\mathbf{p}}^+ = N_+$, and $\sum'''$ is subject to the restriction $\sum_{\mathbf{p}} n_{\mathbf{p}}^- = N_- = N - N_+$. Let $Q_N^{(0)}$ denote the partition function of the ideal Fermi gas of N *spinless* particles of mass m:

$$Q_N^{(0)} \equiv \sum_{\sum n_{\mathbf{p}}=N} \exp\left(-\beta\sum_{\mathbf{p}}\frac{p^2}{2m}n_{\mathbf{p}}\right) \equiv e^{-\beta A(N)} \quad (11.108)$$

Then

$$Q_N = e^{-\beta\mu_0 HN} \sum_{N_+=0}^{N} e^{2\beta\mu_0 HN_+} Q_{N_+}^{(0)} Q_{N-N_+}^{(0)}$$

$$\frac{1}{N}\log Q_N = -\beta\mu_0 H + \frac{1}{N}\log \sum_{N_+=0}^{N} e^{2\beta\mu_0 HN_+ - \beta A(N_+) - \beta A(N_-)} \quad (11.109)$$

There are $N + 1$ positive terms in the sum just given. The logarithm of this sum is equal to the logarithm of the largest term in the sum plus a contribution of the order of $\log N$. Therefore, neglecting a term of order $N^{-1}\log N$, we have

$$\frac{1}{N}\log Q_N = \beta f(\overline{N}_+) \quad (11.110)$$

where

$$f(\overline{N}_+) = \max\left[f(N_+)\right]$$

$$f(N_+) \equiv \mu_0 H\left(\frac{2N_+}{N} - 1\right) - \frac{1}{N}\left[A(N_+) + A(N - N_+)\right] \quad (11.111)$$

Obviously we can interpret $\overline{N}_+$ as the average number of particles with spin up. If $\overline{N}_+$ is known, the magnetization per unit volume can be obtained through the formula

$$\mathcal{M} = \frac{\mu_0(2\overline{N}_+ - N)}{V} \tag{11.112}$$

We now explicitly find $\overline{N}_+$. The condition (11.111) is equivalent to the condition*

$$\left[\frac{\partial f(N_+)}{\partial N_+} \right]_{N_+ = \overline{N}_+} = 0$$

or

$$2\mu_0 H - \left[\frac{\partial A(N')}{\partial N'} \right]_{N' = \overline{N}_+} - \left[\frac{\partial A(N - N')}{\partial N'} \right]_{N' + \overline{N}_+} = 0 \tag{11.113}$$

Let $kT\nu(N)$ be the chemical potential of an ideal Fermi gas of N *spinless* particles:

$$kT\nu(N) = \frac{\partial A(N)}{\partial N} \tag{11.114}$$

Then

$$\left[\frac{\partial A(N')}{\partial N'} \right]_{N' = \overline{N}_+} = kT\nu(\overline{N}_+)$$

$$\left[\frac{\partial A(N - N')}{\partial N'} \right]_{N' = \overline{N}_+} = -\left[\frac{\partial A(N - N')}{\partial (N - N')} \right]_{N - N' = N - \overline{N}_+} = -kT\nu(N - \overline{N}_+)$$

Thus (11.113) becomes

$$kT\left[\nu(\overline{N}_+) - \nu(N - \overline{N}_+) \right] = 2\mu_0 H \tag{11.115}$$

This condition states that at a given temperature the average number of particles with spin up is such that the chemical potential of the particles with spin up is greater than that of the particles with spin down by $2\mu_0 H$. We solve (11.115) in the low-temperature and high-temperature limits.

Let the Fermi energy for the present system be

$$\epsilon_F(N) \equiv \left(\frac{3\pi^2 N}{V} \right)^{2/3} \frac{\hbar^2}{2m} \tag{11.116}$$

In the low-temperature region ($kT \ll \epsilon_F$), we can use the expansion (11.24)[†] for $kT\nu(N)$:

$$kT\nu(N) = \epsilon_F(2N)\left\{ 1 - \frac{\pi^2}{12}\left[\frac{kT}{\epsilon_F(2N)} \right]^2 + \cdots \right\}$$

---

*We should make sure that (11.113) determines a maximum and not a minimum and that $\overline{N}_+$ lies between 0 and N. It can be verified that (11.113) has only one real root that automatically satisfies these requirements.

[†] Note that in (11.24) the symbol ϵ_F stands for the Fermi energy of N spinless particles and does not have the same meaning as ϵ_F here.

Thus (11.115) becomes

$$\epsilon_F(2\overline{N}_+) - \epsilon_F(2N - 2\overline{N}_+) - \frac{\pi^2(kT)^2}{12}$$

$$\times \left[\frac{1}{\epsilon_F(2\overline{N}_+)} - \frac{1}{\epsilon_F(2N - 2\overline{N}_+)} \right] + \cdots = 2\mu_0 H \quad (11.117)$$

Let

$$r \equiv \frac{2\overline{N}_+}{N} - 1 \quad (-1 \le r \le +1) \quad (11.118)$$

Then (11.117) becomes

$$(1 + r)^{2/3} - (1 - r)^{2/3} - \frac{\pi^2}{12}\left(\frac{kT}{\epsilon_F}\right)^2$$

$$\times \left[(1 + r)^{-2/3} - (1 - r)^{-2/3} \right] + \cdots = \frac{2\mu_0 H}{\epsilon_F} \quad (11.119)$$

At absolute zero, r satisfies the equation

$$(1 + r)^{2/3} - (1 - r)^{2/3} = \frac{2\mu_0 H}{\epsilon_F} \quad (11.120)$$

This may be solved graphically, as shown in Fig. 11.16. For $\ll \epsilon_F/2\mu$ an approximate solution is

$$r \approx \frac{3\mu_0 H}{2\epsilon_F}$$

$$\overline{N}_+ \approx \frac{N}{2}\left(1 + \frac{3\mu_0 H}{2\epsilon_F}\right) \quad (11.121)$$

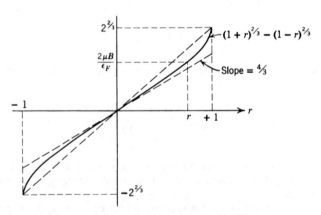

Fig. 11.16 Graphical solution of (11.122).

Thus when $H = 0$ half the particles have spin up, and the other half spin down. When $H > 0$ the balance shifts in favor of spin up. From (11.112) and (11.121) we obtain, for absolute zero,

$$\mathcal{M} = \frac{\mu_0 r}{v} \approx \frac{3\mu_0^2}{2\epsilon_F v}$$

$$\chi \approx \frac{3\mu_0^2 H}{2\epsilon_F v} \tag{11.122}$$

For $0 < kT \ll \epsilon_F$ and $\mu \ll \epsilon_F$ we can solve (11.119) by expanding the left side in powers of r, and we obtain

$$r \approx \frac{3\mu_0 H}{2\epsilon_F}\left[1 - \frac{\pi^2}{12}\left(\frac{kT}{\epsilon_F}\right)^2\right]$$

$$\chi \approx \frac{3\mu_0^2}{2\epsilon_F v}\left[1 - \frac{\pi^2}{12}\left(\frac{kT}{\epsilon_F}\right)^2\right] \tag{11.123}$$

For high temperatures $(kT \gg \epsilon_F)$ we use (11.12):

$$\nu(N) \approx \log\left(\frac{N\lambda^3}{V}\right)$$

Hence (11.115) gives

$$\log\left[\frac{\lambda^3(1 + r)}{v}\right] - \log\left[\frac{\lambda^3(1 - r)}{v}\right] = \frac{2\mu_0 H}{kT}$$

or

$$r = \tanh\frac{\mu_0 H}{kT} \approx \frac{\mu_0 H}{kT} \tag{11.124}$$

The magnetic susceptibility per unit volume is then given by

$$\chi \approx \frac{\mu_0^2}{kTv} \tag{11.125}$$

A qualitative plot of $kT\chi$ is shown in Fig. 11.17.

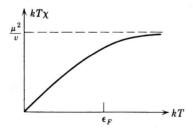

Fig. 11.17 Pauli paramagnetism.

11.7 MAGNETIC PROPERTIES OF AN IMPERFECT GAS

How would interparticle interaction affect the magnetic properties of an electron gas? Qualitatively speaking, the effect of a repulsive interaction will enhance the paramagnetism, because two electrons would prefer to be in a spatially antisymmetric wave function to minimize the repulsive energy. But an antisymmetric spatial wave function requires a symmetric spin wave function, which is a triplet state. Thus the repulsion tends to align the spins of the electrons.

We demonstrate this effect by considering an imperfect gas of spin-$\frac{1}{2}$ fermions at very low temperatures, so that the repulsive interaction can be characterized by a single parameter, the scattering length a, or effective hard-sphere diameter. To first order in a, the energy for an N-particle system is given by (A.41) in the Appendix. We take our model to be defined by (11.102) plus the interaction energy:

$$E_n = \sum_{\mathbf{p}} \left(n_{\mathbf{p}}^+ + n_{\mathbf{p}}^- \right) \frac{p^2}{2m} + \frac{4\pi a \hbar^2}{mV} N_+ N_- - (N_+ - N_-)\mu_0 H \quad (11.126)$$

The condition for the validity of this formula is that $k_F|a| \ll 1$ where k_F is the wave number of a particle at the Fermi level: $k_F^2 = (3\pi^2 n)^{2/3}$. Thus, the condition for validity is low density, i.e., $na^3 \ll 1$.

The partition function is

$$Q_N = \sum_{\{n_{\mathbf{p}}^+\},\{n_{\mathbf{p}}^-\}}' \exp\left\{ -\beta \left[\sum_{\mathbf{p}} \left(n_{\mathbf{p}}^+ + n_{\mathbf{p}}^- \right) \frac{p^2}{2m} - \mu_0 H(N_+ - N_-) + \frac{4\pi a \hbar^2 N_+ N_-}{mV} \right] \right\}$$

$$(11.127)$$

The notation is the same as that of (11.106). Proceeding in the same way as in the evaluation of (11.106), we obtain

$$\frac{1}{N} \log Q_N = \beta g(\overline{N}_+) \quad (11.128)$$

where

$$g(\overline{N}_+) = \max\left[g(N_+) \right]$$

$$g(N_+) \equiv \mu_0 H\left(\frac{2N_+}{N} - 1 \right) - \frac{4\pi a \hbar^2}{mV} N_+(N - N_+)$$

$$- \frac{1}{N}\left[A(N_+) + A(N - N_+) \right] \quad (11.129)$$

Thus $\overline{N}_+$ is the root of the equations

$$\left[\frac{\partial g(N_+)}{\partial N_+} \right]_{N_+ = \overline{N}_+} = 0$$

$$\left[\frac{\partial^2 g(N_+)}{\partial N_+^2} \right]_{N_+ = \overline{N}_+} < 0 \quad (11.130)$$

It must be noted that (11.130) locates the point at which the curve $g(N_+)$ passes through a maximum. It is conceivable (and in fact true) that $\overline{N}_+$ may occur not at a maximum of the curve $g(N_+)$ but at the boundary of the range of N_+, i.e., at $\overline{N}_+ = 0$ or $\overline{N}_+ = N$. We keep this in mind as we proceed. With $\nu(N)$ defined as in (11.114), we rewrite (11.130) as

$$kT\left[\nu\left(\overline{N}_+\right) - \nu\left(N - \overline{N}_+\right)\right] = 2\mu H + \frac{4\pi a\hbar^2}{mV}\left(2\overline{N}_+ - N\right)$$

(11.131)

$$kT\left[\nu'\left(\overline{N}_+\right) + \nu'\left(N - \overline{N}_+\right)\right] - \frac{8\pi a\hbar^2}{mV} > 0$$

where $\nu'(N) \equiv \partial\nu(N)/\partial N$. Let

$$r \equiv \frac{2\overline{N}_+}{N} - 1 \qquad (-1 \le r \le +1)$$

(11.132)

Then (13.131) becomes

$$kT\left\{\nu\left[\frac{N}{2}(1 + r)\right] - \nu\left[\frac{N}{2}(1 - r)\right]\right\} = 2\mu_0 H + \frac{a\lambda^2}{v}2kTr$$

(11.133)

$$\frac{\partial}{\partial r}\left\{N\left[\frac{N}{2}(1 + r)\right] - \nu\left[\frac{N}{2}(1 - r)\right]\right\} - \frac{2a\lambda^2}{v} > 0$$

where $\nu[x] \equiv \nu(x)$ and $\lambda = \sqrt{2\pi\hbar^2/mkT}$, the thermal wavelength. The low-temperature and high-temperature approximations for $\nu(Nx/2)$ are obtainable from (11.24) and (11.12), respectively. They are

$$kT\nu\left(\frac{N}{2}x\right) \approx x^{2/3}\epsilon_F\left[1 - \frac{\pi^2}{12}\left(\frac{kT}{\epsilon_F}\right)^2\frac{1}{x^{4/3}}\right] \qquad \left(\frac{kT}{\epsilon_F} \ll 1\right)$$

(11.134)

$$kT\nu\left(\frac{N}{2}x\right) \approx \log\frac{\lambda^2}{2v} \qquad \left(\frac{kT}{\epsilon_F} \gg 1\right)$$

Spontaneous Magnetization

We first consider the case $H = 0$. At absolute zero, (11.134) reduces to

$$(1 + r)^{2/3} - (1 - r)^{2/3} = \zeta r$$

$$\frac{1}{2}\left[\frac{1}{(1 + r)^{1/3}} + \frac{1}{(1 - r)^{1/3}}\right] > \frac{3}{4}\zeta$$

(11.135)

where

$$\zeta \equiv \frac{8}{3\pi}k_F a$$

(11.136)

Equation (11.135) is invariant under a change of sign of r. This is to be expected; in the absence of field, no absolute meaning can be attached to "up" or "down."

Thus it is sufficient consider $r \geq 0$. We may solve (11.135) graphically by referring to Fig. 11.16, where $(1 + r)^{2/3} - (1 - r)^{2/3}$ is plotted against r. We need only obtain the intersection between the curve in Fig. 11.16 and the straight line ζr. It is seen that for $\zeta < \frac{4}{3}$, $r = 0$ is the only intersection. If ζ is such that

$$\tfrac{4}{3} < \zeta < 2^{2/3} \tag{11.137}$$

then there is an additional intersection $r > 0$, and the value $r > 0$ corresponds to a maximum, whereas the value $r = 0$ corresponds to a minimum. If $\zeta > 2^{2/3}$, then (11.135) has no solution. In this case the maximum of $g(N_+)$ must occur either at $N_+ = 0$ or at $N_+ = N$, unless $g(N_+)$ is a constant. Since $g(N_+)$ is not a constant, and since there is no distinction between $N_+ = 0$ and $N_+ = N$, we can choose to let $\overline{N}_+ = N$, or $r = 1$. The value of r at absolute zero as a function of the repulsive strength ζ, is summarized as follows:

$$r = 0 \qquad (\zeta < \tfrac{4}{3}) \qquad \text{(no spontaneous magnetization)}$$
$$0 < r < 1 \quad (\tfrac{4}{3} < \zeta < 2^{2/3}) \quad \text{(partial spontaneous magnetization)} \tag{11.138}$$
$$r = 1 \qquad (\xi > 2^{2/3}) \qquad \text{(saturated spontaneous magnetization)}$$

That is, if the repulsive strength is sufficiently strong, the system becomes ferromagnetic. The critical value of a at which ferromagnetism first sets in $(\zeta = \frac{4}{3})$ corresponds to

$$k_F a = \frac{\pi}{2} \tag{11.139}$$

The foregoing results hold at absolute zero. At a finite but small temperature we have, instead of (11.135),

$$(1 + r)^{2/3} - (1 - r)^{2/3} - \frac{\pi^2}{12}\left(\frac{kT}{\epsilon_F}\right)^2\left[\frac{1}{(1 + r)^{2/3}} - \frac{1}{(1 - r)^{2/3}}\right] = \zeta r$$

$$\frac{1}{2}\left\{\frac{1}{(1 + r)^{1/3}} + \frac{1}{(1 - r)^{1/3}} + \frac{\pi^2}{12}\left(\frac{kT}{\epsilon_F}\right)^2\left[\frac{1}{(1 + r)^{5/3}} + \frac{1}{(1 - r)^{5/3}}\right]\right\} > \frac{3}{4}\zeta \tag{11.140}$$

Let $r(T)$ be the solution at absolute temperature T. It is easily seen that if $r(0) = 0$, then $r(T) = 0$; if $r(0) > 0$, then $r(T) < r(0)$. Thus, if there is spontaneous magnetization at absolute zero, the magnetization decreases with temperature. The spontaneous magnetization vanishes above a critical temperature T_c (the Curie temperature), which is the value of T at which both equations in (13.45) are satisfied for $\zeta > \frac{4}{3}$ and $r = 0$. We find that

$$\frac{kT_c}{\epsilon_F} = \frac{3}{\pi}\sqrt{\zeta - \tfrac{4}{3}} = \frac{2}{\pi\sqrt{3}}\sqrt{\frac{2}{\pi}k_F a - 1} \tag{11.141}$$

A qualitative plot of the magnetization $\mu r/v$ is shown in Fig. 11.18.

It must be pointed out that the model we have used is a physical model only if $k_F a \ll 1$. Therefore the case of ferromagnetism, which requires $k_F a > \pi/2$, is beyond the domain of validity of the model. It is instructive, however, to see how

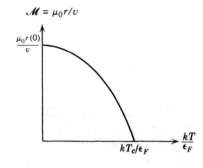

$\mathcal{M} = \mu_0 r/v$

$\dfrac{\mu_0 r(0)}{v}$

$\dfrac{kT_c/\epsilon_F}{}$ $\dfrac{kT}{\epsilon_F}$

Fig. 11.18 Spontaneous magnetization of an imperfect Fermi gas with repulsive interactions.

the spatial repulsion between the fermions can enhance the spin alignment to such an extent that, if we are willing to extrapolate the results of a weak interaction model, ferromagnetism results.

Paramagnetic Susceptibility

We now consider the case of $H > 0$. Let $r_0(T)$ be the value of r for $H = 0$, but for an arbitrary temperature. Putting

$$r = r_0(T) + \frac{\chi v}{\mu_0} H \tag{11.142}$$

and treating $\chi v H/\mu_0$ as a small quantity, we can solve (11.134) and obtain

$$\chi = \frac{2\mu_0^2/\epsilon_F v}{\dfrac{NkT}{2\epsilon F}\left\{ v'\left[\dfrac{N}{2}(1 + r_0)\right] + v'\left[\dfrac{N}{2}(1 - r_0)\right]\right\} - \dfrac{8}{3\pi}k_F a} \tag{11.143}$$

The low- and high-temperature limits are

$$\chi \xrightarrow[T \to 0]{} \frac{3\mu_0^2/\epsilon_F v}{(1 + r_0)^{-1/3} + (1 - r_0)^{-1/3} - (4/\pi)k_F a} \tag{11.144}$$

$$\chi \xrightarrow[T \to \infty]{} \frac{\mu_0^2}{kTv} \tag{11.145}$$

Hence Curie's constant is

$$C = \frac{\mu_0^2}{kv} \tag{11.146}$$

Note that r_0 depends on $k_F a$. It approaches unity when $k_F a$ exceeds a certain value. Thus is can be seen from (11.143) that in general $\chi > 0$. The system is either ferromagnetic or paramagnetic, never diamagnetic.

Consider now the case of paramagnetism, for which we require $r_0 = 0$ for all temperatures. This means that

$$k_F a < \frac{\pi}{2} \tag{11.147}$$

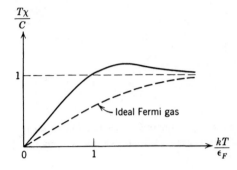

Fig. 11.19 Paramagnetic susceptibility of an imperfect Fermi gas with repulsive interactions. The model used is well founded only for $kT/\epsilon_F \ll 1$.

Here (11.143) becomes

$$\frac{T\chi}{C} = \frac{3kT}{2\epsilon_F}\frac{1}{f - (2/\pi)k_F a} \tag{11.148}$$

where

$$f \equiv \frac{3kT}{2\epsilon_F}\frac{N}{2}\nu'\left(\frac{N}{2}\right) \tag{11.149}$$

The function $(T\chi/C)$ rises linearly at $T = 0$, with a slope given by

$$\frac{\partial}{\partial t}\left(\frac{T\chi}{C}\right)_{T=0} = \frac{3}{2}\frac{1}{1 - (2/\pi)k_F a} \tag{11.150}$$

It reaches a maximum value, which is greater than unity, at $kT/\epsilon_F \approx 1$. Then it approaches unity as $T \to \infty$. A qualitative plot of $T\chi/C$ is shown in Fig. 11.19. If we calculate χ for an ideal Fermi gas endowed with the same magnetic moment, we find the slope

$$\frac{\partial}{\partial T}\left(\frac{T\chi}{C}\right)_{T=0} = \frac{3}{2} \qquad \text{(ideal Fermi gas)} \tag{11.151}$$

The imperfect gas has a steeper slope, as (11.150) shows, which is again a reflection of the enhancement of spin alignment by the repulsive interaction. The result is sometimes described by saying that imperfect gas behaves like an ideal gas with a higher Fermi energy.*

PROBLEMS

11.1 Give numerical estimates for the Fermi energy of

(a) electrons in a typical metal;

(b) nucleons in a heavy nucleus;

(c) He³ atoms in liquid He³ (atomic volume = 46.2 Å³/atom). Treat all the mentioned particles as free particles.

*See, however, Problem 11.7.

11.2 Show that for the ideal Fermi gas the Helmholtz free energy per particle at low temperatures is given by

$$\frac{A}{N} = \tfrac{3}{5}\epsilon_F\left[1 - \frac{5\pi^2}{12}\left(\frac{kT}{\epsilon_F}\right)^2 + \cdots\right]$$

11.3 A collection of free nucleons is enclosed in a box of volume V. The energy of a single nucleon of momentum $\mathbf{p}$ is

$$\epsilon_{\mathbf{p}} = \frac{p^2}{2m} + mc^2$$

where $mc^2 = 1000$ MeV.

(*a*) Pretending that there is no conservation law for the number of nucleons, calculate the partition function of a system of nucleons (which obey Fermi statistics) at temperature T.

(*b*) Calculate the average energy density.

(*c*) Calculate the average particle density.

(*d*) Discuss the necessity for a conservation law for the number of nucleons, in the light of the foregoing calculations.

11.4 (*a*) What is the heat capacity C_V of a three-dimensional cubic lattice of atoms at room temperature? Assume each atom to be bound to its equilibrium position by Hooke's law forces.

(*b*) Assuming that a metal can be represented by such a lattice of atoms plus freely moving electrons, compare the specific heat due to the electrons with that due to the lattice, at room temperature.

11.15 A cylinder is separated into two compartments by a free sliding piston. Two ideal Fermi gases are placed into the two compartments, numbered 1 and 2. The particles in compartment 1 have spin $\tfrac{1}{2}$, while those in compartment 2 have spin $\tfrac{3}{2}$. They all have the same mass. Find the equilibrium relative density of the two gases at $T = 0$ and at $T \to \infty$.

11.6 Consider a two-dimensional electron gas in a magnetic field strong enough so that all particles can be accommodated in the lowest Landau level. Taking into account both orbital and spin paramagnetism, find the magnetization at absolute zero.

11.7 (*a*) Show that for the imperfect Fermi gas discussed in Section 11.6 the specific heat at constant volume is given by

$$\frac{C_V}{N} = -2k\frac{\partial}{\partial T}\left[I(r)T^2\frac{\partial r}{\partial T}\right] + \frac{32\pi a\hbar^2}{mv}T\left[\left(\frac{\partial r}{\partial T}\right)^2 + r\frac{\partial^2 r}{\partial T^2}\right]$$

where

$$I(r) \equiv v\left[\frac{N}{2}(1 + r)\right] - v\left[\frac{N}{2}(1 - r)\right]$$

(*b*) Show that when there is no spontaneous magnetization

$$C_V = (C_V)_{\text{ideal gas}}$$

and hence the interpretation that the imperfect gas behaves like an ideal gas with a higher Fermi energy cannot be consistently maintained.

12

BOSE SYSTEMS

The dominant characteristic of a system of bosons is a "statistical" attraction between the particles. In contradistinction to the case of fermions, the particles like to have the same quantum numbers. When the particle number is conserved, this attraction leads to the Bose-Einstein condensation, which is the basis of superfluidity. In this chapter we illustrate various bose systems, discuss the Bose-Einstein condensation, and introduce the notion of the superfluid order parameter.

12.1 PHOTONS

Consider the equilibrium properties of electromagnetic radiation enclosed in a volume V at temperature T, a system known as a "blackbody cavity." It can be experimentally produced by making a cavity in any material, evacuating the cavity completely, and then heating the material to a given temperature. The atoms in the walls of this cavity will constantly emit and absorb electromagnetic radiation, so that in equilibrium there will be a certain amount of electromagnetic radiation in the cavity, and nothing else. If the cavity is sufficiently large, the thermodynamic properties of the radiation in the cavity should be independent of the nature of the wall. Accordingly we can impose on the radiation field any boundary condition that is convenient.

The Hamiltonian for a free electromagnetic field can be written as a sum of terms, each having the form of a Hamiltonian for a harmonic oscillator of some frequency. This corresponds to the possibility of regarding any radiation field as a linear superposition of plane waves of various frequencies. In quantum theory each harmonic oscillator of frequency ω can only have the energies $(n + \frac{1}{2})\hbar\omega$, where $n = 0, 1, 2, \ldots$. This fact leads to the concept of photons as quanta of the electromagnetic field. A state of the free electromagnetic field is specified by the number n for each of the oscillators. In other words, it is specified by enumerating the number of photons present for each frequency.

According to the quantum theory of radiation, photons are massless bosons of spin $\hbar$. The masslessness implies that a photon always moves with the velocity of light c in free space, and that its spin can have only two independent orientations: parallel and antiparallel to the momentum. A photon in a definite spin state corresponds to a plane electromagnetic wave that is either right- or left-circularly polarized. We may, however, superimpose two photon states with definite spins and obtain a photon state that is linearly polarized but that is not an eigenstate of spin. In the following we consider linearly polarized photons.

For our purpose it is sufficient to know that a photon of frequency ω has the following properties:

$$\text{Energy} = \hbar\omega$$

$$\text{Momentum} = \hbar\mathbf{k}, \qquad |\mathbf{k}| = \frac{\omega}{c} \tag{12.1}$$

$$\text{Polarization vector} = \boldsymbol{\epsilon}, \qquad |\boldsymbol{\epsilon}| = 1, \qquad \mathbf{k} \cdot \boldsymbol{\epsilon} = 0$$

Such a photon corresponds* to a plane wave of electromagnetic radiation whose electric field vector is

$$\mathbf{E}(\mathbf{r}, t) = \boldsymbol{\epsilon}\, e^{i(\mathbf{k}\cdot\mathbf{r} - \omega t)} \tag{12.2}$$

The direction of $\boldsymbol{\epsilon}$ is the direction of the electric field. The condition $\boldsymbol{\epsilon} \cdot \mathbf{k} = 0$ is a consequence of the transversality of the electric field, i.e., $\nabla \cdot \mathbf{E} = 0$. Thus for given $\mathbf{k}$ there are two and only two independent polarization vectors $\boldsymbol{\epsilon}$. If we impose periodic boundary conditions on $\mathbf{E}(\mathbf{r}, t)$ in a cube of volume $V = L^3$, we obtain the following allowed values of $\mathbf{k}$:

$$\mathbf{k} = \frac{2\pi\mathbf{n}}{L} \tag{12.3}$$

$\mathbf{n} = $ a vector whose components are $0, \pm 1, \pm 2, \ldots$

Thus the number of allowed momentum values between k and $k + dk$ is

$$\frac{V}{(2\pi)^3} 4\pi k^2\, dk \tag{12.4}$$

Since atoms can emit and absorb photons, the total number of photons is not a conserved quantity.

The total energy of the state of the electromagnetic field in which there are $n_{\mathbf{k}, \epsilon}$ photons of momentum $\mathbf{k}$ and polarization ϵ is given by

$$E\{n_{\mathbf{k}, \epsilon}\} = \sum_{\mathbf{k}, \epsilon} \hbar\omega n_{\mathbf{k}, \epsilon} \tag{12.5}$$

where
$$\omega = c|\mathbf{k}|$$

$$n_{\mathbf{k}, \epsilon} = 0, 1, 2, \ldots \tag{12.6}$$

---

*For a precise meaning of this statement, we refer the reader to any book on the quantum theory of radiation.

Since the number of photons is indefinite, the partition function is

$$Q = \sum_{\{n_{k,\epsilon}\}} e^{-\beta E\{n_{k,\epsilon}\}} \tag{12.7}$$

with no restriction on $\{n_{k,\epsilon}\}$.* The calculation of Q is trivial:

$$Q = \sum_{\{n_{k,\epsilon}\}} \exp\left(-\beta \sum_{k,\epsilon} \hbar\omega n_{k,\epsilon}\right) = \prod_{k,\epsilon} \sum_{n=0}^{\infty} e^{-\beta\hbar\omega n} = \prod_{k,\epsilon} \frac{1}{1 - e^{-\beta\hbar\omega}}$$

$$\log Q = - \sum_{k,\epsilon} \log(1 - e^{-\beta\hbar\omega}) = -2\sum_{k} \log(1 - e^{-\beta\hbar\omega}) \tag{12.8}$$

The average occupation number for photons of momentum k, regardless of polarization, is

$$\langle n_k \rangle = -\frac{1}{\beta} \frac{\partial}{\partial(\hbar\omega)} \log Q = \frac{2}{e^{\beta\hbar\omega} - 1} \tag{12.9}$$

where the factor 2 comes from the two possible polarizations.

The internal energy is

$$U = -\frac{\partial}{\partial\beta} \log Q = \sum_{k} \hbar\omega \langle n_k \rangle \tag{12.10}$$

To find the pressure, we express Q in the form

$$\log Q = -2\sum_{n} \log(1 - e^{-\beta\hbar c 2\pi|n|V^{-1/3}}) \tag{12.11}$$

from which we obtain

$$P = \frac{1}{\beta} \frac{\partial}{\partial V} \log Q = \frac{1}{3V} \sum_{k} \hbar\omega \langle n_k \rangle$$

Comparison between this equation and (12.10) leads to the equation of state

$$PV = \tfrac{1}{3}U \tag{12.12}$$

We now calculate U in the limit as $V \to \infty$. From (12.10), (12.9), and (12.3) we have

$$U = \frac{2V}{(2\pi)^3} \int_0^{\infty} dk\, 4\pi k^2 \frac{\hbar c k}{e^{\beta\hbar c k} - 1} = \frac{Vh}{\pi^2 c^3} \int_0^{\infty} d\omega \frac{\omega^3}{e^{\beta\hbar\omega} - 1}$$

Hence the internal energy per unit volume is

$$\frac{U}{V} = \int_0^{\infty} d\omega\, u(\omega, T) \tag{12.13}$$

where

$$u(\omega, T) = \frac{\hbar}{\pi^2 c^3} \frac{\omega^3}{e^{\beta\hbar\omega} - 1} \tag{12.14}$$

---

*One could say that the chemical potential is 0, because a photon can disappear into the vacuum.

This is Planck's radiation law, which gives the energy density due to photons of frequency ω, regardless of polarization and direction of momentum. The integral (12.13) can be explicitly evaluated to give

$$\frac{U}{V} = \frac{\pi^2}{15} \frac{(kT)^4}{(\hbar c)^3}$$

(12.15)

It follows that the specific heat per unit volume is

$$c_V = \frac{4\pi^2 k^4 T^3}{15(\hbar c)^3}$$

(12.16)

The specific heat is not bounded as $T \rightarrow \infty$, because the number of photons in the cavity is not bounded.

Both (12.14) and (12.15) can be verified experimentally by opening the black-body cavity to the external world through a small window. Radiation would then escape from the cavity with the velocity c. The amount of energy radiated per second per unit area of the opening, in the form of photons of frequency ω, is

$$I(\omega, T) = c \int \frac{d\Omega}{4\pi} u(\omega, T) \cos \theta = \frac{c}{4} u(\omega, T)$$

(12.17)

where the angular integration extends only over a hemisphere. Integrating over the frequency, we obtain

$$I(T) = \int_0^\infty d\omega \, I(\omega, T) = \sigma T^4$$

$$\sigma = \frac{\pi^2 k^4}{60 \hbar^3 c^3}$$

(12.18)

This is known as Stefan's law, and σ is Stefan's constant. The function $I(\omega, T)$ is shown in Fig. 12.1, showing that the radiation peaks at a frequency that is an increasing function of T. The area under the curves shown in Fig. 12.1 increases like T^4. All these conclusions are in excellent agreement with experiments.

It should be noted that although the form of $u(\omega, T)$ can be arrived at only though quantum theory the equation of state $PV = U/3$ and the fact that $U \propto T^4$ can be derived in classical physics.

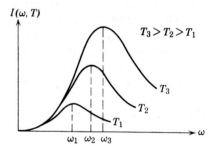

$I(\omega, T)$

$T_3 > T_2 > T_1$

T_3

T_2

T_1

ω_1 ω_2 ω_3

ω

Fig. 12.1 Planck's radiation law.

The equation of state may be derived as follows. Consider first a plane wave whose electric and magnetic field vectors are **E** and **B**. The average energy density is

$$\tfrac{1}{2}\overline{(E^2 + B^2)} = \overline{E^2}$$

The radiation pressure, which is equal to the average momentum flux, is

$$\overline{|\mathbf{E} \times \mathbf{B}|} = \overline{E^2}$$

Thus the energy density is numerically equal to the radiation pressure. Now consider an amount of isotropic radiation contained in a cubical box. The radiation field in the box may be considered an *incoherent* superposition of plane waves propagating in all directions. The relative intensities of the plane waves depend only on the temperature as determined by the walls of the box. The radiation pressure on any wall of the box is one-third of the energy density in the box, because, whereas all the plane waves contribute to the energy density, only one-third of the plane waves contribute to the radiation pressure on any wall of the box.

To derive $U \propto T^4$, recall that the second law of thermodynamics implies the following relation, which holds for all systems:

$$\left(\frac{\partial U}{\partial V}\right)_T = T\left(\frac{\partial P}{\partial T}\right)_V - P \tag{12.19}$$

From $PV = U/3$ and the fact that P depends on temperature alone we have

$$\left(\frac{\partial U}{\partial V}\right)_T = 3P = \frac{U}{V} \equiv u(T) \tag{12.20}$$

Using (12.19) we have

$$u = \frac{T}{3}\frac{du}{dT} - \tfrac{1}{3}u$$

$$\frac{du}{u} = 4\frac{dT}{T}$$

Hence

$$u = CT^4 \tag{12.21}$$

The constant C cannot be obtained through classical considerations.

If the photon had a finite rest mass, no matter how small, then it would have three independent polarizations instead of two.* There would be, in addition to transverse photons, longitudinal photons. If this were so, Planck's radiation formula (12.14) would be altered by a factor of $\tfrac{3}{2}$. The fact that (12.14) has been experimentally verified means that either the photon has no rest mass, or if it does the coupling between longitudinal photons and matter is so small that

---

*If the photon had a finite rest mass, it could be transformed to rest by a Lorentz transformation. We could then make a second Lorentz transformation in an arbitrary direction, so that the spin would lie neither parallel nor antiparallel to the momentum.

thermal equilibrium between longitudinal photons and matter cannot be established during the course of any of our experiments concerned with Planck's radiation law.

12.2 PHONONS IN SOLIDS

Phonons are quanta of sound waves in a macroscopic body. Mathematically they emerge in a similar way that photons arise from the quantization of the electromagnetic field. For low-lying excitations, the Hamiltonian for a solid, which is made up of atoms arranged in a crystal lattice, may be approximated by a sum of terms, each representing a harmonic oscillator, corresponding to a normal mode of lattice oscillation.* Each normal mode is classically a wave of distortion of the lattice planes—a sound wave. In quantum theory these normal modes give rise to quanta called phonons. A quantum state of a crystal lattice near its ground state may be specified by enumerating all the phonons present. Therefore at a very low temperature a solid can be regarded as a volume containing a gas of noninteracting phonons.

Since a phonon is a quantum of a certain harmonic oscillator, it has a characteristic frequency ω_i and an energy $\hbar\omega_i$. The state of the lattice in which one phonon is present corresponds to a sound wave of the form

$$\boldsymbol{\epsilon}\, e^{i(\mathbf{k}\cdot\mathbf{r}-\omega t)} \tag{12.22}$$

where the propagation vector $\mathbf{k}$ has the magnitude

$$|\mathbf{k}| = \frac{\omega}{c} \tag{12.23}$$

in which c is the velocity of sound.[†] The polarization vector $\boldsymbol{\epsilon}$ can have three independent directions, corresponding to one longitudinal mode of compression wave and two transverse modes of shear wave. Since an excited state of a harmonic oscillator may contain any number of quanta, the phonons obey Bose statistics, with no conservation of their total number.

If a solid has N atoms, it has $3N$ normal modes. Therefore there will be $3N$ different types of phonon with the characteristic frequencies

$$\omega_1, \omega_2, \ldots, \omega_{3N} \tag{12.24}$$

The values of these frequencies depend on the nature of the lattice. In the Einstein model of a lattice they are taken to be equal to one another. An improved model is that of Debye, who assumed that for the purpose of finding the frequencies (12.24), one may consider the solid as an elastic continuum of volume V. The frequencies (12.24) are then taken to be the lowest $3N$ normal frequencies of such a system. Since an elastic continuum has a continuous

---

*In as much as anharmonic forces between atoms, which at high temperatures allow the lattice to melt, can be neglected.

[†] We assume an isotropic solid, for which c is independent of the polarization vector $\boldsymbol{\epsilon}$.

distribution of normal frequencies we shall be interested in the number of normal modes whose frequency lies between ω and $\omega + d\omega$. To find this number we must know the boundary conditions on a sound wave in the elastic medium. Taking periodic boundary conditions, we find as usual that $\mathbf{k} = (2\pi/L)\mathbf{n}$, where $L = V^{1/3}$ and $\mathbf{n}$ has the components $0, \pm 1, \pm 2, \ldots$. The number we seek is then

$$f(\omega)\, d\omega \equiv \begin{matrix} \text{no. of normal modes with} \\ \text{frequency between } \omega \text{ and } \omega + d\omega \end{matrix} = \frac{3V}{(2\pi)^3} 4\pi k^2\, dk \quad (12.25)$$

where the factor 3 comes from the three possible polarizations. Since $k = \omega/c$ we have

$$f(\omega)\, d\omega = V \frac{3\omega^2}{2\pi^2 c^3}\, d\omega \quad (12.26)$$

The maximum frequency ω_m is obtained by the requirement that

$$\int_0^{\omega_m} f(\omega)\, d\omega = 3N \quad (12.27)$$

which gives, with $v = V/N$,

$$\omega_m = c \left(\frac{6\pi^2}{v} \right)^{1/3} \quad (12.28)$$

The wavelength corresponding to ω_m is

$$\lambda_m = \frac{2\pi c}{\omega_m} = \left(\tfrac{4}{3}\pi v \right)^{1/3} \approx \text{interparticle distance} \quad (12.29)$$

This is a reasonable criterion because for wavelengths shorter than λ_m a wave of displacements of atoms becomes meaningless.

We now calculate the equilibrium properties of a solid at low temperatures by calculating the partition function for an appropriate gas of phonons. The energy of the state in which there are n_i phonons of the ith type is*

$$E\{n_i\} = \sum_{i=1}^{3N} n_i \hbar \omega_i \quad (12.30)$$

The partition function is

$$Q = \sum_{\{n_i\}} e^{-\beta E\{n_i\}} = \prod_{i=1}^{3N} \frac{1}{1 - e^{-\beta \hbar \omega_i}}$$

Hence

$$\log Q = -\sum_{i=1}^{3N} \log\left(1 - e^{-\beta \hbar \omega_i}\right) \quad (12.31)$$

---

*We should add to (12.30) an unknown constant representing the ground state energy of the solid, but this constant does not affect any subsequent results and hence can be ignored.

The average occupation number is

$$\langle n_i \rangle = -\frac{1}{\beta} \frac{\partial}{\partial(\hbar\omega_i)} \log Q = \frac{1}{e^{\beta\hbar\omega_i} - 1} \tag{12.32}$$

The integral energy is

$$U = -\frac{\partial}{\partial\beta} \log Q = \sum_{i=1}^{3N} \hbar\omega_i \langle n_i \rangle = \sum_{i=1}^{3N} \frac{\hbar\omega_i}{e^{\beta\hbar\omega_i} - 1} \tag{12.33}$$

Passing to the limit $V \to \infty$ we obtain, with the help of (12.26),

$$U = \frac{3V}{2\pi^2 c^3} \int_0^{\omega_m} d\omega\, \omega^2 \frac{\hbar\omega}{e^{\beta\hbar\omega} - 1} \tag{12.34}$$

or

$$\frac{U}{N} = \frac{9(kT)^4}{(\hbar\omega_m)^3} \int_0^{\beta\hbar\omega_m} dt\, \frac{t^3}{e^t - 1} \tag{12.35}$$

We define the Debye function $D(x)$ by

$$D(x) \equiv \frac{3}{x^3} \int_0^x dt\, \frac{t^3}{e^t - 1} = \begin{cases} 1 - \frac{3}{8}x + \frac{1}{20}x^2 + \cdots & (x \ll 1) \\[2mm] \dfrac{\pi^4}{5x^3} + O(e^{-x}) & (x \gg 1) \end{cases} \tag{12.36}$$

and the Debye temperature T_D by

$$kT_D \equiv \hbar\omega_m = \hbar c \left(\frac{6\pi^2}{v}\right)^{1/3} \tag{12.37}$$

Then

$$\frac{U}{N} = 3kTD(\lambda) = \begin{cases} 3kT\left(1 - \dfrac{3}{8}\dfrac{T_D}{T} + \cdots \right) & (T \gg T_D) \\[3mm] 3kT\left[\dfrac{\pi^4}{5}\left(\dfrac{T}{T_D}\right)^3 + O(e^{-T_D/T})\right] & (T \ll T_D) \end{cases} \tag{12.38}$$

where $\lambda \equiv T_D/T$. Then the specific heat is given by

$$\frac{C_V}{Nk} = 3D(\lambda) + 3T\frac{dD(\lambda)}{dT} = 3\left[4D(\lambda) - \frac{3\lambda}{e^\lambda - 1}\right] \tag{12.39}$$

The high- and low-temperature behaviors of C_v are as follows:

$$\frac{C_V}{Nk} = \begin{cases} 3\left[1 - \dfrac{1}{20}\left(\dfrac{T_D}{T}\right)^2 + \cdots \right] & (T \gg T_D) \\[3mm] \dfrac{12\pi^4}{5}\left(\dfrac{T}{T_D}\right)^3 + O(e^{-T_D/T}) & (T \ll T_D) \end{cases} \tag{12.40}$$

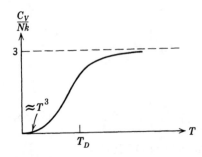

Fig. 12.2 Specific heat of a crystal lattice in Debye's theory.

A plot of the specific heat is shown in Fig. 12.2, which agrees quite well with experimental findings.

At low temperatures C_V vanishes like T^3, verifying the third law of thermodynamics. When the temperature is much greater than the Debye temperature the lattice behaves classically, as indicated by the fact that $C_V \approx 3NK$. For most solids the Debye temperature is of the order of 200 K. This is why the Dulong-Petit law $C_V \approx 3Nk$ holds at room temperatures. At extremely high temperatures the model of noninteracting phonons breaks down because the lattice eventually melts. The melting of the lattice is made possibly by the fact that the forces between the atoms in the lattice are not strictly harmonic forces. In the phonon language the phonons are not strictly free. They must interact with each other, and this interaction becomes strong at very high temperatures.

12.3 BOSE-EINSTEIN CONDENSATION

Equation (8.71) gives the equation of state for the ideal Bose gas of N particles of mass m contained in a volume V. To study in detail the properties of the equation of state we must find the fugacity z as a function of temperature and specific volume by solving the second equation of (8.71), namely

$$\frac{1}{v} = \frac{1}{\lambda^3} g_{3/2}(z) + \frac{1}{V}\frac{z}{1-z} \tag{12.41}$$

where $v = V/N$, and $\lambda = \sqrt{2\pi\hbar^2/mkT}$, the thermal wavelength. To do this, we must first study the properties of the function $g_{3/2}(z)$, which is a special case of a more general class of functions

$$g_n(z) \equiv \sum_{l=1}^{\infty} \frac{z^l}{l^n} \tag{12.42}$$

These functions have been studied* and tabulated[†] in the literature.

It is obvious that for real values of z between 0 and 1, $g_{3/2}(z)$ is a bounded, positive, monotonically increasing function of z. To satisfy (12.41) it is necessary

---

*J. E. Robinson, *Phys. Rev.* **83**, 678 (1951).
†F. London, *Superfluids*, Vol. II (Wiley, New York, 1954), Appendix.

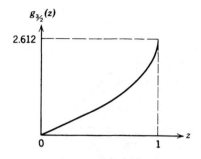

Fig. 12.3 The function $g_{3/2}(z)$.

that
$$0 \le z \le 1$$

For comparison we recall that $0 \le z < \infty$ in the case of Fermi statistics. For small z, the power series (12.42) furnishes a practical way to calculate $g_{3/2}(z)$:

$$g_{3/2}(z) = z + \frac{z^2}{2^{3/2}} + \frac{z^3}{3^{3/2}} + \cdots \tag{12.43}$$

At $z = 1$ its derivative diverges, but its value is finite:

$$g_{3/2}(1) = \sum_{l=1}^{\infty} \frac{1}{l^{3/2}} = \zeta\left(\tfrac{3}{2}\right) = 2.612\ldots \tag{12.44}$$

where $\zeta(x)$ is the Riemann zeta function of x. Thus for all z between 0 and 1,

$$g_{3/2}(z) \le 2.612\ldots \tag{12.45}$$

A graph of $g_{3/2}(z)$ is shown in Fig. 12.3.

Let us rewrite (12.41) in the form

$$\lambda^3 \frac{\langle n_0 \rangle}{V} = \frac{\lambda^3}{v} - g_{3/2}(z) \tag{12.46}$$

This implies that $\langle n_0 \rangle / V > 0$ when the temperature and the specific volume are such that

$$\frac{\lambda^3}{v} > g_{3/2}(1) \tag{12.47}$$

This means that a finite fraction of the particles occupies the level with $\mathbf{p} = 0$. This phenomenon is known as the *Bose-Einstein condensation*. The condition (12.47) defines a subspace of the thermodynamic P-v-T space of the ideal Bose gas, which corresponds to the transition region of the Bose-Einstein condensation. As we see later, in this region the system can be considered to be a mixture of two thermodynamic phases, one phase being composed of particles with $\mathbf{p} = 0$ and the other with $\mathbf{p} \ne 0$. We refer to the region (12.47) as the condensation region. It is separated from the rest of the P-v-T space by the two-dimensional surface

$$\frac{\lambda^3}{v} = g_{3/2}(1) \tag{12.48}$$

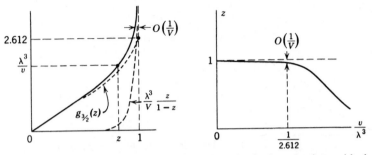

Fig. 12.4 (*a*) Graphical solution of (12.41); (*b*) the fugacity for an ideal Bose gas contained in a finite volume V.

For a given specific volume v, (12.48) defines a critical temperature T_c:

$$\lambda_c^3 = v g_{3/2}(1) \tag{12.49}$$

or

$$kT_c = \frac{2\pi\hbar^2/m}{\left[v g_{3/2}(1)\right]^{2/3}} \tag{12.50}$$

As indicated by (12.49), T_c is the temperature at which the thermal wavelength is of the same order of magnitude as the average interparticle separation. For a given temperature T, (12.48) defines a critical volume v_c:

$$v_c = \frac{\lambda^3}{g_{3/2}(1)} \tag{12.51}$$

In terms of T_c and v_c the region of condensation is the region in which $T < T_c$ or $v < v_c$.

To find z as a function of T and v we solve (12.41) graphically. For a large but finite value of the total volume V the graphical construction in Fig. 12.4a yields the curve for z shown in Fig. 12.4b. In the limit as $V \to \infty$ we obtain

$$z = \begin{cases} 1 & \left(\dfrac{\lambda^3}{v} \geq g_{3/2}(1)\right) \\[4mm] \text{the root of } g_{3/2}(z) = \lambda^3/v & \left(\dfrac{\lambda^3}{v} \leq g_{3/2}(1)\right) \end{cases} \tag{12.52}$$

For $(\lambda^3/v) \leq g_{3/2}(1)$, the value of z must be found by numerical methods. A graph of z is given in Fig. 12.5.

To make these considerations more rigorous the following point must be noted. It is recalled that (12.41) is derived from the condition

$$\frac{N}{V} = \frac{1}{V}\sum_{\mathbf{p}\neq 0}\langle n_{\mathbf{p}}\rangle + \frac{\langle n_0\rangle}{V}$$

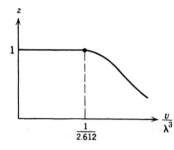

Fig. 12.5 The fugacity for an ideal Bose gas of infinite volume.

by replacing the sum on the right side by an integral. It is clear that this integral is unchanged if we subtract from the sum any *finite* number of terms. More generally, (12.41) should be replaced by the equation

$$\frac{N}{V} = \frac{1}{\lambda^3} g_{3/2}(z) + \frac{\langle n_0 \rangle}{V} + \left(\frac{\langle n_1 \rangle}{V} + \frac{\langle n_2 \rangle}{V} + \cdots \right)$$

where, in the parentheses, there appear any finite number of terms. Every term in the parentheses, however, approaches zero as $V \to \infty$. For example,

$$\frac{\langle n_1 \rangle}{V} = \frac{1}{V} \frac{1}{z^{-1} e^{\beta \epsilon_1} - 1} \leq \frac{1}{V} \frac{1}{e^{\beta \epsilon_1} - 1}$$

where

$$2m\epsilon_1 = (2\pi\hbar)^2 \frac{l_1}{V^{2/3}}$$

l_1 = sum of the squares of three integers not all zero

Hence

$$\frac{\langle n_1 \rangle}{V} \leq \frac{1}{V} \frac{2m\beta V^{2/3}}{(2\pi\hbar)^2 \beta^2 l_1} \xrightarrow[V \to \infty]{} 0 \qquad (12.53)$$

This shows that (12.41) is valid.

By (12.52) and the fact that $\langle n_0 \rangle = z/(1 - z)$ we can write

$$\frac{\langle n_0 \rangle}{N} = \begin{cases} 0 & \left(\frac{\lambda^3}{v} \leq g_{3/2}(1) \right) \\ 1 - \left(\frac{T}{T_c} \right)^{3/2} = 1 - \frac{v}{v_c} & \left(\frac{\lambda^3}{v} \geq g_{3/2}(1) \right) \end{cases} \qquad (12.54)$$

A plot of $\langle n_0 \rangle / N$ is shown in Fig. 12.6. It is seen that when $T < T_c$, a finite fraction of the particles in the system occupy the single level with $\mathbf{p} = 0$. On the other hand (12.53) shows that $\langle n_{\mathbf{p}} \rangle / N$ is always zero for $\mathbf{p} \neq 0$. Therefore we have the following situation: For $T > T_c$ no single level is occupied by a finite fraction of all the particles. The particles "spread thinly" over all levels. For $T < T_c$ a finite fraction $1 - (T/T_c)^{3/2}$ occupies the level with $\mathbf{p} = 0$ while the rest of the particles "spread thinly" over the levels with $\mathbf{p} \neq 0$. At absolute zero all particles occupy the level with $\mathbf{p} = 0$.

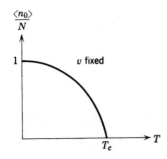

Fig. 12.6 Average occupation number of the level with $p = 0$.

The Bose-Einstein condensation is sometimes described as a "condensation in momentum space." We shall see, however, that its thermodynamic manifestations are those of a first-order phase transition. If we examine the equation of state alone, we discern no difference between the Bose-Einstein condensation and an ordinary gas-liquid condensation. If the particles of the ideal Bose gas are placed in a gravitational field, then in the condensation region there will be a spatial separation of the two phases, just as in a gas-liquid condensation.* The term "momentum-space condensation" merely serves to emphasize the fact that the cause of the Bose-Einstein condensation lies in the symmetry of the wave function and not in any interparticle interaction.

By virtue of (12.52) all thermodynamic functions of the ideal Bose gas will be given by different analytical expressions for the region of condensation and for the complement of that region. Only in the condensation region will these analytical expressions be simple. In the other region numerical computations would be necessary to obtain explicit formulas.

Throughout the remainder of this section let z be defined only for the region $(\lambda^3/v) \leq g_{3/2}(1)$. Some equivalent definitions of z are

$$g_{3/2}(z) = \frac{\lambda^3}{v}$$

$$\frac{g_{3/2}(z)}{g_{3/2}(1)} = \frac{v_c}{v} \tag{12.55}$$

$$\frac{g_{3/2}(z)}{g_{3/2}(1)} = \left(\frac{T_c}{T}\right)^{3/2}$$

In the region $(\lambda^3/v) > g_{3/2}(1)$, z need not be mentioned because $z = 1$.

The equation of state can be obtained from (8.71):

$$\frac{P}{kT} = \begin{cases} \dfrac{1}{\lambda^3} g_{5/2}(z) & (v > v_c) \\[2mm] \dfrac{1}{\lambda^3} g_{5/2}(1) & (v < v_c) \end{cases} \tag{12.56}$$

*W. Lamb and A. Nordsieck, *Phys. Rev.* **59**, 677 (1941).

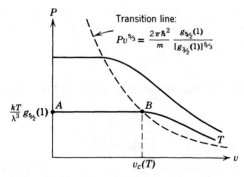

Fig. 12.7 Isotherms of the ideal Bose gas.

where

$$g_{5/2}(1) = \zeta(\tfrac{5}{2}) = 1.342\ldots \tag{12.57}$$

The term $V^{-1} \log(1 - z)$ in (8.71) is zero as $V \to \infty$. For $v > v_c$ this is obvious. For $v < v_c$, it is also true, because $(1 - z) \propto V^{-1}$. It is immediately seen that for $v < v_c$, P is independent of v. The isotherms are shown in Fig. 12.7, and the P-T diagram is shown in Fig. 12.8. We may, as in the case of a gas-liquid condensation, interpret the horizontal portion of an isotherm to mean that in that region the system is a mixture of two phases. In the present case these two phases correspond to the two points labeled A and B in Fig. 12.7. We refer to these respectively as the condensed phase and the gas phase. The horizontal portion of the isotherm is the region of phase transition between the two phases. The vapor pressure is

$$P_0(T) = \frac{kT}{\lambda^3} g_{5/2}(1) \tag{12.58}$$

Differentiation of this equation leads to

$$\frac{dP_0(T)}{dT} = \frac{5}{2} \frac{k g_{5/2}(1)}{\lambda^3} = \frac{1}{Tv_c} \left[\frac{5}{2} kT \frac{g_{5/2}(1)}{g_{3/2}(1)} \right] \tag{12.59}$$

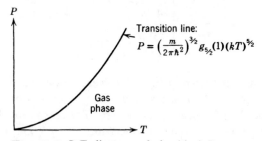

Fig. 12.8 P-T diagram of the ideal Bose gas. Note that the space above the transition curve does not correspond to anything. The condensed phase lies on the transition line itself.

When the two phases coexist the gas phase has the specific volume v_c, whereas the condensed phase has the specific volume 0. Hence the difference in specific volume between the two phases is

$$\Delta v = v_c \qquad (12.60)$$

In fact (12.59) is the Clapeyron equation, and the latent heat of transition per particle is

$$L = \frac{g_{5/2}(1)}{g_{3/2}(1)} \frac{5}{2} kT \qquad (12.61)$$

Therefore the Bose-Einstein condensation is a first-order phase transition.

Other thermodynamic functions for the ideal Bose gas are given in the following. For each thermodynamic function the upper equation refers to the region $v > v_c$ (or $T > T_c$) and the lower equation refers to the region $v < v_c$ (or $T < T_c$):

$$\frac{U}{N} = \tfrac{3}{2} Pv = \begin{cases} \dfrac{3}{2} \dfrac{kTv}{\lambda^3} g_{5/2}(z) \\[2mm] \dfrac{3}{2} \dfrac{kTv}{\lambda^3} g_{5/2}(1) \end{cases} \qquad (12.62)$$

$$-\frac{A}{NkT} = \begin{cases} \dfrac{v}{\lambda^3} g_{5/2}(z) - \log z \\[2mm] \dfrac{v}{\lambda^3} g_{5/2}(1) \end{cases} \qquad (12.63)$$

$$\frac{G}{NkT} = \begin{cases} \log z \\ 0 \end{cases} \qquad (12.64)$$

$$\frac{S}{Nk} = \begin{cases} \dfrac{5}{2} \dfrac{v}{\lambda^3} g_{5/2}(z) - \log z \\[2mm] \dfrac{5}{2} \dfrac{v}{\lambda^3} g_{5/2}(1) \end{cases} \qquad (12.65)$$

$$\frac{C_V}{Nk} = \begin{cases} \dfrac{15}{4} \dfrac{v}{\lambda^3} g_{5/2}(z) - \dfrac{9}{4} \dfrac{g_{3/2}(z)}{g_{1/2}(z)} \\[2mm] \dfrac{15}{4} \dfrac{v}{\lambda^3} g_{5/2}(1) \end{cases} \qquad (12.66)$$

The specific heat is shown in Fig. 12.9. Near absolute zero, C_V vanishes like $T^{3/2}$. This behavior is to be contrasted with a photon gas or a phonon gas, for which C_V vanishes like T^3 near absolute zero. The reason for this difference lies in the difference between the particle spectrum $\epsilon_p = p^2/2m$ and the photon or phonon spectrum $\epsilon_p = cp$. At the same energy the particle spectrum has a higher density of states than the photon or phonon spectrum. Consequently there are more modes of excitation available for a particle, and the specific heat is greater.

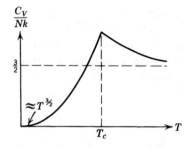

Fig. 12.9 Specific heat of the ideal Bose gas.

From (12.65) we see that $S = 0$ at $T = 0$, in accordance with the third law of thermodynamics. This means that the condensed phase (which exists at $T = 0$) has no entropy. At any finite temperature the total entropy is entirely due to the gas phase. The fraction of particles in the gas phase in the transition region is v/v_c, or $(T/T_c)^{3/2}$. If we rewrite S in the transition region in the form

$$\frac{S}{N} = \left(\frac{T}{T_c}\right)^{3/2} s = \left(\frac{v}{v_c}\right) s \tag{12.67}$$

we find that

$$s = \frac{g_{5/2}(1)}{g_{3/2}(1)} \frac{5}{2} k \tag{12.68}$$

which is the entropy per particle of the gas phase. The difference in specific entropy between the gas phase and the condensed phase is

$$\Delta s = s = \frac{g_{5/2}(1)}{g_{3/2}(1)} \frac{5}{2} k \tag{12.69}$$

Comparing this with (12.61), we find that

$$L = T\Delta s \tag{12.70}$$

This shows that the interpretation of the Bose-Einstein condensation as a first-order phase transition is self-consistent.

The only Bose system known to exist at low temperatures is liquid He^4. At a temperature of 2.18 K, He^4 exhibits the remarkable λ transition, at which the specific heat becomes logarithmically infinite. Since He^4 atoms obey Bose statistics, it is natural to suppose that this transition is the Bose-Einstein condensation modified by intermolecular interactions. This is supported by the fact that no such transition occurs in liquid He^3, whose atoms obey Fermi statistics. Furthermore, substituting the mass of He^4 and the density of liquid helium into (12.50) leads to the transition temperature $T_c = 3.14$ K, which is of the right order of magnitude.

Finally, we must re-emphasize that Bose-Einstein condensation can occur only when the particle number is conserved. For example, photons do not condense. They have a simpler alternative, namely, to disappear into the vacuum.

We have pointed out in Section 7.5 that heavy-particle conservation as physically observed is a low-energy approximation to the real conservation law, which says that the conserved quantity is the number of particles minus the number of antiparticles. Thus, any discussion of the Bose-Einstein condensation for a relativistic Bose gas must take antiparticles into account.*

12.4 AN IMPERFECT BOSE GAS

The ideal Bose gas is an artificial example in that the particles condense into a highly idealized phase with infinite compressibility. That is, the Bose-Einstein condensate is unphysical and uninteresting. We now study an interacting Bose gas in a crude approximation, to see how the nature of the Bose-Einstein condensation changes.

The Energy Levels

We consider a dilute system of N identical spinless bosons of mass m, contained in a box of volume V, at very low temperatures. The bosons interact with one another through binary collisions characterized by the scattering length a which is assumed to be positive. The energy levels to the first order in a may be obtained from (10.124) through the use of first-order perturbation theory.

Let the unperturbed wave functions be free-particle wave functions Φ_n, labeled by the occupation numbers $\{\ldots, n_p, \ldots\}$, where n_p is the number of bosons with momentum p. The energy levels to the first order in a are

$$E_n \equiv (\Phi_n, \mathscr{H}'\Phi_n) = \sum_p \frac{p^2}{2m} n_p + \frac{4\pi a \hbar^2}{m} \left(\Phi_n, \sum_{i<j} \delta(r_i - r_j) \Phi_n \right) \quad (12.71)$$

The second term is calculated in (A.36) of the Appendix. With that, we have

$$E_n = \sum_p \frac{p^2}{2m} n_p + \frac{4\pi a \hbar^2}{mV} \left(N^2 - \tfrac{1}{2} \sum_p n_p^2 \right) \quad (12.72)$$

This formula is valid only under the conditions

$$\frac{a}{v^{1/3}} \ll 1 \quad (12.73)$$

$$ka \ll 1$$

where k is the relative wave number of any pair of particles. Thus (12.72) becomes invalid if there are excited particles of high momentum.

Let us first study the implications of (12.72). The ground state energy per particle is obtained from (12.72) by setting all $n_p = 0$ for $p \neq 0$, and $n_0 = N$:

$$\frac{E_0}{N} = \frac{2\pi a \hbar^2}{mv} = \left(\frac{\hbar}{m} \right)^2 2\pi a \rho \quad (12.74)$$

*H. E. Haber and H. A. Weldon, *Phys. Rev. Lett.* **46**, 1497 (1981).

where ρ is the mass density. It is proportional to the scattering length a and to the mass density, and it may be interpreted to be the energy shift of an average particle in the "optical approximation," whereby the effect of the rest of the system is replaced by a medium having an index of refraction. This interpretation can be justified as follows. In the shape-independent approximation we may replace a scattering potential by one of any shape, provided it gives the same scattering length. Let us replace the interparticle potential by a very shallow but very long-ranged square well such that the scattering length is still a. Now a particle moving through the system essentially "sees" a uniform potential of an appropriate depth. This gives (12.74).

For an excited state in which the particles have vanishingly small momenta the energy per particle is

$$\frac{E_n}{N} = \left(\frac{\hbar}{m}\right)^2 4\pi a\rho \left[1 - \tfrac{1}{2}\sum_{\mathbf{p}}\left(\frac{n_p}{N}\right)^2\right] \qquad (12.75)$$

The second term is most negative when all the excited particles are in the same momentum state. Thus we may say that "spatial repulsion leads to momentum space attraction." This is a consequence of the symmetry of the wave function.*

The "momentum space attraction" just mentioned also leads to an "energy gap" in the spectrum (12.72). This may be seen as follows. The energies of the very low excited states of the system are approximately given by

$$E_n \approx \sum_{\mathbf{p}} \frac{p^2}{2m} n_{\mathbf{p}} + N\left(\frac{\hbar}{m}\right)^2 4\pi a\rho \left[1 - \frac{1}{2}\left(\frac{n_0}{N}\right)^2\right] \qquad (12.76)$$

According to this formula, the excitation of one particle from the momentum state $\mathbf{p} = 0$ to a state of infinitesimal momentum changes the energy by the *finite* amount

$$\Delta = \left(\frac{\hbar}{m}\right)^2 2\pi a\rho \qquad (12.77)$$

Thus the single-particle energy spectrum is separated from the zero point of energy by the amount Δ. This "energy gap," however, is a feature only of the lowest-order formula. When the energy levels are calculated to higher orders in perturbation theory,† the energy gap disappears. Instead, there is only a decrease of level density just above the ground state, changing the single-particle spectrum $p^2/2m$ into a phonon spectrum $\hbar cp/2m$, where c is a constant. The "energy gap," which implies that the level density is strictly zero just above the ground state, is a crude approximation to the actual state of affairs.

The foregoing discussions make it clear that the energy levels (12.72), although not exact, possess many qualitative features of the effect of a repulsive interaction among bosons. We use them to calculate the partition function. The

*See Problem 12.7.
†See Section 13.8.

validity of this calculation is discussed as we proceed. We introduce a further simplification, namely, we take the energy levels to be

$$E_n = \sum_p \frac{p^2}{2m} n_p + \frac{4\pi a h^2}{mV}(N^2 - \tfrac{1}{2}n_0^2) \qquad (12.78)$$

The behavior of the model defined by (12.78) should be qualitatively the same as that by (12.72) when the temperature is so low that few particles are excited.[*]

The Equation of State

For the calculation of the partition function, we confine our considerations to the region in which

$$a/\lambda \ll 1, \qquad a\lambda^2/v \ll 1 \qquad (12.79)$$

because these are the only dimensionless parameters in the problem involving a, and our model is valid only to the first order in a.

Let n be an abbreviation for $\{n_k\}$, and let ϵ_n denote the first term of (12.78), the unperturbed energy. Introducing the parameter

$$\xi \equiv \frac{n_0}{N} \qquad (12.80)$$

and denoting, as usual, the thermal wavelength by $\lambda = \sqrt{2\pi\hbar^2/mkT}$, we can write the partition function in the form

$$Q_n = \sum_n e^{-\beta\epsilon_n} e^{-N(a\lambda^2/v)(2-\xi^2)} = Q_N^{(0)}\langle e^{-N(a\lambda^2/v)(2-\xi^2)}\rangle_0 \qquad (12.81)$$

where $Q_N^{(0)}$ is the partition function of the ideal Bose gas, and $\langle\ \rangle_0$ denotes the thermodynamic average with respect to the ideal Bose gas. Hence the free energy per particle is

$$\frac{A}{N} = \frac{A^{(0)}}{N} - \frac{kT}{N}\log\langle e^{-N(a\lambda^2/v)(2-\xi)^2}\rangle_0$$

$$\approx \frac{A^{(0)}}{N} + \frac{kT}{N}\frac{a\lambda^2}{v}\langle(2-\xi)^2\rangle_0 \qquad (12.82)$$

It can be easily verified that the fluctuations of $\langle n_0\rangle$ are small. In fact, for any k

$$\langle n_k^2\rangle_0 - \langle n_k\rangle_0^2 = \langle n_k\rangle_0 \qquad (12.83)$$

Hence the mean-square fluctuation of $\langle\xi\rangle_0$ is of the order $1/N$. Hence

$$\frac{A}{N} = \frac{A^{(0)}}{N} + \frac{h^2}{m}\frac{4\pi a}{v}(1 - \tfrac{1}{2}\xi^2) \qquad (12.84)$$

---

[*]For a detailed study of the equation of state based on the more accurate energy levels (12.72), see K. Huang, C. N. Yang, and J. M. Luttinger, *Phys. Rev.* **105**, 776 (1957).

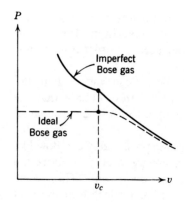

Fig. **12.10** Isotherm of an imperfect Bose gas with repulsive interactions.

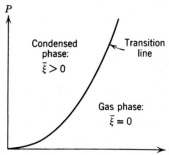

Fig. **12.11** P-T diagram of the imperfect Bose gas. In contradistinction to that of the diagram of the ideal Bose gas, Fig. 12.8, the space above the transition now corresponds to the condensed phase.

where $\bar{\xi} \equiv \langle \xi \rangle_0 = \langle n_0/N \rangle_0$ is given by (12.54). This result is extremely simple, being the free energy of the ideal gas plus the interaction term in (12.78), except that the quantum number n_0 is replaced by its thermodynamic average with respect to the ideal gas, thereby turning it into a thermodynamic parameter.

The pressure can be immediately obtained:

$$P = P^{(0)} + \frac{4\pi a\hbar^2}{m}\left[\frac{1}{v^2}\left(1 - \tfrac{1}{2}\bar{\xi}^2\right) + \frac{1}{v}\bar{\xi}\frac{\partial\bar{\xi}}{\partial v}\right] \tag{12.85}$$

where $P^{(0)}$ is the pressure of the ideal Bose gas. Using (12.54) to evaluate $\bar{\xi}$ and $\partial\bar{\xi}/\partial v$, we obtain

$$P = \begin{cases} P^{(0)} + \dfrac{4\pi a\hbar^2}{mv^2} & (v > v_c, T > T_c) \\[2mm] P^{(0)} + \dfrac{2\pi a\hbar^2}{m}\left(\dfrac{1}{v^2} + \dfrac{1}{v_c^2}\right) & (v < v_c, T < T_c) \end{cases} \tag{12.86}$$

An isotherm is shown in Fig. 12.10, and the P-T diagram is shown in Fig. 12.11. The Bose-Einstein condensation is here a second-order transition. The specific heat decreases across the transition point by the amount

$$\frac{\Delta C_V}{Nk} = \frac{9a}{2\lambda_c}g_{3/2}(1) \tag{12.87}$$

We cannot deduce from these results that an imperfect Bose gas with repulsive interactions generally exhibits a second-order transition. The present model merely shows that the transition appears to be a second-order transition if higher-order effects in a/λ and $a\lambda^2/v$ are neglected.

The model here is more realistic than the ideal Bose gas in that the condensed phase now has finite compressibility. We see from (12.86) that the isothermal compressibility increases discontinuously by a factor of 2 when we go from the gas to the condensed phase.

In a nonequilibrium situation, the condensed phase presumably can flow like a real substance. Since the system has a paucity of energy levels just above the ground state, we expect it to flow with little or no dissipation. This is the phenomenon of superfluidity that we shall discuss at greater length in the next chapter.

12.5 THE SUPERFLUID ORDER PARAMETER

We have suggested that the Bose-Einstein condensate is a "superfluid" in more realistic systems than the ideal Bose gas. This idea will be taken up in the next chapter. Here we shall analyze in greater depth the implications of the existence of a condensate. It will be necessary to use the quantized-field description of a many-body system (sometimes call 1 "second quantization"), as discussed in the appendix.

Identifying the Order Parameter

Let us begin by examining the one-particle density matrix

$$\rho_1(\mathbf{x}, \mathbf{y}) = \langle \psi^\dagger(\mathbf{x}) \psi(\mathbf{y}) \rangle = \frac{1}{V} \sum_{\mathbf{k}, \mathbf{q}} e^{i(\mathbf{k} \cdot \mathbf{x} - \mathbf{q} \cdot \mathbf{y})} \langle a_\mathbf{q}^\dagger a_\mathbf{k} \rangle \qquad (12.88)$$

where $\langle\ \rangle$ denotes ensemble average, and $\psi(\mathbf{x})$ is the quantized boson field operator, which is expanded in terms of annihilation operators $a_\mathbf{k}$ for a plane-wave state of wave vector $\mathbf{k}$ (see (A.65)). Roughly speaking, this is the probability that, having lost a particle at $\mathbf{x}$, you will find one at $\mathbf{y}$.

Consider first a translationally invariant system. Using the fact that the total momentum operator $\mathbf{P}$ commutes with the Hamiltonian, we can verify

$$\left\langle \left[\mathbf{P}, a_\mathbf{q}^\dagger a_\mathbf{k} \right] \right\rangle = 0$$

by writing out the trace and using $\mathrm{Tr}(AB) = \mathrm{Tr}(BA)$. On the other hand, a direct calculation gives

$$\left[\mathbf{P}, a_\mathbf{q}^\dagger a_\mathbf{k} \right] = \hbar(\mathbf{k} - \mathbf{q}) a_\mathbf{q}^\dagger a_\mathbf{k}$$

Hence for a translationally invariant system

$$\langle a_\mathbf{q}^\dagger a_\mathbf{k} \rangle = \delta_{\mathbf{q}, \mathbf{k}} \langle n_\mathbf{k} \rangle \qquad (12.89)$$

where $n_k = a_k^\dagger a_k$. Thus

$$\rho_1(\mathbf{x}, \mathbf{y}) = \frac{1}{V} \sum_k e^{i\mathbf{k}\cdot(\mathbf{x}-\mathbf{y})} \langle n_p \rangle$$

$$= \frac{\langle n_0 \rangle}{V} + \int \frac{d^3k}{(2\pi)^3} e^{i\mathbf{k}\cdot(\mathbf{x}-\mathbf{y})} \langle n_k \rangle \qquad (12.90)$$

where we have separated out the $k = 0$ term before passing to the limit $V \to \infty$. The second term in (12.90) vanishes when $|\mathbf{x} - \mathbf{y}| \to \infty$, because in that limit the integral gets contributions only from the neighborhood of $k = 0$ (see Problem 12.9). Thus

$$\rho_1(\mathbf{x}, \mathbf{y}) \xrightarrow[|\mathbf{x}-\mathbf{y}| \to \infty]{} \frac{\langle n_0 \rangle}{V} \qquad (12.91)$$

This does not imply a positional correlation over infinite distances, as it might seem at first sight. It says only that there is a constant density of zero-momentum particles over the entire system.

Note that in a general interacting system, the single-particle momenta are not good quantum numbers. In particular n_0 does not commute with the Hamiltonian. But $\langle n_0 \rangle/N$ can still be used as a characterization of a Bose-Einstein condensate.

In analogy with (12.91), Penrose and Onsager* proposed the following general criterion for Bose-Einstein condensation:

$$\langle \psi^\dagger(\mathbf{x})\psi(\mathbf{y}) \rangle \xrightarrow[|\mathbf{x}-\mathbf{y}| \to \infty]{} f^*(\mathbf{x})f(\mathbf{y}) \qquad (12.92)$$

To be of practical value, this criterion must be applicable to real systems with nontranslationally invariant geometry, and under nonequilibrium situations. It is then not obvious (in fact, it is somewhat of a mystery) how the criterion can be satisfied. Consider, for example, a Bose fluid contained in two separate tanks connected by a pipe a mile long. (For added realism, have an experimentalist kick the apparatus from time to time.) Suppose $\mathbf{x}$ lies in one tank, and $\mathbf{y}$ the other. It is physically absurd to suppose that there can be any correlation between $\mathbf{x}$ and $\mathbf{y}$. But then how does it come about that these separate points are characterized by the same function f?

The way out is to make f dependent only on local dynamical variables. It is now generally accepted that the correct choice is $f = \langle \psi \rangle$.[†] Thus the Penrose-Onsager criterion takes the more specific form

$$\langle \psi^\dagger(\mathbf{x})\psi(\mathbf{y}) \rangle \xrightarrow[|\mathbf{x}-\mathbf{y}| \to \infty]{} \langle \psi^\dagger(\mathbf{x}) \rangle \langle \psi(\mathbf{y}) \rangle \qquad (12.93)$$

---

*O. Penrose, *Philos. Mag.* **42**, 1373 (1951); O. Penrose and L. Onsager, *Phys. Rev.* **104**, 576 (1956).

[†] The first use of this was implicit in N. N. Bogoliubov, *J. Phys. USSR* **11**, 23 (1947). J. Goldstone, *N. Cim.*, **19**, 154 (1961), clarified its meaning in terms of "broken symmetry." For a review see P. W. Anderson, *Rev. Mod. Phys.* **38**, 298 (1966).

We call the complex number

$$\langle \psi(\mathbf{x}) \rangle = r(\mathbf{x}) \, e^{i\phi(\mathbf{x})} \qquad (12.94)$$

the superfluid order parameter. The fact that $r(\mathbf{x}) > 0$ implies the existence of momentum-space order, i.e., a Bose-Einstein condensate. As we shall see in the next chapter, the phase $\phi(\mathbf{x})$ is the velocity potential for superfluid flow.

Spontaneous Symmetry Breaking

The ensemble average $\langle \psi \rangle$ should be taken in a grand canonical ensemble, because we are interested in open systems, where the number of particles is not definite. Thus,

$$\langle \psi(x) \rangle = \frac{\mathrm{Tr}\left[e^{-\beta \mathscr{E}[\psi]} \psi(x) \right]}{\mathrm{Tr}\, e^{-\beta \mathscr{E}[\psi]}} \qquad (12.95)$$

$$\mathscr{E}[\psi] = \mathscr{H} - \mu \mathscr{N}$$

where $\mathscr{H}$ is the Hamiltonian and $\mathscr{N}$ is the number operator. The notation $\mathscr{E}[\psi]$ indicates that $\mathscr{E}$ is a functional of ψ. The immediate question is why the ensemble average above should not be always zero. There is a trivial and a subtle aspect to this question.

First, the trivial aspect. Since ψ annihilates a particle, its expectation value with respect to any eigenstate of $\mathscr{N}$ is zero. This makes one feel uneasy about taking its ensemble average. In the grand canonical ensemble, however, the relevant basis are not the simultaneous eigenstates of $\mathscr{H}$ and $\mathscr{N}$, but those of $\mathscr{H} - \mu \mathscr{N}$, and the latter need not be eigenstates of $\mathscr{N}$. In the infinite-volume limit, the eigenvalues of $\mathscr{H} - \mu \mathscr{N}$ are highly degenerate: systems with different particle numbers can maintain the same eigenvalue by going into different energy levels. One can form a new basis by superimposing these degenerate states (of different particle numbers), with respect to which the expectation value of ψ will have the form (12.94). The phase ϕ labels the degenerate states.

But, one argues, the ensemble of ψ is still zero, because one has the freedom to calculate the trace using a basis with definite particle numbers. This is true, and is a reflection of the fact that particle number is conserved, which can be expressed formally by saying that the Hamiltonian has a "global gauge invariance"—an invariance under the transformation

$$\psi(\mathbf{x}) \rightarrow e^{i\alpha} \, \psi(\mathbf{x}) \qquad (12.96)$$

where α is an arbitrary real number. Thus, in the ensemble average, every value $\psi = r \exp(i\alpha)$ will be canceled by a value $\psi = r \exp(i\alpha + i\pi)$ of equal weight. This argument is technically correct, and illustrates the need to redefine the ensemble average more carefully. This is the subtle part of the problem.

There is a parallel between $\langle \psi \rangle$ and the spontaneous magnetization of a ferromagnet:

$$\langle M \rangle = \frac{\mathrm{Tr}\left(e^{-\beta \mathscr{H}} M \right)}{\mathrm{Tr}\, e^{-\beta \mathscr{H}}} \qquad (12.97)$$

where M is the total magnetic moment. Since the Hamiltonian $\mathcal{H}$ in the absence of external field is invariant under rotations, the ensemble average of M is always zero because M and $-M$ occur with equal probability. The mathematical correctness of this statement is irrefutable. But we know that it is physically the wrong answer, for we do have ferromagnets in nature.

The resolution of the apparent paradox lies in the recognition that the symmetry of a system may be "spontaneously broken," in that the ground state of a Hamiltonian does not possess the symmetry of the Hamiltonian. This requires that the ground state be degenerate. The symmetry is realized by the fact that any one of the degenerate ground states is equally as good as the physical ground state, and by the existence of characteristic "Goldstone excitations."* For a ferromagnet, the ground state is not rotationally invariant, because the magnetization points along a definite axis in space. The Goldstone excitations in this case are the spin waves.

The essential point in the present context is that, once the system magnetizes along a certain direction it cannot make a transition to another direction, even though doing so requires no expenditure in energy. For to do so requires that all the atomic magnetic moments in the system spontaneously and simultaneously rotate through exactly the same angle. The probability for this to happen is essentially zero for a macroscopic system. (One would have to wait for a time of the order of a Poincaré cycle to see this happen.) The ensemble average has physical significance only if it corresponds to time averages over microscopic relaxation times. One must therefore redefine it in such a way that M and $-M$ are not both included among the configurations. This can be done most simply by placing the system in an external field H pointing along an arbitrary but fixed direction, and calculating the ensemble average in the limit $H \rightarrow 0$. To emphasize the importance of various limiting processes, we explicitly indicate the infinite-volume limit:

$$\frac{\langle M \rangle}{V} \equiv \lim_{H \to 0} \lim_{V \to \infty} \frac{1}{V} \frac{\text{Tr}\left[e^{-\beta(\mathcal{H}-MH)} M\right]}{\text{Tr}\, e^{-\beta(\mathcal{H}-MH)}} \tag{12.98}$$

The thermodynamic limit of (12.97), which is not physically relevant, corresponding a reversal of the limiting process above[†]:

$$\lim_{V \to \infty} \lim_{H \to 0} \frac{\text{Tr}\left[e^{-\beta(\mathcal{H}-MH)} M\right]}{\text{Tr}\, e^{-\beta(\mathcal{H}-MH)}} = 0 \tag{12.99}$$

Returning to the Bose system, we see that Bose-Einstein condensation corresponds to a spontaneous breaking of the global gauge invariance. In analogy with ferromagnetism, we imagine subjecting the system to an external field coupled to $\psi(\mathbf{x})$, calculate the ensemble average of $\psi(\mathbf{x})$ in the thermodynamic

---

*J. Goldstone, *op. cit.* A brief discussion of this phenomenon will be given in Section 16.6.

[†] Note that in calculating the spontaneous magnetization in the model in Section 11.6, we in effect used the correct average (12.98) instead of (12.99), because we ignored the $-M$ solution (by common sense).

limit, and then let the external field go to zero:

$$\langle \psi(\mathbf{x}) \rangle \equiv \lim_{\eta \to 0} \lim_{V \to \infty} \frac{\text{Tr}\left[e^{-\beta \mathscr{E}[\psi, \eta]} \psi(\mathbf{x}) \right]}{\text{Tr}\, e^{-\beta \mathscr{E}[\psi, \eta]}} \tag{12.100}$$

where

$$\mathscr{E}[\psi, \eta] = \mathscr{H} - \mu \mathscr{N} - \int d^3x \left[\psi(\mathbf{x}) \eta(\mathbf{x}) + \psi^\dagger(\mathbf{x}) \eta^\dagger(\mathbf{x}) \right] \tag{12.101}$$

The only essential difference with the ferromagnetic case is that, unlike the magnetic field, the external field $\eta(\mathbf{x})$ here is a mathematical device that cannot be realized experimentally.*

PROBLEMS

12.1 (*a*) Show that the entropy per photon in blackbody radiation is independent of the temperature, and in d spatial dimensions is given by

$$s = (d + 1) \frac{\sum\limits_{n=1}^{\infty} n^{-d-1}}{\sum\limits_{n=1}^{\infty} n^{-d}}$$

(*b*) Show that the answer would have been $d + 1$ if the photons obeyed Boltzman statistics.

12.2 Some experimental values[†] for the specific heat of liquid He^4 are given in the accompanying table. The values are obtained along the vapor pressure curve of liquid He^4, but we may assume that they are not very different from the values of c_V at the same temperatures.

Temperature (K)	Specific Heat ($joule/g\text{-}deg$)
0.60	0.0051
0.65	0.0068
0.70	0.0098
0.75	0.0146
0.80	0.0222
0.85	0.0343
0.90	0.0510
0.95	0.0743
1.00	0.1042

*The Bose-Einstein condensation of the ideal gas has been reanalyzed in terms of the superfluid order parameter by J. D. Gunton and M. J. Buckingham, *Phys. Rev.* **166**, 152 (1968).

† Taken from H. C. Kramers, "Some Properties of Liquid Helium below 1°K," Dissertation, Leiden (1955).

(*a*) Show that the behavior of the specific heat at very low temperatures is characteristic of that of a gas of phonons.

(*b*) Find the velocity of sound in liquid He^4 at low temperature.

12.3 Equation (12.64) states that $G = 0$ for $v < v_c$. Using the formula $S = -(\partial G/\partial T)_P$, we would obtain $S = 0$ for $v < v_c$, in contradiction to (12.65). What is wrong with the previous statement?

12.4 In the neighborhood of $z = 1$ the following expansion may be obtained (F. London, *loc. cit.*):

$$g_{5/2}(z) = 2.363v^{3/2} + 1.342 - 2.612v - 0.730v^2 + \cdots$$

where $v \equiv -\log z$. From this the corresponding expansions for $g_{3/2}$, $g_{1/2}$, and $g_{-1/2}$ may be obtained by the recursion formula $g_{n-1} = -\partial g_n/\partial v$. Using this expansion show that for the ideal Bose gas the discontinuity of $\partial C_V/\partial T$ at $T = T_c$ is given by

$$\left(\frac{\partial}{\partial T}\frac{C_V}{Nk}\right)_{T\to T_c^+} - \left(\frac{\partial}{\partial T}\frac{C_V}{Nk}\right)_{T\to T_c^-} = \frac{3.66}{T_c}$$

12.5 Show that the equation of state of the ideal Bose gas in the gas phase can be written in the form of a virial expansion, i.e.,

$$\frac{Pv}{kT} = 1 - \frac{1}{4\sqrt{2}}\left(\frac{\lambda^3}{v}\right) + \left(\frac{1}{8} - \frac{2}{9\sqrt{3}}\right)\left(\frac{\lambda^3}{v}\right)^2 - \cdots$$

12.6 (*a*) Calculate the grand partition function $\mathcal{Q}(z, V, T)$ for a two-dimensional ideal Bose gas and obtain the limit

$$\lim_{V\to\infty} \frac{1}{V}\log \mathcal{Q}(z, V, T)$$

where $V = L^2$ is the area available to the system.

(*b*) Find the average number of particles per unit area as a function of z and T.

(*c*) Show that there is no Bose-Einstein condensation for a two-dimensional ideal Bose gas.

12.7 Consider two free bosons contained in a box of volume V with periodic boundary conditions. Let the momenta of the two particles be $\mathbf{p}$ and $\mathbf{q}$.

(*a*) Write down the normalized wave function $\psi_{pq}(\mathbf{r}_1, \mathbf{r}_2)$ for both $\mathbf{p} \neq \mathbf{q}$ and $\mathbf{p} = \mathbf{q}$.

(*b*) Show that for $\mathbf{p} \neq \mathbf{q}$

$$\left|\psi_{pq}(\mathbf{r}, \mathbf{r})\right|^2 > \left|\psi_{pp}(\mathbf{r}, \mathbf{r})\right|^2$$

(*c*) Explain the meaning of the statement "spatial repulsion leads to momentum space attraction."

12.8 For the imperfect Bose gas discussed in Section 12.4, show that in the gas phase

$$\frac{Pv}{kT} = 1 + \left(-\frac{1}{4\sqrt{2}} + \frac{2a}{\lambda}\right)\frac{\lambda^3}{v} + \left(\frac{1}{8} - \frac{1}{3\sqrt{3}}\right)\left(\frac{\lambda^3}{v}\right)^2 - \cdots$$

Thus we can conclude that the third and higher virial coefficients, if they depend on a, must involve orders of a^2 or higher.

12.9 Consider an ideal Bose gas. Let $\psi(\mathbf{x})$ be the boson field operator.

(a) Show

$$\langle \psi^\dagger(\mathbf{x})\psi(\mathbf{y}) \rangle = \frac{\langle n_0 \rangle}{V} + f(|\mathbf{x} - \mathbf{y}|)$$

where

$$f(r) = \int \frac{d^3k}{(2\pi)^3}\, e^{i\mathbf{k}\cdot\mathbf{r}} \langle n_k \rangle = \frac{mkT}{2\hbar^2}\, \frac{e^{-r/r_0}}{r}$$

with $r_0 = \hbar/\sqrt{2mkT|\log z|}$

(b) Let $T \to T_c$ from the high-temperature side. Find r_0 as a function of $t = (T - T_c)/T_c$, as $t \to 0$.

(c) The density-density correlation function is defined as

$$\Gamma(x) \equiv \langle \rho(x)\rho(0) \rangle - (N/V)^2$$

where $\rho(\mathbf{x}) = \psi^\dagger(\mathbf{x})\psi(\mathbf{x})$ is the density operator. Show

$$\Gamma(\mathbf{x}) = \frac{1}{V^2} \sum_{k \neq q} e^{i(\mathbf{k}-\mathbf{q})\cdot\mathbf{x}} \langle n_q(n_k + 1) \rangle \xrightarrow[V \to \infty]{} \left| \frac{1}{V} \sum_k e^{i\mathbf{k}\cdot\mathbf{x}} \langle n_k \rangle \right|^2$$

Work out $\Gamma(\mathbf{x})$ more explicitly, using the results of (a) and (b).

C

SPECIAL TOPICS IN STATISTICAL MECHANICS

CHAPTER

13

SUPERFLUIDS

A superfluid is a fluid that flows without dissipation—a phenomenon that never ceases to cause excitement, because it defies common sense and raises the hope that one might capitalize on it. The classic example is liquid He^4 in its low-temperature phase, which can flow through the tiniest cracks without viscosity. Equally remarkable is a superconducting metal in which "supercurrents" can be maintained without an applied emf for months, and presumably indefinitely in principle. In this case one thinks of the charge carriers (the Cooper pairs) as a superfluid. Other examples include spin-aligned atomic hydrogen, which experimenters can now create in the laboratory.

In this chapter, we shall concentrate on liquid He^4 as the prototype of a superfluid. The essence of superfluidity, as we shall see, is the existence of a condensate characterizable by a nonvanishing superfluid order parameter.

13.2 LIQUID HELIUM

Why Helium Does Not Solidify

Helium is the only substance known to resist solidification under atmospheric pressure down to the lowest temperatures observed, and presumably to absolute zero. To solidify it requires an external pressure of at least 25 atm. This happy circumstance furnishes us with two naturally occurring quantum liquids, He^4 and its less abundant isotope He^3.

The qualitative reasons for the fluidity of helium are two: (a) The molecular interactions between He atoms are weak, as evidenced by the fact that He is a noble gas; (b) the mass of He is the smallest among the noble gases. These circumstances lead to a large zero-point motion of the He atoms, so that it becomes impossible to localize the atoms at well-defined lattice sites. To understand these reasons, we must first have a few facts.

The potential energy $v(r)$ between two He atoms separated by a distance r has been calcuiᵤted by Slater and Kirkwood* on the basis of the electronic

---

*J. C. Slater and J. G. Kirkwood, *Phys. Rev.* **37**, 682 (1931).

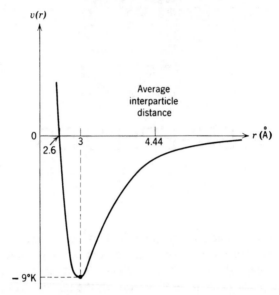

Fig. 13.1 Potential energy between two He atoms separated by distance r.

structure of the He atom. They find that

$$v(r) = (5.67 \times 10^6) \, e^{-21.5(r/\sigma)} - 1.08\left(\frac{\sigma}{r}\right)^6 \qquad (13.1)$$

where $\sigma = 4.64$ Å and $v(r)$ is in kelvins. A graph of $v(r)$ is shown in Fig. 13.1.

From the slope dP/dT of the experimental vapor pressure curve we can deduce the latent heat of vaporization of liquid helium. Extrapolation of experimental data shows that $dP/dT > 0$ at absolute zero. Hence liquid helium has a nonvanishing binding energy per atom at absolute zero. That is, the ground state of liquid helium is an N-body bound state that has a self-determined equilibrium density in the absence of external pressure.

Consider a collection of He atoms at absolute zero under no external pressure. The most probable configuration of the atoms is determined by the ground state wave function, which, according to the variational principle, must be such as to minimize the total energy of the system, with no external constraint imposed. Hence energy consideration alone determines the most probable configuration. We can then make the following qualitative argument. If a He atom is to have a well-defined location, it must be confined to within a distance Δx that is small compared to the range of the potential, say $\Delta x \approx 0.5$ Å. By the uncertainty principle, we would then expect an uncertainty in energy (in units of

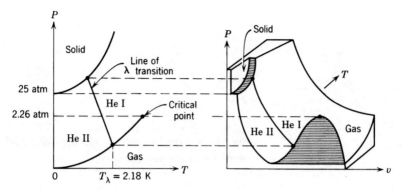

Fig. 13.2 Equation of state of He^4 (not to scale).

Boltzmann's constant) of the order of

$$\Delta E \approx \frac{1}{2m}\left(\frac{\hbar}{\Delta x}\right)^2 \approx 10 \text{ K} \tag{13.2}$$

This is comparable to the depth of the potential well. Hence the localization is impossible. The fact that no other noble element can remain in liquid form down to very low temperatures is explained by their much greater masses. The fact that H_2, although lighter than He, solidifies at a finite temperature is explained by the strong molecular interactions between H_2 molecules. The argument we have given is independent of statistics and also explains why both He^4 and He^3 remain liquid down to absolute zero.

The λ Transition

The equation of state of He^4 is shown in Fig. 13.2. In the liquid phase there is a second-order (i.e., non-first-order) phase transition dividing the liquid further into He I and He II. The former is a normal-type liquid, but He II exhibits superfluid behavior, and will be the focus of our attention. The transition from He I to He II is called the λ *transition*. Along the vapor pressure curve it occurs at the temperature T_λ and the specific volume v_λ, given by

$$T_\lambda = 2.18 \text{ K}$$
$$v_\lambda = 46.2 \text{ Å}^3/\text{atom} \tag{13.3}$$

The specific heat along the vapor curve becomes logarithmically infinite as the point of λ transition is approached from either side,* as shown in Fig. 13.3. The shape of the specific-heat curve near T_λ gives rise to the name λ transition.

It is significant that He^4 exhibits the λ transition and He^3 does not. Apart from a difference in atomic masses, the only difference between these two

---

*W. M. Fairbank, M. J. Buckingham, and C. F. Kellers, in *Low Temperature Physics and Chemistry*, edited by J. R. Dillinger (University of Wisconsin Press, Madison, 1958), p. 50.

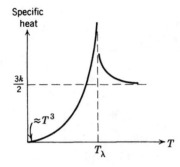

Fig. 13.3 Experimental specific heat of liquid He^4 along the vapor pressure curve.

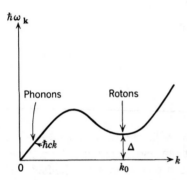

Fig. 13.4 Energy spectrum of elementary excitations in liquid He^4.

substances is that He^4 atoms are bosons, whereas He^3 atoms are fermions. Hence it is natural to assume that the λ transition is the Bose-Einstein condensation, modified, of course, by molecular interactions. In fact, an ideal Bose gas with the same mass and density as liquid He^4 would undergo the Bose-Einstein condensation at 3.14 K, which is of the same order of magnitude as T_λ.

The isotope He^3 undergoes a phase transition to a superfluid phase at much lower temperature (about 10^{-3} K.) The physical mechanism is similar to that in superconductivity, in which bound fermion pairs form a condensate. We shall not discuss this case, and will consider He^4 exclusively from now on.

Elementary Excitations

The lowest excitations of liquid He^4 can be found experimentally by neutron scattering.* If the temperature is sufficiently low, the scattering proceeds predominantly through the creation of an elementary excitation whose energy and momentum is equal to the energy and momentum transferred to the system by the neutron. In this manner, one finds the spectrum of elementary excitations ω_k shown schematically in Fig. 13.4. The linear portion near $k = 0$ represents phonons. The portion near $k = k_0$ corresponds to "rotons," which behaves like a

---

*Palevsky, Otnes, and Larsson, *Phys. Rev.* **112**, 11 (1958); Yarnell, Arnold, Bendt, and Kerr, *Phys. Rev.* **113**, 1379 (1959). The measurements are made at about 1 K.

particle with finite mass σ, and requires a minimal energy Δ for its creation. (It is an excitation with an energy gap). The experimentally determined parameters are

$$
\hbar\omega_k = \begin{cases} \hbar c k & (k \ll k_0) \\ \Delta + \dfrac{\hbar^2(k - k_0)^2}{2\sigma} & (k \approx k_0) \end{cases}
$$

$$
\begin{aligned}
c &= (239 \pm 5) \text{ m/s} \\
\frac{\Delta}{k} &= (8.65 \pm 0.04) \text{ K} \\
k_0 &= (1.92 \pm 0.01) \text{ Å}^{-1} \\
\frac{\sigma}{m} &= 0.16 \pm 0.01
\end{aligned}
\tag{13.4}
$$

The parameter c is the velocity of sound, and m is the mass of a helium atom.

The noteworthy fact is that phonons constitute the only type of low-lying excitations. While one expects this to be true of a crystal lattice, it is somewhat surprising for a liquid. As we shall see, the reason for this has to do with the Bose statistics of the system.

13.2 TISZA'S TWO-FLUID MODEL

When Kamerlingh Onnes first liquified helium in 1908, some unusual properties of this liquid below 2.18 K became obvious. It was found that liquid helium will crawl up walls and coat all inside surfaces of its container (unless, of course, part of the container is above the temperature T_λ). If placed in a open beaker, the liquid will crawl up the wall, across the rim, and out of the beaker. It is obviously "super" fluid. Some other unusual features can be described in terms of a phenomenological model proposed by Tisza, known as the two-fluid model.

Tisza postulates that the phase He II is made up of two components called the *normal fluid* and the *superfluid*. In contradistinction, the phase He I is pure normal fluid. It is imagined that we may attribute characteristic mass densities ρ_n and ρ_s, respectively, to the normal fluid and the superfluid. If the liquid flows, we may attribute characteristic velocity fields $\mathbf{v}_n$ and $\mathbf{v}_s$, respectively, to the normal and the superfluid. The mass density ρ and the velocity field $\mathbf{v}$ of He II are assumed to be given by

$$
\begin{aligned}
\rho &= \rho_n + \rho_s \\
\rho\mathbf{v} &= \rho_n\mathbf{v}_n + \rho_s\mathbf{v}_s
\end{aligned}
\tag{13.5}
$$

The normal fluid is supposed to behave like an ordinary classical fluid, whereas the superfluid has the unusual properties that

(a) its entropy is zero;

(b) it flows with no resistance through channels of extremely small diameters (10^{-2} cm or even smaller).

With no further assumptions many strange properties of He II can be qualitatively understood.

First, an extremely small opening in a tank of He II acts as a filter for the superfluid component because the latter can pass through the opening, leaving the normal-fluid component behind. Suppose two tanks of He II are connected by an extremely thin tube, and suppose some superfluid is made to flow from tank A to tank B by the establishment of a pressure differential. Then, since the superfluid has no entropy, the entropy per unit mass in tank A will increase whereas that of tank B will decrease. Therefore tank A will warm up while tank B cools. This is known as the *mechanocaloric effect*.

The inverse effect, that of the creation of a pressure differential by heating, is known as the *fountain effect*.

In the two fluid model a sound wave in He II must mean a sinusoidal oscillation of ρ_n and ρ_s in phase with each other, because only when they oscillate in phase can the total mass density vary sinusoidally. We can also imagine a new mode of oscillation, however, in which the normal fluid and the superfluid oscillate out of phase by 180°. This would not be a sound wave, for the total mass density would be constant throughout the liquid, but it would represent a sinusoidal variation of the entropy per unit mass, because the superfluid has no entropy. Therefore, we may hope to excite this new mode of oscillation by local heating of He II. The temperature gradient established would not propagate by diffusion, as in heat conduction, but would propagate like a wave, with a characteristic velocity. This phenomenon is called the *second sound*.

All these phenomena are observed in He II. Applying purely thermodynamic considerations, we can describe them quantitatively in terms of the two-fluid model.

In the two-fluid model the relative amount of normal fluid and superfluid present may be deduced from an experiment of Andronikashvili. A pile of disks spaced 0.2 mm apart are mounted on a shaft, and the assembly is made to rotate in He II, as shown in Fig. 13.5. The moment of inertia of the assembly is measured as a function of the temperature, when the He II in equilibrium with its own vapor. By assuming that the superfluid remains completely unaffected by the rotation disks, whereas the normal fluid between the disks is dragged into rotation, we see that the moment of inertia must be proportional to ρ_n/ρ. The proportionality constant is determined by the fact that at the transition tempera-

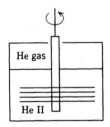

Fig. 13.5 Experiment of Andronikashvili.

ture $\rho_n/\rho = 1$. The experimental results fit the formula

$$\frac{\rho_n}{\rho} = \begin{cases} \left(\dfrac{T}{T_\lambda}\right)^{5.6} & (T < T_\lambda) \\ 1 & (T > T_\lambda) \end{cases} \tag{13.6}$$

We shall give a microscopic explanation of the two-fluid model later.

13.3 THE BOSE-EINSTEIN CONDENSATE

Penrose and Onsager* have given an estimate of the fraction of particles in liquid He^4 having zero momentum, by making use of (12.92) for a translationally invariant liquid. An equivalent expression for the one-particle density matrix in terms of the wave functions of the system is (cf. (A.53))

$$\rho_1(\mathbf{x} - \mathbf{y}) \equiv \langle \psi^\dagger(\mathbf{x}) \psi(\mathbf{y}) \rangle$$

$$= N \sum_n \int d^3 r_2 \cdots d^3 r_N \, \Psi_n(\mathbf{y}, \mathbf{r}_2, \ldots, \mathbf{r}_N) \, e^{-\beta \mathcal{H}} \Psi_n^*(\mathbf{x}, \mathbf{r}_2, \ldots, \mathbf{r}_N) \tag{13.7}$$

which at absolute zero reduces to

$$\rho_1(\mathbf{r}) = N \int d^3 r_2 \cdots d^3 r_N \, \Psi_0(\mathbf{r}, \mathbf{r}_2, \ldots, \mathbf{r}_N) \Psi_0(0, \mathbf{r}_2, \ldots, \mathbf{r}_N)$$

where Ψ_0 is the ground state wave function. It is real, positive, and unique, as shown in the Appendix. Intuitively we expect Ψ_0 to be approximately constant except when two He atoms "touch." Then it is essentially zero. The main effect of the attractive part of the potential is to give the whole system a self-determined equilibrium density. It is plausible that this effect is not important for our considerations, as long as we assume the liquid to have the observed density of liquid He^4. We assume that

$$\Psi_0(\mathbf{r}_1, \mathbf{r}_2, \ldots, \mathbf{r}_N) = \frac{1}{\sqrt{Z_N}} F_N(\mathbf{r}_1, \ldots, \mathbf{r}_N) \tag{13.8}$$

where Z_N is a normalization constant, and

$$F_N(\mathbf{r}_1, \ldots, \mathbf{r}_N) = \begin{cases} 0 & (|\mathbf{r}_i - \mathbf{r}_j| \leq a, \text{ any } i \neq j) \\ 1 & (\text{otherwise}) \end{cases} \tag{13.9}$$

$$a = 2.56 \text{ Å}$$

The constant $Z_N/N!$ is numerically equal to the configuration integral of a classical hard-sphere gas. We note that

$$F_N(\mathbf{r}, \mathbf{r}_2, \ldots, \mathbf{r}_N) F_N(0, \mathbf{r}_2, \ldots, \mathbf{r}_N) = F_{N+1}(\mathbf{r}, 0, \mathbf{r}_2, \ldots, \mathbf{r}_N).$$

*O. Penrose and L. Onsager, *Phys. Rev.* **104**, 576 (1956).

Hence

$$\rho_1(\mathbf{r}) = \frac{N}{Z_N} \int d^3 r_2 \cdots d^3 r_N \, F_{N+1}(\mathbf{r}, 0, \mathbf{r}_2, \ldots, \mathbf{r}_N) \qquad (13.10)$$

which is proportional to the pair correlation function for $N + 1$ classical hard spheres (see Problem 9.3):

$$\rho_2(\mathbf{r}) \equiv \frac{N(N+1)}{Z_{N+1}} \int d^3 r_2 \cdots d^3 r_N \, F_{N+1}(\mathbf{r}, 0, \mathbf{r}_2, \ldots, \mathbf{r}_N) \qquad (13.11)$$

which tends to $(N/V)^2$ as $r \to \infty$. We can now write

$$\rho_1(\mathbf{r}) = z\rho_2(\mathbf{r})$$

$$z = \frac{N!}{(N+1)!} \frac{Z_{N+1}}{Z_N} \qquad (13.12)$$

As $r \to \infty$, we have $\lim \rho_1(\mathbf{r}) = z(N/V)^2$. Hence by (12.91)

$$\frac{\langle n_0 \rangle}{N} = \frac{zN}{V} \qquad (13.13)$$

showing that Bose-Einstein condensation exists at absolute zero, if z is finite. Let

$$\lim_{N \to \infty} \frac{1}{N} \log \frac{Z_N}{N!} = f(v) \qquad (13.14)$$

where $v \equiv V/N$. Then

$$\log z = f(v) - v \frac{\partial f(v)}{\partial v} \qquad (13.15)$$

The atomic volume of liquid He^4 is $v = 46.2 \, \text{Å}^3$, which implies an average interatomic distance of 4.44 Å. Since this is somewhat larger than the effective hard-sphere radius of 1.28 Å, we may calculate $f(v)$ by making a virial expansion as discussed in Problem 7.6. The result is

$$f(v) = \log v - \frac{2\pi}{3} \frac{a^3}{v} \qquad (13.16)$$

which leads to

$$\frac{\langle n_0 \rangle}{N} = e^{-[1+(4\pi/3)(a^3/v)]} = 0.08 \qquad (13.17)$$

Thus the modulus of the superfluid order parameter at absolute zero is $|\langle \psi \rangle|$ $\cong \sqrt{0.08}$. One has to be careful to distinguish this number from the "superfluid" fraction ρ_s/ρ in Tisza's phenomenological model. The latter is by definition 1 at absolute zero. The difference is made clearer in the next section.

13.4 LANDAU'S THEORY

We now describe a modernized version of Landau's theory.* We assume that there are no other excitations near the ground state except the elementary excitations whose spectrum is illustrated in Fig. 13.4. Thus the quantum states immediately above the ground state can be described as a gas of noninteracting elementary excitations, with energy

$$E_n = E_0 + \sum_{\mathbf{k}} \hbar \omega_{\mathbf{k}} n_{\mathbf{k}} \tag{13.18}$$

where $\hbar \omega_{\mathbf{k}}$ is the energy of an elementary excitation of wave number $\mathbf{k}$ and $\{n_{\mathbf{k}}\}$ is a set of occupation numbers. It is assumed that the elementary excitations are bosons, so that $n_{\mathbf{k}} = 0, 1, 2, \ldots$. Clearly, there is some restriction on the occupation numbers, just as in the Debye theory of the crystal lattice. This restriction is unknown; but it cannot be important for the low-temperature properties of the liquid, because only a few excitations would be present. Therefore we may take each $n_{\mathbf{k}}$ to be an independent number. The thermodynamic average of $n_{\mathbf{k}}$ is then given by

$$\langle n_{\mathbf{k}} \rangle = \frac{1}{e^{\hbar \beta \omega_{\mathbf{k}}} - 1} \tag{13.19}$$

and the internal energy of the liquid is

$$U = E_0 + \sum_{\mathbf{k}} \hbar \omega_{\mathbf{k}} \langle n_{\mathbf{k}} \rangle = E_0 + \frac{V}{2\pi^2} \int_0^\infty dk \, \frac{k^2 \hbar \omega_{\mathbf{k}}}{e^{\beta \hbar \omega_{\mathbf{k}}} - 1} \tag{13.20}$$

where V is the volume occupied by the liquid. The specific heat at constant volume is

$$\frac{1}{Nk} C_V = \frac{1}{Nk} \frac{\partial U}{\partial T} \tag{13.21}$$

where N is the number of He^4 atoms in the liquid. For low temperatures, only the phonon part and the roton part contribute significantly to the integral in (13.20). The specific heat can therefore be written approximately as a sum of two terms, the phonon contribution and the roton contribution.

$$\frac{1}{Nk} C_{\text{phonon}} = \frac{2\pi^2 v (kT)^3}{15 \hbar^3 c^3} \tag{13.22}$$

where $v = V/N$ and m is the mass of a He^4 atom. The roton part may be calculated by treating kT/Δ as a small quantity. We obtain

$$\frac{1}{Nk} C_{\text{roton}} \approx \frac{2\sqrt{\sigma} \, k_0^2 \, \Delta^2 v \, e^{-\Delta/kT}}{(2\pi)^{3/2} \hbar (kT)^{3/2}} \tag{13.23}$$

The total specific heat is the sum of (13.22) and (13.23). Fitting these to specific

*L. D. Landau, *J. Phys. USSR* **5**, 71 (1941).

heat data, one recovers the parameters (13.4), in good agreement with the directly measured values given there.

The Landau theory is important because it provides an explicit construction of the two-fluid model near absolute zero. Let us consider the wave function of the liquid. At absolute zero, it is the function Ψ_0 given approximately in (13.8). This is by definition the pure superfluid, in absolute thermal equilibrium. As the temperature increases, there will be elementary excitations. Let us superimpose the wave functions of elementary excitations of slightly different momenta, so as to form a wave packet $f(\mathbf{r})$. Then the wave function in which there is one wave packet is

$$\Psi_f = \sum_{j=1}^{N} f(\mathbf{r}_j)\Psi_0 \tag{13.24}$$

where the symmetrization over particle coordinates is required by Bose statistics. The wave functions for two packets $f(\mathbf{r})$ and $g(\mathbf{r})$ present can be represented by

$$\Psi_{fg} \equiv \sum_{i=1}^{N} \sum_{j=1}^{N} f(\mathbf{r}_i) g(\mathbf{r}_j)\Psi_0 \tag{13.25}$$

We can define Ψ_{fgh}, etc., in a similar manner. At very low temperatures the wave function of the liquid is an *incoherent* superposition of all such wave functions, with different numbers of wave packets (incoherent, because we are considering a quantum ensemble). The coefficients in this superposition depend on the temperature, in such a manner that the average number of excitations of approximate wave vector $\mathbf{k}$ is $\langle n_{\mathbf{k}} \rangle$.

In this ensemble, the factor Ψ_0 in (13.24) and (13.25) is the same for all wave functions, as long as $\langle n_{\mathbf{k}} \rangle / N \ll 1$. Thus, at a finite but small temperature, the superfluid continues to be represented by the ground state wave function Ψ_0, while the normal fluid is the gas of elementary excitations.

Near absolute zero, then, the normal fluid component of liquid helium consists of a very dilute gas of elementary excitations, in the form of phonons. Landau argues that at such low temperatures, He II can flow past a wall without function, as long as the flow velocity is less than c.

To understand this we need only consider what happens if an external object with a velocity v_e is dragged through a stationary liquid. Since the only excitations of the liquid are phonons, the only ways energy and momentum can be imparted to the liquid are by (a) excitation of new phonons and (b) scattering of existing phonons. Suppose an amount of energy ΔE is transferred through excitation of a number of phonons, specified by the occupation numbers $\{n_{\mathbf{k}}\}$:

$$\Delta E = \sum_{\mathbf{k} \neq 0} c|\mathbf{k}|n_{\mathbf{k}} \tag{13.26}$$

The momentum imparted to the system is necessarily

$$\Delta \mathbf{P} = \sum_{\mathbf{k} \neq 0} \mathbf{k} n_{\mathbf{k}} \tag{13.27}$$

Therefore

$$|\Delta\mathbf{P}| \le \sum_{\mathbf{k} \neq 0} |\mathbf{k}| n_{\mathbf{k}} \tag{13.28}$$

or

$$c|\Delta\mathbf{P}| \le \Delta E \tag{13.29}$$

On the other hand, if the external object loses the amount of energy ΔE and momentum $\Delta\mathbf{P}$, we must have

$$\Delta E = \mathbf{v}_e \cdot \Delta\mathbf{P} \tag{13.30}$$

This is impossible unless $|\mathbf{v}_e| > c$. At low temperatures the transfer of energy and momentum through scattering of existing phonons can be neglected because as $T \to 0$ the number of phonons becomes zero. It is noted that the argument depends on the linearity of the phonon energy spectrum and does not apply to an ideal gas of bosons.

The foregoing argument is valid at absolute zero, but it breaks down at a higher temperature when there are many phonons present. The external object may now transfer energy and momentum by scattering the phonons. Experiments on the flow of He II past a wall at 1 K indicate that the critical velocity is orders of magnitude smaller than c.

13.5 SUPERFLUID VELOCITY

Galilean Transformation

Thus far we have considered only states for which the total energy and total momentum are given by

$$\begin{aligned} E_n &= E_0 + \sum_{\mathbf{k} \neq 0} \hbar\omega_{\mathbf{k}} n_{\mathbf{k}} \\ \mathbf{P}_n &= \sum_{\mathbf{k} \neq 0} \hbar\mathbf{k} n_{\mathbf{k}} \end{aligned} \tag{13.31}$$

The total momenta of these states come exclusively from the elementary excitations (the normal fluid), while the superfluid remains uniform and stationary. It is clear that there are states not included above in which the superfluid also contributes to the total momentum. Such states become important if we want to describe the flow of the superfluid.

To generate the states for which the superfluid is flowing uniformly, all we have to do is to multiply the ground state wave function, which is a common factor to all the states we are considering, by $\exp(i\mathbf{P} \cdot \mathbf{R}/\hbar)$ where $\mathbf{P}$ is the total momentum and $\mathbf{R}$ the center-of-mass coordinate:

$$\Psi_0 \to e^{i\mathbf{P} \cdot \mathbf{R}/\hbar}\Psi_0 \tag{13.32}$$

The new ground state wave function corresponds to a superfluid moving uniformly with velocity $\mathbf{v}_s = \mathbf{P}/Nm$. In a frame moving with the superfluid, all operators undergo a Galilean transformation

$$S = e^{-iG}, \qquad G = \frac{m}{\hbar} \mathbf{v}_s \cdot \sum_{j=1}^{N} \mathbf{r}_j \tag{13.33}$$

The Hamiltonian and the total momentum of the liquid are given by

$$\mathcal{H} = -\frac{\hbar^2}{2m} \sum_{j=1}^{N} \nabla_j^2 + \sum_{i<j} v(|\mathbf{r}_i - \mathbf{r}_j|)$$

$$\mathbf{P}_{0p} = \frac{\hbar}{i} \sum_{j=1}^{N} \nabla_j \tag{13.34}$$

Under the Galilean transformation they become*

$$e^{-iG} \mathcal{H} e^{iG} = \mathcal{H} + \hbar \mathbf{v}_s \cdot \mathbf{P}_{0p} + \tfrac{1}{2} N m v_s^2$$

$$e^{-iG} \mathbf{P}_{0p} e^{iG} = \mathbf{P}_{0p} + N m \mathbf{v}_s \tag{13.35}$$

with eigenvalues

$$E_n(\mathbf{v}_s) = E_0 + \sum_{\mathbf{k} \neq 0} \hbar(\omega_{\mathbf{k}} + \mathbf{v}_s \cdot \mathbf{k}) n_{\mathbf{k}} + \tfrac{1}{2} N m v_s^2$$

$$P_n(\mathbf{v}_s) = \sum_{\mathbf{k} \neq 0} \hbar \mathbf{k} n_{\mathbf{k}} + N m \mathbf{v}_s \tag{13.36}$$

which include (13.31) as a special case when $\mathbf{v}_s = 0$. This shows that an elementary excitation of momentum $\hbar \mathbf{k}$ transforms according to

$$\omega_{\mathbf{k}} \rightarrow \omega_{\mathbf{k}} + \mathbf{v}_s \cdot \mathbf{k} \tag{13.37}$$

Superfluid Velocity Potential

Let us apply the Galilean transformation to the superfluid order parameter $\langle \psi(\mathbf{x}) \rangle$. For this purpose we re-expressed the transformation in quantized field language as

$$G = \frac{m}{\hbar} \mathbf{v}_s \cdot \int d^3 x \, \psi^\dagger(\mathbf{x}) \mathbf{x} \psi(\mathbf{x}) \tag{13.38}$$

Then

$$e^{-iG} \psi(\mathbf{x}) e^{iG} = e^{im\mathbf{v}_s \cdot \mathbf{x}/\hbar} \psi(\mathbf{x}) \tag{13.39}$$

which shows that

$$\langle \psi(\mathbf{x}) \rangle \rightarrow e^{im\mathbf{v}_s \cdot \mathbf{x}/\hbar} \langle \psi(\mathbf{x}) \rangle \tag{13.40}$$

Writing $\langle \psi \rangle = r e^{i\phi}$ in general, we can obtain the superfluid velocity as

$$\mathbf{v}_s = \frac{\hbar}{m} \nabla \phi \tag{13.41}$$

---

*Proof. $e^{-iG} \mathcal{H} e^{iG} = \mathcal{H} - i[G, \mathcal{H}] - \dfrac{1}{2!}[G, [G, \mathcal{H}]] + \cdots$ The series terminates after the third term. ∎

Therefore the phase of $\langle \psi \rangle$ is the superfluid velocity potential. It follows from (13.40) that the flow the superfluid is irrotational (see, however, Section 13.6):

$$\nabla \times \mathbf{v}_s = 0 \qquad (13.42)$$

Superfluidity: v_s As an Independent Degree of Freedom

In absolute thermodynamic equilibrium $\mathbf{v}_s$ is not independent of the total momentum $\mathbf{P}$ of the system. In fact, as we show below,

$$\mathbf{P} = Nm\langle \mathbf{v}_s \rangle \qquad \text{(Absolute equilibrium)} \qquad (13.43)$$

which means that *there is no relative motion between the centers of mass of normal fluid and superfluid.* In nonequilibrium situations, (13.43) need not be true, and the question is one of how long it will take for equilibrium relation (13.43) to be established. We argue that it will take a macroscopically long time. In fact, to the extend that the energy levels (13.36) are valid, the absolute equilibrium can never be established. The reason is as follows. To establish the relation (13.43) momentum must be transferred from the gas of excitations to somewhere else. According to (13.36), however, the excitations are stable, and hence there is no mechanism for momentum transfer. In a more accurate treatment, the elementary excitations should interact with one another and have finite lifetimes. Only in such an accurate treatment can we discuss the approach to absolute equilibrium.

We assume that it is a good approximation to take the lifetime of an elementary excitation to be infinite. The kinetic theory which results from this assumption is analogous to the zero-order approximation in the classical kinetic theory of gases and leads to nonviscous hydrodynamics. Specifically, the assumption is that $\mathbf{P}$ and $\mathbf{v}_s$ are *independent* variables. Such a treatment does not correspond to the situation of absolute equilibrium; it corresponds instead to situations of *quasi-equilibrium.*

By this assumption, the liquid is endowed with a new degree of freedom, namely, the relative motion between the normal fluid and the superfluid. This new degree of freedom is the essence of the transport phenomena in He II known collectively as superfluidity.

The new degree of freedom $\mathbf{v}_s$ leads to a two-fluid hydrodynamics, in which the superfluid moves independently of the normal fluid.*

To complete the discussion, we supply a proof of (13.43).

Let us calculate the partition function of the liquid with the energy levels (13.36), subject to the condition that

$$\mathbf{P} = \sum_{\mathbf{k} \neq 0} \hbar \mathbf{k} n_{\mathbf{k}} + Nm\mathbf{v}_s \qquad (13.44)$$

where $\mathbf{P}$ is a given vector. The partition function will be a function of $\mathbf{P}$ and also

---

*See K. Huang in *Studies in Statistical Mechanics*, Vol. II, J. De Boer and G. E. Uhlenbeck, eds. (North-Holland, Amsterdam, 1964).

of N, v, and T, but we leave the latter variables understood. Thus

$$Q(\mathbf{P}) = e^{-\beta E_0} \sum_{\mathbf{v}_s,\{n\}}' \exp\left\{-\beta\left[\tfrac{1}{2}Nmv_s^2 + \sum_{\mathbf{k}\neq 0} \hbar(\omega_\mathbf{k} + \mathbf{k}\cdot\mathbf{v}_s)n_\mathbf{k}\right]\right\} \quad (13.45)$$

where the sum $\sum'$ extends over all sets $\{n_\mathbf{q}\}$ and all values of $\mathbf{v}_s$ that satisfy (13.44). Define the following generating function:

$$\mathscr{G}(\mathbf{w}) \equiv \sum_{\mathbf{P}} e^{\beta\mathbf{w}\cdot\mathbf{P}} Q(\mathbf{P}) \quad (13.46)$$

where the sum over $\mathbf{P}$ extends over all vectors of the form

$$\mathbf{P} = \frac{2\pi\mathbf{n}\hbar}{L} \quad (13.47)$$

where $L = V^{1/3}$ and $\mathbf{n}$ is a vector whose components take on the values $0, \pm 1, \pm 2, \ldots$. One easily finds

$$\mathscr{G}(\mathbf{w}) = V e^{-\beta E_0 + (1/2)N\beta m w^2} \int \frac{d^3k}{(2\pi)^3} e^{-N[\beta\hbar^2 k^2/2m + g(\mathbf{k})]}$$

where

$$(13.48)$$

$$g(\mathbf{k}) = v\int \frac{d^3q}{(2\pi)^3} \log\left[1 - e^{-\beta\hbar(\omega_\mathbf{q} + \mathbf{k}\cdot\mathbf{q})}\right]$$

As $N \to \infty$ we can evaluate $\mathscr{G}(\mathbf{w})$ by the method of saddle-point integration:

$$\frac{1}{N} \log \mathscr{G}(\mathbf{w}) = -\frac{\beta E_0}{N} + \tfrac{1}{2}\beta m w^2 - v\int \frac{d^3k}{(2\pi)^3} \log\left(1 - e^{-\beta\hbar\omega_\mathbf{k}}\right) \quad (13.49)$$

We now recover $Q(\mathbf{P})$ through

$$Q(\mathbf{P}) = \frac{1}{(2\pi i)^3} \oint dt_1 \oint dt_2 \oint dt_3 \frac{\mathscr{G}(\mathbf{w})}{t_1^{n_1+1}t_2^{n_2+1}t_3^{n_3+1}} \quad (13.50)$$

where $t_j = \exp(2\pi\beta w_j/L)$, $n_j = (L/2\pi)P_j$, and the contours of integration are circles about $t_j = 0$. We can also write

$$Q(\mathbf{P}) = \frac{(-i\beta)^3}{v} \int_0^{2\pi} dw_1 \int_0^{2\pi} dw_2 \int_0^{2\pi} dw_3\, e^{-\beta\mathbf{P}\cdot\mathbf{w}} \mathscr{G}(\mathbf{w}) \quad (13.51)$$

The result is

$$\frac{1}{N} \log Q(\mathbf{P}) = \frac{1}{N} \log Q_0 - \frac{\beta}{2m}\left(\frac{\mathbf{P}}{N}\right)^2 \quad (13.52)$$

where Q_0 is the partition function for $\mathbf{v}_s = 0$. Thus, $\langle n_\mathbf{k}\rangle$ is still given by (13.19). Taking the ensemble average of (13.44) and noting that $\sum \mathbf{k}n_\mathbf{k} = 0$, we obtain

$$\mathbf{P} = Nm\langle\mathbf{v}_s\rangle \qquad\qquad \blacksquare$$

13.6 SUPERFLUID FLOW

Anderson's Equations

Consider pure superfluid flow in a hydrodynamic regime, in which the phase ϕ, and therefore the superfluid velocity v_s, varies slowly on an atomic scale. We shall perform coarse graining, by imagining the superfluid (i.e., liquid He^4 in its ground state), being divided into cells containing a large number of atoms, but which are small on a macroscopic scale. We assume that both the phase ϕ and the particle number N vary only slightly from cell to cell, and average their values over a cell. These average values will be used as macroscopic thermodynamic variables.

First let us construct a state of definite phase ϕ by superposing states of definite particle number N. As a model we choose

$$|\phi\rangle = \sum_{N=0}^{\infty} B_N e^{iN\phi}|N\rangle$$

$$B_N = \frac{e^{-(1/2)(N-\bar{N})^2/\Delta N}}{\sqrt{2\pi\Delta N}} \tag{13.53}$$

which gives

$$\langle\phi|\psi|\phi\rangle = e^{i\phi}\sum_{N=0}^{\infty} B_N B_{N+1}\langle N|\psi|N+1\rangle \tag{13.54}$$

It is assumed that the phase of $|N\rangle$ has been so chosen such that the matrix elements of ψ are real. The important feature to note is that states of different N have definite relative phases (instead of random relative phases as in an ensemble).

The states labeled by different ϕ are not independent, however, because they are not orthogonal to each other:

$$\langle\phi|\phi'\rangle = \sum_{N=0}^{\infty} B_N^2 e^{iN\Delta\phi} \propto e^{-(\Delta N\Delta\phi)^2/4} \tag{13.55}$$

where $\Delta\phi = \phi - \phi'$. But states satisfying*

$$\Delta N\Delta\phi \geq 1 \tag{13.56}$$

are essentially orthogonal, and thus may be considered independent. Accordingly, in each of our coarse-grained cells, the values of ϕ and N are to be averaged over respective intervals $\Delta\phi$ and ΔN, which are small on a macroscopic scale, and at the same time satisfy (13.56). The variables ϕ and N are then semiclassical coordinates for the coarse-grained cell, which we call the "superfluid coordinates." Their values are defined only to accuracies within the "phase-space" cells shown in Fig. 13.6.

When the superfluid coordinates are uniform throughout the system, the superfluid is in equilibrium. Any variations in the coordinates induce a flow of

---

*This is an uncertainty relation similar to that between energy and time ($\Delta E\Delta t \geq \hbar$), there being no operator whose eigenvalue is ϕ.

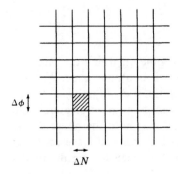

$\Delta\phi\updownarrow$

$\xleftrightarrow{\hspace{1cm}}$
ΔN

Fig. 13.6 Number of particles N and superfluid phase ϕ obey uncertainty relation $\Delta N\Delta\phi \geq 1$. Shown here are minimal cells in "phase space."

the superfluid, which tends to restore uniformity. This is because any nonuniformity in density will obviously cause the energy to rise above its absolute minimum. A nonuniform phase ϕ will also make the energy rise, because, while the system is invariant under a global gauge transformation (i.e., constant ϕ), it is not invariant under a local one (i.e., space-dependent ϕ.)

If the number of particles in a cell changes by ΔN during the time interval Δt, then the energy of the cell is uncertain by an amount $\Delta E \approx \hbar/\Delta t$. Thus

$$\Delta E\,\Delta t \approx \hbar\,\Delta\phi\,\Delta N \tag{13.57}$$

from which we obtain

$$\hbar\frac{d\phi}{dt} = \frac{\partial E}{\partial N} = \mu$$

$$\hbar\frac{dN}{dt} = \frac{\partial E}{\partial \phi} \tag{13.58}$$

where μ is the chemical potential. Another way to derive these equations is to note that the energy $E(\phi, N)$ of a cell is an effective classical Hamiltonian, where $\hbar\phi$ and N are canonically conjugate coordinates. The equations above are the Hamiltonian equations of motion. These are Anderson's equations* governing superfluid flow. Although we have derived them at absolute zero, they continue to apply at finite temperatures, with the partial derivatives taken to be adiabatic derivatives.

Vortex Motion

The superfluid order parameter $\langle\psi(\mathbf{x})\rangle$ must be a continuous function of $\mathbf{x}$. In particular, its phase $\phi(\mathbf{x})$ must change by at most an integer multiple of 2π when we traverse a closed loop in the superfluid. This means that the vorticity in

*P. W. Anderson, *Rev. Mod. Phys.* **38**, 298 (1966). This paper contains applications of these equations.

superfluid flow must be quantized:

$$\oint ds \cdot \nabla \phi = 2\pi n$$

$$\oint ds \cdot v_s = 2\pi \hbar n / m \qquad (n = 0, 1, 2, \dots)$$

(13.59)

Obviously $n = 0$ if the closed loop can be continuously shrunken to a point. The values $n \neq 0$ occur if the superfluid flows in a multiply connected region, for example, between two cylinders.

The multiply connected region may also be created by the liquid itself: The superfluid can be expelled from a self-formed tube filled with normal fluid, and will flow around this tube with nonzero vorticity, i.e., $n \neq 0$ in (13.59). The tube, called the vortex core, could be a line whose ends terminate on boundary surfaces or it could form a ring. The excitation is called a vortex line in the first case, and a vortex ring in the second. Both types of vortex motion have been observed experimentally.[*]

The coarse-grained description of the superfluid obviously breaks down in the vicinity of the vortex core. The core diameter is a phenomenological parameter that can be calculated only by considering the microscopic dynamics.

In the approximation of incompressible superfluid flow in the presence of vorticity, the equations for the superfluid velocity are

$$\nabla \cdot v_s = 0$$

$$\nabla \times v_s = j$$

(13.60)

where j is the vorticity density, the vorticity per unit cross-sectional area of the vortex core. Note that v_s in this case is mathematically analogous to the magnetic field set up by a current density distribution j.

The energy associated with superfluid flow is just the kinetic energy of the fluid elements:

$$K = \int d^3r \, \tfrac{1}{2} \rho v_s^2$$

(13.61)

where ρ is the density of the superfluid, and the integral extends over the superfluid. Likewise, the momentum density is that due to the fluid elements. Many of the properties of classical hydrodynamics can be carried over. For example, the properties of a vortex ring can be taken straight from Lamb's classic book.[†] A vortex ring of radius R and vorticity κ (the line integral of v_s around the vortex core) has a definite energy $E(R)$ and translational velocity normal to the

---

[*]For vortex lines see W. F. Vinen in *Progress in Low Temperature Physics*, Vol. III, C. J. Gorter, ed. (North-Holland, Amsterdam, 1961). For vortex rings see G. W. Rayfield and F. Reif, *Phys. Rev.* **136**, A1194 (1964).

[†]H. Lamb, *Hydrodynamics* (Dover, New York, 1945).

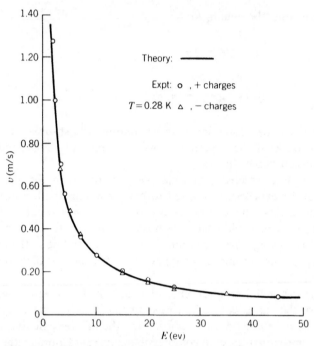

Fig. 13.7 Translational velocity vs. energy for a quantized vortex ring of one quantum of circulation h/m in superfluid He^4. The solid curve is theory. The data points are those of Rayfield and Reif, who trap ions in the vortex ring (in the core, presumably), and drag the excitation in an electric field, measuring the velocity by a time-of-flight device.

the plane of the vortex ring $v(R)$:

$$E(R) = \tfrac{1}{2}\rho\kappa^2 R\left(\log \frac{8R}{b} - \frac{7}{4}\right)$$

$$v(R) = \frac{\kappa}{4\pi R}\left(\log \frac{8R}{b} - \frac{1}{4}\right)$$

(13.62)

where b is the core radius and, according to (13.59), κ is an integer multiple of h/m. This has the usual feature that the velocity of translation decreases as the energy increases. Using $\rho = 0.145$ g/cm^3, and $b = 10^{-8}$ cm, we eliminate R from (13.62) to obtain the $E - v$ curve in Fig. 13.7 for a vortex ring of unit quantized vorticity. The data points of Rayfield and Reif* are superimposed on the theoretical curve.

*op. cit.

13.7 THE PHONON WAVE FUNCTION

In this section we shall show that for liquid He^4, excited states immediately above the ground state have the form

$$\Psi_{\mathbf{k}} = \text{Const.} \sum_{j=1}^{N} e^{i\mathbf{k}\cdot\mathbf{r}_j} \Psi_0 \qquad (13.63)$$

where Ψ_0 is the ground state wave function. The momentum and energy of this state are, respectively, $\hbar\mathbf{k}$ and $\hbar ck$, where c is the velocity of sound. (13.63) represents a one-phonon wave function. From the experimental fact that the specific heat is proportional to T^3 near $T = 0$, we know that the low-lying excitations are exclusively phonons, which are longitudinal density waves. Thus it is not altogether surprising that we have a wave function of the form (13.63), in which the ground state wave function is multiplied by the kth Fourier component of the density.

Let us first picture the wave function physically, before getting down to the mathematical business of deriving it. A wave function of the form (13.63) does not automatically describe a sound wave. It depends on the property of the ground state wave function Ψ_0.

For example, for an ideal Bose gas, for which Ψ_0 is a constant, $\Psi_{\mathbf{k}}$ certainly does not represent a collective sound wave mode; it represents a single-particle excitation instead. Its modulus is peaked at isolated points in the configuration space, when every $\mathbf{r}_j$ is such that $\mathbf{k} \cdot \mathbf{r}_j$ is 2π times an integer. Otherwise the wave function is practically zero.

In liquid helium the situation is quite different: the ground state wave function is not constant, but achieves maximum amplitude when the particles are uniformly spread apart, because of the hard-core repulsion between the atoms. To obtain an idea of what $\Psi_{\mathbf{k}}$ looks like here, consider only one space dimension. The real part of $\Psi_{\mathbf{k}}$ is

$$\sum_{j=1}^{N} \cos(kx_j)\Psi_0(x_1,\ldots,x_N)$$

The behavior of this function is markedly different for $kr_0 \ll 1$ and for $kr_0 \gg 1$, where r_0 is the average interparticle distance. Plot $\cos(kx)$ as a function of x, and place N points on the abscissa representing the positions of the N particles. Then each term $\cos(kx_j)$ can be read off the graph. For $kr_0 \ll 1$, such a graph is shown in Fig. 13.8. The factor $\sum_j \cos(kx_j)$ is zero when the particles are uniformly distributed, for then kx_j tends to be random, but the factor Ψ_0 becomes zero when any two particles "touch." Thus there is an opposing tendency between the first factor and the second. The former tends to bunch the particles together, whereas the latter tends to spread them out uniformly. The magnitude of the wave function is largest when the particles are distributed in space in such a way that the number of particles in the region $\cos kx > 0$ exceeds those in the region $\cos kx < 0$, or vice versa. This corresponds to a sinusoidal

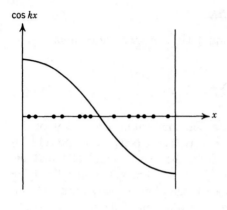

Fig. 13.8 To find $\sum_j \cos(kx_j)$, mark off the values of $x_1, \ldots, x_N$ on the abscissa and read off each term $\cos(kx_j)$ from this graph, which is drawn for $kr_0 \ll 1$, where $r_0 =$ average interparticle distance.

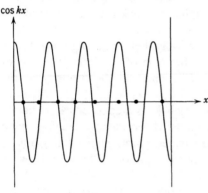

Fig. 13.9 This is the same graph as 13.8, but for $kr_0 \gg 1$.

spatial variation of density and hence to a sound wave. When $kr_0 \gg 1$, the situation becomes that illustrated in Fig. 13.9, and Ψ_k looks like that of the ideal gas (i.e., single-particle excitation).

We now derive (13.63).* The Hamiltonian is taken to be (with $\hbar = 1$):

$$\mathscr{H} = \frac{1}{2m} \int d^3x \, |\nabla \psi(\mathbf{x})|^2 + \frac{1}{2} \int d^3x \, d^3y \, \psi^\dagger(\mathbf{x}) \psi^\dagger(\mathbf{y}) v(\mathbf{x} - \mathbf{y}) \psi(\mathbf{y}) \psi(\mathbf{x}) - E_0$$

(13.64)

where $\psi(\mathbf{x})$ is the usual quantized field operator, and E_0 is the ground state energy for an N-particle system. In the subspace of N-particle states, let $|\alpha\rangle$ be an eigenstate of $\mathscr{H}$, and let $|0\rangle$ be the lowest eigenstate:

$$\mathscr{H}|0\rangle = 0$$
$$\mathscr{H}|\alpha\rangle = \omega_\alpha |\alpha\rangle$$

(13.65)

We assume that $|0\rangle$ has zero total momentum, and is invariant under translations and rotations.

*K. Huang and A. Klein, *Ann. Phys. (N.Y.)* **30**, 203 (1964). The result was first shown using a different method by R. P. Feynman, *Phys. Rev.* **94**, 262 (1954).

The density operator is

$$\rho(\mathbf{x}) = \psi^\dagger(\mathbf{x})\psi(\mathbf{x}) \tag{13.66}$$

with Fourier transform

$$\rho_\mathbf{k} = V^{-1/2} \int d^3x\, e^{i\mathbf{k}\cdot\mathbf{x}} \rho(\mathbf{x}), \qquad \rho_{-\mathbf{k}} = \rho_\mathbf{k}^\dagger \tag{13.67}$$

Our purpose is to show that $\rho_\mathbf{k}|0\rangle$ is an eigenstate of the Hamiltonian with momentum $\mathbf{k}$ and energy ck. To this end, consider the function

$$S(k,\omega) = n^{-1}\langle 0|\rho_\mathbf{k}^\dagger\, \delta(\mathcal{H} - \omega)\, \rho_\mathbf{k}|0\rangle = n^{-1}\sum_\alpha \delta(\omega_\alpha - \omega)\,|\langle\alpha|\rho_\mathbf{k}|0\rangle|^2$$

$$\tag{13.68}$$

which depends on the magnitude and not the direction of $\mathbf{k}$ by the assumed rotational invariance of $|0\rangle$. Since the eigenvalues of $\mathcal{H}$ are all positive, $S(k,\omega) = 0$ for $\omega < 0$. It is easily verified that the matrix element $\langle\alpha|\rho_\mathbf{k}|0\rangle$ vanishes unless $|\alpha\rangle$ has momentum $\mathbf{k}$. Thus a knowledge of $S(k,\omega)$ tells us something about the density of states of momentum $\mathbf{k}$. Integrating (13.68) over ω, we obtain

$$S_k = \int_0^\infty d\omega\, S(k,\omega) = n^{-1}\langle 0|\rho_\mathbf{k}^\dagger\rho_\mathbf{k}|0\rangle \tag{13.69}$$

which is called the *liquid structure factor*. Both $S(k,\omega)$ and S_k are accessible to direct measurements through neutron or light scattering from liquid He[4].[*]

We now make use of the sum rules satisfied by $S(k,\omega)$. These are actually independent of the statistics of the particles, and are proved in the Appendix:

$$n^{-1}\langle 0|\rho_\mathbf{k}^\dagger\rho_\mathbf{k}|0\rangle = S_k \tag{13.70}$$

$$n^{-1}\langle 0|\rho_\mathbf{k}^\dagger\mathcal{H}\rho_\mathbf{k}|0\rangle = k^2/2m \tag{13.71}$$

$$\lim_{k\to 0} n^{-1}\langle 0|\rho_\mathbf{k}^\dagger\mathcal{H}^{-1}\rho_\mathbf{k}|0\rangle = 1/2mc^2 \tag{13.72}$$

where n is the number density of liquid helium. Of these, the first is merely the definition of S_k. These can be rewritten as follows:

$$\int_0^\infty d\omega\, S(k,\omega) = S_k \tag{13.73}$$

$$\int_0^\infty d\omega\, \omega S(k,\omega) = \frac{k^2}{2m} \tag{13.74}$$

$$\lim_{k\to 0}\int_0^\infty \frac{d\omega}{\omega} S(k,\omega) = \frac{1}{2mc^2} \tag{13.75}$$

Let

$$\begin{aligned} \nu &= \omega/ck, \\ R(k,\nu) &= 2mc^2 S(k, ck\nu) \end{aligned} \tag{13.76}$$

[*]For references and analyses of data, see A. Miller, D. Pines, and P. Nozierres, *Phys. Rev.* **127**, 1452 (1962).

Then

$$\int_0^\infty dv\, R(k,v) = (2mc/k)S_k \tag{13.77}$$

$$\int_0^\infty dv\, vR(k,v) = 1 \tag{13.78}$$

$$\lim_{k \to 0} \int_0^\infty \frac{dv}{v} R(k,v) = 1 \tag{13.79}$$

By taking a suitable linear combination of (13.77)–(13.79) we obtain

$$\int_0^\infty \frac{dv}{v}(v-1)^2 R(K,v) \xrightarrow[k \to 0]{} 2\left(1 - \frac{2mc}{k}S_k\right) \tag{13.80}$$

Since both $R(k,v)$ and S_k are by definition positive definite, we conclude that

$$\lim_{k \to 0} \frac{2mc}{k} S_k \leq 1 \tag{13.81}$$

Neutron scattering experiments have yielded $S(k, \omega)$ down to very small k. They show that $S(k, \omega)$ is strongly peaked about $\omega = ck$. For given k, the integral over ω yields S_k, which is the area under the peak. Experiments are consistent with the assumptions that S_k lies at the limit of the inequality (13.81):

$$S_k \cong k/2mc \tag{13.82}$$

Thus the right side of (13.80) is zero. Since $R(k,v)$ is positive definite we must have

$$R(k,v) \xrightarrow[k \to 0]{} \delta(v-1) \tag{13.83}$$

to satisfy both (13.80) and (13.77). This implies that as $k \to 0$, the state $\rho_k|0\rangle$ becomes an eigenstate of $\mathcal{H}$. The momentum of the state is $\mathbf{k}$, as stated previously. The excitation energy is given by

$$\omega_{\mathbf{k}} = \frac{\langle 0|\rho_{\mathbf{k}}^\dagger \mathcal{H} \rho_{\mathbf{k}}|0\rangle}{\langle 0|\rho_{\mathbf{k}}^\dagger \rho_{\mathbf{k}}|0\rangle} = \frac{k^2}{2mS_k} \tag{13.84}$$

which, by virtue of (13.82), reduces for small k to

$$\omega_{\mathbf{k}} \xrightarrow[k \to 0]{} ck \tag{13.85}$$

In this derivation we have appealed to the experimental fact (13.82). Actually one can prove the latter theoretically and show that, because of Bose statistics, the low-lying excitations consist solely of phonons.*

Although the wave function (13.63) was established only in the limit $\mathbf{k} \to 0$, it gives a fair representation of the excitation even for finite $\mathbf{k}$. To indicate this, we show in Fig. 13.10 the experimental data for S_k. Substituting it into (13.84)

---

*Huang and Klein, *op. cit.*

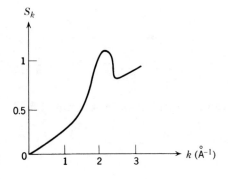

Fig. 13.10 Experimental liquid structure factor (Fourier transform of pair correlation function) of liquid He^4.

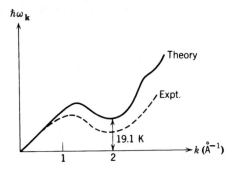

Fig. 13.11 Comparison between the calculated excitation spectrum (solid curve) and measurements (dashed curve).

gives an excitation energy as function of k shown in Fig. 13.11, which reproduces the roton part of the spectrum qualitatively.*

13.8 DILUTE BOSE GAS

Dilute Bose gases have become accessible to experiments, furnishing us with simpler examples of superfluids. Among these are spin-aligned atomic hydrogen gas,[†] and liquid He^4 adsorbed in porous Vycor glass.[‡] At sufficiently low temperatures, the interparticle potential may be summarized by only one parameter, the scattering length, which is a kind of effective hard-sphere diameter. The low-lying energy levels of a dilute Bose gas then correspond to those of a hard-sphere gas, which can be treated through the methods of pseudopotentials introduced in Section 10.5.[§]

*An improved wave function that fits the data better is given by R. P. Feynman and M. Cohen, *Phys. Rev.* **102**, 1189 (1956).

[†] T. J. Greytak and D. Kleppner, in *New Trends in Atomic Physics, Vol. II (Les Houches 38)* G. Greyberg and R. Stora, eds. (North-Holland, Amsterdam, 1984).

[‡] B. C. Crooker, B. Hebral, E. N. Smith, and J. D. Reppy, *Phys. Rev. Lett.* **51**, 666 (1983).

[§] T. D. Lee, K. Huang, and C. N. Yang, *Phys. Rev.* **106**, 1135 (1957); see also K. Huang, *op. cit.*

Effective Hamiltonian

The Hamiltonian is

$$\mathcal{H} = -\frac{1}{2m} \sum_{j=1}^{N} \nabla_j^2 + \frac{4\pi a}{m} \sum_{i<j} \delta(\mathbf{r}_i - \mathbf{r}_j) \frac{\partial}{\partial r_{ij}} r_{ij} \tag{13.86}$$

where a is the hard-sphere diameter. We can re-express it in the quantized field representation as

$$\begin{aligned}
\mathcal{H} &= -\frac{1}{2m} \int d^3r \, \psi^\dagger(\mathbf{r}) \nabla^2 \psi(\mathbf{r}) \\
&\quad + \frac{2\pi a}{m} \int d^3r_1 \, d^3r_2 \, \psi^\dagger(\mathbf{r}_1) \psi^\dagger(\mathbf{r}_2) \, \delta(\mathbf{r}_1 - \mathbf{r}_2) \frac{\partial}{\partial r_{12}} [r_{12} \psi(\mathbf{r}_1) \psi(\mathbf{r}_2)] \\
&= \frac{1}{2m} \sum k^2 a_k^\dagger a_k + \frac{2\pi a}{mV} \sum_{\mathbf{p},\mathbf{q}} a_\mathbf{p}^\dagger a_\mathbf{q}^\dagger \frac{\partial}{\partial r} \left[r \sum_\mathbf{k} e^{i\mathbf{k}\cdot\mathbf{r}} a_{\mathbf{p}+\mathbf{k}} a_{\mathbf{q}-\mathbf{k}} \right]_{r=0}
\end{aligned} \tag{13.87}$$

where

$$\psi(\mathbf{r}) = \sum_\mathbf{k} a_\mathbf{k} \frac{e^{i\mathbf{k}\cdot\mathbf{r}}}{\sqrt{V}}, \qquad [a_\mathbf{k}, a_{\mathbf{k}'}^\dagger] = \delta_{\mathbf{k}\mathbf{k}'} \tag{13.88}$$

and we have imposed periodic boundary conditions:

$$\mathbf{k} = \frac{2\pi \mathbf{n}}{L} \tag{13.89}$$

when $\mathbf{n}$ is a vector whose components are independently $0, \pm 1, \pm 2, \ldots$. The appearance of the operator $(\partial/\partial r)r$ seems to complicate things, but actually the effect is quite simple. The operator has the property that

$$\begin{aligned}
\frac{\partial}{\partial r}[rf(r)]_{r=0} &= f(0) \qquad \text{(if } f \text{ is regular at } r=0\text{)} \\
\frac{\partial}{\partial r}[rf(r)]_{r=0} &= 0 \qquad \left(\text{if } f \xrightarrow[r\to 0]{} \frac{1}{r} \right)
\end{aligned} \tag{13.90}$$

The operator $(\partial/\partial r)r$ weeds out, so to speak, any singularity of the $1/r$ type in the function to which it is applied.[*]

We treat the interaction as a perturbation on the ideal gas. The validity of the Hamiltonian is restricted by the conditions

$$a/v^{1/3} \ll 1, \qquad ha \ll 1 \tag{13.91}$$

where v is the specific volume, and k is any wave number of an excited particle in the unperturbed state. For low-lying excited states, we assume

$$n_0 \approx N, \qquad \sum_{\mathbf{k}\neq 0} n_k/N \ll 1 \tag{13.92}$$

[*]cf. Problem 10.7.

First we decompose the interaction into a part that has only diagonal matrix elements with respect to the unperturbed free-particle states and a part that has only off-diagonal elements. The part with diagonal matrix elements has been worked out and given in (A.36) of the Appendix, and discussed in Section 10.5. For the off-diagonal elements, we isolate a part that contains terms proportional to $a_0 a_0$ or $a_0^\dagger a_0^\dagger$. What is left contains terms both proportional to and independent of a_0 or $a_0^\dagger$. The matrix elements of the first part are of order N larger than those of the remaining part, which will be neglected. Thus we have

$$2m\mathcal{H} = \sum_k k^2 a_k^\dagger a_k + \frac{8\pi a N}{v}\left(1 - \frac{1}{2N^2}\sum_k n_k^2\right)$$
$$+ \frac{4\pi a}{v}\sum_k{}' \left(a_k^\dagger a_k^\dagger a_0 a_0 + a_k a_k a_0^\dagger a_0^\dagger\right) \tag{13.93}$$

Now we replace both $a_0 a_0$ and $a_0^\dagger a_0^\dagger$ by N, and obtain an effective Hamiltonian:

$$2m\mathcal{H}_{\text{eff}} = \frac{4\pi a N}{v} + \sum_{k \neq 0}{}'\left[\left(k^2 + \frac{8\pi a}{v}\right)a_k^\dagger a_k + \frac{4\pi a}{v}\left(a_k^\dagger a_{-k}^\dagger + a_k a_{-k}\right)\right] \tag{13.94}$$

The sum $\sum'$ in (13.93) and (13.94) is defined by

$$\sum_{k \neq 0}{}' f_k \equiv \lim_{r \to 0} \frac{\partial}{\partial r}\left(r \sum_{k \neq 0} f_k\right) \tag{13.95}$$

In practice, this operation affects only the ground state energy, and has the effect of subtracting out the term proportional to a^2.

Energy Levels

To diagonalize (13.94) we introduce a linear transformation first used by Bogolubov*:

$$a_k = \frac{1}{\sqrt{1 - \alpha_k^2}}\left(b_k - \alpha_k b_{-k}^\dagger\right)$$
$$a_k^\dagger = \frac{1}{\sqrt{1 - \alpha_k^2}}\left(b_k^\dagger - \alpha_k b_{-k}\right) \tag{13.96}$$

or

$$b_k = \frac{1}{\sqrt{1 - \alpha_k^2}}\left(a_k + \alpha_k a_{-k}^\dagger\right)$$
$$b_k^\dagger = \frac{1}{\sqrt{1 - \alpha_k^2}}\left(a_k^\dagger + \alpha_k a_{-k}\right) \tag{13.97}$$

where α is assumed to be a real number than one. It is clear that b_k and $b_k^\dagger$

*N. N. Bogolubov, *J. Phys. USSR* **11**, 23 (1947).

satisfy the same commutation rules as a_k and $a_k^\dagger$, namely

$$[b_k, b_{k'}] = [b_k^\dagger, b_{k'}^\dagger] = 0$$
$$[b_k, b_{k'}^\dagger] = \delta_{kk'} \qquad (13.98)$$

Therefore b_k and $b_k^\dagger$ can be interpreted, respectively, as annihilation and creation operators, just as a_k and $a_k^\dagger$. If we substitute (13.96) into the effective Hamiltonian (13.94), we find that $\mathscr{H}_{\text{eff}}$ is diagonalized by choosing

$$\alpha_k = 1 + x^2 - x\sqrt{x^2 + 2}, \qquad x^2 \equiv \frac{k^2}{8\pi a/v} \qquad (13.99)$$

Clearly α_k is real and is less than one, as we assumed earlier.

The Hamiltonian in diagonal form reads

$$2m\mathscr{H}_{\text{eff}} = \frac{4\pi aN}{v} - \frac{1}{2}\sum_{k\neq 0}\left[\frac{8\pi a}{v} + k^2 - k\sqrt{k^2 + \frac{16\pi a}{v}} - \frac{1}{2}\left(\frac{8\pi a}{v}\right)^2 \frac{1}{k^2}\right]$$
$$+ \sum_{k\neq 0} k\sqrt{k^2 + \frac{16\pi a}{v}}\, b_k^\dagger b_k \qquad (13.100)$$

where the term $-\frac{1}{2}(8\pi a/v)^2 k^{-2}$ is the subtraction required by (13.95). It cancels the a^2 term in the expansion of $k\sqrt{k^2 + (16\pi a/v)}$. If we had not made this subtraction, the sum over k would be divergent. It is now of course finite and can be converted into an integral as $V \to \infty$:

$$-\frac{V}{(2\pi)^2}\left(\frac{8\pi a}{v}\right)^{5/2}\int_0^\infty dx\, x^2\left(1 + x^2 - x\sqrt{x^2 + 2} - \frac{1}{2x^2}\right) = \frac{4\pi aN}{v}\frac{128}{15}\sqrt{\frac{a^3}{\pi v}} \qquad (13.101)$$

The Hamiltonian now reads

$$\mathscr{H}_{\text{eff}} = \frac{2\pi aN}{mv}\left(1 + \frac{128}{15}\sqrt{\frac{a^3}{\pi v}}\right) + \sum_{k\neq 0}\frac{k}{2m}\sqrt{k^2 + \frac{16\pi a}{v}}\, b_k^\dagger b_k \qquad (13.102)$$

It follows from the commutation rules (13.98) that the eigenvalues of $b_k^\dagger b_k$ are $0, 1, 2, \dots$. Therefore the lowest eigenvalue of $\mathscr{H}_{\text{eff}}$ is

$$E_0 = \frac{2\pi aN}{mv}\left(1 + \frac{128}{15}\sqrt{\frac{a^3}{\pi v}}\right) \qquad (13.103)$$

In a more elaborate calculation, Wu* obtains the ground state energy as

$$E_0 = \frac{2\pi aN}{mv}\left[1 + \frac{128}{15}\sqrt{\frac{a^3}{\pi v}} + 8\left(\frac{4\pi}{3} - \sqrt{3}\right)\frac{a^3}{v}\log\frac{a^3}{v} + \text{const.}\left(\frac{a^3}{v}\right)\right] \qquad (13.104)$$

*T. T. Wu, *Phys. Rev.* **115**, 1390 (1959).

The excited states are characterized by the occupation numbers $b_k^\dagger b_k = 0, 1, 2, \ldots$ of noninteracting excitations whose energies are given by

$$\epsilon_k = (k/2m)\sqrt{k^2 + 16\pi a/v} \, .$$

For very small k, these energies become $(k/2m)\sqrt{16\pi a/v}$. These excitations are therefore phonons. The velocity of sound for very long lengths ($k \to 0$) is

$$c = \sqrt{\frac{4\pi a}{m^2 v}} \tag{13.105}$$

We may check the consistency of this interpretation by calculating the sound velocity independently from the compressibility of the system at absolute zero:

$$c = \frac{1}{\sqrt{\rho \kappa_s}} = \sqrt{\frac{v}{m} \frac{\partial P_0}{\partial v}} \tag{13.106}$$

where P_0 is the pressure at absolute zero:

$$P_0 = -\frac{\partial}{\partial v} \frac{E_0}{N} \tag{13.107}$$

Using (13.103) we find that

$$c = \sqrt{\frac{4\pi a}{m^2 v}} \left(1 + 16\sqrt{\frac{a^3}{\pi v}}\right) \tag{13.108}$$

We see that the first term agrees with (13.105). The next term is of the order $16\sqrt{a^3/v}$ and is beyond the accuracy of the calculation from which (13.105) is obtained.

Wave Functions

We now calculate the wave functions of the system, both for the ground state and for states with phonons.

The eigenstates of (13.102) may be labeled by phonon occupation numbers σ_k, which can independently assume the values $0, 1, 2, \ldots$. We denote an eigenstate by $| \ldots, \sigma_k, \ldots \rangle$. It has the properties that

$$b_k^\dagger b_k | \ldots, \sigma_k, \ldots \rangle = \sigma_k | \ldots, \sigma_k, \ldots \rangle$$

$$b_k | \ldots, \sigma_k, \ldots \rangle = \sqrt{\sigma_k} | \ldots, \sigma_k - 1, \ldots \rangle \tag{13.109}$$

$$b_k^\dagger | \ldots, \sigma_k, \ldots \rangle = \sqrt{\sigma_k + 1} | \ldots, \sigma_k + 1, \ldots \rangle$$

The ground state is denoted by

$$|\Psi_0\rangle \equiv |0, 0, 0, \ldots \rangle$$

It is defined by

$$b_k |\Psi_0\rangle = 0 \tag{13.110}$$

Let $|n_1 m_1; n_2 m_2; \dots\rangle$ be the unperturbed state in which there are

$$n_k \quad \text{particles of momentum } \mathbf{k}$$

$$m_k \quad \text{particles of momentum } -\mathbf{k} \quad (\mathbf{k} > 0)$$

Then we may expand $|\Psi_0\rangle$ as follows:

$$|\Psi_0\rangle = \sum_{\substack{n_1=0 \\ m_1=0}}^{\infty} \sum_{\substack{n_2=0 \\ m_2=0}}^{\infty} \cdots \left(C_{n_1 m_1} C_{n_2 m_2} \cdots \right) |n_1 m_1; n_2 m_2; \dots\rangle \quad (13.111)$$

Only the amplitudes for $\mathbf{k}, -\mathbf{k}$ are coupled together because the transformation (13.97) has this property. Substituting (13.111) into (13.110) and using (13.97) we obtain the following equation for C_{nm}:

$$\sum_{n=0}^{\infty} \sum_{m=0}^{\infty} \left[C_{nm}\sqrt{n}\,|n-1, m\rangle + \alpha C_{nm}\sqrt{m+1}\,|n, m+1\rangle \right] = 0$$

where

$$|n, m\rangle \equiv |n_1 m_1; n_2 m_2; \dots\rangle$$

$$\alpha \equiv \alpha_k$$

By changing the indices of summation, we can write

$$\sum_{n=0}^{\infty} \sum_{m=0}^{\infty} \left[C_{n+1, m}\sqrt{n+1} + \alpha C_{n, m-1}\sqrt{m} \right]|n, m\rangle = 0$$

which implies

$$C_{n+1, m}\sqrt{n+1} + \alpha C_{n, m-1}\sqrt{m} = 0$$

From this we can deduce that

$$C_{nm} = 0 \quad (n \neq m)$$

for this is obviously true for $m = 0$, and the general case can be proved by induction. Hence it is sufficient to consider C_{mm}, which satisfies the equation

$$C_{mm} + \alpha C_{m-1, m-1} = 0 \quad (13.112)$$

The solution is

$$C_{mm} = (-\alpha)^m C_{00} \quad (13.113)$$

where C_{00} is to be determined by normalizing the total wave function. Therefore the unperturbed states that appear in the expansion of $|\Psi_0\rangle$ are states in which pairs of particles $\mathbf{k}, -\mathbf{k}$, are excited. We denote such a state by $|l_1, l_2, \dots\rangle$, in which there are l_k particles with momentum $\mathbf{k}$, and the same number of particles with momentum $-\mathbf{k}$. Thus

$$|\Psi_0\rangle = Z \sum_{l_1=0}^{\infty} \sum_{l_2=0}^{\infty} \cdots \left[(-\alpha_1)^{l_1} (-\alpha_2)^{l_2} \cdots \right]|l_1, l_2, \dots\rangle \quad (13.114)$$

where α_k is defined by (19.51) and one factor α_k appears for each $\mathbf{k} > 0$. The normalization constant Z can be shown to be

$$Z = \prod_{k>0} \sqrt{1 - \alpha_k^2} = \exp\left[-\tfrac{4}{9}N(3\pi - 8)\sqrt{\frac{a^3}{\pi v}}\right] \qquad (13.115)$$

In (13.114) the term with all $l_k = 0$ corresponds to the unperturbed ground state $|0\rangle$. Hence Z is none other than $\langle 0|\Psi_0\rangle$, the probability amplitude of finding the unperturbed ground state in the perturbed ground state. According to (13.115) these two states become orthogonal to each other as $N \to \infty$.

The wave functions for excited states can be easily calculated. For example, for the state with one phonon of momentum k, the normalized wave function is defined by

$$|\Psi_k\rangle = b_k^\dagger|\Psi_0\rangle \qquad (13.116)$$

By a straightforward calculation we obtain

$$|\Psi_k\rangle = \sqrt{1 - \alpha_k^2}\, a_k^\dagger|\Psi_0\rangle \qquad (13.117)$$

Thus the one-phonon state is a superposition of unperturbed states in which there are any number of particles $p, -p$ for all p, *plus* an additional particle of momentum k.

The average number of particles that have momentum k in the perturbed ground state is

$$\langle n_k\rangle = \langle\Psi_0|a_k^\dagger a_k|\Psi_0\rangle = \frac{\alpha_k^2}{1 - \alpha_k^2} \qquad (k \neq 0) \qquad (13.118)$$

Therefore the total number of excited particles in the perturbed ground state is

$$\sum_{k\neq 0}\langle n_k\rangle = \sum_{k\neq 0}\frac{\alpha_k^2}{1 - \alpha_k^2} = \frac{8}{3}\sqrt{\frac{a^3}{\pi v}}\,N \qquad (13.119)$$

The number of particles of zero momentum in the perturbed ground state is

$$\langle n_0\rangle = N\left[1 - \frac{8}{3}\sqrt{\frac{a^3}{\pi v}}\right] \qquad (13.120)$$

which shows that the approximation (13.92) is justified.

It is instructive to calculate the wave functions in configuration space. In each term of the sum in (13.114), the number $n \equiv \sum_{k>0} l_k$ is half the total number of particles with nonzero momentum. Thus we must have $N \geq 2n$. We rewrite (13.114) as follows:

$$|\Psi_0\rangle = Z\sum_{n=0}^{N/2}\sum_{\substack{l_1, l_2, \ldots \\ \Sigma l_k = n}}\left[(-\alpha_1)^{l_1}(-\alpha_2)^{l_2}\cdots\right]|l_1, l_2, \ldots\rangle \qquad (13.121)$$

We imagine $N \to \infty$ at the end of the calculation. The normalized configuration space wave function for an unperturbed state specified by the occupation numbers $\{n_0, n_1, \ldots\}$ is

$$\langle \mathbf{r}_1, \ldots, \mathbf{r}_N | n_0, n_1, \ldots \rangle \equiv \frac{1}{V^{N/2}} \frac{1}{\sqrt{N! \prod(n_k!)}} \sum_P P \, e^{i(\mathbf{p}_1 \cdot \mathbf{r}_1 + \cdots + \mathbf{p}_N \cdot \mathbf{r}_N)} \quad (13.122)$$

where among the N momenta $\mathbf{p}_1, \ldots, \mathbf{p}_N$, n_0 are 0, n_1 are $\mathbf{k}_1$, etc. The symbol P denotes a permutation of $\mathbf{r}_1, \ldots, \mathbf{r}_N$. Hence

$$\langle \mathbf{r}_1, \ldots, \mathbf{r}_N | l_1, l_2, \ldots \rangle$$

$$= \frac{1}{V^{N/2}} \frac{1}{\sqrt{N!(N-2n)!}} \frac{1}{\prod(l_k!)} \sum_P P \, e^{i[\mathbf{p}_1 \cdot (\mathbf{r}_1 - \mathbf{r}_2) + \cdots + \mathbf{p}_n \cdot (\mathbf{r}_3 - \mathbf{r}_4)]} \quad (13.123)$$

where among the n vectors $\mathbf{p}_1, \ldots, \mathbf{p}_n$, l_1 are $\mathbf{k}_1$, l_2 are $\mathbf{k}_2$, etc.

Now consider the sum appearing in (13.121):

$$|\chi_n\rangle \equiv \sum_{\substack{l_1, l_2, \ldots \\ \Sigma l_k = n}} \left[(-\alpha_1)^{l_1} (-\alpha_2)^{l_2} \cdots \right] | l_1, l_2, \ldots \rangle$$

This may be rewritten as

$$|\chi_n\rangle = \sum_{k_1 \leq k_2 \leq \cdots} \left[(-\alpha_1)(-\alpha_2)(-\alpha_3) \cdots \right] | l_1, l_2, \ldots \rangle$$

$$= \frac{\prod(l_k!)}{n!} \sum_{k_1 > 0} \sum_{k_2 > 0} \cdots \left[(-\alpha_1)(-\alpha_2)(-\alpha_3) \cdots \right] | l_1, l_2, \ldots \rangle \quad (13.124)$$

where, in the first line, $k_1 \leq k_2 \leq \cdots$ denotes any ordering of the momenta. In the second line each momentum independently ranges through half of momentum space excluding $\mathbf{k} = 0$ (as denoted by $\mathbf{k} > 0$). The configuration space representation of (13.124) is obtained by substituting (13.123) for $|l_1, l_2, \ldots \rangle$:

$$\chi_n(\mathbf{r}_1, \ldots, \mathbf{r}_N) = \frac{1}{V^{N/2}} \frac{1}{n! \sqrt{N!(N-2n)!}} \sum_P P$$

$$\times \left[\sum_{k_1 > 0} (-\alpha_1) e^{i\mathbf{k}_1 \cdot (\mathbf{r}_1 - \mathbf{r}_2)} \cdots \sum_{k_n > 0} (-\alpha_n) e^{i\mathbf{k}_n \cdot (\mathbf{r}_3 - \mathbf{r}_4)} \right]$$

Let us define

$$f(ij) \equiv f(\mathbf{r}_i - \mathbf{r}_j)$$

$$f(\mathbf{r}) = -\frac{2}{N} \sum_{k > 0} \alpha_k e^{i\mathbf{k} \cdot \mathbf{r}} = -\frac{v}{8\pi^3} \int d^3k \, \alpha_k e^{i\mathbf{k} \cdot \mathbf{r}} \quad (13.125)$$

Since α_k depends only on $|\mathbf{k}|$, it follows that $f(\mathbf{r})$ depends only on $|\mathbf{r}|$. We can then write

$$\chi_n(\mathbf{r}_1, \ldots, \mathbf{r}_N) = \frac{1}{V^{N/2}} \frac{1}{n! \sqrt{N!(N-2n)!}} \left(\frac{N}{2} \right)^n \sum_P P[f(12) \cdots f(34)]$$

in which there are n factors $f(ij)$.

The number of distinct ways of choosing the arguments of $f(ij)$ from the N coordinates $\mathbf{r}_1, \ldots, \mathbf{r}_N$ can be found by filling the n boxes in the following with N balls. The boxes are identical, each holding two balls, and the order of the two balls is irrelevant.

It is evident that there are

$$\frac{N!}{(N-2n)!\,n!\,2^n}$$

distinct ways to fill the boxes, whereas there are $N!$ permutations of $\mathbf{r}_1, \ldots, \mathbf{r}_N$. Hence

$$\chi_n(\mathbf{r}_1,\ldots,\mathbf{r}_N) = \frac{1}{V^{N/2}}\sqrt{\frac{(N-2n)!}{N!}}\,N^n\sum[f(12)\cdots f(34)] \quad (13.126)$$

where the sum is extended over all distinct ways of filling the "boxes." Therefore

$$\Psi_0(\mathbf{r}_1,\ldots,\mathbf{r}_N) = Z\sum_{n=0}^{N/2}\chi_n(\mathbf{r}_1,\ldots,\mathbf{r}_N)$$

$$= \frac{Z}{V^{N/2}}\{1 + [f(12) + f(34) + \cdots]$$

$$+ [f(12)f(34) + f(12)f(56) + \cdots]$$

$$+ [f(12)f(34)f(56) + \cdots] + \cdots\} \quad (13.127)$$

It differs from the function

$$\Psi_0'(\mathbf{r}_1,\ldots,\mathbf{r}_N) \equiv \frac{Z}{V^{N/2}}\prod_{i<j}[1 + f(ij)] \quad (13.128)$$

only in that (13.128) contains extra terms of the type

$$f(12)f(13), \; f(12)f(13)f(34), \ldots$$

in which the same particle appears in more than one "box." These terms may be shown to belong to a higher order in the calculation than we have considered. Therefore we may take (13.128) to be the wave function to the order $\sqrt{a^3/v}$. The function $f(\mathbf{r})$ defined in (13.125) has the following asymptotic behavior:

$$f(\mathbf{r}) \approx -\frac{a}{r} \qquad (r \ll r_0)$$

$$f(\mathbf{r}) \approx -32\sqrt{\frac{a^3}{\pi v}}\left(\frac{r_0}{r}\right)^4 \qquad (r \gg r_0) \qquad (13.129)$$

where

$$r_0 = \sqrt{\frac{v}{8\pi a}} \qquad (13.130)$$

Thus although (13.128) vanishes at $r_{ij} = a$, (13.127) does so only approximately, but its value at $r_{ij} = a$ is of a higher order than $\sqrt{a^3/v}$.

The one-phonon wave function (13.116) is easily shown to have the following form in configuration space:

$$\Psi_k(r_1, \ldots, r_N) = \text{const.} \sum_{j=1}^{N} e^{ik \cdot r_j} \Psi_0(r_1, \ldots, r_N) \qquad (13.131)$$

which verifies the result of the last section.

The extension of the present results to a dilute Bose gas with attractive interaction, which has a more realistic phase diagram, has also been worked out.[*]

PROBLEMS

13.1 The most visible "super" quality of the fluidity of liquid He^4 is its ability to crawl up walls and out of a containing vessel. The reason is that it coats the wall with a thick film. At a height z above the free liquid surface (in equilibrium with its own vapor), the film adhering to a vertical wall has a thickness proportional to $z^{-1/2}$. Derive this result by the following steps:

(a) Suppose the temperature is such that the superfluid fraction is f. Consider an element of the adhering film at height z, of thickness d, vertical dimension Δz, and unit width in

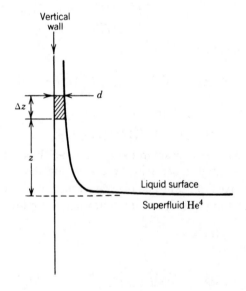

*K. Huang, *op. cit.*

the other direction (see sketch). Regard the superfluid component in this volume element as a free Bose gas. Write down its energy, with contributions from potential energy due to gravity and kinetic energy due to its being confined to a thickness d.

(**b**) Minimize the energy with respect to d.

13.2 A pure superfluid contained in an infinite cylinder of radius a develops a vortex line along the axis of the cylinder, of core size b.

(**a**) Find the superfluid velocity v_s as a function of normal distance r from the axis of the cylinder.

(**b**) Find the energy per unit length of the cylinder.

(**c**) Find the angular momentum per unit length of the cylinder.

13.3 Consider a channel connecting two reservoirs A and B of superfluid He^4. Let

$$\Delta\mu = \mu_A - \mu_B$$

$$\Delta\phi = \phi_A - \phi_B$$

be the differences in chemical potential and superfluid phases between A and B. In equilibrium $\Delta\mu = \Delta\phi = 0$.

Now suppose vortex liens are created on one wall of the channel, migrate across the channel as indicated by the arrow in the accompanying top-view sketch, and are annihilated on the opposite wall. With the help of Anderson's equation

$$\hbar\frac{d\phi}{dt} = \mu$$

show that this mechanism maintains a finite chemical potential difference $\Delta\mu$. Find the relation between $\Delta\mu$ and N, the rate of vortex creation.

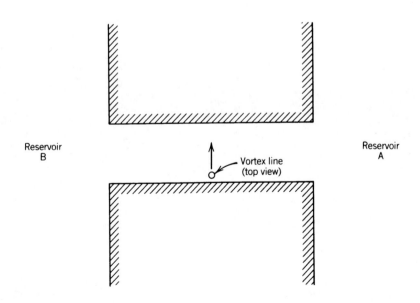

Reservoir B

Vortex line (top view)

Reservoir A

13.4 The first-order perturbation for the ground state energy due to hard-sphere interactions is calculated for a Bose system in (A.36) of the Appendix using configuration-space methods. Repeat the calculation using the method of quantized fields. (You will find it much neater.)

CHAPTER

14

THE ISING MODEL

14.1 DEFINITION OF THE ISING MODEL

One of the most interesting phenomena in the physics of the solid state is ferromagnetism. In some metals, e.g., Fe and Ni, a finite fraction of the spins of the atoms becomes spontaneously polarized in the same direction, giving rise to a macroscopic magnetic field. This happens, however, only when the temperature is lower than a characteristic temperature known as the Curie temperature. Above the Curie temperature the spins are oriented at random, producing no net magnetic field. As the Curie temperature is approached from both sides the specific heat of the metal approaches infinity.

The Ising model is a crude attempt to simulate the structure of a physical ferromagnetic substance.* Its main virtue lies in the fact that a two-dimensional Ising model yields to an exact treatment in statistical mechanics. It is the only nontrivial example of a phase transition that can be worked out with mathematical rigor.

In the Ising model[†] the system considered is an array of N fixed points called lattice sites that form an n-dimensional periodic lattice ($n = 1, 2, 3$). The geometrical structure of the lattice may (for example) be cubic or hexagonal. Associated with each lattice site is a spin variable s_i ($i = 1, \ldots, N$) which is a *number* that is either $+1$ or -1. There are no other variables. If $s_i = +1$, the ith site is said to have spin up, and if $s_i = -1$, it is said to have spin down. A given set of numbers $\{s_i\}$ specifies a configuration of the whole system. The energy of the system in the configuration specified by $\{s_i\}$ is defined to be

$$E_I\{s_i\} = -\sum_{\langle ij \rangle} \epsilon_{ij} s_i s_j - H \sum_{i=1}^{N} s_i$$

---

*More accurately, the Ising model simulates a "domain" in a ferromagnetic substance. A discussion of the physical properties and the atomic structure of ferromagnets (and antiferromagnets) is beyond the scope of the present discussion. For information in these aspects consult a book on solid-state physics.

[†] E. Ising, *Z. Phys.* **31**, 253 (1925).

where the subscript I stands for Ising and the symbol $\langle ij \rangle$ denotes a nearest-neighbor pair of spins. There is no distinction between $\langle ij \rangle$ and $\langle ji \rangle$. Thus the sum over $\langle ij \rangle$ contains $\gamma N/2$ terms, where γ is the number of nearest neighbors of any given site. For example,

$$\gamma = \begin{cases} 4 \ \text{(two-dimensional square lattice)} \\ 6 \ \text{(three-dimensional simple cubic lattice)} \\ 8 \ \text{(three-dimensional body-centered cubic lattice)} \end{cases}$$

The interaction energy ϵ_{ij} and the external magnetic field H are given constants. The geometry of the lattice enters the problem through γ and ϵ_{ij}. For simplicity we specialize the model to the case of isotropic interactions, so that all ϵ_{ij} are equal to a given number ϵ. Thus the energy will be taken as

$$E_I\{s_i\} = -\epsilon \sum_{\langle ij \rangle} s_i s_j - H \sum_{i=1}^{N} s_i \tag{14.1}$$

The case $\epsilon > 0$ corresponds to ferromagnetism and the case $\epsilon < 0$ to antiferromagnetism. We consider only the case $\epsilon > 0$. The partition function is

$$Q_I(B, T) = \sum_{s_1} \sum_{s_2} \cdots \sum_{s_N} e^{-\beta E_I\{s_i\}} \tag{14.2}$$

where each s_i ranges independently over the values ± 1. Hence there are 2^N terms in the summation. The thermodynamic functions are obtained in the usual manner from the Helmholtz free energy:

$$A_I(H, T) = -kT \log Q_I(H, T) \tag{14.3}$$

Some of the interesting ones are

$$U_I(H, T) = -kT^2 \frac{\partial}{\partial T}\left(\frac{A_I}{kT}\right) \qquad \text{(internal energy)} \tag{14.4}$$

$$C_I(H, T) = \frac{\partial U_I}{\partial T} \qquad \text{(heat capacity)} \tag{14.5}$$

$$M_I(H, T) = -\frac{\partial}{\partial H}\left(\frac{A_I}{kT}\right) = \left\langle \sum_{i=1}^{N} s_i \right\rangle \qquad \text{(magnetization)} \tag{14.6}$$

where $\langle \ \rangle$ denotes ensemble average. The quantity $M_I(0, T)$ is called the spontaneous magnetization. If it is nonzero the system is said to be ferromagnetic.

Although a configuration of the system is specified by the N numbers $s_1, \ldots, s_N$, the energy value (14.1) is in general degenerate. There is another way of writing (14.1) that makes this manifest. In any given configuration of the lattice, let

$$N_+ = \text{total number of up spins}$$
$$N_- = \text{total number of down spins}$$
$$= N - N_+$$

Each nearest-neighbor pair is one of the three types $(+ +)$, $(- -)$, or $(+ -)$, where $(+ -)$ is not distinguished from $(- +)$. Let the respective number of such pairs be denoted by N_{++}, N_{--}, N_{+-}. These numbers are not independent of each other, nor of N_+, N_-. A relation among them may be found as follows. Choose a particular lattice site with spin up and draw a line connecting it to all its nearest neighbors. There should be γ lines drawn. Repeat this procedure for another site with spin up, and continue to do so until this is done for all sites with spin up. After the procedure is completed the total number of lines drawn is γN_+. A construction of the kind described is illustrated in Figure 14.1 for a two-dimensional square lattice. The total number of lines drawn can also be counted by noting that between every $(+ +)$ pair there are two lines, between every $(+ -)$ pair there is one line, and between every $(- -)$ pair there is no line. Hence $\gamma N_+ = 2N_{++} + N_{+-}$. This relation remains valid if we interchange $+$ and $-$. Therefore we have the set of relations

$$\gamma N_+ = 2N_{++} + N_{+-}$$
$$\gamma N_- = 2N_{--} + N_{+-} \qquad (14.7)$$
$$N_+ + N_- = N$$

from which any three of the five numbers $N_+, N_-, N_{++}, N_{--}, N_{+-}$ can be eliminated. If we choose to eliminate N_{+-}, N_{--}, N_-, we have

$$N_{+-} = \gamma N_+ - 2N_{++}$$
$$N_- = N - N_+ \qquad (14.8)$$
$$N_{--} = \frac{\gamma}{2}N + N_{++} - \gamma N_+$$

We further note that

$$\sum_{\langle ij \rangle} s_i s_j = N_{++} + N_{--} - N_{+-} = 4N_{++} - 2\gamma N_+ + \frac{\gamma}{2}N$$

$$\sum_{i=1}^{N} s_i = N_+ - N_- = 2N_+ - N \qquad (14.9)$$

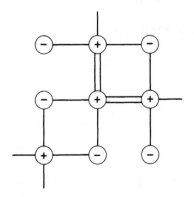

Fig. 14.1 Construction for the derivation of (14.7).

Substituting (14.9) into (14.1) and using (14.8) we obtain

$$E_I(N_+, N_{++}) = -4\epsilon N_{++} + 2(\epsilon\gamma - H)N_+ - (\tfrac{1}{2}\gamma\epsilon - H)N \quad (14.10)$$

Thus although a configuration of the system depends on N numbers the energy of a state depends only on two numbers. The partition function can also be written as

$$e^{-\beta A_I(H,T)} = e^{N\beta(\frac{1}{2}\gamma\epsilon - H)} \sum_{N_+=0}^{N} e^{-2\beta(\epsilon\gamma - H)N_+} \sum_{N_{++}}{}' g(N_+, N_{++}) e^{4\beta\epsilon N_{++}} \quad (14.11)$$

where $g(N_+, N_{++})$ is the number of configurations that has a given set of values (N_+, N_{++}). The sum $\sum'$ extends over all values of N_{++} consistent with the fact that there are N spins of which N_+ are up. Since $g(N_+, N_{++})$ is a complicated function, the form (14.11) is not a simplification over (14.2) for actual calculations.

14.2 EQUIVALENCE OF THE ISING MODEL TO OTHER MODELS

By a change of names the Ising model can be made to simulate systems other than a ferromagnet. Among these are a lattice gas and a binary alloy.

Lattice Gas

A lattice gas is a collection of atoms whose positions can take on only discrete values. These discrete values form a lattice of given geometry with γ nearest neighbors to each lattice site. Each lattice site can be occupied by at most one atom. Figure 14.2 illustrates a configuration of a two-dimensional lattice gas in which the atoms are represented by solid circles and the empty lattice sites by open circles. We neglect the kinetic energy of an atom and assume that only nearest neighbors interact, and the interaction energy for a pair of nearest neighbors is assumed to be a constant $-\epsilon_0$. Thus the potential energy of the system is equivalent to that of a gas in which the atoms are located only on lattice sites and interact through a two-body potential $v(|\mathbf{r}_i - \mathbf{r}_j|)$ with

$$v(r) = \begin{cases} \infty & (r = 0) \\ -\epsilon_0 & (r = \text{nearest-neighbor distance}) \\ 0 & (\text{otherwise}) \end{cases} \quad (14.12)$$

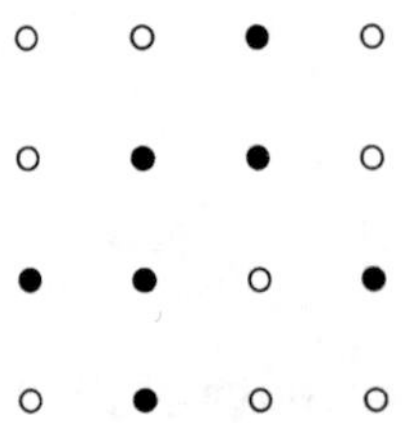

Fig. 14.2 A configuration of the lattice gas.

Let

$$N = \text{total no. of lattice sites}$$
$$N_a = \text{total no. of atoms} \tag{14.13}$$
$$N_{aa} = \text{total no. of nearest-neighbor pairs of atoms}$$

The total energy of the lattice gas is

$$E_G = -\epsilon_0 N_{aa} \tag{14.14}$$

and the partition function is

$$Q_G(N_a, T) = \frac{1}{N_a!} \sum^a e^{\beta \epsilon_0 N_{aa}} \tag{14.15}$$

where the sum $\sum^a$ extends over all ways of distributing N_a distinguishable atoms over N lattice sites. If the volume of a unit cell of the lattice is chosen to be unity, then N is the volume of the system. The grand partition function is

$$\mathcal{Q}_G(z, N, T) = \sum_{N_a=0}^{\infty} z^{N_a} Q_G(N_a, T) \tag{14.16}$$

The equation of state is given, as usual, by

$$\begin{cases} \beta P_G = \dfrac{1}{N} \log \mathcal{Q}_G(z, N, T) \\[2mm] \dfrac{1}{v} = \dfrac{1}{N} z \dfrac{\partial}{\partial z} \log \mathcal{Q}_G(z, N, T) \end{cases} \tag{14.17}$$

To establish a correspondence between the lattice gas and the Ising model, let occupied sites correspond to spin up and empty sites to spin down. Then $N_a \leftrightarrow N_+$. In the Ising model a set of N numbers $\{s_1, \ldots, s_N\}$ uniquely defines a configuration. In the lattice gas an enumeration of the occupied sites determines not one, but $N_a!$ configurations. The difference arises from the fact that the atoms are supposed to be able to move from site to site. This difference, however, is obliterated by the adoption of "correct Boltzmann counting." Hence

$$Q_G(N_a, T) = \sideset{}{'}\sum_{N_{++}} g(N_+, N_{++}) e^{\beta \epsilon_0 N_{++}} \tag{14.18}$$

where the function $g(N_+, N_{++})$ and the sum $\sum'$ are identical with those appearing in (14.11). The grand partition function is

$$e^{\beta N P_G} = \mathcal{Q}_G(z, N, T) = \sum_{N_+=0}^{\infty} z^{N_+} \sideset{}{'}\sum_{N_{++}} g(N_+, N_{++}) e^{\beta \epsilon_0 N_{++}} \tag{14.19}$$

A comparison between (14.19) and (14.11) yields the accompanying table of correspondence. Hence a solution of the Ising model can be immediately transcribed to be a solution of the lattice gas.

Ising Model		*Lattice Gas*
N_+	$\leftrightarrow$	N_a
N_-	$\leftrightarrow$	$N - N_a$
4ϵ	$\leftrightarrow$	ϵ_0
$e^{2\beta(\epsilon\gamma - H)}$	$\leftrightarrow$	z
$-\left(\dfrac{A_I}{N} + \dfrac{1}{2}\gamma\epsilon - H\right)$	$\leftrightarrow$	P_G
$\dfrac{1}{2}\left(\dfrac{M_I}{N} + 1\right)$	$\leftrightarrow$	$\dfrac{1}{v}$

The lattice gas does not directly correspond to any real system in nature. If we allow the lattice constant to approach zero, however, and then add to the resulting equation of state the pressure of an ideal gas, the model corresponds to a real gas of atoms interacting with one another through a zero-range potential. Thus it may be interesting to study the phase transition of a lattice gas.

The lattice gas has also been used as a model for the melting of a crystal lattice. When it is so used, however, the lattice constant must be kept finite. The kinetic energy of the atoms in the crystal lattice is appended in some ad hoc fashion. Such a model would only have a mathematical interest, because it is not clear that it describes melting.

Binary Alloy

Before introducing a model for the binary alloy, let us describe some of the salient features of an actual binary alloy, β-brass, which is a body-centered cubic lattice made up of Zn and Cu atoms. A unit cell of this lattice in its completely ordered state, which exists only at absolute zero, is shown in Fig. 14.3, where an open circle indicates a Zn atom and a solid circle a Cu atom. As the temperature is increased some Zn atoms will exchange positions with Cu atoms, but the probability of finding a Zn atom in the "right" place is greater than $\frac{1}{2}$. Above the critical temperature of 742 K, however, the Zn and Cu atoms are thoroughly mixed, and the probability of finding a Zn atom in the "right" place becomes exactly $\frac{1}{2}$. This transition can be discovered experimentally through the Bragg

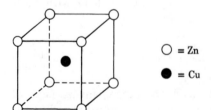

$\bigcirc$ = Zn

$\bullet$ = Cu

Fig. 14.3 Body-centered cubic lattice of β-brass.

reflection of X rays from the crystal. In the ordered state X-ray reflection will reveal that there are two sets of atomic planes with spacing d, whereas in the state of disorder there is only one set of atomic planes with spacing $d/2$. It is observed experimentally that the specific heat c_P approaches infinity as the temperature approaches the critical temperature from both sides.

A model for a binary alloy follows. Let there be two kinds of atoms, called 1 and 2, of which there are N_1 and N_2, respectively. Let their positions be confined to the lattice sites of a given lattice, with γ nearest neighbors to each lattice site. At each lattice site there shall be one and only one atom. Thus the total number of sites is $N = N_1 + N_2$. There are three types of nearest-neighbor pairs: (11), (22), and (12). The pair (12) is not distinguished from the pair (21). Let a configuration of the system be such that the number of pairs of each type present is, respectively, N_{11}, N_{22}, N_{12}. Neglecting the kinetic energy of the atoms and all but nearest-neighbor interactions, we take the energy of the system to be

$$E_A(N_{11}, N_{22}, N_{12}) = \epsilon_1 N_{11} + \epsilon_2 N_{22} + \epsilon_{12} N_{12} \qquad (14.20)$$

where the subscript A stands for "alloy." Obviously E_A is in general degenerate. Moreover, the numbers N_{11}, N_{22}, N_{12} are not independent of one another. By analogy with (14.7) we have the relations

$$\gamma N_1 = 2N_{11} + N_{12}$$
$$\gamma N_2 = 2N_{22} + N_{12} \qquad (14.21)$$
$$N_1 + N_2 = N$$

Thus

$$N_{12} = \gamma N_1 - 2N_{11}$$
$$N_{22} = \tfrac{1}{2}\gamma N + N_{11} - \gamma N_1 \qquad (14.22)$$

Hence the energy depends only on one variable, N_{11}:

$$E_A(N_{11}) = (\epsilon_1 + \epsilon_2 - 2\epsilon_{12})N_{11} + \left[\gamma(\epsilon_{12} - \epsilon_2)N_1 + \tfrac{1}{2}\gamma\epsilon_2 N\right] \qquad (14.23)$$

where the term in brackets is a constant. A correspondence between the binary alloy and the lattice gas may be established by identifying N_1 with N_a, which in turn is identified with N_+ of the Ising model. Comparison of (14.23) and the exponent in (14.19) immediately leads to the correspondence summarized in the accompanying table:

Lattice Gas		*Binary Alloy*
N_a	↔	N_1
$N - N_a$	↔	$N - N_1 = N_2$
$-\epsilon_0$	↔	$\epsilon_1 + \epsilon_2 - 2\epsilon_{12}$
(Helmholtz free energy)	↔	(Helmholtz free energy)
		$+[\gamma(\epsilon_{12} - \epsilon_2)N_1 + \tfrac{1}{2}\gamma\epsilon_2 N]$

14.3 SPONTANEOUS MAGNETIZATION

The energy (14.1) of the Ising model in the absence of external field is invariant when all the spins change sign simultaneously: $s_i \rightarrow -s_i$. For $H = 0$, therefore, the magnetization as defined by (14.6) is zero, because the contribution of every spin configuration $\{s_i\}$ is canceled by that of $\{-s_i\}$. As we have discussed in Section 12.5, however, the correct way to calculate the spontaneous magnetization, which is a case of spontaneous symmetry breaking, is to take the thermodynamic limit of (14.1) in the presence of a arbitrarily small H, and then let $H \rightarrow 0$. Figure 14.4 illustrates the typical behavior of the magnetization for a system exhibiting spontaneous magnetization. The magnetization per unit volume M/V is plotted as a function of temperature T for a range of values of the external field H. As $H \rightarrow 0^{\pm}$, M/V approaches one or the other branch of the $H = 0$ curve. It would be physically incorrect to average the two branches (thus getting 0), for reasons discussed in Section 12.5.

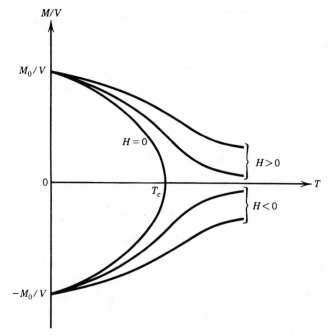

Fig. 14.4 Magnetization per unit volume as a function of temperature for various external fields H, typical of a ferromagnet. Below T_c for $H = 0$ there is spontaneous symmetry breaking. The mathematical limit that corresponds to physical observations consists of first taking the thermodynamic limit, and then the limit $H \rightarrow 0^{\pm}$. For $T < T_c$ this gives either the upper or the lower branch of the curve, but not the average of the two.

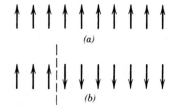

Fig. 14.5 (a) A one-dimensional array of spins. (b) A domain wall in one dimension.

We now show that the existence of spontaneous magnetization in the Ising model depends on the dimensionality of the model.

Absence of Spontaneous Magnetization in One Dimension

When all spins are aligned, as shown in Fig. 14.5a, the energy is at the absolute minimum, but the entropy is zero. Now create a "domain wall" by flipping all the spins to the right of some site, as shown in Fig. 14.5b. The energy increases by 2ϵ, but the entropy increases by $k \log(N - 1)$, since we have $N - 1$ choices to place the wall. Thus the free energy associated with the creation of one domain wall is

$$\Delta A = 2\epsilon - kT \log(N - 1) \tag{14.24}$$

For $T > 0$ and $N \to \infty$, the creation of a domain wall lowers the free energy. Hence more will be created until the spins are randomized. Therefore there can be no spontaneous magnetization at $T > 0$ in one-dimensional models.

In an antiferromagnetic model, where $\epsilon < 0$, the analog of spontaneous magnetization is a staggered magnetization, in which the neighboring spins are alternately ± 1. Again, there can be no staggered magnetization in one-dimensional antiferromagnetic models at $T > 0$. We can see this by mentally separating the antiferromagnetic lattice into two interpenetrating sublattices, one consisting of every other spin, the other consisting of the remaining spins. Existence of staggered magnetization means that each sublattice is spontaneously magnetized. This is not possible for $T > 0$ because the proof we gave for the ferromagnetic case applies to each sublattice separately.

It should be emphasized that the discussion above is valid only for $T > 0$. Phase transitions in one-dimensional systems can occur at absolute zero, and can actually be seen in the laboratory.*

Existence of Spontaneous Magnetization in Two Dimensions

Peierls[†] has shown that in two dimensions there will be spontaneous magnetization below a finite temperature. We follow the proof given by Griffiths.[‡]

*For a more detailed discussion, see R. Birgeneau and G. Shirane, *Phys. Today* (Dec. 1978).
[†] R. Peierls, *Proc. Camb. Philos. Soc.* **32**, 477 (1936).
[‡] R. B. Griffiths, *Phys. Rev. A*, **136**, 437 (1964).

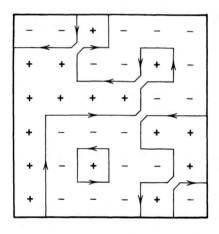

Fig. 14.6 Domain walls in the two-dimensional Ising model.

Consider an arbitrary configuration in a two-dimensional square lattice. A domain wall is a continuous line drawn between up ($+$) spins and down ($-$) spins. The length b of the wall is the number of lattice spacings it traverses. To make the definition unique, we draw a domain wall in a particular sense, such that the $-$ spins always lie to the right of the wall, and the $+$ spins to the left. Where there is still an ambiguity, the domain wall will bend to the right. An example is shown in Fig. 14.6. By this definition, no two domain walls ever cross. Some are close curves, while others begin and end on the boundary of the lattice. In the set of all possible domain walls, two walls of the same shape but differ in location are counted as distinct. Also distinct are two walls of identical shape and location but of different sense. (In the latter case, both cannot occur in the same configuration, of course.)

Now we impose the boundary condition that all spins on the boundary are $+$. This simulates an external field, whose influence should become arbitrarily weak in the limit of an infinite lattice. The boundary condition destroys the up-down symmetry of the model. To show that spontaneous magnetization exists, we shall show that, at a sufficiently low but finite temperature, the average fraction of down spins is less than $\frac{1}{2}$.

With our boundary condition, every domain wall is a closed curve. Consider the set of all closed domain walls. We classify them according to length b, and within a class of given length we give each a number i. Thus any domain wall is uniquely characterized by a label (b, i). Let us first state some properties of the domain walls.

A wall of length b is a closed polygon made up of b vertical or horizontal segments of unit length. If these segments were rearranged so that the closed figure resembles a square as much as possible, it will fit inside a square of side $b/4$. Therefore the area enclosed by a domain wall of length b is less than $b^2/16$.

Let $m(b)$ be the number of domain walls of length b. To find a bound for $m(b)$, imagine constructing a continuous line on a large square lattice of N sites, by successively laying down b sticks of unit length, either vertically or horizontally. The number of distinct ways to do this is greater than $m(b)$ because we did not require the line be closed. The first stick can go in N positions, and subsequent ones have three choices corresponding to the three possible directions: up, down, or horizontal. (For large N we can ignore constraints imposed by the boundary condition.) Therefore

$$m(b) \leq 3^{b-1}N \qquad (14.25)$$

Now, because the spins on the boundary are all $+$, every $-$ spin is enclosed by at least one domain wall. In a particular configuration let

$$X(b, i) = \begin{cases} 1 & \text{(if the domain wall } (b, i) \text{ occurs in that configuration)} \\ 0 & \text{(otherwise)} \end{cases} \qquad (14.26)$$

Then in that configuration the number of $-$ spins satisfy

$$N_- \leq \sum_b (b^2/16) \sum_{i=1}^{m(b)} X(b, i) \qquad (14.27)$$

We now calculate the thermal average of $X(b, i)$:

$$\langle X(b, i) \rangle = \frac{\sum_{\{s\}}' e^{-\beta E\{s\}}}{\sum_{\{s\}} e^{-\beta E\{s\}}} \qquad (14.28)$$

where the prime in the numerator indicates that the sum is restricted to those configurations in which (b, i) occurs. If C is a configuration in which (b, i) occurs, let $\tilde{C}$ be the configuration obtained from C by reversing every spin inside the domain wall (i.e., every $+$ becomes $-$, and vice versa.) An example is shown in Fig. 14.7. Their energies are related by

$$E_C = E_{\tilde{C}} + 2\epsilon b \qquad (14.29)$$

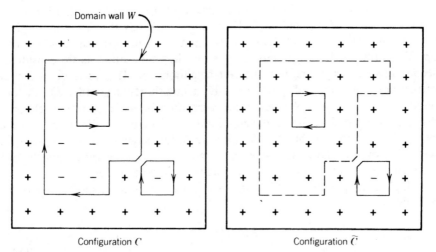

Configuration C Configuration $\tilde{C}$

Fig. 14.7 The configuration C is one that contains a specific domain wall W. The configuration $\tilde{C}$ is obtained by reversing every spin inside W.

The denominator in (14.28) can only decrease, if we restrict the sum to those configurations $\tilde{C}$ corresponding to a C. Therefore

$$\langle X(b, i) \rangle \le e^{-2\beta\epsilon b} \tag{14.30}$$

We now take the thermal average of (14.27) to obtain

$$\langle N_- \rangle \le \frac{1}{48} \sum_{b=4,6,8,\dots} b^2 3^b e^{-\beta\epsilon b}$$

$$= \frac{x^2}{3(1-x)^3} \left(1 - \tfrac{3}{4}x + \tfrac{1}{4}x^2\right), \qquad x = 9\,e^{-2\beta\epsilon} \tag{14.31}$$

It is clear that for sufficiently large but finite β the above ratio is less than $1/2$. ∎

The above proof can be extended to more than two dimensions. Griffiths has also shown the existence of spontaneous magnetization without imposing the special boundary condition, but we shall forego the proof.

It should be noted that the existence of long-range order in two-dimensions is a special feature of the Ising model. In more realistic models with a continuous order parameter, one can show that long-range order is not tenable in two-dimensions, because of large fluctuations (see Section 16.7).

14.4 THE BRAGG-WILLIAMS APPROXIMATION

In the Ising model the energy of a configuration of the spin lattice depends not on the detailed distribution of spins over lattice sites but only on the two numbers N_+ and N_{++}. The number N_+/N is said to be a measure of the "long-range order" in the lattice, and $N_{++}/(\gamma N/2)$ is said to be a measure of the "short-range order." The reason for this terminology is as follows. Let us imagine that the distribution of spin is random, except for the restriction that it possesses the given values of N_+ and N_{++}. If we know definitely that a given spin is up, then the number $N_{++}/(\gamma N/2)$ is the fraction of its nearest neighbors with spin up. This number, however, imposes less and less correlation as we consider the second-nearest neighbors, third-nearest neighbors, etc. It is therefore a measure of the local correlation of spins; hence the name short-range order. On the other hand, the number N_+/N requires no correlation between nearest neighbors. It does, however, require that in the entire lattice a fraction N_+/N of all spins must be up. Thus, if the number N_+/N is known in the neighborhood of a given spin, we will know that no matter how far we go away from the given spin the order measured by it is the same. Hence the name long-range order.

Define the parameter of long-range order L and that of short-range order σ as follows:

$$\frac{N_+}{N} \equiv \frac{1}{2}(L + 1) \qquad (-1 \le L \le +1)$$

$$\frac{N_{++}}{\tfrac{1}{2}\gamma N} \equiv \tfrac{1}{2}(\sigma + 1) \qquad (-1 \le \sigma \le +1) \tag{14.32}$$

From (14.9) we see that

$$\sum_{\langle ij \rangle} s_i s_j = \tfrac{1}{2}\gamma N(2\sigma - 2L + 1)$$

$$\sum_{i=1}^{N} s_i = NL \tag{14.33}$$

Thus the ensemble average of the long-range order is the magnetization per particle. The energy per spin is, by (14.1),

$$\frac{1}{N}E_I(L,\sigma) = -\tfrac{1}{2}\epsilon\gamma(2\sigma - 2L + 1) - HL \tag{14.34}$$

The Bragg-Williams approximation* is contained in the statement that "there is no short-range order apart from that which follows from long-range order." More precisely, the approximation consists of putting $N_{++}/(\tfrac{1}{2}\gamma N) \approx (N_+/N)^2$ or

$$\sigma \approx \tfrac{1}{2}(L + 1)^2 - 1 \tag{14.35}$$

In this approximation the energy becomes

$$\frac{1}{N}E_I(L) \approx -\tfrac{1}{2}\epsilon\gamma L^2 - HL \tag{14.36}$$

This approximation clearly has heuristic value, but it is difficult to estimate the error involved.

With (14.36) the partition function (14.2) becomes

$$Q_I(H,T) = \sum_{\{s_i\}} e^{\beta N(\tfrac{1}{2}\epsilon\gamma L^2 + HL)} \tag{14.37}$$

The sum extends over all sets $\{s_i\}$, but the summand depends only on L. Hence we want to find the number of sets $\{s_i\}$ that share the same L. According to (14.32), L is determined by N_+. The number we seek is the number of ways to pick N_+ things out of N, namely $N!/N_+!(N - N_+)!$. Therefore

$$Q_I(H,T) = \sum_{L=-1}^{+1} \frac{N!}{[\tfrac{1}{2}N(1 + L)]![\tfrac{1}{2}N(1 - L)]!} e^{\beta N(\tfrac{1}{2}\epsilon\gamma L^2 + HL)} \tag{14.38}$$

As $N \to \infty$ the logarithm of Q_I is equal to the logarithm of the largest term in the summand. Using Sterling's approximation for $N!$ we find that

$$\frac{1}{N} \log Q_I(H,T) = \beta\left(\tfrac{1}{2}\epsilon\gamma\overline{L}^2 + B\overline{L}\right) - \frac{1 + \overline{L}}{2}\log\frac{1 + \overline{L}}{2} - \frac{1 - \overline{L}}{2}\log\frac{1 - \overline{L}}{2} \tag{14.39}$$

where $\overline{L}$ is the value of L that maximizes the summand of (14.38). We easily find that $\overline{L}$ is the root of the equation

$$\log\frac{1 + \overline{L}}{1 - \overline{L}} = 2\beta H + 2\beta\epsilon\gamma\overline{L} \tag{14.40}$$

*W. L. Bragg and E. J. Williams, *Proc. Soc. London Ser. A* **145**, 699 (1934).

which is equivalent to

$$\bar{L} = \tanh\left(\frac{H}{kT} + \frac{\gamma\epsilon\bar{L}}{kT}\right) \tag{14.41}$$

Thus (14.39) can also be rewritten

$$\frac{1}{N}A_I(H, T) = -\frac{kT}{N}\log Q_I(H, T) = \frac{\epsilon\gamma}{2}\bar{L}^2 + \frac{kT}{2}\log\frac{1 - \bar{L}^2}{4} \tag{14.42}$$

We consider the case of no external magnetic field ($H = 0$). Then (14.41) becomes

$$\bar{L} = \tanh\left(\frac{\gamma\epsilon\bar{L}}{kT}\right) \tag{14.43}$$

which may be solved graphically as illustrated in Fig. 14.8. The main feature of the solution is that

$$\bar{L} = 0 \qquad \left(\frac{\gamma\epsilon}{kT} < 1\right)$$

$$\bar{L} = \begin{cases} L_0 \\ 0 \\ -L_0 \end{cases} \qquad \left(\frac{\gamma\epsilon}{kT} > 1\right)$$

In the second case the root $\bar{L} = 0$ must be rejected, because substituting it into (14.39) shows that it corresponds to a minimum instead of a maximum. If $\epsilon > 0$, there exists a critical temperature T_c, given by

$$kT_c = \gamma\epsilon \tag{14.44}$$

such that

$$\bar{L} = \begin{cases} 0 & (T > T_c) \\ \pm L_0 & (T < T_c) \end{cases} \tag{14.45}$$

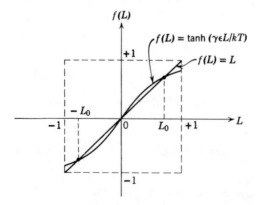

Fig. 14.8 Graphical solution of (14.43).

where L_0 is the root of (14.43) that is greater than zero. Since $\overline{L}$ is the magnetization per particle, we immediately see that for $T < T_c$ the system is a ferromagnet, whereas for $T > T_c$ it has no magnetization. The temperature T_c is the Curie temperature of the system. The degeneracy $\overline{L} = \pm L_0$ arises from the fact that in the absence of an external magnetic field there is no intrinsic distinction between "up" and "down." This degeneracy has no effect on the free energy, which is an even function of $\overline{L}$.

In general L_0 must be computed numerically, but near $T = 0$ and $T = T_c$ an approximation can be easily worked out:

$$
\begin{aligned}
L_0 &\approx 1 - 2\,e^{-2T_c/T} & \left(\frac{T_c}{T} \ll 1\right) \\
L_0 &\approx \sqrt{3\left(1 - \frac{T}{T_c}\right)} & \left(0 < 1 - \frac{T}{T_c} \ll 1\right)
\end{aligned}
\tag{14.46}
$$

A graph of L_0 is shown in Fig. 14.9.

The thermodynamic functions are summarized next:

$$
\frac{1}{N}A_I(0,T) = \begin{cases} 0 & (T > T_c) \\ \dfrac{\gamma\epsilon}{2}L_0^2 + \dfrac{kT}{2}\log\dfrac{1 - L_0^2}{4} & (T < T_c) \end{cases}
\tag{14.47}
$$

$$
\frac{1}{N}M_I(0,T) = \begin{cases} 0 & (T > T_c) \\ L_0 & (T < T_c) \end{cases}
\tag{14.48}
$$

$$
\frac{1}{N}U_I(0,T) = \begin{cases} 0 & (T > T_c) \\ -\dfrac{\epsilon\gamma}{2}L_0^2 & (T < T_c) \end{cases}
\tag{14.49}
$$

$$
\frac{1}{Nk}C_I(0,T) = \begin{cases} 0 & (T > T_c) \\ -\dfrac{\epsilon\gamma}{2}\dfrac{dL_0^2}{dT} & (T < T_c) \end{cases}
\tag{14.50}
$$

Using (14.46) we obtain

$$
\frac{1}{Nk}C_I(0,T_c) = \tfrac{3}{2}
\tag{14.51}
$$

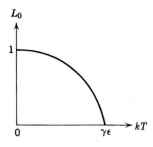

Fig. 14.9 Spontaneous magnetization in the Bragg-Williams approximation.

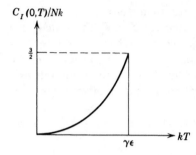

Fig. 14.10 Specific heat in the Bragg-Williams approximation.

A graph of C_I is shown in Fig. 14.10. Above the critical temperature the specific heat vanishes. This is a consequence of the fact that both the long-range and the short-range order vanish in that region in the present approximation.

We now turn to the lattice gas. The equation of state for the lattice gas in the Bragg-Williams approximation can be immediately obtained through the use of the table of correspondence in Section 14.2:

$$
\begin{cases}
P_G = H - \dfrac{\epsilon_0 \gamma}{8}(1 + \bar{L}^2) - \dfrac{kT}{2}\log\left(\dfrac{1 - \bar{L}^2}{4}\right) \\
\dfrac{1}{v} = \dfrac{1}{2}(1 + \bar{L})
\end{cases}
\tag{14.52}
$$

where H is a free parameter, related to the fugacity z by the equation

$$
H = \frac{\epsilon_0 \gamma}{4} - \frac{kT}{2}\log z
\tag{14.53}
$$

and $\bar{L}$ is a function of H and T to be found by solving (14.41). To find P_G as a function of T and v we must eliminate H from (14.52). The explicit solution of the Ising model in the absence of external field ($H = 0$) corresponds only to a restricted region in the P–v diagram.

For $H = 0$ we have

$$
\bar{L} =
\begin{cases}
0 & (T > T_c) \\
\pm L_0 & (T < T_c)
\end{cases}
\tag{14.54}
$$

where $T_c = \gamma \epsilon_0/4$ and L_0 is a function of temperature alone, being the nonzero root of the equation

$$
L_0 = \tanh\left(\frac{T_c}{T}L_0\right)
\tag{14.55}
$$

If $L_0 \neq 0$ is a solution, so is $-L_0$. Thus fixing the temperature determines two values of v but only one value of P_G (since P_G is an even function of L_0). Without going into the details we present the results in the P–v diagram of Fig. 14.11. The solid curves indicate the points on the equation of state surface that correspond to $H = 0$. The two points marked T_1 represent two points on the

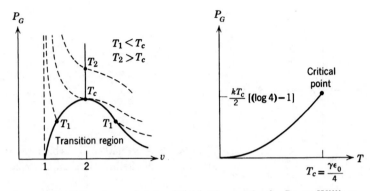

Fig. 14.11 Equations of state of the lattice gas in the Bragg-Williams approximation.

same isotherm with $T_1 < T_c$, whereas the point marked T_2 lies on an isotherm with $T_2 > T_c$. Thus with $H = 0$ it is only possible to obtain two points on each isotherm for $T < T_c$ and one point for $T > T_c$. To obtain a complete isotherm we must consider the case $B \neq 0$. We then obtain the isotherms represented by the dotted curves. In the Bragg-Williams approximation, the transition region is empty. This indicates that the Bragg-Williams approximation is not satisfactory. We can see, however, there is a first-order transition in the lattice gas and that T_c is the critical temperature of the transition region. It is to be noted that $v = 1$ is the smallest specific volume possible, because the volume of a unit cell of the lattice has been chosen to be unity. The profile of the transition region is shown in the P–T diagram of Fig. 14.11.

The results of the Bragg-Williams approximation can be obtained by other methods. One way is through the Gibbs variational principle, as suggested in Problem 10.6. Another way is suggested in Problem 14.3. It is an example of a general type of approximation known as "mean-field theory," a general formulation of which will be given in Section 17.4.

14.5 THE BETHE-PEIERLS APPROXIMATION

The Bethe-Peierls approximation is an improvement over the Bragg-Williams approximation, in that the former takes into account specific short-range order.

In the Bragg-Williams approximation the assumption $N_{++}/\frac{1}{2}\gamma N = (N_+/N)^2$ ignores the possibility of local correlation between spins. The Bethe-Peierls approximation replaces it by a better assumption. An outline of the method follows. We try to find a more accurate relation between N_{++} and N_+ by focusing our attention not on the entire lattice, but only on a sublattice composed of any lattice site plus its γ nearest neighbors. We form the mental picture of this sublattice being "immersed" in the background provided by the rest of the lattice, just as a small-volume element in a liquid is "immersed" in the

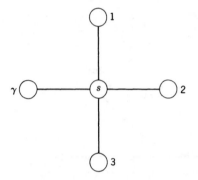

Fig. 14.12 A sublattice on which we focus our attention in the Bethe-Peierls approximation.

background provided by the rest of the liquid. It is assumed that the background influences the sublattice only through a single parameter that is similar to the fugacity in a liquid. The relation between N_{++} and N_+ will be obtained for the sublattice through heuristic arguments. It is then assumed that the same relation holds throughout the entire lattice.

We consider only the case $H = 0$. To begin, consider a sublattice consisting of any lattice site whose spin state is denoted by s, together with its γ nearest neighbors, as shown in Fig. 14.12. Let $P(s, n)$ be the probability that n of the nearest neighbors are in a spin-up state while the center site has the spin state s. If $s = +1$, then $P(s, n)$ refers to configurations of the sublattice in which there are n pairs $(+ +)$ and $\gamma - n$ pairs $(+ -)$. If $s = -1$, then $P(s, n)$ refers to configurations in which there are n pairs $(+ -)$ and $n - \gamma$ pairs $(- -)$. For a given n there are $\binom{\gamma}{n}$ ways to decide which of the γ neighbors are the n spins in question. Thus we *assume* that

$$P(+1, n) = \frac{1}{q}\binom{\gamma}{n} e^{\beta \epsilon (2n - \gamma)} z^n \tag{14.56}$$

$$P(-1, n) = \frac{1}{q}\binom{\gamma}{n} e^{\beta \epsilon (\gamma - 2n)} z^n \tag{14.57}$$

where q is a normalization factor and z is introduced to represent the effect of the background formed by the rest of the lattice. Because of the similarity between z and the fugacity, this method is also known as the *quasi-chemical method*. To determine q we require that

$$\sum_{n=0}^{\gamma} [P(+1, n) + P(-1, n)] = 1 \tag{14.58}$$

This leads to

$$q = \sum_{n=0}^{\gamma} \binom{\gamma}{n} [(z e^{2\beta \epsilon})^n e^{-\beta \epsilon \gamma} + (z e^{-2\beta \epsilon})^n e^{\beta \epsilon \gamma}]$$

$$= (e^{\beta \epsilon} + z e^{-\beta \epsilon})^\gamma + (z e^{\beta \epsilon} + e^{-\beta \epsilon})^\gamma \tag{14.59}$$

It follows from the meaning given to $P(+1, n)$ that

$$\frac{1 + L}{2} \equiv \frac{N_+}{N} = \sum_{n=0}^{\gamma} P(+1, n) = \frac{1}{q}(e^{\beta \epsilon} + z e^{-\beta \epsilon})^{\gamma} \tag{14.60}$$

$$\frac{1 + \sigma}{2} \equiv \frac{N_{++}}{\frac{1}{2}\gamma N} = \frac{1}{\gamma} \sum_{n=0}^{\gamma} nP(+1, n) = \frac{z}{q} e^{\beta \epsilon} (e^{-\beta \epsilon} + z e^{\beta \epsilon})^{\gamma - 1} \tag{14.61}$$

These equations express L and σ in terms of a single variable z. Since the energy of the Ising lattice depends on L and σ, we have an expression of the energy in terms of a single parameter z if we assume that (14.60) and (14.61) hold throughout the lattice. We may then use the resulting expression for the energy to calculate the partition function. This completes the statement of the Bethe-Peierls approximation.

It is not necessary to calculate the partition function, because there is a simpler procedure to obtain the magnetization. The interpretation that (14.56) and (14.57) are probabilities leads to the interpretation

$$\sum_{n=0}^{\gamma} P(+1, n) = \text{probability of finding an up spin at the center}$$

$$\frac{1}{\gamma} \sum_{n=0}^{\gamma} n[P(+1, n) + P(-1, n)]$$

$$= \text{probability of finding an up spin among the neighbors}$$

Since these probabilities are not conditioned by the knowledge of anything else, they must be equal to each other for the interpretation to be consistent. Thus we require that

$$\sum_{n=0}^{\gamma} P(+1, n) = \frac{1}{\gamma} \sum_{n=0}^{\gamma} n[P(+1, n) + P(-1, n)] \tag{14.62}$$

This condition determines z. Using (14.56) and (14.57) we find that

$$(e^{-\beta \epsilon} + z e^{\beta \epsilon})^{\gamma} = \frac{z}{\gamma} \frac{\partial}{\partial z} [(e^{-\beta \epsilon} + z e^{\beta \epsilon})^{\gamma} + (e^{\beta \epsilon} + z e^{-\beta \epsilon})^{\gamma}]$$

$$= z[(e^{-\beta \epsilon} + z e^{\beta \epsilon})^{\gamma - 1} e^{\beta \epsilon} + (e^{\beta \epsilon} + z e^{-\beta \epsilon})^{\gamma - 1} e^{-\beta \epsilon}]$$

or

$$z = \left(\frac{1 + z e^{2\beta \epsilon}}{z + e^{2\beta \epsilon}} \right)^{\gamma - 1} \tag{14.63}$$

After solving this equation for z, we can obtain $\bar{L}$ and $\bar{\sigma}$ from (14.60) and (14.61):

$$\bar{L} = \frac{z^x - 1}{z^x + 1}, \qquad\qquad x \equiv \frac{\gamma}{\gamma - 1} \tag{14.64}$$

$$\bar{\sigma} = \frac{2z^2}{(1 + z e^{-2\beta \epsilon})(1 + z^x)} - 1, \tag{14.65}$$

The internal energy of the Ising lattice in the absence of a magnetic field is then given by

$$\frac{1}{N} U_I(0, T) = -\tfrac{1}{2}\epsilon\gamma(2\bar{\sigma} - 2\bar{L} + 1) \tag{14.66}$$

Thus it only remains to solve (14.63).

We note that

(a) $z = 1$ is always a solution of (14.63);

(b) if z is a solution of (14.63), then $1/z$ is also a solution of (14.63);

(c) interchanging z and $1/z$ interchanges $\bar{L}$ and $-\bar{L}$;

(d) $z = 1$ corresponds to $\bar{L} = 0$; $z = \infty$ corresponds to $\bar{L} = 1$.

The actual solution of (14.63) may be found graphically, as shown in Fig. 14.13. The slope of the right side of (14.63) at $z = 1$ is

$$c = \frac{(\gamma - 1)(e^{4\beta\epsilon} - 1)}{(1 + e^{2\beta\epsilon})^2} \tag{14.67}$$

Thus if $c < 1$ the only solution is $z = 1$. If $c > 1$, there are three solutions $z = 1$, z_0, $1/z_0$, of which $z = 1$ is discarded by comparison with the solution in the Bragg-Williams approximation. The solution $1/z_0$ does not lead to anything new since it merely means interchanging up spin with down spin. It too will be ignored.

Let us define the critical temperature T_c by the equation

$$\frac{(\gamma - 1) e^{4\epsilon/kT_c}}{(1 + e^{2\epsilon/kT_c})^2} = 1 \tag{14.68}$$

which leads to the explicit expression

$$kT_c = \frac{2\epsilon}{\log\left[\gamma/(\gamma - 2)\right]} \tag{14.69}$$

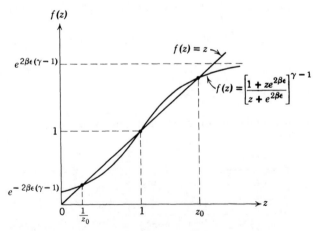

Fig. 14.13 Graphical solution of (14.63).

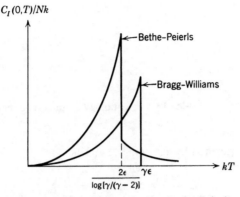

Fig. 14.14 Specific heat in the Bethe-Peierls approximation and in the Bragg-Williams approximation.

For $T > T_c$ we have

$$z = 1$$
$$\bar{L} = 0 \tag{14.70}$$

$$\bar{\sigma} = \frac{1}{2(1 + e^{-2\beta\epsilon})}$$

For $T < T_c$ we have

$$z > 1$$
$$\bar{L} > 0 \tag{14.71}$$

In this case we have spontaneous magnetization. Of the thermodynamic functions we only consider the specific heat, which can be shown to be

$$\frac{C_I(0, T)}{Nk} = \frac{1}{Nk}\frac{d}{dT}U_I(0, T) = -\frac{\epsilon\gamma}{Nk}\left(\frac{d\bar{\sigma}}{dT} - \frac{d\bar{L}}{dT}\right) \tag{14.72}$$

and which, in contradistinction the Bragg-Williams result, does not vanish for $T > T_c$:

$$\frac{C_I(0, T)}{Nk} = \frac{2\gamma\epsilon^2}{(kT)^2}\frac{e^{2\epsilon/kT}}{(1 + e^{2\epsilon/kT})^2} \qquad (T > T_c) \tag{14.73}$$

A more detailed calculation yields the graph shown in Fig. 14.14 for the specific heat. For comparison the Bragg-Williams result is also shown.

14.6 THE ONE-DIMENSIONAL ISING MODEL

The one-dimensional Ising model is a chain of N spins, each spin interacting only with its two nearest neighbors and with an external magnetic field. The energy for

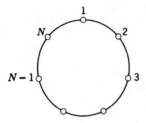

Fig. 14.15 Topology of the one-dimensional Ising lattice.

the configuration specified by $\{s_1, s_2, \ldots, s_N\}$ is

$$E_I = -\epsilon \sum_{k=1}^{N} s_k s_{k+1} - H \sum_{k=1}^{N} s_k \qquad (14.74)$$

We impose the periodic boundary condition

$$s_{N+1} \equiv s_1 \qquad (14.75)$$

making the topology of the chain that of a circle, as shown in Fig. 14.15. The partition function is

$$Q_I(H, T) = \sum_{s_1} \sum_{s_2} \cdots \sum_{s_N} \exp\left[\beta \sum_{k=1}^{N} (\epsilon s_k s_{k+1} + H s_k)\right] \qquad (14.76)$$

where each s_k independently assumes the values ± 1.

The partition function can be expressed in terms of matrices.* Let us write

$$Q_I(H, T) = \sum_{s_1} \sum_{s_2} \cdots \sum_{s_N}$$

$$\times \exp\left\{\beta \sum_{k=1}^{N} \left[\epsilon s_k s_{k+1} + \tfrac{1}{2}H(s_k + s_{k+1})\right]\right\} \qquad (14.77)$$

which is equivalent to (14.76) by virtue of (14.75). Let a 2×2 matrix **P** be so defined that its matrix elements are given by

$$\langle s|\mathbf{P}|s'\rangle = e^{\beta[\epsilon s s' + \tfrac{1}{2}H(s+s')]} \qquad (14.78)$$

where s and s' may independently take on the values ± 1. A list of all the matrix elements is

$$\langle +1|\mathbf{P}|+1\rangle = e^{\beta(\epsilon + H)}$$
$$\langle -1|\mathbf{P}|-1\rangle = e^{\beta(\epsilon - H)} \qquad (14.79)$$
$$\langle +1|\mathbf{P}|-1\rangle = \langle -1|\mathbf{P}|+1\rangle = e^{-\beta \epsilon}$$

Thus an explicit representation for **P** is

$$\mathbf{P} = \begin{bmatrix} e^{\beta(\epsilon + H)} & e^{-\beta \epsilon} \\ e^{-\beta \epsilon} & e^{\beta(\epsilon - H)} \end{bmatrix} \qquad (14.80)$$

*The matrix formulation of the Ising model in general is due to H. A. Kramers and G. H. Wannier, *Phys. Rev.* **60**, 252 (1941).

With these definitions we may rewrite (14.77) in the form

$$Q_I(H, T) = \sum_{s_1} \sum_{s_2} \cdots \sum_{s_N} \langle s_1|P|s_2\rangle\langle s_2|P|s_3\rangle \cdots \langle s_N|P|s_1\rangle$$

$$= \sum_{s_1} \langle s_1|P^N|s_1\rangle = \text{Tr} P^N = \lambda_+^N + \lambda_-^N \qquad (14.81)$$

where λ_+ and λ_- are the two eigenvalues of P, with $\lambda_+ \geq \lambda_-$. The fact that Q_I is the trace of the Nth power of a matrix is a consequence of the periodic boundary condition (14.75). Because of the way P enters in (14.81), it is called the *transfer matrix*.

By an easy calculation we find that the two eigenvalues $\lambda_\pm$ are

$$\lambda_\pm = e^{\beta\epsilon}\left[\cosh(\beta H) \pm \sqrt{\sinh^2(\beta H) + e^{-4\beta\epsilon}}\right] \qquad (14.82)$$

Thus $\lambda_+ > \lambda_-$ for all H. As $N \to \infty$ only the larger one of the eigenvalues λ_+ is relevant, because

$$\frac{1}{N}\log Q_I(H, T) = \log\lambda_+ + \frac{1}{N}\log\left[1 + \left(\frac{\lambda_-}{\lambda_+}\right)^N\right] \xrightarrow[N \to \infty]{} \log\lambda_+ \quad (14.83)$$

Thus the Helmholtz free energy per spin is given by

$$\frac{1}{N}A_I(H, T) = -\epsilon - kT\log\left[\cosh(\beta H) + \sqrt{\sinh^2(\beta H) + e^{-4\beta\epsilon}}\right] \quad (14.84)$$

The magnetization per spin is

$$\frac{1}{N}M_I(H, T) = \frac{\sinh(\beta H)}{\sqrt{\sinh^2(\beta H) + e^{-4\beta\epsilon}}} \qquad (14.85)$$

Graphs of $N^{-1}M_I(H, T)$ for various temperatures are shown in Fig. 14.16.
For all $T > 0$,

$$\frac{1}{N}M_I(0, T) = 0$$

in agreement with the general result stated in Section 14.3.

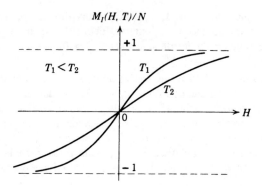

Fig. 14.16 Magnetization of the one-dimensional Ising model. There is no spontaneous magnetization.

PROBLEMS

14.1 Consider a one-dimensional Ising-like spin system with pairwise interactions, such that every pair of spins (s_1, s_2) separated by r lattice spacings contribute to the energy a term $s_1 s_2 / r^\sigma$, where σ is a positive constant. Show that the argument for the absence of spontaneous magnetization fails when $\sigma < 2$.

14.2 Modify the argument for the existence of spontaneous magnetization in a two-dimensional Ising model to show the same result in three dimensions.

14.3 Heuristic Mean-Field Theory of Ferromagnetism. Here is yet another way to obtain the results of the Bragg-Williams approximation. (One alternative was suggested in Problem 10.6.) We shall state the problem in a physical context. Consider a solid containing N electrons localized at lattice sites. Each electron has spin $\hbar/2$, with magnetic moment $g\mu_0$, where μ_0 is the Bohr magneton ($\mu_0 = e\hbar/2mc$.) In a magnetic field H each electron can exist in only one of two states, with energies $\pm\mu_0 H$.

(*a*) Make the assumptions that each electron sees the same effective magnetic field H_{eff} and that in thermal equilibrium the total magnetic moment of the solid is given by

$$M = N\mu_0 \tanh\left(\mu_0 H_{\text{eff}}/kT\right)$$

(*b*) Assume that H_{eff} consists of an externally applied field H, plus a local field (the Weiss field) $H_{\text{loc}} = qM$. Write down the self-consistency condition that determines M. Compare the result with (14.41).

(*c*) Show that there is spontaneous magnetization ($H = 0$) below a critical temperature T_c. Obtain T_c in terms of the constants given in the problem.

(*d*) For $H = 0$ show that the magnetization just below the critical temperature behaves like

$$\frac{M}{N\mu_0} \approx \sqrt{3}\left(\frac{T_c - T}{T_c}\right)^{\frac{1}{2}}$$

(*e*) Show that the magnetic susceptibility $\chi = (\partial M/\partial H)_{H=0}$ diverges as $T \to T_c$ from above, in accordance with the Curie-Weiss law

$$\chi \approx \frac{N\mu_0^2}{k(T - T_c)}$$

14.4 Consider a square lattice of Ising spins in any dimension, with energy given by (14.1). Show that in the absence of an external field ($H = 0$) the free energy at a given temperature is the same for the ferromagnetic case ($\epsilon > 0$) and the antiferromagnetic case ($\epsilon < 0$).

    Solution. Consider any lattice site. Designate it as a site of sublattice A. Designate its nearest neighbors as sites belonging to sublattice B. The sublattices are completely defined by the rule that the nearest neighbors of any site in sublattice B are sites of sublattice A, and vice versa. In a square lattice, the nearest neighbors are situated one lattice spacing away, along mutually orthogonal directions. Therefore both A and B are square lattices. Denote the spin variables in sublattice A by s_{Ai}, and those in sublattice B by s_{Bj}. Only spins in different sublattices interact. Thus the partition function can be written in the form

$$Q = \sum_{\{s_A\}} \sum_{\{s_B\}} \exp\left(J\sum_{\langle ij \rangle} s_{Ai} s_{Bj}\right)$$

where $J = \beta\epsilon$. Since $\{s_A\}$ is being summed over, the above is invariant under $s_{Ai} \rightarrow -s_{Ai}$. Therefore it is invariant under $\epsilon \rightarrow -\epsilon$. ■ Note that the proof fails for a triangular lattice.

14.5 Duality of Two-Dimensional Ising model (H. A. Kramers and G. H. Wannier, *Phys. Rev.* **60**, 252 (1941))

(*a*) Let $Q(N, \beta)$ denote the partition function for a two-dimensional Ising model of N sites on a square lattice at temperature $kT = 1/\beta$, with no external field. Show that in the limit $N \rightarrow \infty$

$$\frac{\log Q(N, \beta)}{N} = \frac{\log Q(N, \beta^*)}{N} - \sinh(2\beta^*)$$

where $\beta^* = -\frac{1}{2}\log\tanh\beta$, or

$$\sinh(2\beta)\sinh(2\beta^*) = 1$$

Note that β^* is a decreasing function of β. Thus the high-temperature properties of the system are explicitly related to its low-temperature properties.

(*b*) We know through the Peierls argument that the system exhibits spontaneous magnetization. Assuming that the critical temperature T_c is unique, we conclude $\beta_c = \beta_c^*$. Show

$$\beta_c = \frac{1}{2}\log(1 + \sqrt{2})$$

This is how the transition temperature of the two-dimensional Ising model was obtained before Onsager's explicit solution.

Solution.

(1)

$$Q = \sum_{\{s\}} \exp\left(J\sum_{\langle ij\rangle} s_i s_j\right) = \sum_{\{s\}} \prod_{\langle ij\rangle} \exp(Js_i s_j)$$

$$= \sum_{\{s\}} \prod_{\langle ij\rangle} (\cosh J + s_i s_j \sinh J) = \sum_{\{s\}} \prod_{\langle ij\rangle} \sum_{k=0}^{1} c_k(s_i s_j)^k$$

where $J = \beta\epsilon$, $c_0 = \cosh J$, $c_1 = \sinh J$. We have used the identity $\exp(ax) = \cosh x + a\sinh x$, where $a^2 = 1$.

(2) Each $\langle ij\rangle$ corresponds to a link b between the two sites. The last expression associates with each link an integer $k = 0, 1$. Denote the set of k's by $\{k\} = \{k_1, k_2, \dots\}$, where k_b refers to the bth link. Note that

$$\prod_{\langle ij\rangle} \sum_{k=0}^{1} c_k(s_i s_j)^k = \sum_{\{k\}} (c_{k_1} c_{k_2} \cdots) \prod_{\langle ij\rangle} (s_i s_j)^{k_{ij}}$$

where $k_{ij} = k_{ji} \equiv k_b$, with b referring to the link that joins $\langle ij\rangle$. To derive this, consider $\prod_{\langle ij\rangle}(c_0 + c_1 s_i s_j)$, which is a product of F factors (F = no. of links). Expand the product into a sum of terms, each made up of F factors, obtained by choosing either the c_0 or c_1 term.

(3) $$\prod_{\langle ij\rangle} (s_i s_j)^{k_{ij}} = \prod_i \left[\prod_j{}'(s_i)^{k_{ij}}\right] = \prod_i (s_i)^{n_i}, \qquad n_i \equiv \sum_{b \supset i} k_0$$

where the prime on $\prod_j$ means product over sites j that are nearest neighbors of i. The sum $\sum_{b \supset i}$ denotes a sum over all links b that meet at the site i. To derive the first equality, note that both sides represent different ways of writing the same product. Convince yourself of this by writing out each form explicitly.

(4)

$$Q = \sum_{\{s\}} \sum_{\{k\}} (c_{k_1} c_{k_2} \cdots) \prod_i (s_i)^{n_i} = \sum_{\{k\}} (c_{k_1} c_{k_2} \cdots) \sum_{\{s\}} \left[(s_1)^{n_1} (s_2)^{n_2} \cdots \right]$$

$$= \sum_{\{k\}} (c_{k_1} c_{k_2} \cdots) \prod_i \sum_{s=-1}^{1} s^{n_i} = \sum_{\{k\}} (c_{k_1} c_{k_2} \cdots) \prod_i \left[1 + (-1)^{n_i} \right]$$

Note $1 + (-1)^{n_i} = 0$ for n_i odd, 2 for n_i even. Since $\{n_i\}$ depends on $\{k_b\}$, only that subset of $\{k_b\}$ corresponding to all n_i even will contribute:

$$Q = 2^N \sum_{\{k\}}' (c_{k_1} c_{k_2} \cdots)$$

where the prime on the sum denotes the constraint

$$\sum_{b \supset i} k_b = 0 \,(\text{mod } 2), \qquad \text{for all } i$$

That is, associate with each link b an integer $k_b = 0, 1$, such that $k_1 + k_2 + k_3 + k_4 = 0 \,(\text{mod } 2)$ whenever the links $1, 2, 3, 4$ all meet at one site.

(5) Now comes the key step. We solve the constraint through a geometrical construction. Define the "dual lattice" as the lattice whose sites are located at centers of each square of the original lattice. In the accompanying sketch, the dual lattice

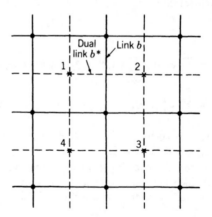

sites are marked with an $\times$, and the links of the dual lattice are shown dotted. Each original site i is contained in a square of the dual lattice, which we call a "plaquette." Attach to each site of the dual lattice a dual spin variable σ_i, which can assume only the values ± 1. If $1, 2, 3, 4$ are dual sites at successive corners of a plaquette (say in clockwise order), then

$$\sigma_1 \sigma_2 + \sigma_2 \sigma_3 + \sigma_3 \sigma_4 + \sigma_4 \sigma_1 = 0 \,(\text{mod } 4)$$

To see this, note that the left side can be written in the form $(\sigma_1 + \sigma_3)(\sigma_2 + \sigma_4) = 0, \pm 4$. Now each link b of the original lattice cuts a unique link b^* of the dual lattice. The constraint is solved by taking

$$k_b = \tfrac{1}{2}(1 - \sigma_1 \sigma_2) \,,$$

where $1, 2$ mark the dual sites at the ends of b^*. This is so because, first, $k_b = 0, 1$ as required, and second whenever four links meet at the same site, they cut the sides of the plaquette containing the site. We note in passing that an equally acceptable definition of k_b can be obtained by changing the $-$ sign to $+$. This shows that in this case the ferromagnetic model and the antiferromagnetic model have the same free energy, in agreement with the general theorem stated in the last problem.

(**6**) Since there is a one-to-one correspondence between links on the original and on the dual lattice, we can associate with a dual link b^* the integer $k_{b^*} \equiv k_b$. For large N, the number of dual sites is also N. We can associate each dual site with two dual links (say, the ones pointing north and east). Thus there are $2N$ integers k_{b^*}, of which N are independent (because they satisfy N constraints). Thus the partition function is obtained by replacing the constraint sum $\sum\limits_{\{k\}}'$ by the unconstraint sum $\frac{1}{2} \sum\limits_{\{\sigma\}}$:

$$Q = 2^{N-1} \sum_{\{\sigma\}} \left(c_{k_1} c_{k_2} \cdots \right)$$

where the c's are to be re-expressed in terms of the σ's.

(**7**)

$$c_k = k \sinh J + (1 - k) \cosh J = \tfrac{1}{2}(1 - \sigma_1\sigma_2) \sinh J + \tfrac{1}{2}(1 + \sigma_1\sigma_2) \cosh J$$
$$= \tfrac{1}{2} e^J \left(1 + \sigma_1\sigma_2 e^{-2J} \right) = \tfrac{1}{2} e^J \left(1 + \sigma_1\sigma_2 \tanh J^* \right)$$

where we have defined J^* by $\tanh J^* \equiv e^{-2J}$, with a view to re-expressing the above expression as an exponential:

$$c_k = \frac{e^J}{2 \cosh J^*} (\cosh J^* + \sigma_1\sigma_2 \sinh J^*)$$
$$= [2 \sinh (2J)]^{-1/2} \exp (J^* \sigma_1\sigma_2)$$

Thus finally

$$Q = \tfrac{1}{2}(\sinh 2J^*)^{-N} \sum_{\{\sigma\}} \exp \left(J^* \sum_{\langle ij \rangle} \sigma_i \sigma_j \right) \qquad \blacksquare$$

15

THE ONSAGER SOLUTION

15.1 FORMULATION OF THE TWO-DIMENSIONAL ISING MODEL

Matrix Formulation

We formulate the two-dimensional Ising model in terms of matrices as a preliminary step toward an exact solution of the model. Consider a square lattice of $N = n^2$ spins consisting of n rows and n columns, as shown in Fig. 15.1. Let us imagine the lattice to be enlarged by one row and one column with the requirement that the configuration of the $(n + 1)$th row and column be identical with that of the first row and column, respectively. This boundary condition endows the lattice with the topology of a torus, as depicted in Fig. 15.2. Let μ_α $(\alpha = 1, \ldots, n)$ denote the collection of all the spin coordinates of the αth row:

$$\mu_\alpha \equiv \{s_1, s_2, \ldots, s_n\}_{\alpha\text{th row}} \tag{15.1}$$

The toroidal boundary condition implies the definition

$$\mu_{n+1} \equiv \mu_1 \tag{15.2}$$

A configuration of the entire lattice is then specified by $\{\mu_1, \ldots, \mu_n\}$. By assumption, the αth row interacts only with the $(\alpha - 1)$th and the $(\alpha + 1)$th row. Let $E(\mu_\alpha, \mu_{\alpha+1})$ be the interaction energy between the αth and the $(\alpha + 1)$th row. Let $E(\mu_\alpha)$ be the interaction energy of the spins within the αth row plus their interaction energy with an external magnetic field. We can write

$$E(\mu, \mu') = -\epsilon \sum_{k=1}^{n} s_k s_k'$$

$$E(\mu) = -\epsilon \sum_{k=1}^{n} s_k s_{k+1} - H \sum_{k=1}^{n} s_k \tag{15.3}$$

where μ and μ' respectively denote the collection of spin coordinates in two

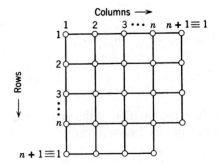

Fig. 15.1 Two-dimensional Ising lattice.

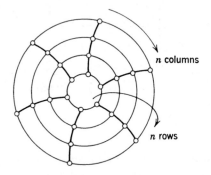

n columns

n rows

Fig. 15.2 Topology of the two-dimensional Ising lattice.

neighboring rows:

$$\mu \equiv \{s_1, \ldots, s_n\}$$
$$\mu' \equiv \{s'_1, \ldots, s'_n\} \tag{15.4}$$

The toroidal boundary condition implies that in each row

$$s_{n+1} \equiv s_1 \tag{15.5}$$

The total energy of the lattice for the configuration $\{\mu_1, \ldots, \mu_n\}$ is then given by

$$E_I\{\mu_1, \ldots, \mu_n\} = \sum_{\alpha=1}^{n} \left[E(\mu_\alpha, \mu_{\alpha+1}) + E(\mu_\alpha) \right] \tag{15.6}$$

The partition function is

$$Q_I(H, T) = \sum_{\mu_1} \cdots \sum_{\mu_n} \exp\left\{ -\beta \sum_{\alpha=1}^{n} \left[E(\mu_\alpha, \mu_{\alpha+1}) + E(\mu_\alpha) \right] \right\} \tag{15.7}$$

Let a $2^n \times 2^n$ matrix* **P** be so defined that its matrix elements are

$$\langle \mu | \mathsf{P} | \mu' \rangle \equiv e^{-\beta[E(\mu, \mu') + E(\mu)]} \tag{15.8}$$

*From now on all $2^n \times 2^n$ matrices are denoted by sans serif letters: P, Γ.

Then

$$Q_I(H, T) = \sum_{\mu_1} \cdots \sum_{\mu_n} \langle \mu_1|P|\mu_2\rangle\langle\mu_2|P|\mu_3\rangle \cdots \langle\mu_n|P|\mu_1\rangle$$

$$= \sum_{\mu_1} \langle\mu_1|P^n|\mu_1\rangle = \operatorname{Tr} P^n \qquad (15.9)$$

Since the trace of a matrix is independent of the representation of the matrix, the trace in (15.9)* may be evaluated by bringing P into its diagonal form:

$$P = \begin{bmatrix} \lambda_1 & & & \\ & \lambda_2 & & \\ & & \ddots & \\ & & & \lambda_{2^n} \end{bmatrix} \qquad (15.10)$$

where $\lambda_1, \lambda_2, \ldots, \lambda_{2^n}$ are the 2^n eigenvalues of P. The matrix P^n is then also diagonal, with the diagonal matrix elements $(\lambda_1)^n, (\lambda_2)^n, \ldots, (\lambda_{2^n})^n$. Therefore

$$Q_I(H, T) = \sum_{\alpha=1}^{2^n} (\lambda_\alpha)^n \qquad (15.11)$$

From the form of (15.8) we expect that the eigenvalues of P are in general of the order of e^n when n is large, since $E(\mu, \mu')$ and $E(\mu)$ are of the order of n. If λ_{max} is the largest eigenvalue of P, we expect that

$$\lim_{n\to\infty} \frac{1}{n} \log \lambda_{max} = \text{finite number} \qquad (15.12)$$

If this is true and if all the eigenvalues λ_α are positive, then

$$(\lambda_{max})^n \le Q_I \le 2^n(\lambda_{max})^n$$

or

$$\frac{1}{n} \log \lambda_{max} \le \frac{1}{n^2} \log Q_I \le \frac{1}{n} \log \lambda_{max} + \frac{1}{n} \log 2 \qquad (15.13)$$

Therefore

$$\lim_{N\to\infty} \frac{1}{N} \log Q_I = \lim_{n\to\infty} \frac{1}{n} \log \lambda_{max} \qquad (15.14)$$

where $N = n^2$. It will turn out that (15.12) is true and that all the eigenvalues λ_α are positive. Thus it is sufficient to find the largest eigenvalue of P. The remaining parts of this section are devoted to the description of an explicit representation for P.

---

*The fact that Q_I is of the form of a trace is a consequence of (15.2) alone and does not require (15.5). In other words, to make Q_I a trace of some matrix it is only necessary to fold the two-dimensional lattice into a cylinder. The purpose of (15.5), which turns the cylinder into a torus, is to facilitate the actual diagonalization of P.

The Matrix P

From (15.8) and (15.3) we may obtain the matrix elements of P in the form

$$\langle s_1, \ldots, s_n | P | s_1', \ldots, s_n' \rangle = \prod_{k=1}^{n} e^{\beta H s_k} \, e^{\beta \epsilon s_k s_{k+1}} \, e^{\beta \epsilon s_k s_k'} \tag{15.15}$$

Let us define three $2^n \times 2^n$ matrices V_1', V_2, and V_3 whose matrix elements are respectively given by

$$\langle s_1, \ldots, s_n | V_1' | s_1', \ldots, s_n' \rangle \equiv \prod_{k=1}^{n} e^{\beta \epsilon s_k s_k'} \tag{15.16}$$

$$\langle s_1, \ldots, s_n | V_2 | s_1', \ldots, s_n' \rangle \equiv \delta_{s_1 s_1'} \ldots \delta_{s_n s_n'} \prod_{k=1}^{n} e^{\beta \epsilon s_k s_{k+1}} \tag{15.17}$$

$$\langle s_1, \ldots, s_n | V_3 | s_1', \ldots, s_n' \rangle \equiv \delta_{s_1 s_1'} \cdots \delta_{s_n s_n'} \prod_{k=1}^{n} e^{\beta H s_k} \tag{15.18}$$

where $\delta_{ss'}$ is the Kronecker symbol. Thus V_2 and V_3 are diagonal matrices in the present representation. It is easily verified that

$$P = V_3 V_2 V_1' \tag{15.19}$$

in the usual sense of matrix multiplication, namely

$$\langle s_1, \ldots, s_n | P | s_1', \ldots, s_n' \rangle$$
$$= \sum_{s_1'', \ldots, s_n''} \sum_{s_1''', \ldots, s_n'''} \langle s_1, \ldots, s_n | V_3 | s_1'', \ldots, s_n'' \rangle$$
$$\times \langle s_1'', \ldots, s_n'' | V_2 | s_1''', \ldots, s_n''' \rangle \langle s_1''', \ldots, s_n''' | V_1' | s_1', \ldots, s_n' \rangle$$

Direct Product of Matrices

Before describing a convenient way to represent the matrices V_3, V_2, and V_1 we introduce the notion of a direct product of matrices. Let A and B be two $m \times m$ matrices whose matrix elements are respectively $\langle i | A | j \rangle$ and $\langle i | B | j \rangle$, where i and j independently take on the values $1, 2, \ldots, m$. Then the direct product $A \times B$ is the $m^2 \times m^2$ matrix whose matrix elements are

$$\langle ii' | A \times B | jj' \rangle \equiv \langle i | A | j \rangle \langle i' | B | j' \rangle \tag{15.20}$$

This definition can be immediately extended to define the direct product $A \times B \times \cdots \times C$ of any number of $m \times m$ matrices $A, B, \ldots, C$:

$$\langle ii' \cdots i'' | A \times B \times \cdots \times C | jj' \cdots j'' \rangle$$
$$\equiv \langle i | A | j \rangle \langle i' | B | j' \rangle \cdots \langle i'' | C | j'' \rangle \tag{15.21}$$

If AB denotes the product of the matrices A and B under ordinary matrix multiplication, then

$$(A \times B)(C \times D) = (AC) \times (BD) \tag{15.22}$$

To prove this, take matrix elements of the left side:

$$\langle ii'|(A \times B)(C \times D)|jj'\rangle = \sum_{kk'} \langle ii'|A \times B|kk'\rangle\langle kk'|C \times D|jj'\rangle$$

$$= \sum_{k}\langle i|A|k\rangle\langle k|C|j\rangle \cdot \sum_{k'}\langle i'|B|k'\rangle\langle k'|C|j'\rangle$$

$$= \langle i|AC|j\rangle\langle i'|BD|j'\rangle$$

$$= \langle ii'|(AC) \times (BD)|jj'\rangle$$

A generalization of (15.22) can be proved in the same way:

$$(A \times B \times \cdots \times C)(D \times E \times \cdots \times F) = (AD) \times (BE) \times \cdots \times (CF) \tag{15.23}$$

Spin Matrices

We now introduce some special matrices in terms of which V_1', V_2, and V_3 may be conveniently expressed. Let the three familiar 2×2 Pauli spin matrices be denoted by X, Y, and Z:

$$X \equiv \begin{pmatrix} 0 & 1 \\ 1 & 0 \end{pmatrix}, \qquad Y \equiv \begin{pmatrix} 0 & -i \\ i & 0 \end{pmatrix}, \qquad Z \equiv \begin{pmatrix} 1 & 0 \\ 0 & -1 \end{pmatrix} \tag{15.24}$$

The following properties are easily verified

$$X^2 = 1, \quad Y^2 = 1, \qquad\qquad Z^2 = 1$$
$$XY + YX = 0, \quad YZ + ZY = 0, \quad ZX + XZ = 0 \tag{15.25}$$
$$XY = iZ, \quad YZ = iX, \qquad ZX = iY$$

Let three sets of $2^n \times 2^n$ matrices $X_\alpha, Y_\alpha, Z_\alpha$ ($\alpha = 1, \ldots, n$) be defined as follows[*]:

$$X_\alpha \equiv 1 \times 1 \times \cdots \times X \times \cdots \times 1 \qquad (n \text{ factors})$$
$$Y_\alpha \equiv 1 \times 1 \times \cdots \times Y \times \cdots \times 1 \qquad (n \text{ factors}) \tag{15.26}$$
$$Z_\alpha \equiv 1 \times 1 \times \cdots \times Z \times \cdots \times 1 \qquad (n \text{ factors})$$

$$\uparrow$$
$$\alpha\text{th factor}$$

For $\alpha \neq \beta$ we can easily verify that

$$[X_\alpha, X_\beta] = [Y_\alpha, Y_\beta] = [Z_\alpha, Z_\beta] = 0$$
$$[X_\alpha, Y_\beta] = [X_\alpha, Z_\beta] = [Y_\alpha, Z_\beta] = 0 \tag{15.27}$$

For any given α the $2^n \times 2^n$ matrices $X_\alpha, Y_\alpha, Z_\alpha$ formally satisfy all the relations (15.25).

[*]The matrices $X_\alpha, Y_\alpha, Z_\alpha$ are familiar in quantum mechanics. For example, for a system of n nonrelativistic electrons the spin matrices for the αth electron are precisely X_α, Y_α, and Z_α.

The following identity holds for any matrix X whose square is the unit matrix

$$e^{\theta X} = \cosh\theta + X \sinh\theta \tag{15.28}$$

where θ is a number. The proof is as follows. Since $X^n = 1$ if n is even, and $X^n = X$ if n is odd,

$$e^{\theta X} = \sum_{n=0}^{\infty} \frac{\theta^n}{n!} X^n = \sum_{n \text{ even}} \frac{\theta^n}{n!} + X \sum_{n \text{ odd}} \frac{\theta^n}{n!} = \cosh\theta + X \sinh\theta$$

In particular (15.28) is satisfied separately by X, Y, Z and by $X_\alpha, Y_\alpha, Z_\alpha$ ($\alpha = 1, \ldots, n$).

The Matrices V_1', V_2, and V

By inspection of (15.16) it is clear that V_1' is a direct product of n 2×2 identical matrices:

$$V_1' = a \times a \times \cdots \times a \tag{15.29}$$

where

$$\langle s | a | s' \rangle = e^{\beta\epsilon ss'} \tag{15.30}$$

Therefore

$$a = \begin{bmatrix} e^{\beta\epsilon} & e^{-\beta\epsilon} \\ e^{-\beta\epsilon} & e^{\beta\epsilon} \end{bmatrix} = e^{\beta\epsilon} + e^{-\beta\epsilon}X \tag{15.31}$$

Using (15.28) we obtain

$$a = \sqrt{2 \sinh(2\beta\epsilon)} \; e^{\theta X} \tag{15.32}$$

where

$$\tanh\theta \equiv e^{-2\beta\epsilon} \tag{15.33}$$

Hence

$$V_1' = [2 \sinh(2\beta\epsilon)]^{n/2} e^{\theta X} \times e^{\theta X} \times \cdots \times e^{\theta X} \tag{15.34}$$

The following identity can be verified by a direct calculation of matrix elements:

$$e^{\theta X} \times e^{\theta X} \times \cdots \times e^{\theta X} = e^{\theta X_1} e^{\theta X_2} \cdots e^{\theta X_n} = e^{\theta(X_1 + X_2 + \cdots + X_n)} \tag{15.35}$$

Applying (15.35) to (15.34) we obtain

$$V_1' = [2 \sinh(2\beta\epsilon)]^{n/2} V_1 \tag{15.36}$$

$$V_1 = \prod_{\alpha=1}^{n} e^{\theta X_\alpha}, \qquad \tanh\theta \equiv e^{-2\beta\epsilon} \tag{15.37}$$

A straightforward calculation of matrix elements shows that

$$V_2 = \prod_{\alpha=1}^{n} e^{\beta\epsilon Z_\alpha Z_{\alpha+1}} \tag{15.38}$$

$$V_3 = \prod_{\alpha=1}^{n} e^{\beta B Z_\alpha}, \qquad Z_{n+1} \equiv Z_1 \tag{15.39}$$

Therefore

$$P = [2 \sinh (2\beta\epsilon)]^{n/2} V_3 V_2 V_1 \tag{15.40}$$

For the case $H = 0$, $V_3 = 1$. This completes the formulation of the two-dimensional Ising model.

15.2 MATHEMATICAL DIGRESSION

The following study of a general class of matrices is relevant to the solution of the two-dimensional Ising model in the absence of magnetic field ($H = 0$).

Let $2n$ matrices Γ_μ ($\mu = 1, \ldots, 2n$) be defined as a set of matrices satisfying the following anticommutation rule

$$\Gamma_\mu \Gamma_\nu + \Gamma_\nu \Gamma_\mu = 2\delta_{\mu\nu} \qquad \begin{matrix} (\mu = 1, \ldots, 2n) \\ (\nu = 1, \ldots, 2n) \end{matrix} \tag{15.41}$$

The following properties of $\{\Gamma_\mu\}$ are stated without proof.*

(*a*) The dimensionality of Γ_μ cannot be smaller than $2^n \times 2^n$.

(*b*) If $\{\Gamma_\mu\}$ and $\{\Gamma'_\mu\}$ are two sets of matrices satisfying (15.41), there exists a nonsingular matrix S such that $\Gamma_\mu = S\Gamma'_\mu S^{-1}$. The converse is obviously true.

(*c*) Any $2^n \times 2^n$ matrix is a linear combination of the unit matrix, the matrices Γ_μ (chosen to be $2^n \times 2^n$), and all the independent products $\Gamma_\mu \Gamma_\nu, \Gamma_\mu \Gamma_\nu \Gamma_\lambda, \ldots$.

For $n = 1$, (15.41) defines two of the 2×2 Pauli spin matrices from which the third can be obtained as their product. It is obvious that any 2×2 matrix is a linear combination of the unit matrix and the Pauli spin matrices. For $n = 2$, (15.41) defines the four 4×4 Dirac matrices γ_μ.

A possible representation of $\{\Gamma_\mu\}$ by $2^n \times 2^n$ matrices is

$$\begin{matrix} \Gamma_1 = Z_1 & \Gamma_2 = Y_1 \\ \Gamma_3 = X_1 Z_2 & \Gamma_4 = X_1 Y_2 \\ \Gamma_5 = X_1 X_2 Z_3 & \Gamma_6 = X_1 X_2 Y_3 \\ \vdots & \vdots \end{matrix} \tag{15.42}$$

That is,

$$\begin{matrix} \Gamma_{2\alpha-1} = X_1 X_2 \cdots X_{\alpha-1} Z_\alpha & (\alpha = 1, \ldots, n) \\ \Gamma_{2\alpha} = X_1 X_2 \cdots X_{\alpha-1} Y_\alpha & (\alpha = 1, \ldots, n) \end{matrix} \tag{15.43}$$

An equally satisfactory representation is obtained by interchanging the roles of

---

*These general properties are not necessary for future developments, since we work with an explicit representation. A general study of (15.41) was made by R.Brauer and H. Weyl, *Am. J. Math.* **57**, 425 (1935).

X_α and Z_α ($\alpha = 1, \ldots, n$). It is also obvious that given any representation, such as (15.43), an equally satisfactory representation is obtained by an arbitrary permutation of the numbering of $\Gamma_1, \ldots, \Gamma_{2n}$.

It will presently be revealed that V_1 and V_2 are matrices that transform one set of $\{\Gamma_\mu\}$ into another equivalent set.

Let a definite set $\{\Gamma_\mu\}$ be given and let ω be the $2n \times 2n$ matrix describing a linear orthogonal transformation among the members of $\{\Gamma_\mu\}$:

$$\Gamma_\mu' = \sum_{\nu=1}^{2n} \omega_{\mu\nu} \Gamma_\nu \qquad (15.44)$$

where $\omega_{\mu\nu}$ are complex numbers satisfying

$$\sum_{\mu=1}^{2n} \omega_{\mu\nu} \omega_{\mu\lambda} = \delta_{\nu\lambda} \qquad (15.45)$$

This may be written in matrix form as

$$\omega^T \omega = 1 \qquad (15.46)$$

where ω^T is the transpose of ω. If Γ_μ is regarded as a component of a vector in a $2n$-dimensional space, then ω induces a rotation in that space:

$$\begin{bmatrix} \Gamma_1' \\ \Gamma_2' \\ \vdots \\ \Gamma_{2n}' \end{bmatrix} = \begin{bmatrix} \omega_{11} & \omega_{12} & \cdots & \omega_{1,2n} \\ \omega_{31} & \omega_{22} & \cdots & \omega_{2,2n} \\ \vdots & & & \\ \omega_{2n,1} & \omega_{2n,2} & \cdots & \omega_{2n,2n} \end{bmatrix} \begin{bmatrix} \Gamma_1 \\ \Gamma_2 \\ \vdots \\ \Gamma_{2n} \end{bmatrix} \qquad (15.47)$$

Substitution of (15.44) into (17.41) shows that the set $\{\Gamma_\mu'\}$ also satisfies (15.41), because of (15.45). Therefore

$$\Gamma_\mu' = S(\omega) \Gamma_\mu S^{-1}(\omega) \qquad (15.48)$$

where $S(\omega)$ is a nonsingular $2^n \times 2^n$ matrix. The existence of $S(\omega)$ will be demonstrated by explicit construction. Thus there is a correspondence

$$\omega \leftrightarrow S(\omega) \qquad (15.49)$$

which establishes $S(\omega)$ as a $2^n \times 2^n$ matrix representation of a rotation in a $2n$-dimensional space. Combining (15.48) and (15.44) we have

$$S(\omega) \Gamma_\mu S^{-1}(\omega) = \sum_{\nu=1}^{2n} \omega_{\mu\nu} \Gamma_\nu \qquad (15.50)$$

We call ω a *rotation* and $S(\omega)$ the *spin representative* of the rotation ω. It is obvious that if ω_1 and ω_2 are two rotations then $\omega_1 \omega_2$ is also a rotation. Furthermore

$$S(\omega_1 \omega_2) = S(\omega_1) S(\omega_2) \qquad (15.51)$$

We now study some special rotations ω and their corresponding $S(\omega)$. Consider a rotation in a two-dimensional plane of the $2n$-dimensional space. A rotation in the plane $\mu\nu$ through the angle θ is defined by the transformation.

$$\begin{cases} \Gamma'_\lambda = \Gamma_\lambda & (\lambda \neq \mu, \lambda \neq \nu) \\ \Gamma'_\mu = \Gamma_\mu \cos\theta - \Gamma_\nu \sin\theta & (\mu \neq \nu) \\ \Gamma'_\nu = \Gamma_\mu \sin\theta + \Gamma_\nu \cos\theta & (\mu \neq \nu) \end{cases} \qquad (15.52)$$

where θ is a complex number. The rotation matrix, denoted by $\omega(\mu\nu|\theta)$, is explicitly given by

$$\omega(\mu\nu|\theta) = \begin{bmatrix} & \overset{\mu\text{th}}{\underset{\text{column}}{\vdots}} & & \overset{\nu\text{th}}{\underset{\text{column}}{\vdots}} & \\ \cdots & \cos\theta & \cdots & \sin\theta & \cdots \\ & \vdots & & \vdots & \\ \cdots & -\sin\theta & \cdots & \cos\theta & \cdots \\ & \vdots & & \vdots & \end{bmatrix} \begin{matrix} \\ \mu\text{th row} \\ \\ \nu\text{th row} \\ \end{matrix} \qquad (15.53)$$

where the matrix elements not displayed are unity along the diagonal and zero everywhere else. Now $\omega(\mu\nu|\theta)$ is called *the plane rotation in the plane* $\mu\nu$. It is easily verified that

$$\begin{aligned} \omega(\mu\nu|\theta) &= \omega(\nu\mu|-\theta) \\ \omega^T(\mu\nu|\theta)\omega(\mu\nu|\theta) &= 1 \end{aligned} \qquad (15.54)$$

The properties of ω and $S(\omega)$ that are relevant to the solution of the Ising model are summarized in the following lemmas.*

LEMMA 1

If $\omega(\mu\nu|\theta) \leftrightarrow S_{\mu\nu}(\theta)$, then

$$S_{\mu\nu}(\theta) = e^{-1/2\theta\Gamma_\mu\Gamma_\nu} \qquad (15.55)$$

Proof Since $\Gamma_\mu\Gamma_\nu = -\Gamma_\nu\Gamma_\mu$ for $\mu \neq \nu$, $(\Gamma_\mu\Gamma_\nu)^2 = \Gamma_\mu\Gamma_\nu\Gamma_\mu\Gamma_\nu = -1$. An identity analogous to (15.28) is

$$e^{-1/2\theta\Gamma_\mu\Gamma_\nu} = \cos\frac{\theta}{2} - \Gamma_\mu\Gamma_\nu \sin\frac{\theta}{2}$$

Since $(\Gamma_\mu\Gamma_\nu)(\Gamma_\nu\Gamma_\mu) = (\Gamma_\nu\Gamma_\mu)(\Gamma_\mu\Gamma_\nu) = 1$, we have

$$e^{1/2\theta\Gamma_\mu\Gamma_\nu}e^{-1/2\theta\Gamma_\mu\Gamma_\nu} = e^{1/2\theta\Gamma_\mu\Gamma_\nu}e^{1/2\theta\Gamma_\nu\Gamma_\mu} = e^{1/2\theta(\Gamma_\mu\Gamma_\nu + \Gamma_\nu\Gamma_\mu)} = 1$$

Hence

$$S_{\mu\nu}^{-1}(\theta) = e^{1/2\theta\Gamma_\mu\Gamma_\nu} \qquad (15.56)$$

---

*It will be noted that the proofs of these lemmas make use only of the general property (15.41) and the special representation (15.42) of $\{\Gamma_\mu\}$.

A straightforward calculation shows that

$$S_{\mu\nu}(\theta)\Gamma_\lambda S_{\mu\nu}^{-1}(\theta) = \Gamma_\lambda \qquad (\lambda \neq \mu, \lambda \neq \nu)$$

$$S_{\mu\nu}(\theta)\Gamma_\mu S_{\mu\nu}^{-1}(\theta) = \Gamma_\mu \cos\theta + \Gamma_\nu \sin\theta$$

$$S_{\mu\nu}(\theta)\Gamma_\nu S_{\mu\nu}^{-1}(\theta) = \Gamma_\mu \sin\theta - \Gamma_\nu \cos\theta \qquad \blacksquare$$

LEMMA 2

The eigenvalues of $\omega(\mu\nu|\theta)$ are 1 ($2n - 2$-fold degenerate), and $e^{\pm i\theta}$ (nondegenerate). The eigenvalues of $S_{\mu\nu}(\theta)$ are $e^{\pm i\theta/2}$ (each 2^{n-1}-fold degenerate).

Proof The first part is trivial. The second part can be proved by choosing a special representation for $\Gamma_\mu\Gamma_\nu$, since the eigenvalues of $S_{\mu\nu}(\theta)$ are independent of the representation. As a representation for Γ_μ and Γ_ν we use (15.43) with X and Z interchanged. Since the number of the Γ_μ in (15.43) is not unique we are free to choose any two to be Γ_μ and Γ_ν. What we choose for the remaining $2n - 2$ matrices is irrelevant for the proof. We choose

$$\Gamma_\mu = Z_1 X_2$$

$$\Gamma_\nu = Z_1 Y_2$$

Then

$$\Gamma_\mu\Gamma_\nu = X_2 Y_2 = iZ_2 = 1 \times \begin{pmatrix} i & 0 \\ 0 & -i \end{pmatrix} \times 1 \times \cdots \times 1$$

Therefore

$$S_{\mu\nu}(\theta) = \cos\frac{\theta}{2} - \Gamma_\mu\Gamma_\nu \sin\frac{\theta}{2} = 1 \times \begin{pmatrix} e^{-i\theta/2} & 0 \\ 0 & e^{i\theta/2} \end{pmatrix} \times 1 \times \cdots \times 1$$

The matrix elements of $S_{\mu\nu}(\theta)$ in this representation are

$$\langle s_1, \ldots, s_n | S_{\mu\nu}(\theta) | s_1', \ldots, s_n' \rangle = e^{1/2 i\theta s_2} \prod_{k=1}^{n} \delta_{s_k s_k'}$$

Thus $S_{\mu\nu}(\theta)$ is diagonal. The diagonal elements are either $e^{i\theta/2}$ or $e^{-i\theta/2}$, each appearing the same number of times, i.e., 2^{n-1} times each. $\blacksquare$

LEMMA 3

Let ω be a product of n commuting plane rotations:

$$\omega = \omega(\alpha\beta|\theta_1)\omega(\gamma\delta|\theta_2) \cdots \omega(\mu\nu|\theta_n) \qquad (15.57)$$

where $\{\alpha, \beta, \ldots, \mu, \nu\}$ is a permutation of the set of integers $\{1, 2, \ldots, 2n - 1, 2n\}$, and $\theta_1, \ldots, \theta_n$ are complex numbers. Then

(a) $\omega \leftrightarrow S(\omega)$, with

$$S(\omega) = e^{-1/2\theta_1\Gamma_\alpha\Gamma_\beta} e^{-1/2\theta_2\Gamma_\gamma\Gamma_\delta} \cdots e^{-1/2\theta_n\Gamma_\mu\Gamma_\nu} \qquad (15.58)$$

(b) The $2n$ eigenvalues of ω are

$$e^{\pm i\theta_1}, e^{\pm i\theta_2}, \ldots, e^{\pm i\theta_n} \tag{15.59}$$

(c) The 2^n eigenvalues of $S(\omega)$ are the values

$$e^{\frac{1}{2}i(\pm\theta_1 \pm \theta_2 \pm \cdots \pm \theta_n)} \tag{15.60}$$

with the signs $\pm$ chosen independently.

Proof This lemma is an immediate consequence of lemmas 1 and 2 and the fact that $[\Gamma_\mu\Gamma_\nu, \Gamma_\alpha\Gamma_\beta] = 0$.   ∎

By this lemma, the eigenvalues of $S(\omega)$ can be immediately obtained from those of ω, if the eigenvalues of ω are of the form (15.59).

The usefulness of these lemmas rests on the fact that V_2V_1 can be expressed in terms of $S(\omega)$.

15.3 THE SOLUTION*

In the absence of an external magnetic field the formulas (15.14) and (15.40) lead to the following:

$$\lim_{N \to \infty} \frac{1}{N} \log Q_I(0, T) = \tfrac{1}{2} \log \left[2 \sinh(2\beta\epsilon) \right] + \lim_{n \to \infty} \frac{1}{n} \log \Lambda \tag{15.61}$$

where

$$\Lambda = \text{largest eigenvalue of } V \tag{15.62}$$

and

$$V = V_1V_2 \tag{15.63}$$

where V_1 is given by (15.37) and V_2 by (15.38). These formulas are valid if all eigenvalues of V are positive and if $\lim n^{-1} \log \Lambda$ exists. Our main task is to diagonalize the matrix V. *Throughout the present section, all matrices are understood to be matrices in a definite representation.*

Expression of V in Terms of Spin Representatives

Using the representation (15.42), we note that

$$\Gamma_{2\alpha}\Gamma_{2\alpha-1} = Y_\alpha Z_\alpha = iX_\alpha \qquad (\alpha = 1, \ldots, n) \tag{15.64}$$

---

*L. Onsager, *Phys. Rev.* **65**, 117 (1944); B. Kaufmann, *Phys. Rev.* **76**, 1232 (1949). The present account follows Kaufmann's treatment.

From (15.37) we immediately have

$$V_1 = \prod_{\alpha=1}^{n} e^{\theta X_\alpha} = \prod_{\alpha=1}^{n} e^{-i\theta \Gamma_{2\alpha} \Gamma_{2\alpha-1}} \tag{15.65}$$

Thus V_1 is a spin representative of a product of commuting plane rotations.

Again, from (15.42),

$$\begin{aligned}\Gamma_{2\alpha+1}\Gamma_{2\alpha} &= X_\alpha Z_{\alpha+1} Y_\alpha = iZ_\alpha Z_{\alpha+1} \quad (\alpha = 1,\ldots, n-1)\\ \Gamma_1\Gamma_{2n} &= Z_1(X_1 \cdots X_{n-1})Y_n = -iZ_1 Z_n(X_1 \cdots X_n)\end{aligned} \tag{15.66}$$

By (15.38),

$$V_2 = \left[\prod_{\alpha=1}^{n-1} e^{\beta\epsilon Z_\alpha Z_{\alpha+1}}\right] e^{\beta\epsilon Z_n Z_1}$$

The last factor commutes with the bracket. Therefore we can write

$$V_2 = e^{\beta\epsilon Z_n Z_1}\left[\prod_{\alpha=1}^{n-1} e^{\beta\epsilon Z_\alpha Z_{\alpha+1}}\right] = e^{i\beta\epsilon U \Gamma_1 \Gamma_{2n}} \sum_{\alpha=1}^{n-1} e^{-i\beta\epsilon \Gamma_{2\alpha+1} \Gamma_{2\alpha}} \tag{15.67}$$

where

$$U \equiv X_1 X_2 \cdots X_n \tag{15.68}$$

Were it not for the first factor in (15.67), V_2 would also be the spin representative of a product of commuting plane rotations. This factor owes its existence to the toroidal boundary condition imposed on the problem (i.e., the condition that $s_{n+1} \equiv s_1$ in every row of the lattice). At first sight this condition seems to be an unnecessary and artificial complication, but it actually simplifies our future task.

Substituting (15.67) and (15.65) into (15.63) we obtain

$$V \equiv V_2 V_1 = e^{i\phi U \Gamma_1 \Gamma_{2n}}\left[\prod_{\alpha=1}^{n-1} e^{-i\phi \Gamma_{2\alpha+1}\Gamma_{2\alpha}}\right]\left[\prod_{\lambda=1}^{n} e^{-i\theta \Gamma_{2\lambda}\Gamma_{2\lambda-1}}\right] \tag{15.69}$$

where

$$\phi = \beta\epsilon, \qquad \epsilon > 0, \qquad \theta \equiv \tanh^{-1} e^{-2\phi}$$

Some relevant properties of U are

(a) $U^2 = 1, \qquad U(1+U) = 1+U, \qquad U(1-U) = -(1-U)$
$$\tag{15.70}$$

(b) $U = i^n \Gamma_1 \Gamma_2 \cdots \Gamma_{2n} \tag{15.71}$

(c) U commutes with a product of an even number of Γ_μ and anticommutes with a product of an odd number of Γ_μ.* A simple calculation shows that

$$\begin{aligned}e^{i\phi\Gamma_1\Gamma_{2n}U} &= \left[\tfrac{1}{2}(1+U) + \tfrac{1}{2}(1-U)\right]\left[\cosh\phi + i\Gamma_1\Gamma_{2n}U\sinh\phi\right]\\ &= \tfrac{1}{2}(1+U)\left[\cosh\phi + i\Gamma_1\Gamma_{2n}\sinh\phi\right]\\ &\quad + \tfrac{1}{2}(1-U)\left[\cosh\phi - i\Gamma_1\Gamma_{2n}\sinh\phi\right]\\ &= \tfrac{1}{2}(1+U)e^{i\phi\Gamma_1\Gamma_{2n}} + \tfrac{1}{2}(1-U)e^{-i\phi\Gamma_1\Gamma_{2n}}\end{aligned} \tag{15.72}$$

---

*This is an immediate consequence of (15.71) and (15.41). For $n = 2$, U is commonly denoted by $\gamma_5 \equiv \gamma_1\gamma_2\gamma_3\gamma_4$.

Substituting this result into (15.69) we obtain

$$V = \tfrac{1}{2}(1 + U)V^+ + \tfrac{1}{2}(1 - U)V^- \tag{15.73}$$

where

$$V^\pm \equiv e^{\pm i\phi\Gamma_1\Gamma_{2n}} \left[\prod_{\alpha=1}^{n-1} e^{-i\phi\Gamma_{2\alpha+1}\Gamma_{2\alpha}} \right] \left[\prod_{\lambda=1}^{n} e^{-i\theta\Gamma_{2\lambda}\Gamma_{2\lambda-1}} \right] \tag{15.74}$$

Thus both V^+ and V^- are spin representatives of rotations.

Representation in Which U Is Diagonal

It is obvious that the three matrices U, V^+, and V^- commute with one another. Hence they can be simultaneously diagonalized. We first transform V into the representation in which U is diagonal (but in which $V^\pm$ are not necessarily diagonal):

$$RVR^{-1} \equiv \tilde{V} = \tfrac{1}{2}(1 + \tilde{U})\tilde{V}^+ + \tfrac{1}{2}(1 - \tilde{U})\tilde{V}^- \tag{15.75}$$

$$\tilde{U} \equiv RUR^{-1} \tag{15.76}$$

$$\tilde{V}^\pm \equiv RV^\pm R^{-1} \tag{15.77}$$

Since $U^2 = 1$, the eigenvalues of U are either $+1$ or -1. From (15.68) it is seen that U can also be written in the form $U = X \times X \times \cdots \times X$. Therefore a diagonal form of U is $Z \times Z \times \cdots \times Z$, and the eigenvalues $+1$ and -1 occur with equal frequency. Other diagonal forms of U may be obtained by permuting the relative positions of the eigenvalues along the diagonal. We shall choose R in such a way that all the eigenvalues $+1$ are in one submatrix, and -1 in the other, so that the matrix $\tilde{U}$ can be represented in the form

$$\tilde{U} = \begin{pmatrix} 1 & 0 \\ 0 & -1 \end{pmatrix} \tag{15.78}$$

where 1 is the $2^{n-1} \times 2^{n-1}$ unit matrix. Since $\tilde{V}^\pm$ commute with $\tilde{U}$, they must have the forms

$$\tilde{V}^\pm = \begin{pmatrix} \mathfrak{A}^\pm & 0 \\ 0 & \mathfrak{B}^\pm \end{pmatrix} \tag{15.79}$$

where $\mathfrak{A}^\pm$ and $\mathfrak{B}^\pm$ are $2^{n-1} \times 2^{n-1}$ matrices are not necessarily diagonal. It is now clear that $\tfrac{1}{2}(1 + \tilde{U})$ annihilates the lower submatrix and $\tfrac{1}{2}(1 - \tilde{U})$ annihilates the upper submatrix:

$$\tfrac{1}{2}(1 + \tilde{U})\tilde{V}^+ = \begin{pmatrix} \mathfrak{A}^+ & 0 \\ 0 & 0 \end{pmatrix} \tag{15.80}$$

$$\tfrac{1}{2}(1 - \tilde{U})\tilde{V}^- = \begin{pmatrix} 0 & 0 \\ 0 & \mathfrak{B}^- \end{pmatrix} \tag{15.81}$$

Therefore

$$\tilde{V} = \begin{pmatrix} \mathfrak{A}^+ & 0 \\ 0 & \mathfrak{B}^- \end{pmatrix} \tag{15.82}$$

To diagonalize V, it is sufficient to diagonalize $\tilde{V}$, which has the same set of eigenvalues as V. To diagonalize V it is sufficient to diagonalize (15.80) and (15.81) *separately and independently*, for each of them has only n nonzero eigenvalues. The combined set of their nonzero eigenvalues constitutes the set of eigenvalues of V.

To diagonalize (15.80) and (15.81), we first diagonalize $\tilde{V}^+$ and $\tilde{V}^-$ separately and independently, thereby obtaining twice too many eigenvalues for each. To obtain the eigenvalues of (15.80) and (15.81), we would then decide which eigenvalues so obtained are to be discarded. This last step will not be necessary, however, for we shall show that as $n \to \infty$ a knowledge of the eigenvalues of $\tilde{V}^+$ and $\tilde{V}^-$ suffices to determine the largest eigenvalue of V. The set of eigenvalues of $\tilde{V}^\pm$, however, is respectively equal to the set of eigenvalues of $V^\pm$. Therefore we shall diagonalize V^+ and V^- *separately and independently*.

Eigenvalues of V^+ and V^-

To find the eigenvalues of V^+ and V^- we first find the eigenvalues of the rotations, of which V^+ and V^- are spin representatives. These rotations shall be respectively denoted by Ω^+ and Ω^-, which are both $2n \times 2n$ matrices:

$$V^\pm \leftrightarrow \Omega^\pm \tag{15.83}$$

From (15.74) we immediately have

$$\Omega^\pm = \omega(1, 2n| \mp 2i\phi)\left[\prod_{\alpha=1}^{n-1} \omega(2\alpha + 1, 2\alpha| - 2i\phi)\right]\left[\prod_{\lambda=1}^{n} \omega(2\lambda, 2\lambda - 1| - 2i\theta)\right] \tag{15.84}$$

where $\omega(\mu\nu|\alpha) = \omega(\nu\mu| - \alpha)$ is the plane rotation in the plane $\mu\nu$ through the angle α and is defined by (15.53). The eigenvalues of $\Omega^\pm$ are clearly the same as that of

$$\omega^\pm \equiv \Delta\Omega^\pm \Delta^{-1} \tag{15.85}$$

where Δ is the square root of the last factor in (15.84):

$$\Delta \equiv \sqrt{\prod_{\lambda=1}^{n} \omega(2\lambda, 2\lambda - 1| - 2i\theta)} = \prod_{\lambda=1}^{n} \omega(2\lambda, 2\lambda - 1| - i\theta) \tag{15.86}$$

Thus

$$\omega^\pm = \Delta\chi^\pm \Delta$$
$$\Delta = \omega(12|i\theta)\omega(34|i\theta) \cdots \omega(2n - 1, 2n|i\theta)$$
$$\chi^\pm = \omega(1, 2n| \pm 2i\phi)[\omega(23|2i\phi)\omega(45|2i\phi) \cdots \omega(2n - 2, 2n - 1|2i\phi)] \tag{15.87}$$

Explicitly,

$$
\Delta = \begin{bmatrix} \boxed{J} & \begin{matrix} 0 & 0 \\ 0 & 0 \end{matrix} & \cdots \\ \begin{matrix} 0 & 0 \\ 0 & 0 \end{matrix} & \boxed{J} & \\ \begin{matrix} \vdots & \vdots \end{matrix} & & \ddots \\ & & & \boxed{J} \end{bmatrix}, \qquad \boxed{J} \equiv \begin{pmatrix} \cosh\theta & i\sinh\theta \\ -i\sinh\theta & \cosh\theta \end{pmatrix}
$$

$$(15.88)$$

$$
\chi^{\pm} = \begin{bmatrix} a & 0 & 0 & \cdots & & & \pm b \\ 0 & \boxed{K} & & & & & \\ 0 & & & \boxed{K} & & & \\ \vdots & & & & \ddots & & \vdots \\ & & & & & \boxed{K} & \begin{matrix} 0 \\ 0 \end{matrix} \\ \mp b & & \cdots & 0 & 0 & & a \end{bmatrix},
\qquad
\begin{aligned} \boxed{K} &\equiv \begin{pmatrix} \cosh 2\phi & i\sinh 2\phi \\ -i\sinh 2\phi & \cosh 2\phi \end{pmatrix} \\ a &\equiv \cosh 2\phi \\ b &\equiv i\sinh 2\phi \end{aligned}
$$

$$(15.89)$$

Performing the matrix multiplication $\Delta\chi^{\pm}\Delta$ in a straightforward way, we obtain

$$
\omega^{\pm} = \begin{bmatrix} A & B & 0 & 0 & \cdots & 0 & \mp B^* \\ B^* & A & B & 0 & & 0 & 0 \\ 0 & B^* & A & B & & & \\ \vdots & & & & & & \vdots \\ 0 & 0 & & & & & \\ \mp B & 0 & & \cdots & & B^* & A \end{bmatrix} \qquad (15.90)
$$

where A and B are 2×2 matrices given by

$$
A \equiv \begin{pmatrix} \cosh 2\phi \cosh 2\theta & -i\cosh 2\phi \sinh 2\theta \\ i\cosh 2\phi \sinh 2\theta & \cosh 2\phi \cosh 2\theta \end{pmatrix} \qquad (15.91)
$$

$$
B \equiv \begin{pmatrix} -\tfrac{1}{2}\sinh 2\phi \sinh 2\theta & i\sinh 2\phi \sinh^2\theta \\ -i\sinh 2\phi \cosh^2\theta & -\tfrac{1}{2}\sinh 2\phi \sinh 2\theta \end{pmatrix} \qquad (15.92)
$$

and B^* is the Hermitian conjugate of B.

To find the eigenvalues of $\omega^{\pm}$, try the following form for an eigenvector of $\omega^{\pm}$:

$$
\psi = \begin{bmatrix} zu \\ z^2 u \\ \vdots \\ z^n u \end{bmatrix} \qquad (15.93)
$$

where z is a number and u is a two-component vector

$$u = \begin{pmatrix} u_1 \\ u_2 \end{pmatrix} \tag{15.94}$$

The requirement that

$$\omega^{\pm}\psi = \lambda\psi \tag{15.95}$$

leads to the following eigenvalue equations:

$$(zA + z^2B \mp z^nB^*)u = z\lambda u$$
$$(z^2A + z^3B + zB^*)u = z^2\lambda u$$
$$(z^3A + z^4B + z^2B^*)u = z^3\lambda u$$
$$\vdots$$
$$(z^{n-1}A + z^nB + z^{n-2}B^*)u = z^{n-1}\lambda u$$
$$(z^nA \mp zB + z^{n-1}B^*)u = z^n\lambda u$$

The second through the $(n-1)$th equations are identical. Hence there are only three independent equations:

$$(A + zB \mp z^{n-1}B^*)u = \lambda u$$
$$(A + zB + z^{-1}B^*)u = \lambda u \tag{15.96}$$
$$(A \mp z^{1-n}B + z^{-1}B^*)u = \lambda u$$

These equations are solved by putting

$$z^n = \mp 1 \tag{15.97}$$

The three equations (15.96) then become the same one:

$$(A + zB + z^{-1}B^*)u = \lambda u \tag{15.98}$$

where the sign $\mp$ in (15.97) is associated with $\omega^{\pm}$. Thus, for ω^+ and for ω^-, there are n values of z:

$$z_k = e^{2i\pi k/n} \qquad (k = 0, 1, \ldots, 2n - 1) \tag{15.99}$$

where

$$k = 1, 3, 5, \ldots, 2n - 1 \quad \text{(for } \omega^+) $$
$$k = 0, 2, 4, \ldots, 2n - 2 \quad \text{(for } \omega^-) \tag{15.100}$$

For each k, two eigenvalues λ_k are determined by the equation

$$(A + z_kB + z_k^{-1}B^*)u = \lambda_k u \tag{15.101}$$

and λ_k is to be associated with $\omega^{\pm}$ according to (15.100). This determines $2n$ eigenvalues each for $\omega^{\pm}$.

To find λ_k, note that according to (15.91) and (15.92)

$$\det |A| = 1, \qquad \det |B| = \det |B^*| = 0$$

$$\det |A + z_kB + z_k^{-1}B^*| = 1$$

Therefore the two values of λ_k must have the forms

$$\lambda_k = e^{\pm \gamma_k} \qquad (k = 0, 1, \ldots, 2n - 1) \qquad (15.102)$$

The value of γ_k may be found from the equation

$$\tfrac{1}{2}\operatorname{Tr}\left(A + z_k B + z_k^{-1} B^*\right) = \tfrac{1}{2}\left(e^{\gamma_k} + e^{-\gamma_k}\right) = \cosh \gamma_k \qquad (15.103)$$

Evaluating the trace with the help of (15.91), (15.92), and (15.99) we obtain

$$\cosh \gamma_k = \cosh 2\phi \cosh 2\theta - \cos \frac{\pi k}{n} \sinh 2\phi \sinh 2\theta$$

$$(k = 0, 1, \ldots, 2n - 1) \quad (15.104)$$

if γ_k is a solution to (15.104) then $-\gamma_k$ is also a solution. But this possibility has already been taken into account in (15.102). Therefore we define γ_k to be the positive solution of (15.104).

It is easily verified that

$$\gamma_k = \gamma_{2n-k}$$
$$0 < \gamma_0 < \gamma_1 < \cdots < \gamma_n \qquad (15.105)$$

The first is obvious. The second can be seen by noting that $\partial \gamma_k / \partial k = (\pi/n)\sin(\pi k/n)/\sin \gamma_k$, which is positive for $k \le n$. A plot of γ_k as a function of ϕ is shown in Fig. 15.3. As $n \to \infty$ these curves merge into a continuum.

The eigenvalues of $\Omega^\pm$ are the same as those of $\omega^\pm$, respectively. Therefore $\Omega^\pm$ are products of commuting plane rotations, although this fact is not obvious from (15.84).

The 2^n eigenvalues of $V^\pm$ can now be written down immediately with the help of lemma 3 of the last section:

$$\text{eigenvalues of } V^- \text{ are } e^{1/2(\pm \gamma_0 \pm \gamma_2 \pm \gamma_4 \pm \cdots \pm \gamma_{2n-2})} \qquad (15.106)$$

$$\text{eigenvalues of } V^+ \text{ are } e^{1/2(\pm \gamma_1 \pm \gamma_3 \pm \gamma_5 \pm \cdots \pm \gamma_{2n-1})} \qquad (15.107)$$

where all possible choices of the signs $\pm$ are to be made independently.

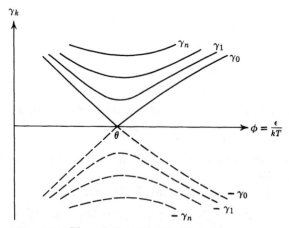

Fig. 15.3 The solutions of (15.104).

EIGENVALUES OF V

As we have explained, the set of eigenvalues of V consists of one-half the set of eigenvalues of V^+ and one-half that of V^-. The eigenvalues of $V^{\pm}$ are all positive and of order e^n. Therefore all eigenvalues of V are positive and of order e^n. This justifies the formula (15.61). To find explicitly the set of eigenvalues of V, it would be necessary to decide which half of the set of eigenvalues of V^+ and V^- should be discarded. We are only interested, however, in the largest eigenvalue of V. For such a purpose it is not necessary to carry out this task.

Suppose the $2^n \times 2^n$ matrix F transforms (15.80) into diagonal form and the $2^n \times 2^n$ matrix G transforms (15.81) into diagonal form:

$$F\left[\tfrac{1}{2}(1 + \tilde{U})\tilde{V}^+\right]F^{-1} = V_D^+ \tag{15.108}$$

$$G\left[\tfrac{1}{2}(1 - \tilde{U})\tilde{V}^-\right]G^{-1} = V_D^- \tag{15.109}$$

where $V_D^{\pm}$ are diagonal matrices with half the eigenvalues of (15.106) and (15.107), respectively, appearing along the diagonal. It is possible to choose F and G in such a way that $F\tilde{U}F^{-1}$ and $G\tilde{U}G^{-1}$ remain diagonal matrices. Then F and G merely permute the eigenvalues of $\tilde{U}$ along the diagonal. But the convention has been adopted that $\tilde{U}$ has the form (15.78). Hence F and G either leave $\tilde{U}$ unchanged or simply interchange the two submatrices 1 and -1 in (15.78). That is, F and G either commute or anitcommute with $\tilde{U}$. This means that

$$V_D^+ = \tfrac{1}{2}(1 \pm \tilde{U})F\tilde{V}^+F^{-1} \tag{15.110}$$

$$V_D^- = \tfrac{1}{2}(1 \pm \tilde{U})G\tilde{V}^-G^{-1} \tag{15.111}$$

where the signs $\pm$ can be definitely determined by an explicit calculation.* For our purpose this determination is not necessary.

We may write

$$\tfrac{1}{2}(1 \pm \tilde{U}) = \tfrac{1}{2}(1 \pm Z_1 Z_2 \cdots Z_n) \tag{15.112}$$

$$F\tilde{V}^+F^{-1} = \prod_{k=1}^{n} e^{1/2\gamma_{2k-1}Z_{Pk}} \tag{15.113}$$

$$G\tilde{V}^-G^{-1} = \prod_{k=1}^{n} e^{1/2\gamma_{2k-2}Z_{Qk}} \tag{15.114}$$

where P and Q are two definite permutations (though so far unknown) of the integers $1, 2, \ldots, n$. The permutation P sends k into Pk and Q sends k into Qk. The forms (15.113) and (15.114) are arrived at by noting that $F\tilde{V}^+F^{-1}$ and $G\tilde{V}^-G^{-1}$ must respectively have the same eigenvalues as V^+ and V^-, except for possible different orderings of the eigenvalues. Since the eigenvalues of Z_k are ± 1, we have

$$\tfrac{1}{2}(1 \pm \tilde{U}) = \begin{cases} 1 & \text{(if an even number of } Z_k \text{ are } \pm 1) \\ 0 & \text{(if an odd number of } Z_k \text{ are } \pm 1) \end{cases} \tag{15.115}$$

*B. Kaufmann (*loc. cit.*) shows that the plus sign is to be chosen in both (15.110) and (15.111).

This condition is invariant under any permutation that sends $\{Z_k\}$ into $\{Z_{Pk}\}$. Therefore the eigenvalues of (15.110) consist of those eigenvalues (15.106) for which an even (odd) number of $-$ signs appear in the exponents, if the $+(-)$ sign is chosen in (15.110). From (15.106) and (15.105), we conclude that the

$$\text{largest eigenvalue of } V_D^+ = e^{1/2(\pm \gamma_0 + \gamma_2 + \gamma_4 + \cdots + \gamma_{2n-2})}$$

where the $\pm$ sign corresponds to the $\pm$ sign in (15.109). As $n \to \infty$, these two possibilities give the same result, for γ_0 is negligible compared to the entire exponent in the last equation. A similar conclusion can be reached for V_D^-. Therefore we conclude that as $n \to \infty$ the

$$\text{largest eigenvalue of } V_D^+ = e^{1/2(\gamma_0 + \gamma_2 + \gamma_4 + \cdots + \gamma_{2n-2})}$$
$$\text{largest eigenvalue of } V_D^- = e^{1/2(\gamma_1 + \gamma_3 + \gamma_5 + \cdots + \gamma_{2n-1})} \tag{15.116}$$

The largest eigenvalue of V is the larger one of (15.116), which by (15.105) is that for V_D^-. Therefore the largest eigenvalue of V is

$$\Lambda = e^{1/2(\gamma_1 + \gamma_3 + \gamma_5 + \cdots + \gamma_{2n-1})} \tag{15.117}$$

The Largest Eigenvalue of V

It is now necessary to evaluate explicitly the largest eigenvalue of V. Using (15.117), we obtain

$$\mathcal{L} \equiv \lim_{n \to \infty} \frac{1}{n} \log \Lambda = \lim_{n \to \infty} \frac{1}{2n} (\gamma_1 + \gamma_3 + \gamma_5 + \cdots + \gamma_{2n-1}) \tag{15.118}$$

Let

$$\gamma(\nu) \equiv \gamma_{2k-1}$$
$$\nu \equiv \frac{\pi}{n}(2k-1) \tag{15.119}$$

As $n \to \infty$, ν becomes a continuous variable, and we have

$$\sum_{k=1}^{n} \gamma_{2k-1} \to \frac{n}{2\pi} \int_0^{2\pi} d\nu\, \gamma(\nu)$$

Therefore

$$\mathcal{L} = \frac{1}{4\pi} \int_0^{2\pi} d\nu\, \gamma(\nu) = \frac{1}{2\pi} \int_0^{\pi} d\nu\, \gamma(\nu) \tag{15.120}$$

where the last step results from (15.105), which states that $\gamma(\nu) = \gamma(2\pi - \nu)$. To express $\mathcal{L}$ in a more convenient form, we recall that $\gamma(\nu)$ is the positive solution of the equation

$$\cosh \gamma(\nu) = \cosh 2\phi \cosh 2\theta - \cos \nu \sinh 2\phi \sinh 2\theta \tag{15.121}$$

with

$$\phi \equiv \beta\epsilon, \quad \epsilon > 0$$
$$\theta \equiv \tanh^{-1} e^{-2\phi} \tag{15.122}$$

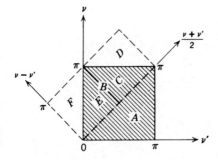

Fig. 15.4 Region of integration in (15.127).

A straightforward calculation shows that

$$\sinh 2\theta = \frac{1}{\sinh 2\phi} \tag{15.123}$$

$$\cosh 2\theta = \coth 2\phi$$

Hence (15.121) can also be written as

$$\cosh \gamma(\nu) = \cosh 2\phi \coth 2\phi - \cos \nu \tag{15.124}$$

We find the following identity helpful in the reduction of (15.120):

$$|z| = \frac{1}{\pi} \int_0^{\pi} dt \, \log \left(2 \cosh z - 2 \cos t \right) \tag{15.125}$$

With its help we see immediately that $\gamma(\nu)$ has the integral representation

$$\gamma(\nu) = \frac{1}{\pi} \int_0^{\pi} d\nu' \, \log \left(2 \cosh 2\phi \coth 2\phi - 2 \cos \nu - 2 \cos \nu' \right) \tag{15.126}$$

Therefore

$$\mathscr{L} = \frac{1}{2\pi^2} \int_0^{\pi} d\nu \int_0^{\pi} d\nu' \, \log \left[2 \cosh 2\phi \coth 2\phi - 2(\cos \nu + \cos \nu') \right] \tag{15.127}$$

The double integral in (15.127) extends over the shaded square in the $\nu\nu'$ plane shown in Fig. 15.4. It is obvious* that the integral remains unchanged if we let the region of integration be the rectangle shown in dotted lines, which corresponds to the range of integration

$$0 \le \frac{\nu + \nu'}{2} \le \pi \tag{15.128}$$

$$0 \le \nu - \nu' \le \pi$$

Let

$$\delta_1 \equiv \frac{\nu + \nu'}{2}, \tag{15.129}$$

$$\delta_2 \equiv \nu - \nu'$$

---

*Since the integrand is symmetric in ν and ν', it is clear that $\int_A = \int_B$ in Fig. 15.4. Noting that $\int_C = \int_D$ and $\int_E = \int_F$, we arrive at the conclusion.

Then

$$\mathscr{L} = \frac{1}{2\pi^2} \int_0^\pi d\delta_1 \int_0^\pi d\delta_2 \log\left(2\cosh 2\phi \coth 2\phi - 4\cos\delta_1 \cos\tfrac{1}{2}\delta_2\right)$$

$$= \frac{1}{\pi^2} \int_0^\pi d\delta_1 \int_0^{\pi/2} d\delta_2 \log\left(2\cosh 2\phi \coth 2\phi - 4\cos\delta_1 \cos\delta_2\right)$$

$$= \frac{1}{\pi^2} \int_0^\pi d\delta_1 \int_0^{\pi/2} d\delta_2 \log\left(2\cos\delta_2\right)$$

$$+ \frac{1}{\pi^2} \int_0^\pi d\delta_1 \int_0^{\pi/2} d\delta_2 \log\left(\frac{D}{\cos\delta_2} - 2\cos\delta_1\right)$$

$$= \frac{1}{\pi} \int_0^{\pi/2} d\delta_2 \log\left(2\cos\delta_2\right) + \frac{1}{\pi} \int_0^{\pi/2} d\delta_2 \cosh^{-1}\frac{D}{2\cos\delta_2}$$

where

$$D \equiv \cosh 2\phi \coth 2\phi \tag{15.130}$$

and where the identity (15.125) has been used once more. Since $\cosh^{-1} x = \log\left[x + \sqrt{x^2 - 1}\right]$, we can write

$$\mathscr{L} = \frac{1}{2\pi} \int_0^\pi d\delta \log\left[D\left(1 + \sqrt{1 + \kappa^2 \cos^2\delta}\right)\right]$$

where

$$\kappa \equiv \frac{2}{D} \tag{15.131}$$

It is clear that in the last integral $\cos^2\delta$ may be replaced by $\sin^2\delta$ without altering the value of the integral. Therefore

$$\mathscr{L} = \tfrac{1}{2}\log\left(\frac{2\cosh^2 2\beta\epsilon}{\sinh 2\beta\epsilon}\right) + \frac{1}{2\pi} \int_0^\pi d\phi \log\tfrac{1}{2}\left(1 + \sqrt{1 - \kappa^2 \sin^2\phi}\right) \tag{15.132}$$

Thermodynamic Functions

From (15.6), (15.118), and (15.132) we obtain the Helmholtz free energy per spin $a_I(0, T)$:

$$\beta a_I(0, T) = -\log\left(2\cosh 2\beta\epsilon\right) - \frac{1}{2\pi} \int_0^\pi d\phi \log\tfrac{1}{2}\left(1 + \sqrt{1 - \kappa^2 \sin^2\phi}\right) \tag{15.133}$$

The internal energy per spin is

$$u_I(0, T) = \frac{d}{d\beta}\left[\beta a_I(0, T)\right] = -2\epsilon\tanh 2\beta\epsilon + \frac{\kappa}{2\pi}\frac{d\kappa}{d\beta}\int_0^\pi d\phi\frac{\sin^2\phi}{\Delta(1 + \Delta)} \tag{15.134}$$

where $\Delta \equiv \sqrt{1 - \kappa^2 \sin^2 \phi}$. It is easily seen that

$$\int_0^\pi d\phi \, \frac{\sin^2 \phi}{\Delta(1 + \Delta)} = -\frac{\pi}{\kappa^2} + \frac{1}{\kappa^2} \int_0^\pi \frac{d\phi}{\Delta}$$

Therefore

$$u_I(0, T) = -2\epsilon \tanh 2\beta\epsilon + \frac{1}{2\kappa} \frac{d\kappa}{d\beta} \left[-1 + \frac{1}{\pi} \int_0^\pi \frac{d\phi}{\sqrt{1 - \kappa^2 \sin^2 \phi}} \right] \quad (15.135)$$

From (15.131) we obtain

$$\frac{1}{\kappa} \frac{d\kappa}{d\beta} = -2\epsilon \coth 2\beta\epsilon (2 \tanh^2 2\beta\epsilon - 1) \quad (15.136)$$

$$-2\epsilon \tanh 2\beta\epsilon - \frac{1}{2\kappa} \frac{d\kappa}{d\beta} = -\epsilon \coth 2\beta\epsilon \quad (15.137)$$

Thus finally

$$\mu_I(0, T) = -\epsilon \coth 2\beta\epsilon \left[1 + \frac{2}{\pi} \kappa' K_1(\kappa) \right] \quad (15.138)$$

where $K_1(\kappa)$ is a tabulated function, the complete elliptic integral of the first kind*:

$$K_1(\kappa) \equiv \int_0^{\pi/2} \frac{d\phi}{\sqrt{1 - \kappa^2 \sin^2 \phi}} \quad (15.139)$$

and

$$\kappa \equiv \frac{2 \sinh 2\beta\epsilon}{\cosh^2 2\beta\epsilon} \quad (15.140)$$

$$\kappa' \equiv 2 \tanh^2 2\beta\epsilon - 1 \quad (15.141)$$

$$\kappa^2 + \kappa'^2 = 1 \quad (15.142)$$

Graphs of κ and κ' are shown in Fig. 15.5.

The specific heat $c_I(0, T)$ is readily shown to be

$$\frac{1}{k} c_I(0, T) = \frac{2}{\pi} (\beta\epsilon \coth 2\beta\epsilon)^2 \left\{ 2K_1(\kappa) - 2E_1(\kappa) - (1 - \kappa') \left[\frac{\pi}{2} + \kappa' K_1(\kappa) \right] \right\} \quad (15.143)$$

where $E_1(\kappa)$ is a tabulated function, the complete elliptic integral of the second kind:

$$E_1(\kappa) \equiv \int_0^{\pi/2} d\phi \sqrt{1 - \kappa^2 \sin^2 \phi} \quad (15.144)$$

*See, for example, H. Hancock, *Elliptic Integrals* (Dover, New York, 1958).

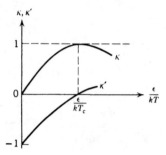

Fig. 15.5 The functions κ and κ'.

The elliptic integral $K_1(\kappa)$ has a singularity at $\kappa = 1$ (or $\kappa' = 0$), in which neighborhood

$$K_1(\kappa) \approx \log \frac{4}{\kappa'}, \qquad \frac{dK_1(\kappa)}{d\kappa} \approx \frac{\pi}{2}$$

$$E_1(\kappa) \approx 1 \tag{15.145}$$

Thus all thermodynamic functions have a singularity of some kind at $T = T_c$, where T_c is such that

$$2 \tanh^2 \frac{2\epsilon}{kT_c} = 1$$

$$\frac{\epsilon}{kT_c} = 0.440\,686\,8 \tag{15.146}$$

$$kT_c = (2.269\,185)\epsilon$$

Other relations satisfied by T_c are

$$e^{-\epsilon/kT_c} = \sqrt{2} - 1$$

$$\cosh \frac{2\epsilon}{kT_c} = \sqrt{2} \tag{15.147}$$

$$\sinh \frac{2\epsilon}{kT_c} = 1$$

Thus near $T = T_c$

$$\frac{1}{k} c_I(0, T) \approx \frac{2}{\pi} \left(\frac{2\epsilon}{kT_c} \right)^2 \left[-\log \left| 1 - \frac{T}{T_c} \right| + \log \left(\frac{kT_c}{2\epsilon} \right) - \left(1 + \frac{\pi}{4} \right) \right] \tag{15.148}$$

It approaches infinity logarithmically as $|T - T_c| \to 0$. A graph of the specific heat is shown in Fig. 15.6, together with the results in the Bragg-Williams and the Bethe-Peierls approximations for comparison. It is seen from (15.138) and (15.145) that the internal energy is continuous at $T = T_c$. Thus the phase transition at $T = T_c$ involves no latent heat.

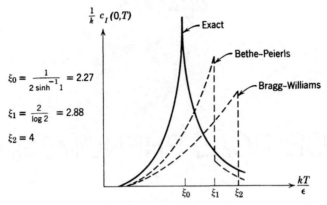

$\frac{1}{k} c_I(0,T)$

Exact

Bethe–Peierls

Bragg–Williams

$\xi_0 = \dfrac{1}{2\sinh^{-1}1} = 2.27$

$\xi_1 = \dfrac{2}{\log 2} = 2.88$

$\xi_2 = 4$

$\xi_0 \quad \xi_1 \quad \xi_2 \qquad \dfrac{kT}{\epsilon}$

Fig. 15.6 Specific heat of the two-dimensional Ising model.

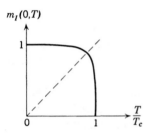

$m_I(0,T)$

$\dfrac{T}{T_c}$

Fig. 15.7 Spontaneous magnetization of the two-dimensional Ising model. The curve is invariant under a reflection about the dotted line.

To justify calling the phenomenon at $T = T_c$ a phase transition, we must examine the long-range order, i.e., the spontaneous magnetization. This cannot be done within the calculations so far outlined, since we have set $B = 0$ from the beginning. To calculate the spontaneous magnetization we have to calculate the derivative of the free energy with respect to H at $H = 0$. This calculation, carried out by Yang,* is as complicated as the one we have presented. The result, however, is simple. Yang shows that the spontaneous magnetization per spin is

$$m_I(0,T) = \begin{cases} 0 & (T > T_c) \\ \left\{1 - \left[\sinh(2\beta\epsilon)\right]^{-4}\right\}^{\frac{1}{8}} & (T < T_c) \end{cases} \qquad (15.149)$$

A graph of the spontaneous magnetization is shown in Fig. 15.7.

*C. N. Yang, *Phys. Rev.* **85**, 809 (1952).

16

CRITICAL PHENOMENA

16.1 THE ORDER PARAMETER

The term critical phenomena refers to the thermodynamic properties of systems near the critical temperature T_c of a second-order phase transition, or near the critical point of a gas-liquid transition. The term "second-order" is used here in the sense of "not first-order."

We shall model after the Curie point in a ferromagnetic system. The two phases on either side of the critical temperature have different spatial symmetries. Above the critical temperature, where there is no magnetization, the system is rotationally invariant. Below the critical temperature, when spontaneous magnetization occurs, the magnetization vector defines a preferred direction in space, destroying the rotational invariance. Since a symmetry is either present or absent, the two phases must be described by different functions of the thermodynamic variables, which cannot be continued analytically across the critical point.

Because of a reduction of the symmetry, an extra parameter is needed to describe the thermodynamics of the low-temperature phase. The extra parameter is called the "order parameter," denoted by M, which is usually an extensive thermodynamic variable accessible to measurements. In the ferromagnetic example M is the magnetization vector, with three components. For the gas-liquid critical point, we can use as the order parameter the volume difference of the coexisting phases, which tends to zero at the critical point. In this case M is a single-component quantity. There is no obvious symmetry change in the gas-liquid case, but the theory works, so we include it in the general scheme.

The basic idea is that near the critical point, the order parameter is the only important thermodynamic quantity. For simplicity let us work with a single-component order parameter M, which we think of as a scalar magnetization. When the order parameter changes by dM, the work done on the system is written in the form

$$dW = H \, dM \qquad (16.1)$$

This defines the "conjugate field" H, which is usually an intensive thermodynamic variable. In the ferromagnetic case H is just the external magnetic field.

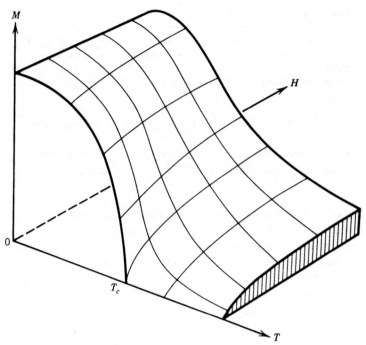

Fig. 16.1 Equation of state surface for a magnetic system with a second-order phase transition at zero field ($H = 0$). The surface is symmetric under $M \rightarrow -M$. The lower half is not shown.

When $H \neq 0$, the order parameter is a regular function of temperature; but it can develop a discontinuous derivative at a critical temperature when $H = 0$. The presence of an external field suppresses the phase transition, as illustrated by the equation of state depicted qualitatively in Fig. 16.1.

Using H and the temperature T as independent thermodynamic variables, we can derive all thermodynamic functions from the Gibbs free energy $G(H, T)$, given by

$$Q(H, T) = e^{-G(H,T)/kT} = \text{Tr } e^{-\mathscr{H}/kT} \tag{16.2}$$

where $\mathscr{H}$ is an appropriate Hamiltonian. Some useful formulas are given below:

$$\text{Magnetization:} \quad M = -\frac{\partial G}{\partial H} \tag{16.3}$$

$$\text{Susceptibility:} \quad \chi = \frac{1}{V}\frac{\partial M}{\partial H} \tag{16.4}$$

$$\text{Internal energy:} \quad U = G - T\frac{\partial G}{\partial T} \tag{16.5}$$

$$\text{Heat capacity:} \quad C = T^2\frac{\partial^2 G}{\partial T^2} \tag{16.6}$$

Table 16.1 Order Parameters

Transition	Order Parameter	Conjugate Field
Ferromagnetic	M	H
Antiferromagnetic	Staggered M	Staggered H
gas-liquid	$V_G - V_L$	$P - P_c$
Superfluidity	$\int d^3r \langle \psi \rangle$	Unobservable
Superconductivity	$\int d^3r \langle \psi_\uparrow \psi_\downarrow \rangle$	Unobservable

In Table 16.1 we give some examples of order parameters. The choice of the order parameter is a phenomenological matter and is not always obvious. In the case of superfluidity and superconductivity, the correct order parameter was discovered only after the phenomena were understood theoretically, long after they had been observed experimentally. In some cases, such as spin-glass transitions, we are not certain what the order parameter is, or indeed whether the concept itself is appropriate.

16.2 THE CORRELATION FUNCTION AND THE FLUCTUATION-DISSIPATION THEOREM

In addition to the thermodynamic quantities given above, the correlation function contains important information about a phase transition. We assume there is an order-parameter density $m(\mathbf{r})$, such that the order parameter can be written as

$$M = \left\langle \int d^3r\, m(\mathbf{r}) \right\rangle \tag{16.7}$$

where $\langle\ \rangle$ denotes ensemble average. The correlation function is defined as

$$\Gamma(\mathbf{r}) = \langle m(\mathbf{r})m(0) \rangle - \langle m(\mathbf{r}) \rangle \langle m(0) \rangle \tag{16.8}$$

It measures the "persistence of memory" of spatial variations in the order-parameter density. For a translationally invariant system the last term may be rewritten as $\langle m(0) \rangle^2$, because $\langle m(\mathbf{r}) \rangle = \langle m(0) \rangle$.

Denote the Fourier transform by a tilde ~. For example,

$$m(\mathbf{r}) = \int \frac{d^3k}{(2\pi)^3} e^{i\mathbf{k}\cdot\mathbf{r}} \tilde{m}(\mathbf{k})$$

$$\tilde{m}(\mathbf{k}) = \int d^3x\, e^{-i\mathbf{k}\cdot\mathbf{r}} m(\mathbf{r}) \tag{16.9}$$

We have $\tilde{m}^*(\mathbf{k}) = \tilde{m}(-\mathbf{k})$, since $m(\mathbf{r})$ is real. Taking the Fourier transform of both sides of (16.8), we obtain

$$\tilde{\Gamma}(\mathbf{k}) = \langle \tilde{m}(\mathbf{k})m(0) \rangle - \langle m(0) \rangle^2 (2\pi)^3 \delta(\mathbf{k}) \qquad (16.10)$$

Consider $H = 0$ and $T > T_c$. The last term vanishes because $\langle m(0) \rangle = 0$. Substituting the expression

$$m(0) = (2\pi)^{-3} \int d^3k\, \tilde{m}(\mathbf{k})$$

into (16.10), and noting that

$$\langle \tilde{m}(\mathbf{k})\tilde{m}(\mathbf{p}) \rangle = (2\pi)^3 \delta(\mathbf{k} + \mathbf{p}) |\tilde{m}(\mathbf{k})|^2$$

we obtain the useful formula

$$\tilde{\Gamma}(\mathbf{k}) = \langle |\tilde{m}(\mathbf{k})|^2 \rangle \qquad (16.11)$$

The mean-square average in (16.11) can be calculated through the following intuitive argument. The fluctuations should be isotropic, since there is no intrinsic direction. Thus to lowest order in the order-parameter density, the free energy should be of the form

$$G = \int d^3r \left[c_1 |\nabla m(\mathbf{r})|^2 + c_2 m^2(\mathbf{r}) \right] = \int \frac{d^3k}{(2\pi)^3} \left(c_1 k^2 + c_2 \right) |\tilde{m}(\mathbf{k})|^2 \quad (16.12)$$

where c_1 and c_2 are parameters that may depend on the temperature. The free energy residing in the kth Fourier mode is therefore

$$G(\mathbf{k}) = \left(c_1 k^2 + c_2 \right) |\tilde{m}(\mathbf{k})|^2 \qquad (16.13)$$

whose average value should be kT, by equipartition of energy. Hence

$$\langle |\tilde{m}(\mathbf{k})|^2 \rangle = \frac{kT}{c_1 k^2 + c_2} \qquad (16.14)$$

This gives the Fourier transform of the correlation function in "Ornstein-Zernike form." Taking the inverse Fourier transform we obtain

$$\Gamma(\mathbf{r}) = \frac{e^{-r/\xi}}{r} \qquad (16.15)$$

where $\xi = \sqrt{c_1/c_2}$ is the "correlation length," a measure of the spatial memory. Experimentally ξ diverges at the critical point. The correlation function then becomes $1/r$, which does not contain any characteristic length. We shall derive Γ in d dimensions in the next chapter, under essentially the same assumptions in different guises (see (17.46) and (17.91)).

From (16.7) we can rewrite the order parameter more explicitly in the form

$$\frac{M}{V} = \frac{1}{V} \int d^3r \frac{\text{Tr}\left[m(\mathbf{r})\, e^{-\mathcal{H}/kT} \right]}{\text{Tr}\, e^{-\mathcal{H}/kT}} = \frac{\text{Tr}\left[m(0)\, e^{-\mathcal{H}/kT} \right]}{\text{Tr}\, e^{-\mathcal{H}/kT}} \qquad (16.16)$$

The last relation is obtained under the assumption that the system is translationally invariant. Assume that the magnetic field H is coupled linearly to the order parameter:

$$\mathcal{H} = \mathcal{H}_0 - H \int d^3r\, m(\mathbf{r}) \tag{16.17}$$

where $\mathcal{H}_0$ is the Hamiltonian for $H = 0$. Differentiating (16.16) with respect to H, we obtain

$$\chi = \frac{1}{kT} \int d^3r \left[\langle m(\mathbf{r})m(0) \rangle - \langle m(0) \rangle^2 \right] \tag{16.18}$$

Comparison with the definition of the correlation function gives

$$\chi = \frac{1}{kT} \int d^3r\, \Gamma(\mathbf{r}) \tag{16.19}$$

This is a special case of a general relation known as the "fluctuation-dissipation theorem." We have encountered examples of this in the relation between the heat capacity and energy fluctuations (Eq. (7.14)), and in the relation between the compressibility and density fluctuations (Eq. (7.43)).

16.3 CRITICAL EXPONENTS

The critical exponents describe the nature of the singularities in various measurable quantities at the critical point. Six are commonly recognized, denoted by (at this point) a rather dull list of Greek letters: $\alpha, \beta, \gamma, \delta, \eta, \nu$. Denote the critical temperature by T_c, and introduce the quantity

$$t = \frac{T - T_c}{T_c} \tag{16.20}$$

We suppose that, in the limit $t \to 0$, any thermodynamic quantity can be decomposed into a "regular" part, which remains finite (but not necessary continuous), plus a "singular" part that may be divergent, or have divergent derivatives. The singular part is assumed to be proportional to some power of t, generally fractional.

The first four critical exponents are defined as follows:

Heat capacity:	$C \sim |t|^{-\alpha}$	(16.21)
Order parameter:	$M \sim |t|^{\beta}$	(16.22)
Susceptibility:	$\chi \sim |t|^{-\gamma}$	(16.23)
Equation of state ($t = 0$):	$M \sim H^{1/\delta}$	(16.24)

Here $\sim$ means "has a singular part proportional to." Since the first three relations all refer to a phase transition, it is understood that $H = 0$. The last one, on the other hand, specifically refers to the case $H \neq 0$.

We should keep in mind that these behaviors refer only to the singular part. For example, $\alpha = 0$ means that the heat capacity has no singular part; but it may still have a finite discontinuity at $t = 0$.

The definitions above implicitly assume that the singularities are of the same type, whether we approach the critical point from above or from below. This has been borne out both theoretically and experimentally, except for M, which is identically zero above the critical point by definition. Thus obviously (16.22) makes sense only for $t < 0$. We shall not bother to make this qualification every time.

The last two in the Greek alphabet soup of exponents concern the correlation function, which we shall assume to have the Ornstein-Zernike form

$$\Gamma(r) \xrightarrow[t \to 0]{} r^{-p} e^{-r/\xi} \qquad (16.25)$$

Then, ν and η are defined as follows:

$$\text{Correlation length:} \qquad \xi \sim |t|^{-\nu} \qquad (16.26)$$

$$\text{Power-law decay at } t = 0: \qquad p = d - 2 + \eta \qquad (16.27)$$

The significance of the critical exponents lie in their universality. As experiments have shown, widely different systems, with critical temperatures differing by orders of magnitudes, approximately share the same critical exponents. Their definitions have been dictated by experimental convenience. Some other linear combinations of them have more fundamental significance, as we shall see later.

Only two of the six critical exponents defined in the preceding discussion are independent, because of the following "scaling laws":

$$\text{Fisher:} \qquad \gamma = \nu(2 - \eta) \qquad (16.28)$$

$$\text{Rushbrooke:} \qquad \alpha + 2\beta + \gamma = 2 \qquad (16.29)$$

$$\text{Widom:} \qquad \gamma = \beta(\delta - 1) \qquad (16.30)$$

$$\text{Josephson:} \qquad \nu d = 2 - \alpha \qquad (16.31)$$

where, in the last relation, d is the dimensionality of space.

Table 16.2 summarizes the experimental values of the critical exponents as well as the results from some theoretical models. We can see that the scaling laws seem to be universal, but the individual exponents show definite deviations from truly universal behavior. The theory based on the renormalization group, as we shall discuss in Chapter 18, suggests that systems fall into "universality classes," and that the critical indices are the same only within a universality class.

To give actual examples of universality, we show in Fig. 16.2 a plot of the reduced temperature T/T_c vs. reduced density n/n_c for eight substances in the gas-liquid coexistence region. The critical data are quite varied, as Table 16.3 shows; but the reduced data points fall on a universal curve. The right branch refers to the liquid phase, and the left branch to the gas phase, and both come together at the critical point. The solid curve is Guggenheim's fit,* which

---

*E. A. Guggenheim, *J. Chem. Phys.* **13**, 253 (1945).

Table 16.2 Critical Exponents[a,b]

Exponent	TH	EXPT	MFT	ISING2	ISING3	HEIS3
α		0–0.14	0	0	0.12	−0.14
β		0.32–0.39	1/2	1/8	0.31	0.3
γ		1.3–1.4	1	7/4	1.25	1.4
δ		4–5	3	15	5	
ν		0.6–0.7	1/2	1	0.64	0.7
η		0.05	0	1/4	0.05	0.04
$\alpha + 2\beta + \gamma$	2	2.00 $\pm$ 0.01	2	2	2	2
$(\beta\delta - \gamma)/\beta$	1	0.93 $\pm$ 0.08	1	1	1	
$(2 - \eta)\nu/\gamma$	1	1.02 $\pm$ 0.05	1	1	1	1
$(2 - \alpha)/\nu d$	1		4/d	1	1	1

[a] TH, theoretical values (from scaling laws); EXPT, experimental values (from a variety of systems); MFT, mean field theory; ISINGd, Ising model in d dimension; HEIS3, classical Heisenberg model, $d = 3$.

[b] For more details and documentation see A. Z. Patashinskii and V. L. Pokrovskii, *Fluctuation Theory of Phase Transitions* (Pergamon, Oxford, 1979), Table 3, pp. 42–43.

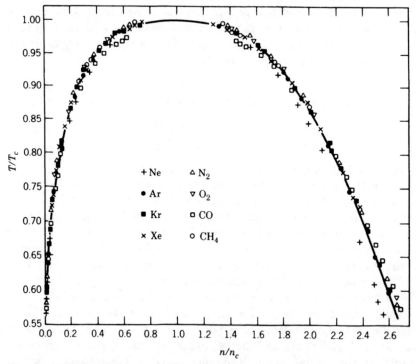

Fig. 16.2 Reduced temperature vs. reduced density in the gas-liquid coexistence region, for eight different substances.

Table 16.3 Critical Data

	T_c (°C)	P_c (atm)
Ne	− 228.7	26.9
Ar	1122.3	48
Kr	− 63.8	54.3
Xe	16.6	58
N_2	− 147	33.5
O_2	− 118.4	50.1
CO	− 140	34.5
CH_4	− 82.1	45.8

reproduces the argon data to 1 part in 10^3:

$$\frac{n_L + n_G}{n_c} = 1 + \frac{3}{4}\left(1 - \frac{T}{T_c}\right)$$

$$\frac{n_L - n_G}{n_c} = \frac{7}{2}\left(1 - \frac{T}{T_c}\right)^{1/3}$$

(16.32)

The first equation is known as the "law of rectilinear diameter." The second immediately shows that the exponent associated with the order parameter has the universal value $\beta = 1/3$.

All physical phenomena take place in three-dimensional space, of course. But there are situations in which the system under study is effectively a two-dimensional film, or a one-dimensional chain. The quantity d then denotes the effective dimensionality. We shall see that it plays a very important role in determining the nature of phase transitions. From now on we shall work in d spatial dimensions, with the spatial coordinate vector denoted by x, which has Cartesian components x_i $(i = 1, \ldots, d)$. The magnitude of x, and the spatial integration, will be respectively denoted by

$$|x| \equiv \left[\sum_{i=1}^{d} x_i^2\right]^{1/2}$$

$$\int (dx) \equiv \int d^d x$$

(16.33)

16.4 THE SCALING HYPOTHESIS

As the name implies, the scaling hypothesis has something to do with how various quantities change under a change of length scale. The value of a quantity with dimension must be expressed in terms of a standard unit of length, and it changes when that standard is changed. Thus, a dimensionless quantity will be invariant; others will change according to their dimensions. Important scaling laws concerning the thermodynamic functions can be derived from the simple (but strong) assumption that, near the critical point, the correlation length ξ is the only characteristic length of the system, in terms of which all other lengths must be measured. This is the "scaling hypothesis." We shall express it in concrete form in different ways, incorporating slightly different additional assumptions.

Dimensional Analysis

One way to implement the scaling hypothesis is to assume that a quantity of dimension (length)$^{-D}$ is proportional to ξ^{-D} near the critical point.

Let us first determine the dimensions of various quantities of interest. First we note that G/kT is dimensionless. Hence $g = G/kTV$ is of dimension (length)$^{-d}$, and we indicate this fact in the following notation:

$$[g] = L^{-d} \tag{16.34}$$

(We consider g instead of G/kT because we want to deal with finite quantities in the thermodynamic limit.) Normalizing the correlation function according to (16.25), we have

$$[\Gamma(x)] = L^{2-d-\eta} \tag{16.35}$$

By definition this has the same dimension as $\langle m(0) \rangle^2$. Hence

$$[M/V] = L^{(2-d-\eta)/2} \tag{16.36}$$

By the fluctuation-dissipation theorem, we have

$$[kT\chi] = L^{2-\eta} \tag{16.37}$$

The dimension of the conjugate field can be obtained from the relation $M = -\partial G/\partial H$ as $[H/kT] = [g]/[M/V]$, or

$$[H/kT] = L^{(2+d-\eta)/2} \tag{16.38}$$

These results are summarized in Table 16.4. The exponent of L is denoted by $-D$, and D is called the "dimension" (the minus sign being introduced such that a quantity changes by a factor b^D when the unit of length is increased by a factor b).

Now replace the length L in the formulas above by ξ. Using $\xi \sim t^{-\nu}$, we obtain all the critical exponents. The results are listed in Table 16.4. Comparing them with their definitions leads to the relations

$$2 - \alpha = \nu d$$
$$\beta = -\nu(2 - d - \eta)/2$$
$$\gamma = \nu(2 - \eta)$$
$$\beta\delta = \nu(2 + d - \eta)/2$$

Table 16.4 Dimensions and Exponents[a]

Function F	Dimension D	Exponent Scaling Hypothesis	Definition
G/kTV	d	νd	$2 - \alpha$
M/V	$(d - 2 - \eta)/2$	$-\nu(2 - d - \eta)/2$	β
$kT\chi$	$\eta - 2$	$-\nu(2 - \eta)$	$-\gamma$
H/kT	$(2 + d - \eta)/2$	$\nu(2 + d - \eta)/2$	$\beta\delta$

[a] $[F] = (\text{length})^{-D}$.

The first is the Josephson law, and the third is the Fisher law. By subtraction and addition, we derive from the second and the fourth the Rushbrooke and the Widom laws.

Scaling Forms

Another way to implement the scaling hypothesis is to assume that, in the absence of external field, the correlation function near $t = 0$ has the functional form

$$\Gamma(x) \xrightarrow[r \to 0^\pm]{} |x|^{-p}\mathscr{F}_{\pm}(x/\xi), \qquad (H = 0) \qquad (16.39)$$

where $p = d - 2 + \eta$. In the presence of an external field H the above is generalized to

$$\Gamma(x, H) \xrightarrow[t \to 0^\pm]{} |x|^{-p}\mathscr{R}_{\pm}(x/\xi, H\xi^y) \qquad (16.40)$$

where y is some constant. This may be justified physically as follows: The magnetic moments in the system are strongly correlated within a correlation length, which tends to infinity as $t \to 0$. Thus, there is a natural tendency for very large blocks of magnetic moments to line up. The effect of an external field is thereby magnified, with a magnification factor proportional to some power of the correlation length. Hence we expect H to occur only in the combination $H\xi^y$. Making the replacement

$$\xi^y \to |t|^{-\nu y}$$

we can write

$$\Gamma(x, H) \xrightarrow[t \to 0^\pm]{} |x|^{-p}\mathscr{R}_{\pm}(x/\xi, H/|t|^\Delta), \qquad \Delta = \nu y \qquad (16.41)$$

This is called a "scaling form" of the correlation function. The quantity Δ is sometimes called the "gap exponent," because (16.41) gains a factor t^Δ when integrated with respect to H.

It is instructive to derive the scaling laws again from the scaling forms. By the fluctuation-dissipation theorem and (16.39), we have

$$\chi = \frac{1}{kT} \int (dx) \Gamma(x) = \frac{1}{kT} \int (dx) |x|^{-p} \mathscr{F}_\pm(x/\xi)$$

By changing the variable of integration to x/ξ, we obtain

$$\chi = \text{const. } \xi^{d-p} \sim |t|^{-\nu(2-\eta)} \tag{16.42}$$

which yields $\gamma = \nu(2 - \eta)$, Fisher's scaling law.

In a similar manner, when $H = 0$, we use the fluctuation-dissipation theorem together with (16.41) to deduce

$$\chi \sim \xi^{2-\eta} \mathscr{X}_\pm (H \xi^y) \sim |t|^{-\gamma} \mathscr{X}_\pm (H/|t|^\Delta) \tag{16.43}$$

Integrating this with respect to H gives the scaling form of the equation of state:

$$M \sim |t|^{\Delta - \gamma} \mathscr{M}_\pm (H/|t|^\Delta) \tag{16.44}$$

which, for $H = 0$, says $M \sim |t|^{\Delta - \gamma}$. Therefore the gap exponent is given by

$$\Delta = \beta + \gamma \tag{16.45}$$

Integrating (16.44) with respect to H yields the scaling form of the Gibbs free energy:

$$G \sim |t|^{2\Delta - \gamma} \mathscr{G}_\pm (H/|t|^\Delta) \tag{16.46}$$

At $H = 0$, we have $G \sim |t|^{2\Delta - \gamma}$, and hence

$$C \sim |t|^{2\Delta - \gamma - 2} \sim |t|^{2\beta + \gamma - 2} \tag{16.47}$$

which gives $\alpha + 2\beta + \gamma = 2$, Rushbrooke's law.

At $t = 0$ and $H \neq 0$, the order parameter is assumed to be finite. Therefore the function $\mathscr{M}_\pm$ in (16.44) must have the following behavior:

$$\mathscr{M}_\pm (H/|t|^\Delta) \xrightarrow[t \to 0]{} |t|^{\alpha + \Delta - 2} = |t|^{-\beta}$$

$$\mathscr{M}_\pm (x) \xrightarrow[x \to 0]{} x^{\beta/\Delta} \tag{16.48}$$

We have used the relation $2 - \alpha - \Delta = \beta$, obtainable from (16.45), and the Rushbrooke scaling law. Using (16.44), we now find

$$M \xrightarrow[t \to 0]{} |t|^\beta (H/|t|^\Delta)^{\beta/\Delta} = H^{\beta/\Delta} \tag{16.49}$$

which says $\delta = \Delta/\beta$. Use of (16.45) then leads to the Widom scaling law $\gamma = \beta(\delta - 1)$.

Finally, to derive the Josephson scaling law from the scaling form, we have to make an extra assumption known as "hyperscaling," that at $H = 0$ the amount of free energy residing in a spatial volume of linear size ξ is of the order of kT. This is consistent with the idea that ξ is the only length scale, so that there can be no fluctuations with wavelengths shorter than ξ. Accordingly the total free energy is of the order of kTV/ξ^d. As $t \to 0$ we have

$$G \sim \xi^{-d} \sim |t|^{\nu d} \tag{16.50}$$

Comparison with (16.46) at $H = 0$ gives $\nu d = 2\Delta - \gamma$, which by (16.45) leads to $\nu d = 2 - \alpha$, the Josephson scaling law.

Widom's Scaling Form

If there are conjugate-variable pairs in the theory other than M and H, say ϕ_i and J_i, then the scaling form of the free energy may be generalized to the following:

$$T \sim |t|^{2-\alpha} \mathscr{G}_{\pm} \left(\frac{H}{|t|^{\beta\delta}}, \frac{J_1}{|t|^{\Delta_1}}, \frac{J_2}{|t|^{\Delta_2}}, \cdots \right) \tag{16.51}$$

which is known as Widom's scaling form. The fields $H, J_1, J_2, \ldots$ are called scaling fields. The associated exponents $\Delta, \Delta_1, \Delta_2, \ldots$, called "crossover exponents," control the relative importance of the fields near $t = 0$. For example, if $\Delta_i < 0$, then the dependence on J_i drops out near $t = 0$, and the field is said to be "irrelevant"; if $\Delta_i > 0$, the field J_i is "relevant"; while if $\Delta_i = 0$, we would have a "marginal" case. We should bear in mind that the scaling form above specifically refers to the neighborhood of a particular critical point. A system may have more than one critical point, and a form like (16.51) is supposed to hold near each of them, with different sets of crossover exponents.

16.5 SCALE INVARIANCE

The scaling hypothesis states that ξ is the only characteristic length of a system in the neighborhood of $t = 0$. When combined with the experimental observation that ξ diverges at $t = 0$, it leads to the conclusion that the system has no characteristic length, and is therefore invariant under scale transformations.

Intuitively speaking, scale invariance means that if part of a system is magnified until it is as large as the original system, one would not be able to tell the difference between the magnified part and the original system (see Fig. 16.3).

More precisely, we define a scale-invariant system to be one in which all thermodynamic functions are homogeneous functions, as typified by the correlation function:

$$\Gamma(x) \sim x^{-p} \qquad (16.52)$$

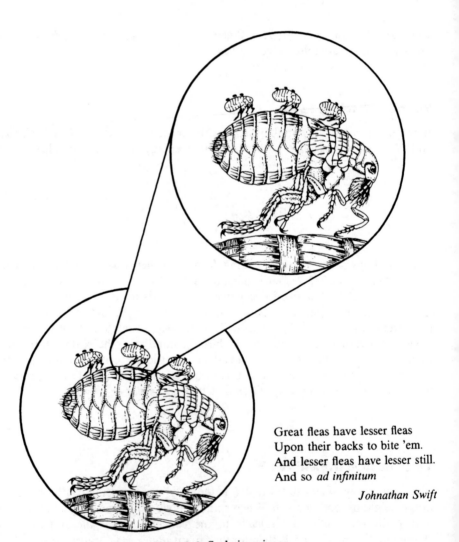

Great fleas have lesser fleas
Upon their backs to bite 'em.
And lesser fleas have lesser still.
And so *ad infinitum*

Johnathan Swift

Fig. 16.3 Scale invariance.

where $p = d - 2 + \eta$ is the dimension of $\Gamma(x)$. When the unit of length is increased by a factor b, the coordinate transforms as $x \to x' = x/b$. Thus the correlation function transforms as

$$\Gamma(x/b) = b^p \Gamma(x) \qquad (16.53)$$

This is a homogeneity rule that would have been impossible to satisfy had Γ contained an exponential factor. Modeling after this, we define a homogeneous function by the rule

$$f(q') = b^{D_f} f(q), \qquad q' = b^{D_q} q \qquad (16.54)$$

The dimensions of various quantities are given in terms of the critical exponents in Table 16.4.

The free energy density $g = G/kTV$ has dimension d. Hence in a scale-invariant system it has the homogeneity property

$$g(h', t') = b^d g(h, t), \qquad (h = H/kT)$$

or

$$g(h, t) = b^{-d} g(b^{D_h} h, b^{D_t} t) \qquad (16.55)$$

The value of D_t has not been given before, but we can deduce it as follows. Since, by assumption, ξ is the only length in the problem, changing ξ is tantamount to changing the unit of length. From the relation $\xi \sim t^{-\nu}$, we then have*

$$D_t = 1/\nu \qquad (16.56)$$

Widom's scaling form is a special case of (16.55) corresponding to the choice $|t'| = 1$, or

$$b = |t|^{-1/D_t} \qquad (16.57)$$

Thus we have,

$$g(h, t) = t^{d/D_t} \mathcal{G}_\pm \left(\frac{h}{t^{D_h/D_t}} \right) \qquad (16.58)$$

---

*It might seem strange to assign a dimension to a dimensionless quantity t. The implicit assumption is that wherever t occurs, it is always multiplied by the same factor that converts it into $\xi^{-1/\nu}$.

where $\mathscr{G}_{\pm}(x) = g(x, \pm 1)$. From a theoretical point of view, the dimensions D_h, D_t are more fundamental than the exponents α, β, etc. The former tell us how thermodynamic variables behave under a scale transformation; the latter are combinations dictated purely by experimental convenience.

The critical exponents can all be expressed in terms of D_h and D_t as follows:

$$\alpha = 2 - \frac{d}{D_t} \qquad \delta = \frac{D_h}{d - D_h}$$

$$\beta = \frac{d - D_h}{D_t} \qquad \nu = \frac{1}{D_t} \qquad (16.59)$$

$$\gamma = \frac{2D_h - d}{D_t} \qquad \eta = 2 + d - 2D_h$$

The first four relations can be obtained from (16.58) by using (16.3)–(16.6) and the definitions (16.21)–(16.24). The last two came from (16.56) and the Fisher scaling law.

We have derived the scaling laws via various routes, using different ways to realize the scaling hypothesis. None of these are satisfactory, for they rely on ad hoc assumptions. However, they do suggest plausible reasons for the experimental success of the scaling laws. Motivated by these discussions, we set the goal of theory as a derivation of scale invariance directly from more basic premises, and calculation of the dimensions of various thermodynamic quantities. This goal will be attained with the help of the renormalization-group method, as will be discussed in Chapter 18.

1.6 GOLDSTONE EXCITATIONS

We must comment on that most remarkable fact, which underlies Landau's original conception of the order parameter, that macroscopic systems generally have a lesser degree of symmetry at low temperatures than at high temperatures. The symmetry manifested at high temperatures is usually a property of the system's microscopic Hamiltonian. As such, it cannot cease to exist, even when it appears to be violated. The question is, where does it go?

For example, the microscopic Hamiltonian of a ferromagnet is rotationally invariant. Lowering the temperature of the system surely does not change that. What is changed, however, is the mode in which the symmetry is expressed. One might find it natural to assume (in the best tradition of Aristotelian logic) that in the most perfect expression of the symmetry, the ground state should be invariant under the symmetry operation. Nature, alas, is not perfect. In most instances, the system possesses many equivalent ground states that transform into one another under the symmetry operation. But, since the system can actually exist in only one of these states, the symmetry appears to be broken. This phenomenon, that the ground state of the system does not possess the symmetry of the Hamiltonian, is called "spontaneous symmetry breaking."

We have encountered the simplest manifestation of spontaneous symmetry breaking in the two-dimensional Ising model. The total energy is invariant under a simultaneous sign change of all the spins. Yet, the lowest-energy configuration is that in which all spins are aligned. In this case the symmetry is expressed through the fact that the spins could also have aligned themselves along the opposite direction. A more subtle example is the nonvanishing of the superfluid order parameter, as discussed in Chapter 12.

When a continuous symmetry is spontaneously broken in a quantum mechanical system, interesting consequences follow. In this case, there is a noncountable infinity of ground states (with energy taken to be zero) orthogonal to one another. The system must choose one of these. As a consequence of the degeneracy, there emerges a type of excited state in which the local ground state changes very gradually over space, so as to form a "wave" of very long wavelength. Such a state is orthogonal to any one ground state, with an energy approaching zero when the wavelength approaches infinity. This is called a "Goldstone excitation." The underlying symmetry is said to be realized in the "Goldstone mode."

Familiar examples of Goldstone excitations include phonons in a crystal lattice, which expresses the broken translational invariance, and spin waves in a ferromagnetic, which realizes the basic rotational invariance.

A new twist occurs when the system is coupled to a zero-mass vector field such as the photon field. In this case the Goldstone mode gives way to the "Higgs

Table 16.5 Spontaneous Symmetry Breaking[a]

System	Broken symmetry	Goldstone excitation
Crystal	Translational	Phonon
Ferromagnet	Rotational	Spin wave
Superfluid	Global gauge	Phonon
Superconductor	Local gauge	(Higgs mode)
Electro-weak[b]	Local gauge	(Higgs mode)
QCD[c]	Chiral	π mesons

[a] For a discussion of broken symmetry in condensed matter physics, see P. W. Anderson, *Basic Notions of Condensed Matter Physics* (Benjamin-Cummings, Menlo Park, CA, 1984), Chapter 2. For its role in particle physics, see K. Huang, *Quarks, Leptons, and Gauge Fields* (World Scientific, Singapore, 1982), Chapters 3, 4.
[b] Weinberg-Salam model of unified electro-weak interactions.
[c] Quantum chromodynamics with massless quarks.

mode," wherein the "photon" acquires a mass dynamically, and the would-be Goldstone excitation becomes the longitudinal degree of freedom of the massive "photon." An example of this is the Meissner effect in superconductivity, in which the photon mass corresponds to the inverse penetration depth. We offer a list of broken symmetry phenomena in condensed matter physics and particle physics in Table 16.5.

In defining spontaneously broken symmetry, we have assumed that a system can exist in only one ground state, even if it possesses many equivalent ones. This is obvious in classical mechanics, but not in quantum mechanics. In the latter case one can legitimately ask, "Why can't the system exist in a linear combination of different ground states?"

The answer is that, to make a transition from one ground state to another, one needs an operator that simultaneously changes the states of all the particles in the system. Such an operator does not exist in the thermodynamic limit, when the number of particles tends to infinity. In this limit, therefore, the different ground states generate separate Hilbert spaces, and transitions among them are forbidden. One says that there is a "super selection rule" insulating the ground states from one another.

If a superposition is made of two different ground states, then no physical measurement can determine the relative phase of the two states; one would have two separate systems that can never communicate with each other. Thus, such a superposition does not correspond to any experimental situation.

16.7 THE IMPORTANCE OF DIMENSIONALITY

The dimensionality of space plays an important role in phase transitions. We have seen this in simple models: In the Ising model, spontaneous magnetization happens only in more than one spatial dimension. In the ideal Bose gas, Bose-Einstein condensation is impossible in dimensions $d \leq 2$. We collect here qualitative arguments regarding critical dimensions of various kinds.

Role of Goldstone Excitations

One way to find out whether an ordered phase can exist is to see if it is stable against long wavelength fluctuations. This is the Peierls-Landau approach,* which we shall rephrase in a somewhat different form.

For definiteness consider a translationally invariant system of atoms in d spatial dimensions, with volume $V \to \infty$. Assume that it forms a crystal lattice,

---

*See L. D. Landau and E. M. Lifshitz, *Statistical Physics* (Pergamon, London, 1958), Section 125

whose detailed nature we need not specify. In other words, we assume that translational invariance is spontaneously broken. Let $u(x)$ be a d-component displacement vector, which gives the deviation of the atoms in a volume element at x from their supposed equilibrium position. In the harmonic approximation we can decompose $u(x)$ into normal modes:

$$u(x) = \frac{1}{V} \sum_{k,i} e^{ik \cdot x} q_i(k) \tag{16.60}$$

where i labels various types of normal modes (e.g., longitudinal and transverse vibrations). The sum over k is cut off at the upper limit at some characteristic inverse length. The energy residing in a normal mode is given by

$$E_{ki} = \tfrac{1}{2} \omega_i^2(k) |q_i(k)|^2 \tag{16.61}$$

whose average value is kT, by the equipartition of energy. Hence the mean-square amplitude of a normal mode is given by

$$\langle |q_{ki}|^2 \rangle = \frac{2kT}{\omega_i^2(k)} \tag{16.62}$$

and the mean-square total displacement is, in the infinite-volume limit,

$$\langle u^2 \rangle = \sum_i \int (dk) \frac{2kT}{\omega_i^2(k)} \tag{16.63}$$

Since by assumption a continuous symmetry has been broken spontaneously, there must exist among the normal modes Goldstone excitations whose frequencies vanish as $k \to 0$. Assuming $\omega_i(k)$ to be regular at $k = 0$, we expect $\omega_i(k) \to c_i k$, where c_i is a constant. Thus the integral (16.63) behaves near the lower limit like $\int dk\, k^{d-1}/k^2$, which diverges for $d \leq 2$. This means that even if one made a crystal in dimensions $d \leq 2$, the spatial order will be destroyed by long wavelength density fluctuations. Similarly, spontaneous magnetization cannot happen in continuous media in dimensions $d \leq 2$, because the magnetic order would have been destroyed by spin waves. For such systems $d = 2$ is the "lower critical dimension."

In the marginal case $d = 2$, the integral in (16.63) diverges only logarithmically. This means that, although the correlation length for long-ranged order is not infinite, as required of a theoretical crystal, it is large enough to be of macroscopic size. Thus, in practice, we should be able to make two-dimensional crystals of finite but macroscopic size.

Role of Domain Walls

Another approach to the problem is to consider whether long-ranged order is stable against the formation of domain walls, which are boundary layers between regions with different values of the order parameter. Let us discuss this in the context of a magnetic system.* The argument is a generalization of that given in Chapter 14 to show that there can be no phase transition in the Ising model in $d = 1$.

To minimize the free energy $U - TS$, we want the internal energy U to be small, and the entropy S large. (We consider only $T > 0$.) Entropy considerations always favor the creation of domains. Hence the key is the energy cost of one domain.

Suppose the system is completely magnetized except for one domain of linear dimension L, in which the magnetization points in the opposite direction. In the case of an Ising model, the spins are either up or down. Hence the energy cost of the domain is proportional to the number of spins that interact across the domain surface, and thus to the surface area L^{d-1}. For $d \le 1$, the energy cost is nonmacroscopic. Creation of N domains will cost energy of order N, but the entropy will be of order $\log N! \approx N \log N$. Hence lots of domains will be created, destroying the long-ranged order. The lower critical dimension for Ising models is therefore $d = 1$, in agreement with our findings in Chapter 14.

For a system with continuous order parameter $m(x)$, the domain wall is of finite thickness, in which $m(x)$ changes continuously from the value inside the domain to that outside. A crude estimate of the energy cost of the domain wall is

$$U \approx \int_{\text{wall}} (dx)|\nabla m(x)|^2 \propto \frac{(\text{Volume of wall})}{L^2} \qquad (16.64)$$

We used the estimate $|\nabla m|^2 \propto 1/L^2$ on the grounds that L is the only relevant length in the problem. By the same token, the thickness of the wall is also proportional to L. Thus, the volume of the wall is proportional to L^d, and we conclude that

$$U \propto L^{d-2} \qquad (16.65)$$

Reasoning as in the previous case, we conclude that entropy favors domain creation for $d \le 2$. Therefore the lower critical dimension is $d = 2$ for continuous systems, in agreement with an earlier result.

None of the arguments presented above are rigorous. But they give us physical insight into the importance of dimensionality. Rigorous arguments in special models support these conclusions. Hohenberg has shown that the superfluid order parameter must vanish in $d \le 2$. Using a similar argument, Mermin

---

*F. Bloch, Z. Phis. **61**, 206 (1930); C. H. Herring and C. Kittel, Phys. Rev. **81**, 869 (1950).

and Wagner proved that in $d \leq 2$ the Heisenberg model can exhibit neither ferromagnetic nor antiferromagnetic order.*

Dimensional Reduction[†]

Experimentally it is possible to create metals containing randomly distributed magnetic impurities that are out of thermal equilibrium with the host system, with a very long relaxation time. This is called "quenched" randomness (as opposed to "annealed" randomness, which obtains when the impurities are in thermal equilibrium with the system). The host system sees the impurities as sources of external fields, which vary from point to point at random. Experimentally the nonequilibrium status can be shown by checking the concentration of the impurities against known equilibrium values. The randomness of the impurities can be demonstrated by X-ray diffraction.

Suppose a system in the presence of a quenched random external field magnetizes spontaneously. Since the random field varies from point to point, there will be regions in which it would pay for the magnetization to reorient itself, so as to lower the interaction energy in the local external field. Thus a domain will be formed, provided the domain wall does not cost too much energy.

To find the energy balance, assume that the random field has Gaussian fluctuations. The value of the external field averaged over a domain will then be of order $\sqrt{N}$, where N is the number of impurities in the domain. For a domain of linear size L, we have $N \propto L^d$. Hence the volume energy is proportional to $-L^{d/2}$. Adding this to the energy of the domain wall, which is proportional to L^{d-2} as before, we obtain the energy cost of one domain:

$$U = c_1 L^{d-2} - c_2 L^{d/2} \tag{16.66}$$

where c_1, c_2 are positive constants. The lower critical dimension is obtained by equating the powers of the two terms:

$$d - 2 = \frac{d}{2}, \qquad \therefore d = 4 \tag{16.67}$$

For $d < 4$ the energy is negative for sufficiently large L, and hence domain formation is energetically favored. For $d = 4$, the negative term also dominates for sufficiently large L, because it is a volume energy instead of surface energy.

The result suggests that a system in a quenched random magnetic field in d dimensions behaves like a system without magnetic field in $d - 2$ dimensions. In other words, the presence of a quenched random field reduces the effective

---

*P. C. Hohenberg, *Phys. Rev.* **158**, 383 (1967); N. D. Mermin and H. Wagner, *Phys. Rev. Lett.* **17**, 1133 (1966).

[†]Y. Imry and S. K. Ma, *Phys. Rev. Lett.* **35**, 1399 (1975).

dimensionality by 2.* However, experiments indicate that this behavior is not universal, and we have yet to understand the phenomenon more fully.[†]

PROBLEMS

16.1 Consider a fluid near its critical point, with isotherms illustrated qualitatively in the sketch. For $T \geq T_c$, assume that the Gibbs free energy of the fluid has a singular part satisfying the scaling form

$$G \sim t^{2-\alpha} \mathcal{G}_{\pm}(p/t^{\Delta})$$
$$p = (P - P_c)/P_c, \qquad t = (T - T_c)/T_c$$

Approaching the critical point along $p = 0$ (path AC in the accompanying sketch), calculate the exponents $x(q)$ that describe the singular parts of the following quantities q, such that $q \sim t^{x(q)}$:

Entropy S
Internal energy U
Heat capacity C_P
Isothermal compressibility $\kappa_T = V^{-1}(\partial V/\partial P)_T$
Thermal expansion coefficient $a = V^{-1}(\partial V/\partial T)_P$
Heat capacity C_V

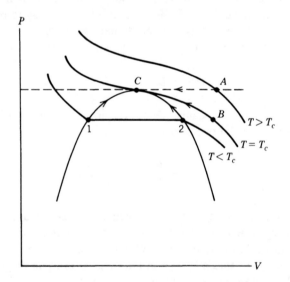

*See G. Parisi and N. Sourlas, *Phys. Rev. Lett.* **43**, 744 (1979) for a formal treatment of this effect.
†For a review and a source of literature see R. J. Birgeneau, Y. Shapira, G. Shirane, R. A. Cowley, and H. Yoshizawa, *Physica B* **137**, 83 (1986).

16.2 Referring to the sketch, approach the critical point along BC (i.e., $t = 0$). Find the equation of state, i.e., the density n as a function of p.

16.3 Continuing with the last problem, approach the critical point separately from the gas and from the liquid phases at $t < 0$ (paths $1C$ and $2C$ in the sketch). The paths are shown in the $P-T$ diagram in the accompanying sketch. Assume that the Gibbs free

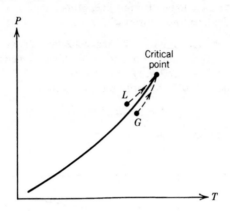

Table 16.6 Equation of State of Nickel[a,b]

T	H	M	T	H	M
678.86	4,160	0.649	614.41	1,820	19.43
	10,070	1.577		3,230	20.00
	17,775	2.776		6,015	20.77
653.31	2,290	0.845		10,070	21.67
	4,160	1.535		14,210	22.36
	6,015	2.215		17,775	22.89
	10,070	3.665	601.14	3,230	25.34
	14,210	5.034		6,015	25.83
	17,775	6.145		10,070	26.40
	21,315	7.211		14,210	26.83
629.53	1,355	6.099		17,775	27.21
	3,230	9.287			
	6,015	11.793			
	10,070	13.945			
	14,210	15.405			
	17,775	16.474			

[a] T, temperature (*K); H, applied field, corrected for demagnetization (oe); M, magnetization number per unit mass (emu/g).
[b] From P. Weiss and R. Forrer, *Ann. Phys. (Paris)* **5**, 153 (1926). The data have been reanalyzed by J. S. Kouvel and M. E. Fisher, *Phys. Rev.* **136**, A1626 (1964). They obtained the following critical exponents: $\gamma = 1.35 \pm 0.02$, $\delta = 4.22$.

energy of the fluid has a singular part satisfying separate scaling forms in the two phases:

$$G \sim t^{2-\alpha} \mathscr{G}_{\pm}\left(\frac{p_{\pm}(t)}{t^{\Delta}}\right)$$

where $p_{\pm}(t)$ are the values of $(P - P_c)/P$ along the paths. Close to transition line and near $t = 0$, we can take them to be the same linear function of t: $p_{\pm}(t) \approx ct$, where c is a constant. Show how the latent heat varies with t.

16.4 The accompanying graph shows the magnetization M vs. magnetic field H for nickel, as measured by Weiss and Forrer in 1926. Some numerical values are listed in Table 16.6. Given that $T_c = 627.2$ K, $\beta = 0.368$, and $\delta = 4.22$, select data from a few temperatures above and below T_c to verify the scaling form

$$M \sim |t|^{\beta} \mathscr{M}_{\pm}\left(H/|t|^{\beta\delta}\right)$$

Answer. See accompanying figure.

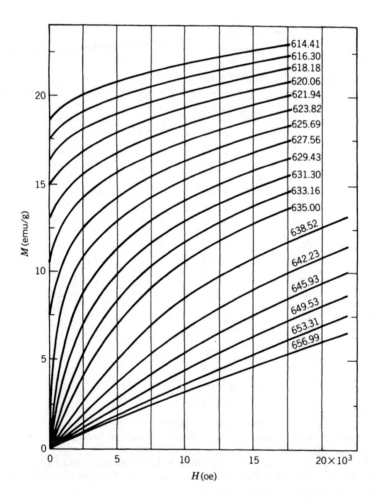

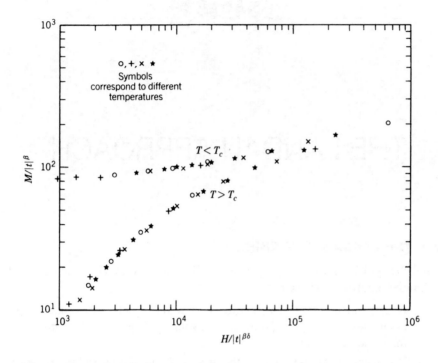

16.5 Derive the relations in (16.59), except the last two, from the scaling form for the Gibbs free energy given in (16.58). Derive the last two in the manner suggested in the text.

THE LANDAU APPROACH

17.1 THE LANDAU FREE ENERGY

To put the subject matter in a proper perspective, let us make clear the nature of the Landau approach* at the outset:

1. It is a phenomenological approach that deals only with macroscopic quantities. The whole point is to avoid dealing with the underlying microscopic structure.

2. It is meant to be used only in the neighborhood of a critical point, where the order parameter is small.

3. It is an inductive approach. We first see that it works for this example and that example, and then try to generalize.

Consider a macroscopic system in d-dimensional space. Assume that there exists a coarse-grained order parameter density $m(x)$. Think of it as a magnetic moment density, averaged over atomic distances. Thus, there are no fluctuations in $m(x)$ with wavelengths smaller than an atomic distance a. Correspondingly $m(x)$ has no Fourier components with wave number greater than a cutoff $\Lambda \sim 1/a$. The partition function, as appropriate for the Gibbs free energy $G(H, T)$, is represented in the form

$$Q(H, T) = e^{-G(H,T)/kT} = \mathcal{N} \int (Dm) \, e^{-E[m, H]} \tag{17.1}$$

where $E[m, H]$, the effective Hamiltonian divided by kT, is a functional of the order parameter density $m(x)$ and its conjugate field $H(x)$:

$$E[m, H] = \int (dx) \psi(m(x), H(x)) \tag{17.2}$$

where (dx) is short for $d^d x$. The dependence on temperature is suppressed.

*L. D. Landau, *Phys. Z. Sowjetunion* **11**, 26 (1937), reprinted in *Collected Papers of L. D. Landau*, D. ter Haar, ed. (Pergamon, London, 1965), p. 193; V. L. Ginzburg and L. D. Landau, *Zh. Eksp. Teor. Fiz.* **20**, 1064 (1950).

We refer to ψ as the "Landau free energy." In the neighborhood of a critical point, where $m(x)$ is small, we can expand it in powers of $m(x)$ and its derivatives:

$$\psi(m(x), H(x)) = \frac{1}{2}|\nabla m(x)|^2 - \frac{1}{kT}m(x)H(x)$$
$$+ \frac{1}{2}r_0m^2(x) + s_0m^3(x) + u_0m^4(x) + \cdots \quad (17.3)$$

where $\nabla m(x)$ is a d-dimensional vector whose components are $\partial m(x)/\partial x$. Terms not shown may involve higher powers of $m(x)$ and its derivatives. The external field $H(x)$ enters only linearly, because we assume it is very weak. We have assumed for simplicity that $m(x)$ is a single real field. More generally it could have many components, or be complex.

The coefficient of $|\nabla m(x)|^2$ is chosen to be $\frac{1}{2}$, to fix the scale of $m(x)$. All other coefficients are phenomenological parameters that may depend on the temperature and the cutoff. In practice we simply choose them to suit our purpose. However, there are constraints to be observed. For example, the term linear in $m(x)$ must have $H(x)$ as a coefficient for the order parameter to be correctly given by $-\partial G/\partial H$. The coefficient s_0 for the cubic term must vanish for systems invariant under reversal of $m(x)$ when $H = 0$, such as a ferromagnet. It is customary to assume that u_0 is independent of T, while

$$r_0 = a_0t, \qquad t = (T - T_c)/T_c \quad (17.4)$$

where a_0 is a positive constant.

The integration in (17.1) is a functional integration which extends over all possible functional forms of $m(x)$ that do not vary over an atomic distance a. The volume element (Dm) is defined only up to a factor, which is absorbed into a normalization constant $\mathcal{N}$ in (17.1). Since it always cancels in ensemble averages, we need not specify it. We shall spell out in greater detail how one might actually do a functional integral in the next section.

We now give an argument to justify the Landau theory. A more rigorous derivation for simple models will be given later. Assume we know how to express $m(x)$ in terms of the microscopic state of the system. We calculate the partition function by first summing over microscopic states holding the form of $m(x)$ fixed, and then integrating over all possible forms of $m(x)$:

$$Q = \mathrm{Tr}\, e^{-\mathcal{H}/kT} = \int (Dm)(\mathrm{Tr}\, e^{-\mathcal{H}/kT})_m \quad (17.5)$$

The trace in the last integral denotes a sum-over-states with the functional form of $m(x)$ held fixed:

$$(\mathrm{Tr}\, e^{-\mathcal{H}/kT})_m = W[m]\, e^{-\epsilon[m]/kT} \quad (17.6)$$

where $\epsilon[m]$ is the energy of the system when $m(x)$ has a particular functional form, and $W[m]$ is the number of microscopic states having that energy. Denoting the "entropy" by $S[m] = \log W[m]$, we have

$$E[m] = \frac{\epsilon[m]}{kT} - S[m] \quad (17.7)$$

From the physical interpretation of $m(x)$ as a magnetic moment density, it is reasonable that the above can be expressed as a spatial integral over some function of $m(x)$. The integrand is then the Landau free energy.

But why do we choose to hold fix $m(x)$, instead of any other variable, while performing the microscopic average? The underlying assumption is that the order parameter is slow to come to thermal equilibrium, long after all other degrees of freedom have become thermalized. In this sense, it is similar to the macroscopic variables of hydrodynamics.

To have a completely macroscopic theory, the cutoff Λ must somehow disappear from final physical answers. Since $1/\Lambda$ is of microscopic scale, the obvious thing to do is to take the limit $\Lambda \to \infty$. That cannot be done in a straightforward manner, however, because it produces divergences in the free energy, (except in the mean-field approximation discussed later.) The correct procedure for taking such a limit involves "renormalization," a scheme originally invented to circumvent divergences of a similar nature in quantum electrodynamics. This will be the subject of discussion in Chapter 18.

That the inspiration for renormalization came from quantum electrodynamics was not accidental. The Landau theory for $d = 4$ is equivalent to a relativistic quantum field theory.* More precisely, the partition function with a spatially varying external field $H(x)$ corresponds to the generating functional for the "vacuum Green's functions" of a relativistic quantum field theory in the presence of an external source $H(x)$. Thus the theory has practical application for $d = 1, 2, 3, 4$:

$$d = 1: \quad \text{Polymers}$$
$$d = 2: \quad \text{Surface physics}$$
$$d = 3: \quad \text{Most physical systems}$$
$$d = 4: \quad \text{Relativistic quantum field theory}$$

17.2 MATHEMATICAL DIGRESSION

Fourier Analysis

We denote the Fourier components of $m(x)$ by $\tilde{m}(k)$. To make them countable, we put the system in a very large d-dimensional cube of volume $V = L^d$, with periodic boundary conditions, and take the limit $L \to \infty$ eventually. Then we have

$$\tilde{m}(k) = \int_V (dx) \, e^{-ik \cdot x} m(x), \qquad k_i = \frac{2\pi n_i}{L}, \quad (n_i = 0, \pm 1, \pm 2, \dots)$$

$$m(x) = \frac{1}{V} \sum_k e^{ik \cdot x} \tilde{m}(k) \xrightarrow[V \to \infty]{} \int (dk)_i e^{ik \cdot x} \tilde{m}(k)$$

(17.8)

---

*See K. Huang, *Quarks, Leptons, and Gauge Fields* (World Scientific, Singapore, 1982), Chapter 7.

where $(dx) = d^dx$, $(dk) = d^kk/(2\pi)^d$. It is useful to remember that

$$\int_V (dx) e^{ik\cdot x} = V\delta_K(k) \xrightarrow[V\to\infty]{} (2\pi)^d \delta(k) \qquad (17.9)$$

where $\delta_K(k)$ is a Kronecker δ.

The Fourier component $\tilde{m}(k)$ is complex. For real $m(x)$ we must require $\tilde{m}(k)^* = \tilde{m}(-k)$. In that case only the vectors in a hemisphere of k space (say, the "upper" hemisphere) are independent.

The following can be easily shown using (17.8). For generality they are stated for a complex field $m(x)$:

$$\int (dx)|\nabla m(x)|^2 = \frac{1}{V}\sum_k k^2 \tilde{m}^*(k)\tilde{m}(k)$$

$$\int (dx)m(x)H^*(x) = \frac{1}{V}\sum_k \tilde{m}(k)\tilde{H}^*(k)$$

$$\int (dx)|m(x)|^2 = \frac{1}{V}\sum_k \tilde{m}^*(k)\tilde{m}(k) \qquad (17.10)$$

$$\int (dx)|m(x)|^4 = \frac{1}{V^3}\sum_{k_1,\ldots,k_4} \delta_K(k_1 + k_2 - k_3 - k_4)$$
$$\times \tilde{m}^*(k_1)\tilde{m}^*(k_2)\tilde{m}(k_3)\tilde{m}(k_4)$$

If $m(x)$ is real, each k sum above should extend only over a hemisphere instead of the entire k space, (or one could take half the sum over the entire k space.)

Functional Integration

A practical way to do the functional integral in (17.1) is to replace the continuous d-dimensional space by a discrete lattice of spacing a, which is of the order of atomic distances. We then regard the value of $m(x)$ at each lattice site x to be an independent number, and integrate them separately:

$$\int (Dm) = \prod_x \int_{-\infty}^{\infty} dm(x) \qquad (17.11)$$

where the product is taken over all the lattice sites x. The lattice constant also serves as a cutoff.

Another way to define the functional integral is to integrate each Fourier component $\tilde{m}(k)$ separately and independently, for $|k| < \Lambda$:

$$\int (Dm) = \prod_{|k|<\Lambda} \int d\tilde{m}(k)\, d\tilde{m}^*(k) \qquad (17.12)$$

The integral $\int d\tilde{m}(k)\, d\tilde{m}^*(k)$ is a two-dimensional integration over the complex plane of $\tilde{m}(k)$, equivalent to integrating the real and imaginary parts independently. The limit $L \to \infty$ here is independent of the more difficult limit $\Lambda \to \infty$.

The product in (17.12) extends over all of k space (subject to the cutoff condition) if $m(x)$ is complex, but only over a hemisphere if $m(x)$ is real. In the latter case, it might be simpler to allow k the full space, and to take the square root of the answer later.

Gaussian Integrals

The only functional integral we can do exactly is of the Gaussian type:

$$I = \int (D\phi) \exp \int (dx) \left[-\tfrac{1}{2}|\nabla\phi(x)|^2 - \tfrac{1}{2}r_0\phi^2(x) + \phi(x)\eta(x) \right]$$

$$= \int (D\phi) \exp \left[-\tfrac{1}{2}(\phi, K, \phi) + (\eta, \phi) \right] \tag{17.13}$$

where the last form rewrites the x integral as an inner product, with $K \equiv r_0 - \nabla^2$. One can obviously do the integral by Fourier analyzing ϕ and using (17.12). There is, however, a shortcut. Consider the elementary integral (done by completing the square)

$$\int_{-\infty}^{\infty} \frac{d\phi}{\sqrt{2\pi}} e^{-(1/2)K\phi^2 + \eta\phi} = \frac{1}{\sqrt{K}} e^{\eta^2/2K} \tag{17.14}$$

This can be immediately generalized to

$$\int (D\phi) \exp \left[-\tfrac{1}{2}(\phi, K\phi) + (\eta, \phi) \right] = (\det K)^{-1/2} \exp \tfrac{1}{2}(\eta, K^{-1}\eta) \tag{17.15}$$

where (A, B) denotes a general inner product between the vectors A and B, and $\int$ has the meaning given in (17.11) or (17.12). This formula is invariant under a unitary transformation of the vector ϕ, which generally transforms K into a nondiagonal matrix.

Following are some examples of the inner product:

Discrete: $(A, B) = \sum_i A_i B_i$

x space: $(A, B) = \int (dx) A(x) B(x)$ (17.16)

k space: $(A, B) = \dfrac{1}{V} \sum_k \tilde{A}(-k) \tilde{B}(k) \xrightarrow[V \to \infty]{} \int (dk) \tilde{A}(-k) \tilde{B}(k)$

17.3 DERIVATION IN SIMPLE MODELS

Model Hamiltonian

The Landau free energy can be calculated from first principles for a class of simple models. The technique relies on expressing certain exponential forms as Gaussian functional integrals. Consider a system whose Hamiltonian $\mathscr{H}$ can be represented in the form

$$\mathcal{H}(s)/kT = -\tfrac{1}{2}(s, Js) - (H, s) \tag{17.17}$$

Using (17.15) to rewrite the factor $\exp \tfrac{1}{2}(s, Js)$, (with $K = J^{-1}$), we can write the partition function as

$$Q = \sum_{(s)} \left\{ (\det J)^{-1/2} \int (D\phi) \exp\left[-\tfrac{1}{2}(\phi, J^{-1}\phi) + (s, \phi)\right] \right\} e^{(H, s)}$$

$$= (\det J)^{-1/2} \int (D\phi) \exp\left[-\tfrac{1}{2}(\phi, J^{-1}\phi)\right] \sum_{(s)} \exp\left(s, (\phi + H)\right)$$

$$= (\det J)^{-1/2} \int (D\phi) \exp\left[-\tfrac{1}{2}((\phi - H), J^{-1}(\phi - H))\right]$$

$$\times \sum_{(s)} \exp(s, \phi) \tag{17.18}$$

which is of Landau form:

$$Q = \mathcal{N} \int (D\phi) \, e^{-E[\phi, H]} \tag{17.19}$$

$$E[\phi, H] = \tfrac{1}{2}((\phi - H), J^{-1}(\phi - H)) - A[\phi] \tag{17.20}$$

$$A[\phi] = \log \sum_{(s)} \exp(s, \phi) \tag{17.21}$$

Ising Model

For the Ising model (in any dimension),

$$s = (s_1, \ldots, s_N), \qquad (s_i = \pm 1)$$

$$J_{ij} = \begin{cases} J & (\text{if } (i, j) \text{ nearest neighbor}) \\ 0 & (\text{otherwise}) \end{cases} \tag{17.22}$$

Thus ϕ is an N-component vector $(\phi_1, \ldots, \phi_N)$, and

$$\sum_{(s)} \exp(s, \phi) = \sum_{(s)} \exp(s_1\phi_1 + \cdots + s_N\phi_N) = \prod_{i=1}^{N} (2 \cosh \phi_i) \tag{17.23}$$

which leads to

$$A[\phi] = N \log 2 + \sum_{i=1}^{N} \log (\cosh \phi_i) \tag{17.24}$$

Therefore

$$Q = \mathcal{N} \prod_{i=1}^{N} \int_{-\infty}^{\infty} d\phi_i \, e^{-E[\phi, H]} \tag{17.25}$$

$$E[\phi, H] = \sum_{i, j} (\phi_i - H_i)(J^{-1})_{ij}(\phi_j - H_j) - \sum_{i=1}^{N} \log (\cosh \phi_i) \tag{12.26}$$

where H_i is the external field at the ith site.

XY **Model**

The *XY* model consists of "magnetic compasses" attached to lattice sites, with nearest neighbor interactions. The lattice can be in any number of spatial dimensions, but the compasses all rotate in the same two-dimensional plane (hence the name *XY*). Thus

$$s = (s_1, \ldots, s_N), \qquad (|s_i|^2 = 1)$$

$$\phi = (\phi_1, \ldots, \phi_N)$$

$$(s, Js) = \sum_{i,j} s_i \cdot J_{ij} s_j \tag{17.27}$$

The matrix J is the same as that given in (17.22) for the Ising model. The vectors s and ϕ lie in the two-dimensional plane of rotation of the compasses. We have

$$\sum_{(s)} \exp(s, \phi) = \prod_{i=1}^{N} \int_0^{2\pi} d\theta_i \exp(\phi_i \cos \theta_i) \tag{17.28}$$

$$\phi_i \equiv |\phi_i|$$

where θ_i is the angle between s_i and ϕ_i. Hence

$$\sum_{(s)} \exp(s, \phi) = \prod_{i=1}^{N} [2\pi I_0(\phi_i)] \tag{17.29}$$

where $I_0(x)$ is a standard Bessel function. Thus

$$a[\phi] = N \log(2\pi) + \sum_{i=1}^{N} \log I_0(\phi_i) \tag{17.30}$$

Therefore

$$Q = \mathcal{N} \prod_{i=1}^{N} \int d^2\phi_i \, e^{-E[\phi, H]} \tag{17.31}$$

$$E[\phi, H] = \sum_{i,j} (\phi_i - H_i) \cdot (J^{-1})_{ij} (\phi_j - H_j) - \sum_{i=1}^{N} \log I_0(\phi_i) \tag{17.32}$$

17.4 MEAN-FIELD THEORY

We have encountered mean-field theory earlier in different guises: the Bragg-Williams approximation in the Ising model, and the Van der Waals theory for the gas-liquid system. Within the context of the Landau theory we can formulate and understand it in a more general and unified way. It consists of making the simplest approximation to the functional integral (17.1), namely, replacing it by the maximum value of the integrand—the saddle-point approximation. For constant H, the functional form of $m(x)$ that maximizes the integrand of (17.1) is clearly a constant $\bar{m}(H, T)$, determined by the conditions

$$\frac{\partial \psi(m, H)}{\partial m}\bigg|_{m=\bar{m}} = 0$$

$$\frac{\partial^2 \psi(m, H)}{\partial m^2}\bigg|_{m=\bar{m}} \geq 0 \tag{17.33}$$

where the Landau free energy is usually expanded as a polynomial in m:

$$\psi(m, H) = -\frac{mH}{kT} + \tfrac{1}{2}r_0 m^2 + s_0 m^3 + u_0 m^4 + \cdots \tag{17.34}$$

The Gibbs free energy density is, up to a constant,

$$\frac{1}{V}G(H, T) = kT\psi(\bar{m}, 0) - \bar{m}H \tag{17.35}$$

The order parameter $M = -\partial G/\partial H$ is therefore given by

$$\frac{M}{V} = \bar{m} \tag{17.36}$$

The mean-field approximation is clearly very crude. It assumes that the only important configuration near the critical point is one of uniform density, whereas we know from experiments that density fluctuations are particularly strong in this region. For this reason, it is not quantitatively correct. However, it is of great value in exploring the qualitative phase structures of a system. Faced with a new situation, one would try it first. Let us illustrate how it can be used to describe second-order transitions—Landau's original goal.

We choose

$$\psi(m, H) = \tfrac{1}{2}r_0 m^2 + u_0 m^4 - \frac{mH}{kT} \tag{17.37}$$

where $u_0 > 0$ and $r_0 = a_0 t$. For $H = 0$ the behavior of ψ is depicted in Fig. 17.1 for $t > 0$ and $t < 0$. It illustrates how a transition from a regime with $\bar{m} = 0$ to one with $\bar{m} \neq 0$ can be described by an underlying continuous function ψ. The Landau free energy for $H = 0$ is invariant under $m \to -m$, but this symmetry is spontaneously broken when $t < 0$. The transition is second order because $\bar{m}$ is continuous at $t = 0$.

The condition determining $\bar{m}$ is

$$r_0 \bar{m} + 4u_0 \bar{m}^3 - \frac{H}{kT} = 0 \tag{17.38}$$

For $H = 0$ the only real solution is

$$\bar{m} = \begin{cases} 0 & (t > 0) \\ \pm(a_0/4u_0)^{1/2}|t|^{1/2} & (t < 0) \end{cases} \tag{17.39}$$

which immediately implies $\beta = \tfrac{1}{2}$.

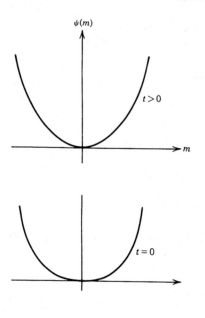

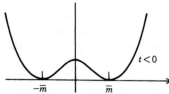

Fig. 17.1 Landau free energy for the m^4 model at different temperatures. Spontaneous symmetry breaking occurs for $t < 0$, giving rise to a second-order phase transition at $t = 0$.

To calculate the susceptibility we differentiate (17.38) with respect to H, obtaining

$$\chi = \frac{1}{kT(r_0 + 12u_0\bar{m}^2)} = \begin{cases} (kTa_0)^{-1}|t|^{-1} & (t > 0) \\ \frac{1}{2}(kTa_0)^{-1}|t|^{-1} & (t < 0) \end{cases} \quad (17.40)$$

Hence $\gamma = 1$.

The equation of state at $t = 0$ is obtained from (17.38) by setting $r_0 = 0$:

$$\bar{m}(0, H) = (4u_0kT_c)^{-1/3}H^{1/3} \quad (17.41)$$

Hence $\delta = 3$.

From (17.37) and (17.39) we have

$$\psi(\bar{m}, 0) = \begin{cases} 0 & (t > 0) \\ (3a_0^2/16u_0)t^2 & (t < 0) \end{cases} \quad (17.42)$$

from which we deduce that the heat capacity has a finite discontinuity at $t = 0$:

$$\frac{C}{V} = \begin{cases} 0 & (t > 0) \\ 3a_0^2kT_c^2/8u_0 & (t < 0) \end{cases} \quad (17.43)$$

Therefore $\alpha = 0$.

To obtain the correlation length in the same approximation we consider the response of $m(x)$ to a "test source" $H(x) = -\lambda \delta(x)$, with $\lambda \to 0$ eventually. Thus

$$E[m] = \int (dx) \left[-\tfrac{1}{2} m \nabla^2 m + \tfrac{1}{2} r_0 m^2 + u_0 m^4 + \lambda m(x) \delta(x) \right] \quad (17.44)$$

Note that we have replaced $|\nabla m|^2$ by $-m \nabla^2 m$ by performing a partial integration, discarding the surface term at infinity on the assumption that $m \to 0$ at infinity. The saddle-point approximation now leads to a partial differential equation:

$$\left(\nabla^2 - r_0 \right) m(x) - 4 u_0 m^3(x) = \lambda \delta(x) \quad (17.45)$$

For $t > 0$, the solution in the absence of source is $m = 0$. Hence we expect $m(x)$ to be of order λ, which is supposed to be the smallest parameter in the equation. Thus we can neglect the cubic term and solve the equation by taking Fourier transforms of both sides, obtaining, in the notation of (17.9),

$$\tilde{m}(k) = \frac{\lambda}{k^2 + r_0}$$

$$\overline{m}(x) = \lambda \int (dk) \frac{e^{ik \cdot x}}{k^2 + r_0} \xrightarrow[x \to \infty]{} x^{2-d} e^{-x/\xi} \quad (17.46)$$

$$\xi = r_0^{-1/2} = (a_0 t)^{-1/2}$$

Thus we have derived the Ornstein-Zernike form. The exponents are given by $\nu = \tfrac{1}{2}$, $\eta = 0$. The neglect of the cubic term in (17.45) is probably a bad approximation when $r_0 = 0$, i.e., at the critical point. (The asymptotic form of $m(x)$ is represented in a sloppy way. See Problem 17.4 for the exact answer.)

For $t < 0$ the cubic term is not negligible, for there is a spontaneous magnetization $\overline{m}$ far from the source. We put $m(x) = \overline{m} + \delta m(x)$, and expand to first order in $\delta m(x)$. The equation is of the same form as before, but with r_0 replaced by $-2r_0$. Thus we get the same exponents, although ξ is a different function:

$$\xi = \begin{cases} (a_0 t)^{-1/2} & (t > 0) \\ (-2 a_0 t)^{-1/2} & (t < 0) \end{cases} \quad (17.47)$$

We summarize the mean-field critical exponents below:

$$\alpha = 0, \quad \beta = \tfrac{1}{2}, \quad \gamma = 1, \quad \delta = 3, \quad \nu = \tfrac{1}{2}, \quad \eta = 0 \quad (17.48)$$

These values are sometimes called "classical," perhaps because they are very old. They are not consistent with experiments, as we can see from Table 16.2, though they are not far wrong.

The mean-field critical exponents have a high degree of universality, for they are independent of the parameters a_0 and u_0, and of the dimesionality of space. The only thing they could depend on is the number of components of the order parameter. Thus, they cannot satisfy Josephson's scaling law in general. (They do satisfy it for $d = 4$, the significance of which we shall discuss later.) The conclusion is that mean-field theory is not reliable quantitatively, and predicts too high a degree of universality. However, we do learn some important lessons from it: First, we learn that it is possible to understand qualitatively, and even to some extent quantitatively, how a perfectly regular function such as the Landau free energy can lead to a singular Gibbs free energy. The point is that, while $\psi(m, H, T)$ is a regular function of its arguments, the value $\bar{m}$ that minimizes ψ may be a singular function of H and T. The mean-field theory shows how such singularities can arise in a physically instructive way.

Second, we learn about universality, even if mean-field theory overstates it. The fact that the critical exponents are independent of the parameters in the Landau free energy suggests the critical phenomena are not sensitive to the details of the microscopic Hamiltonian, but possibly only to general structures, such as the number of "slow" degrees of freedom (i.e., order parameters) and the dimensionality of space. These lessons have given impetus and direction to the more refined treatments that we shall study later.

17.5 THE VAN DER WAALS EQUATION OF STATE

The Van der Waals equation of state can be regarded as a mean-field theory of gas-liquid transitions. That is, we can find a Landau free energy that gives the desired equation upon minimizing with respect to the order parameter. We do this by working backward. The Van der Waals equation in universal form is

$$\left(P + \frac{3}{V^2}\right)\left(V - \frac{1}{3}\right) = \frac{8}{3}T \tag{17.49}$$

where P, V, T respectively stand for pressure, volume, and temperature in units of the corresponding critical quantities. (Thus, for example, $t = T - 1$.) We shall take V as the order parameter and P as the conjugate field, and suppose that (17.49) is the result of minimizing a Landau free energy $\psi(V, P, T)$, i.e., it is the equation $\partial\psi/\partial V = 0$. Thus, by integrating (17.49) at constant T and P, we obtain

$$\psi(V, P, T) = PV - \frac{3}{V} - \frac{8}{3}T \log\left(V - \frac{1}{3}\right) \tag{17.50}$$

up to an undetermined function of T. (The P dependence is correct because P couples linearly to the order parameter V.) A qualitative sketch of this function is shown in Fig. 17.2. Minimizing it with respect to V yields

$$P = \frac{8}{3}\frac{T}{V - \frac{1}{3}} - \frac{3}{V^2} = \frac{\partial}{\partial V}\left\{\frac{8}{3}T \log\left(V - \frac{1}{3}\right) + \frac{3}{V}\right\} \tag{17.51}$$

which is just the Van der Waals equation written in a different form. In the region of a first-order transition, ψ must have two equal minima, occurring at the

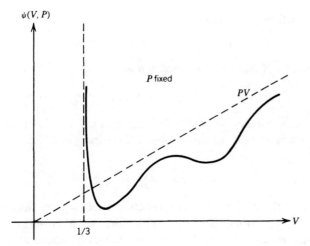

Fig. 17.2 Landau free energy that leads to the Van der Waals equation of state.

volumes V_1 and V_2, satisfying the conditions

$$(\partial\psi/\partial V)_{V=V_1} = (\partial\psi/\partial V)_{V=V_2} = 0$$
$$\psi(V_1) = \psi(V_2)$$

The first of these requires $P(V_1) = P(V_2)$, which leads to

$$\frac{8T}{3}\left(\frac{1}{V_1 - \frac{1}{3}} - \frac{1}{V_2 - \frac{1}{3}}\right) - 3\left(\frac{1}{V_1^2} - \frac{1}{V_2^2}\right) = 0 \qquad (17.52)$$

With the help of (17.50), the second condition can be rewritten

$$\psi(V_1) - \psi(V_2) = P(V_1 - V_2) - \int_{V_2}^{V_1} P \, dV = 0$$
$$P(V_1 - V_2) = \int_{V_2}^{V_1} P \, dV \qquad (17.53)$$

which is the Maxwell construction.

We now calculate the critical exponents. Differentiating (17.49) with respect to V, we have

$$\frac{\partial P}{\partial V} = \frac{6}{V^3} - \frac{8}{3}\frac{T}{\left(V - \frac{1}{3}\right)^2}$$

Approaching the critical point from $T > 1$ along $V = 1$, we find $\partial P/\partial V = -6t$. Hence the susceptibility is

$$\kappa_T = -\frac{1}{V}\left(\frac{\partial V}{\partial P}\right)_T = \frac{1}{6t} \qquad (17.54)$$

which gives $\gamma = 1$.

Near the critical point, put

$$T = 1 + t, \qquad P = 1 + p, \qquad V = 1 + v \tag{17.55}$$

Then the equation of state can be rewritten as

$$v^3 + p(1 + v)^3 - \tfrac{1}{3}(p + 8t)(1 + v)^2 = 0$$

When $t = 0$, this reduces to $p = -3v^3/2$ for small v, which shows that the critical isotherm is an odd function about the critical point. For $t < 0$, the same approximation gives

$$p - \tfrac{8}{3}t = -\tfrac{3}{2}v^3 + \tfrac{16}{3}tv \tag{17.56}$$

which shows that the isotherm is a odd function about a point displaced from the critical point by $8t/3$ along the p axis. Therefore, the Maxwell construction will yield coexisting volumes V_1, V_2 that are symmetrically placed about the critical volume $V = 1$:

$$V_1 = 1 + x, \qquad V_2 = 1 - x \tag{17.57}$$

Substituting these into (17.52) and keeping only terms to order x^2, we obtain

$$x = 2\sqrt{-t} \tag{17.58}$$

Hence $\beta = \tfrac{1}{2}$. The specific heat cannot be calculated because (17.50) is determined only up to a function of the temperature. Assuming it to be such as to give ideal gas behavior at large V, we obtain $\alpha = 0$. In summary, the critical exponents have "classical" values.

If we were interested only in the neighborhood of the critical point, we would write down a mean field theory identical to (17.37), with m identified with $v_1 - v_2$, the specific-volume difference of the coexisting phases, and H with $P - 1$. We would arrive at the same critical exponents, although the model would be quite different. In the model we just discussed, $v_1 - v_2$ is a calculated quantity, and not a free variable. Mean-field theory, as we can see, has many guises, but they lead to the same critical exponents for theories of the same class.

17.6 THE TRICRITICAL POINT

Within the mean-field approximation, consider

$$\psi(m, H) = \tfrac{1}{2}am^2 + \tfrac{1}{4}bm^4 + \tfrac{1}{6}cm^6 - mH/kT \tag{17.59}$$

For simplicity we take c to be a positive fixed constant; but a and b are variable parameters. We shall see that the system has a phase transition at $a = 0$, which is either of first or second order, depending on the sign of b. For definiteness, think of a and b as functions of temperature and pressure. Then the condition $a = b = 0$ determines the critical temperature T_c and critical pressure P_c. Introducing as usual $t = (T - T_c)/T_c$ and $p = (P - P_c)/P_c$, we shall take a and b to be linear in t and p

$$a = At + Bp$$
$$b = Ct + Dp$$
(17.60)

First put $H = 0$. To find the minima of ψ, we need

$$\psi'(m) = am + bm^3 + cm^5$$
$$\psi''(m) = a + 3bm^2 + 5cm^4$$
(17.61)

Setting $\partial\psi/\partial m = 0$, we obtain five roots, as befits a quintic polynomial:

$$\overline{m} = 0, \quad \pm m_+, \quad \pm m_-$$
(17.62)

where

$$m_\pm^2 = \frac{1}{2c}\left(-b \pm \sqrt{b^2 - 4ac}\right)$$
(17.63)

Noting that $\psi(0) = a$, we conclude that $\overline{m} = 0$ is a maximum if $a < 0$. This means that for $a < 0$, the only possible minima are $\pm m_+$. Substituting (17.62) into (17.61), we obtain

$$\psi''(m_+) = \pm\frac{1}{c}\sqrt{b^2 - 4ac}\, m_\pm^2$$
(17.64)

Therefore m_+ corresponds to a minimum, while m_- corresponds to a maximum:

$$\overline{m}^2 = \frac{1}{2c}\left(\sqrt{b^2 - 4ac} - b\right), \quad (a < 0)$$
(17.65)

Consider now $a > 0$. In this region $\overline{m} = 0$ is a minimum. The question is whether it is a lowest minimum. If $a > 0$ and $b > 0$, then we see from (17.63) that m is complex. Hence the only minimum there is $\overline{m} = 0$. Combined with the previous conclusion, we see that the positive b axis is a line of second-order transition, as shown in Fig. 17.3.

If $a > 0$ and $b < 0$, then there are three minima, i.e., $\overline{m} = 0$ and $\overline{m} = \pm m_+$. Since $\psi(0) = 0$, the locus of points in the ab plane where the three minima are equal is determined by the condition $\psi(m_+) = 0$, which leads to

$$m_+^2 = -\frac{4a}{b}$$
(17.66)

Using (17.63) we obtain

$$b = -4\sqrt{ca/3}$$
(17.67)

This is a line of first-order transitions, across which $\overline{m}$ changes discontinuously by the amount

$$\Delta m = \left(-\frac{4a}{b}\right)^{1/2} = \left(\frac{3a}{c}\right)^{1/4}$$
(17.68)

The conclusions we have reached are summarized in the phase diagram in the ab plane, in Fig. 17.3. The point $a = b = 0$ marks the termination of a

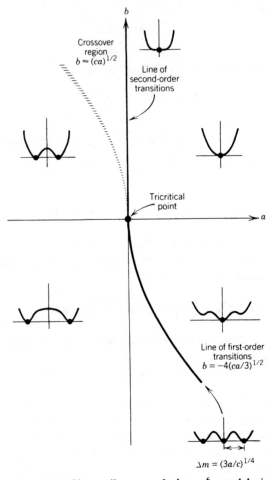

Fig. 17.3 Phase diagram of the m^6 model, in which a is the coefficient of m^2, and b is the coefficient of m^4 in the Landau free energy. The various insets show the forms of the Landau free energy in the particular regions.

second-order phase transition, and the ordinary critical point of a first-order transition. Landau calls it a "critical point of the second-order transition." We adopt the common designation "tricritical point" suggested by Griffiths.*

If we go up sufficiently far along the positive b axis, we should not be aware of the tricritical behavior and the system should undergo a second-order transition of the "classical" type. On the other hand, we expect the "classical" second-order behavior to be overshadowed by tricritical behavior near the origin.

*R. B. Griffiths, *Phys. Rev. Lett.* **24**, 715 (1970).

We therefore expect a "crossover" region in the quadrant $a < 0$, $b > 0$, which may be determined as follows: The root $\overline{m}$ is determined by the condition

$$a + b\overline{m}^2 + c\overline{m}^4 = 0 \tag{17.69}$$

Tricritical behavior results when the $\overline{m}^2$ term can be neglected, or $b\overline{m} \ll a$, which leads to

$$\frac{b^2}{c} \ll |a| \tag{17.70}$$

Classical second-order behavior results when the $\overline{m}^2$ term dominates over the $\overline{m}^4$ term, or $c\overline{m}^4 \ll b\overline{m}^2$, which leads to

$$\frac{b^2}{c} \gg |a| \tag{17.71}$$

Thus the crossover line is qualitatively given by

$$b \approx \left(|a|/c\right)^{1/2} \tag{17.72}$$

This is indicated in Fig. 17.3 by a fuzzy line.

The order parameter $\overline{m}$ as a function of a and b is shown in the three-dimensional plot of Fig. 17.4. Note that since $\psi(m)$ is an even function of m, the solution $-\overline{m}$ is equally good (just as in the Ising model). We arbitrary choose $\overline{m} > 0$.

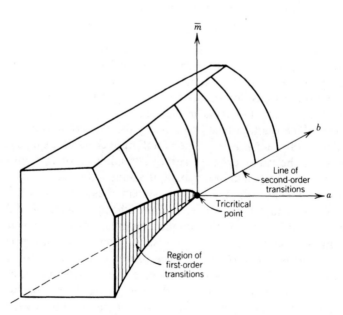

Fig. 17.4 Equation of state of the m^6 model, showing the order parameter as a function of the parameters a and b.

To derive the critical exponents at the tricritical point, restore the external field H and approach the origin of the ab plane along $b = 0$, so that a becomes proportional to t:

$$a = a_0 t, \qquad a_0 = A - \frac{BC}{D} \qquad (17.73)$$

and we shall suppose that $a_0 > 0$. We can then write

$$\psi(m) = \tfrac{1}{2} a_0 t m^2 + \tfrac{1}{6} c m^6 - \frac{mH}{kT}$$

$$\psi'(\overline{m}) = a_0 t \overline{m} + c \overline{m}^5 - \frac{H}{kT} \qquad (17.74)$$

At $H = 0$ we obtain

$$\overline{m} = \begin{cases} 0 & (t > 0) \\ (a_0/c)^{1/4} |t|^{1/4} & (t < 0) \end{cases} \qquad (17.75)$$

Hence $\beta = \tfrac{1}{4}$. Differentiating (17.74) with respect to H to get the susceptibility, we obtain

$$\chi = \begin{cases} (a_0 t)^{-1} & (t > 0) \\ \tfrac{1}{4}(-a_0 t)^{-1} & (t < 0) \end{cases} \qquad (17.76)$$

Hence $\gamma = 1$. At $t = 0$, (17.74) yields $\overline{m} = (H/c)^{1/5}$, implying $\delta = 5$. Finally, putting $\overline{m} \propto t^{1/4}$ in (17.73) for $H = 0$, we have $\psi \propto t^{3/2}$, which leads to $\alpha = \tfrac{1}{2}$. In summary, some of the critical exponents are

$$\alpha = \tfrac{1}{2}, \qquad \beta = \tfrac{1}{4}, \qquad \gamma = 1, \qquad \delta = 5 \qquad (17.77)$$

Note that Rushbrooke's law $\alpha + 2\beta + \gamma = 2$ is satisfied.

A tricritical point has been observed in the He^4–He^3 liquid mixture. The actual data is shown in Fig. 17.5 in the Tx plane, where x is the concentration of He^3. The first-order transition occurs between a He^3-rich phase and a He^3-poor phase, with a discontinuous change in the concentration. This transition ends in a critical point, which is a tricritical point, being also the terminus of the line of the second-order λ transitions. In pure He^4 the latter occurs at a single temperature $T_\lambda = 2.18$ K on the vapor pressure curve. The addition of He^3 into the liquid lowers the transition temperature, making it a function of x.

At fixed T and P, the actual system has two sets of conjugate variables—(x, μ) and (ϕ, η), where μ is the chemical-potential difference between He^3 and He^4, and ϕ is the complex superfluid order parameter, with the generally unobservable conjugate η. We shall identify our order parameter $\overline{m}$ with $x - x_c$, and the field H with μ. The parameters a and b are taken to be functions of t and η. Since only $\eta = 0$ is accessible to experiments, in our model a and b are both functions only of t, and we are going along some line in the ab plane that passes through the origin. We show a comparison between experiments and our mean-field model in the neighborhood of the tricritical point in Fig. 17.6. The resemblance is only qualitative. However, the mean-field prediction $\gamma = 1$ agrees with experiments. This is not accidental, as we shall see later.

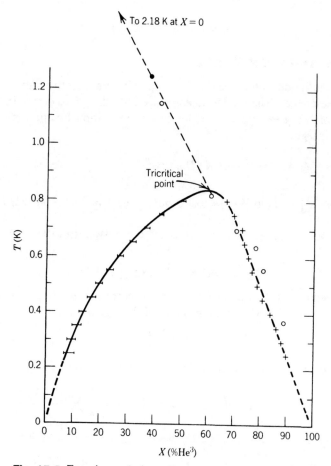

Fig. 17.5 Experimental phase diagram of the liquid He^4–He^3 mixture.

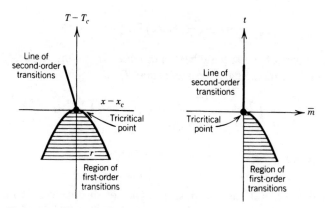

Fig. 17.6 The graph on the left is redrawn from the data of Fig. 17.5 to facilitate comparison with the prediction of mean-field theory, as shown on the right.

17.7 THE GAUSSIAN MODEL

The Gaussian model is obtained by keeping only terms up to order $m^2(x)$ in the Landau free energy. The model makes sense only if $r_0 > 0$, or $t > 0$; but it has the virtue of being exactly soluble, because the partition function is a Gaussian functional integral:

$$Q = e^{-G/kT} = \mathcal{N} \int (Dm)\, e^{-E[m, H]}$$

$$(17.78)$$

$$E[m, H] = \int (dx)\left[-\tfrac{1}{2}m(x)\nabla^2 m(x) + \tfrac{1}{2}r_0 m^2(x) - m(x)h(x)\right]$$

where $h(x) = H(x)/kT$. We can read off the answer from (17.14), with $K = -\nabla^2 + r_0$:

$$Q = \mathcal{N}(\det K)^{-1/2} \exp \tfrac{1}{2}(h, K^{-1}h)$$

$$(17.79)$$

The eigenvectors of K are $\exp(ik \cdot x)$, with corresponding eigenvalue $k^2 + r_0$. The determinant of K is just the product of the eigenvalues:

$$\det K = \prod_k \left(k^2 + r_0\right)$$

$$(17.80)$$

To calculate $(h, K^{-1}h)$, we expand $h(x)$ in a Fourier series as in (17.8), obtaining

$$(h, K^{-1}h) \equiv \int (dx)h(x)\left(-\nabla^2 + r_0\right)^{-1}h(x) = \frac{1}{V}\sum_k \frac{\tilde{h}(-k)\tilde{h}(k)}{k^2 + r_0}$$

$$(17.81)$$

Substituting these results into (17.79), we obtain, up to an additive constant,

$$\log Q = -\frac{1}{2}\sum_k \log\left(k^2 + r_0\right) + \frac{1}{2V}\sum_k \frac{\tilde{h}^*(k)\tilde{h}(k)}{k^2 + r_0}$$

$$(17.82)$$

Let us first set $h = 0$. In the limit $V \to \infty$ the Gibbs free energy is, up to an additive constant, given by

$$\frac{G}{V} = -\frac{kT}{2}\int (dk)\log\left(k^2 + a_0 t\right)$$

$$(17.83)$$

The singular part of the specific heat is obtained by taking the second derivative with respect to t, replacing T by a constant T_c:

$$\frac{C}{V} \sim \int_0^\Lambda dk\, \frac{k^{d-1}}{\left(k^2 + a_0 t\right)^2} \sim t^{(d-4)/2}\int_0^{\Lambda'(t)} ds\, \frac{s^{d-1}}{\left(s^2 + 1\right)^2}$$

$$(17.84)$$

$$\Lambda'(t) = \frac{\Lambda}{\sqrt{a_0 t}}$$

Note that this integral has two possible types of divergences. It may diverge at the upper limit when $\Lambda \to \infty$. This is called an "ultraviolet divergence." The integral may also diverge at the lower limit when $t \to 0$. This would be an "infrared divergence." The critical exponents are determined by the nature of the infrared divergence. As we can see from (17.84), a part of the infrared divergence is already isolated in the factor $t^{(d-4)/2}$. Whether this is the entire contribution depends on how the integral behaves at the upper limit, the ultraviolet domain. By simple power counting, we see that the integral converges for $d < 4$, and diverges for $d \geq 4$. These cases lead to different infrared behavior, and have to be discussed separately:

(a) $d < 4$. The integral converges when $\Lambda \to \infty$:

$$\frac{C}{V} \sim t^{(d-4)/2}$$

(b) $d = 4$. The integral diverges logarithmically when $\Lambda \to \infty$:

$$\frac{C}{V} \sim \text{const.} + \log t \sim \log t$$

(c) $d > 4$. The integral diverges like a power when $\Lambda \to \infty$:

$$\frac{C}{V} \sim \text{const.}$$

These are all consistent with the assignment

$$\alpha = \frac{4 - d}{2}$$

$$D_t = 2$$

$$(17.85)$$

where the last value is obtained from (16.59).

The value of α agrees with the mean-field value $\alpha = 0$ for $d \geq 4$, but disagrees violently for $d < 4$. In Fig. 17.7 we contrast the behavior of the specific heat in mean-field theory with that in the Gaussian model for $d < 4$. The disagreement indicates that in less than four dimensions we cannot ignore the spatial fluctuations of the order parameter.

The free energy (17.83) is ultraviolet-divergent. But the divergence does not affect the nature of the infrared singularities, thanks to the triviality of the model. However, their very presence might make one feel uneasy about the consistency of the model. In the next chapter we shall learn how to rid the free energy of ultraviolet divergences (see Section 18.6).

We now calculate the correlation function. From the definition (16.18), we obtain its Fourier transform as

$$\tilde{\Gamma}(k) = \langle \tilde{m}(k)m(0) \rangle - \langle m(0) \rangle^2 (2\pi)^d \delta(k) \qquad (17.86)$$

Now $m(0)$ is the sum of all Fourier components. When $h = 0$, we have

$$\langle \tilde{m}(k) \rangle = 0, \quad \langle \tilde{m}(k)\tilde{m}(p) \rangle = \delta_K(k + p)$$

because the Landau free energy depends only on the modulus of $\tilde{m}(k)$. Thus

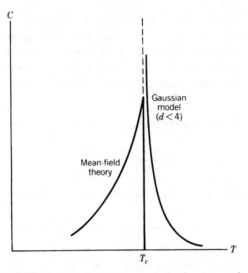

Fig. 17.7 Comparison of the heat capacity in mean-field theory and in the Gaussian model.

when $h = 0$ we have

$$\tilde{\Gamma}(k) = \frac{1}{V}\langle \tilde{m}(-k)\tilde{m}(k)\rangle \qquad (17.87)$$

This can be calculated by temporarily restoring the field h:

$$\tilde{\Gamma}(k) = \left.\left|\frac{\partial^2 \log Q}{\partial \tilde{h}(k)\,\partial \tilde{h}(-k)}\right|\right|_{\tilde{h}=0} \qquad (17.88)$$

Using this with (17.82), we obtain

$$\tilde{\Gamma}(k) = \text{const.}\ \frac{1}{k^2 + r_0} \qquad (17.89)$$

$$\Gamma(x) \sim x^{2-d}e^{-r/\xi}, \qquad \xi = (a_0 t)^{-1/2}$$

This is the Ornstein-Zernike form we obtained earlier in mean-field theory. Therefore $\nu = \frac{1}{2}$, $\eta = 0$, as in mean-field theory. Using $\eta = 0$ in (16.59), we find

$$D_h = \frac{d+2}{2} \qquad (17.90)$$

All other critical exponents can now be found from (16.59), using the known values of D_t and D_h. The results are summarized below:

$$\alpha = \frac{4-d}{2}, \qquad \beta = \frac{d-2}{4}, \qquad \gamma = 1, \qquad \delta = \frac{d+2}{d-2}, \qquad (17.91)$$

$$\nu = \frac{1}{2}, \qquad \eta = 0$$

17.8 THE GINZBURG CRITERION

From the Gaussian model we got the hint that $d = 4$ is a critical dimension, in that we obtain the mean-field exponents if $d > 4$. Why is $d = 4$ special?

The mean-field theory is a saddle-point approximation to a theory characterized by the Landau free energy

$$\psi(m(x)) = \tfrac{1}{2}|\nabla m(x)|^2 + \tfrac{1}{2}r_0 m^2(x) + u_0 m^4(x) \qquad (17.92)$$

To estimate the importance of fluctuations below the critical temperature, we can compare the fluctuations of $m(x)$ over a distance ξ with its mean value $\bar{m}$. For the fluctuations to be small we must have

$$\frac{\Gamma(\xi)}{\bar{m}^2} = \frac{\xi^{2-d}}{(-r_0/4u_0)} \ll 1$$

which, upon using $\xi = (-r_0)^{-1/2}$, leads to the condition

$$u_0[a_0|t|]^{(d-4)/2} \ll 1 \qquad (17.93)$$

This is known as the "Ginzburg criterion." It can be fulfilled in the limit $t \to 0$ only if $d > 4$.

Another way to derive the Ginzburg criterion is to note that there are two length scales in the theory, defined by r_0 and u_0. They may be deduced through dimensional analysis, as follows. The dimension of ψ is $(\text{Length})^{-d}$, since it is the volume integral of the dimensionless energy $E[m]$:

$$[\psi] = L^{-d} \qquad (17.94)$$

The dimensionality of $m(x)$ can be deduced from the fact that $|\nabla m|^2$, which is of dimension $[m]^2/L^2$, must have the same dimension as ψ. The dimensions of r_0 and u_0 are then easily deduced. The results are

$$[m(x)] = L^{(2-d)/2}$$
$$[r_0] = L^{-2} \qquad (17.95)$$
$$[u_0] = L^{d-4}$$

Mean-field theory is good when the m^4 term is dominant. That happens when the characteristic length of the "potential" $m^4(x)$ is much *shorter* than the "Gaussian" correlation length $\xi_G = r_0^{-1/2}$. That is, $u_0/(\xi_G)^{d-4} \ll 1$, which leads to (17.93).

Actually there is a neighborhood ΔT of the critical temperature, called the "critical region," in which fluctuations are always important. In this region the correlation length depends on both r_0 and u_0. Mean-field theory can be valid only outside of the critical region. What the Ginzburg criterion indicates is that $\Delta T/T_c \ll 1$ when $d > 4$. The precise value of ΔT has no universality, and depends on details of the model.

In spatial dimensions less than 4, mean-field theory fails because fluctuations are important. The Gaussian model suggests that the long wavelength fluctuations dominate the critical behavior, for it shows that they are determined by infrared singularities at $t = 0$. Since spatial fluctuations should have wave lengths longer than the correlations, the important wavelengths lie between the correlation length and infinity. But the correlation length diverges when $t \to 0$. Thus, only slightly nonuniform configurations in the neighborhood of the mean-field configuration need be taken into account. Yet it is precisely the failure to include these long wavelength fluctuations that invalidates mean-field theory.*

The critical dimension depends on the type of critical point. For a tricritical point it turns out to be $d = 3$ (see Problem 17.6), which possibly explains why the tricritical mean-field value $\nu = 1$ agrees with experiments, as we mentioned earlier.

17.9 ANOMALOUS DIMENSIONS

The dimension of $m(x)$ derived in (17.95) is in fact valid not only in the Gaussian model, but in Landau theory in general. This being so, one might be puzzled on comparing it with the dimension of its ensemble average, as derived in (16.36) on very general grounds:

$$[m(x)] = L^{-(d-2)/2} \tag{17.96}$$

$$[M/V] \equiv [\langle m(x)\rangle] = L^{-(d-2+\eta)/2} \tag{17.97}$$

Naively we would conclude $\eta = 0$, because ensemble averaging should not change the dimension. But this is contrary to experiments. Hence ensemble averaging must somehow change the dimension of $m(x)$. The naive expectation $(d - 2)/2$ is called the "canonical dimension" of M/V, while the correct answer is called the "anomalous dimension." The question is what causes the two to be different.

The answer lies in the fact that ensemble averaging injects into the problem a hidden length—the ultraviolet cutoff. Let us compare the behaviors of $m(x)$ and $\langle m(x)\rangle$ under a scale transformation $x' = x/b$. The former depends only on x, and hence transforms according to its naive dimension (like a table leg):

$$m(x/b) = b^{D_c}m(x), \qquad D_c = \frac{(d-2)}{2} \tag{17.98}$$

This is "canonical" behavior. On the other hand, the ensemble average $\langle m(x)\rangle = f(x, \Lambda)$ depends on the cutoff Λ as well. Under a scale transformation Λ also changes, contributing an extra term that makes the dimension "anomalous":

$$f(x/b, b\Lambda) = b^D f(x, \Lambda), \qquad D = D_c + \frac{\eta}{2} \tag{17.99}$$

---

*For a more detailed discussion of the Ginsberg criterion, with physical illustrations, see J. Als-Nielsen and R. J. Birgeneau, *Am. J. Phys.* **45**, 554 (1977).

But, one might ask, how can Λ make any difference when it goes to infinity eventually? The answer is that it does not do that quietly, but causes divergences that must be "renormalized" away. The process introduces an arbitrary "renormalization point," a kind of finite effective cutoff. Anomalous dimensions arise from the fact that the renormalization point changes under a scale transformation.

The entanglement with renormalization makes the calculation of anomalous dimensions nontrivial in general. It also explains why we get $\eta = 0$ in simple models like the Gaussian model, in which the ultraviolet divergences are decoupled from the infrared divergences from which one extracts critical exponents. The dimensions D given in Table 16.4 are anomalous dimensions.

PROBLEMS

17.1 The classical Heisenberg model is the same as the XY model defined in (17.27), except that the vector s_i can rotate in three dimensions, instead of being confined to a plane. Put the partition function of the classical Heisenberg model in Landau form, by methods similar to that used for the XY model in the text.

17.2 The infinite range model is an Ising model in which each spin interacts with all other spins. Take the Hamiltonian to be

$$-\mathscr{H}/kT = JS^2 + hS, \qquad S \equiv \sum_{i=1}^{N} s_i$$

where $h = H/kT$, H being the magnetic field. Show that for this model mean-field theory is exact in the limit $N \to \infty$, by the following steps:

(*a*) Write down the partition function, and use (17.14) to express it as an integral over an auxiliary field ϕ. The configuration sum can then be done.

(*b*) Assume $J = \sigma/N$. Use the saddle-point method to evaluate the partition function in the limit $N \to \infty$. Show that the saddle-point condition gives the mean-field theory as described by (14.41).

17.3 Consider mean-field theory with a cubic term in the Landau free energy:

$$\psi(m) = \tfrac{1}{2}r_0 m^2 + s_0 m^3 + u_0 m^4$$

Sketch the shape of $\psi(m)$ for various values of r_0 to show how the lowest minimum $\overline{m}$ of $\psi(m)$ depends on r_0. Show that there is a first-order phase transition at a certain value of r_0. Find that value, and the discontinuity of $\overline{m}$ at the transition.

17.4 In the Gaussian model the Fourier transform of the correlation function is $\tilde{\Gamma}(k) = C(k^2 + r_0)^{-1}$. Take the inverse Fourier transform to calculate $\Gamma(x)$ in d-dimensional space. Show the following for $d \neq 2$:

$$\Gamma(x) = C'|x|^{(2-d)/2}K_{\nu}(z), \qquad \nu = \frac{d-2}{2}, \qquad z = \sqrt{r_0}\,|x|$$

where K_ν is a Bessel function:

$$K_\nu(z) = \int_0^\infty dx\, e^{-z\cosh x} \cosh(\nu x)$$

$$\xrightarrow[z \to \infty]{} \sqrt{\pi/z}\, e^{-z}$$

$$\xrightarrow[z \to 0]{} \tfrac{1}{2}(z/2)^\nu \Gamma(\nu)$$

The asymptotic behaviors of the correlation function are

$$\Gamma(x) \xrightarrow[r \to \infty]{} C'|x|^{(3-d)/2}\, e^{-\sqrt{r_0}\,x}, \qquad (r_0 \neq 0)$$

$$\xrightarrow[r \to \infty]{} C''|x|^{2-d}, \qquad\qquad (r_0 = 0)$$

17.5 Prove (17.47), which gives the correlation length for $T > T_c$ and $T < T_c$.

17.6 Derive the Ginsburg criterion for a tricritical point. What is the critical dimension in this case?

CHAPTER
18

RENORMALIZATION GROUP

18.1 BLOCK SPINS

We have learned that the critical properties of a system depend not on short-distance details, but only on the nature of long-wavelength fluctuations. This suggests that one should do away with the irrelevant degrees of freedom by continuing the coarse-graining procedure (through which the details on an atomic scale had been averaged out) to ever larger distance scales, until one reaches the correlation length.

Kadanoff* first introduced this idea in terms of "block spin" transformations in Ising models. The idea here is that near the critical point the spins should act in concert in large blocks. Thus the important degree of freedom are the average spins of the blocks, rather than the original individual spins. One should describe the system in terms of an effective Hamiltonian involving only the "block spins."

Consider an Ising model in d dimensions, with spin configurations specified by $\{s\} = \{s_1, s_2, \ldots\}$, and a dimensionless Hamiltonian (in units of kT) denoted by

$$E\{s\} = -J \sum_{\langle i, j \rangle} s_i s_j - h \sum_{i=1}^{N} s_i \tag{18.1}$$

where $J = \epsilon/kT$ and $h = H/kT$. Assume the model has a critical point at $T = T_c$ corresponding to a critical coupling J_c. Let

$$t = \frac{(J - J_c)}{J_c} \tag{18.2}$$

The partition function for N spins is

$$Q_N(h, t) \equiv e^{-Ng(h, t)} = \sum_{\{s\}} e^{-E\{s\}} \tag{18.3}$$

where $g(h, t) = G(h, t)/NkT$, $G(h, t)$ being the free energy.

*L. P. Kadanoff, *Physics* **2**, 263 (1966).

Since the coupling constants J and h are dimensionless, the unit of length is set by the lattice spacing (which does not actually appear in the Hamiltonian). For example, the correlation function $\Gamma(i, j)$ refers only to site numbers i and j, the "distance" between sites i and j being defined as the number of lattice spacings between the sites along a minimal path.

Now divide the lattice into blocks of size b^d, and define the block spin $s' = \pm 1$ of a given block as some sort of average of the spins contained in the block. For example, we can use the "majority rule," $s' = \pm 1$, if the majority of the spins in the block is $\pm$. In case of a tie, we arbitrarily define $s' = s_1$, where s_1 is a particular spin, say the one at the upper left corner of the block. Denoting by S the sum of all spins in the block, we have

$$s' = \begin{cases} 1 & (S > 0) \\ -1 & (S < 0) \\ s_1 & (S = 0) \end{cases} \qquad (18.4)$$

After one block-spin transformation, the number of spins in the system is reduced by a factor b^{-d}:

$$N' = b^{-d}N \qquad (18.5)$$

The new lattice spacing is increased b-fold, which means that the unit of length is automatically increased by a factor b. For example, if the site spins are correlated over n lattice spacings, the block spins are correlated over n/b lattice spacings. The process is illustrated in Fig. 18.1.

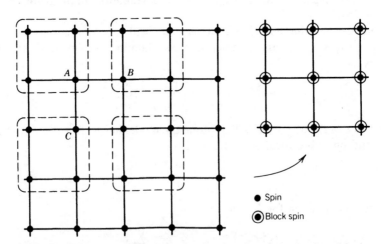

Fig. 18.1 Block-spin transformation: averaging the spins in a block, and then rescaling the lattice to the original size. In more than one dimension, the indirect interaction between B and C gives rise to next-to-nearest-neighbor interactions of the block spins.

Kadanoff assumes that sufficiently close to the critical point one can describe the system by a block-spin Hamiltonian, which again contains only nearest-neighbor interactions:

$$E'\{s'\} = -J' \sum_{\langle i, j \rangle} s_i' s_j' - h' \sum_{i=1}^{N'} s_i' \tag{18.6}$$

The partition function of the block-spin system is thus

$$Q_{N'}(h', t') \equiv e^{-N'g(h', t')} = \sum_{\{s'\}} e^{-E'\{s'\}} \tag{18.7}$$

Note that the free energy g is the same function as in (18.3), because by assumption E' is the same function as E, except for the values of the coupling constants.

Kadanoff now argues that the block-spin system is merely another way of describing the original system, and hence we must have $Q_{N'} = Q_N$. Using (18.5) one can write

$$g(h, t) = b^{-d}g(h', t') \tag{18.8}$$

Now assume

$$t' = b^A t, \qquad h' = b^B h \tag{18.9}$$

Then

$$g(t, h) = b^{-d}g(b^A t, b^B h) \tag{18.10}$$

which is the homogeneity rule (16.55). We have already seen how it leads to Widom's scaling form.

Kadanoff's ideas have great heuristic value, but do not lead to a calculational scheme. There are simply too many ad hoc assumptions, some of them clearly wrong. For example, block spins generally have more complicated interactions than just nearest neighbor. One can see this in the two-dimensional case illustrated in Fig. 18.1. The sites B and C both interact with A. Thus one expects that the block to which B belongs interacts with the block to which C belongs, giving rise to next-to-nearest-neighbor interactions.

18.2 THE ONE-DIMENSIONAL ISING MODEL

A block-spin transformation in one dimension preserves the nearest-neighbor nature of the interactions, as one can see by inspecting Fig. 18.2. Unfortunately the model does not have a critical point, so that its pedagogical value is somewhat limited. Nevertheless we have here the only case in which a block-spin transformation can be carried out analytically. It is worth examining for that alone.

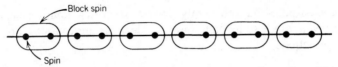

Fig. 18.2 Block-spin transformation in one dimension. The block spins only have nearest-neighbor interactions.

The partition function for an N-spin chain is the trace of the Nth power of the transfer matrix P, as shown in (14.81). If we want to describe the system in terms of 2-spin blocks, the same partition function should be re-expressed as the trace of a new transfer matrix P' raised to the power $N/2$. This is achieved through a trivial rewriting of the partition function:

$$Q_N = \operatorname{Tr} P^N = \operatorname{Tr} \left[(P^2)^{N/2} \right] \tag{18.11}$$

where

$$P = \begin{pmatrix} e^{J+h} & e^{-J} \\ e^{-J} & e^{J-h} \end{pmatrix} \equiv \begin{pmatrix} \dfrac{1}{uv} & u \\ u & \dfrac{v}{u} \end{pmatrix} \tag{18.12}$$

$$u = e^{-J}, \qquad v = e^{-h} \qquad (0 \le u \le 1, \quad 0 \le v \le 1)$$

The limits on u and v come from the conditions $J > 0$ and $h > 0$. The former restricts us to the ferromagnetic case. The second condition occasions no loss in generality because the model is invariant under $h \to -h$.

The transfer matrix for block spins is

$$P' \equiv P^2 = \begin{pmatrix} u^2 + \dfrac{1}{u^2 v^2} & v + \dfrac{1}{v} \\ v + \dfrac{1}{v} & u^2 + \dfrac{v^2}{u^2} \end{pmatrix} \tag{18.13}$$

We demand that P' have the same form as P:

$$P' = C \begin{pmatrix} \dfrac{1}{u'v'} & u' \\ u' & \dfrac{v'}{u'} \end{pmatrix} \tag{18.14}$$

which defines the parameters u' and v' in the block-spin system. A new parameter C must be introduced, because to match (18.13) with (18.14) requires matching three matrix elements, which is generally impossible with only two variables u' and v'. With C, we have three unknowns to satisfy the three conditions

$$Cu' = v + \frac{1}{v}$$

$$\frac{C}{u'v'} = u^2 + \frac{1}{u^2v^2}$$

$$\frac{Cv'}{u'} = u^2 + \frac{v^2}{u^2}$$

(18.15)

The solution is

$$u' = \frac{\left(v + \dfrac{1}{v}\right)^{1/2}}{\left(u^4 + \dfrac{1}{u^4} + v^2 + \dfrac{1}{v^2}\right)^{1/4}}$$

$$v' = \frac{\left(u^4 + v^2\right)^{1/2}}{\left(u^4 + \dfrac{1}{v^2}\right)^{1/2}}$$

(18.16)

$$C = \left(v + \frac{1}{v}\right)^{1/2}\left(u^4 + \frac{1}{u^4} + v^2 + \frac{1}{v^2}\right)^{1/4}$$

The block-spin transformation can be regarded as a mapping $(u, v) \rightarrow (u', v')$ in parameter space. By carrying out the block-spin transformation repeatedly, an initial point (u, v) in parameter space generates a sequence of points, which may be connected to form a trajectory. By doing this for different initial points, we obtain the "flow diagram" of Fig. 18.3.

A salient feature of the mapping are the "fixed points"—values (u, v) that remain unchanged under the transformation:

$$u = 0, \qquad v = 1 \qquad (\infty \text{ interaction, 0 field})$$

$$u = 1, \qquad \text{all } v \qquad (0 \text{ interaction, any field})$$

The fixed point $(0, 1)$ is unstable in the sense that any point in its neighborhood will tend to go away from it under a block-spin transformation. The fixed points on the line $u = 1$ are stable.

At the fixed points the correlation length ξ is invariant under a scale change, and therefore can only be 0 or ∞. The line $u = 0$ corresponds to $T = 0$, with $\xi = \infty$ at $(0, 1)$. The line $u = 0$ corresponds to $T = \infty$, with $\xi = 0$ all along this line. Unfortunately the fixed point $(0, 1)$ is inaccessible. You are either already there or you go away from it. This expresses the fact that the system has no critical point.

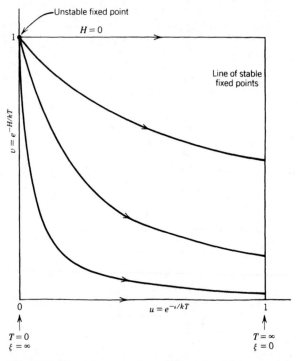

Fig. 18.3 Flow diagram of one-dimensional Ising model, showing how the coupling constant ϵ and the external field H change under successive block-spin transformations.

18.3 RENORMALIZATION-GROUP TRANSFORMATION

The operation of coarse-graining followed by rescaling is called a "renormalization-group" (RG) transformation, of which the block-spin transformation in an Ising model is an example. We now give a formal and general definition of the latter. This is possible owing to the simplicity of the model.* But the results are very instructive, and the concepts illustrated in this case are useful in more general models.

Consider an Ising model with N spin variables $s_i = \pm 1$ defined on sites of a d-dimensional cubic lattice. We will have to include the most general types of interactions, in order that a block-spin transformation, which may generate arbitrarily complicated interactions, can be described as a mapping in parameter space. To this end, let I_α denote an arbitrary set of site labels. Let S_α denote the product of all spins on the sites in I_α:

$$S_\alpha = \prod_{i \in I_\alpha} s_i \tag{18.17}$$

---

*We follow Th. Niemeijer and J. M. J. van Leeuwen, in *Phase Transitions and Critical Phenomena*, Vol. 6, C. Domb and M. S. Green, eds. (Academic Press, New York, 1976), pp. 425–505.

The most general Hamiltonian (in units of kT) is

$$E\{s\} = \sum_\alpha K_\alpha S_\alpha \qquad (18.18)$$

where K_α is a coupling constant for the set of spins in S_α, and the sum over α extends over all possible sets I_α. Note that

$$\sum_{\{s\}} E\{s\} = 0 \qquad (18.19)$$

In principle we can solve for K_α through

$$K_\alpha = 2^{-N} \sum_{\{s\}} S_\alpha E\{s\} \qquad (18.20)$$

For formal manipulations it costs us nothing to regard K_α as completely arbitrary. For practical purposes it suffices for us to think of $E\{s\}$ as

$$E\{s\} = K_1 \sum_{i=1}^N s_i + K_2 \sum_{\langle i,j \rangle} s_i s_j + K_2' \sum_{\langle\langle i,j \rangle\rangle} s_i s_j + K_3 \sum_{\langle i,j,k \rangle} s_i s_j s_k + \cdots \qquad (18.21)$$

where $\langle i, j \rangle$ denote nearest-neighbor pairs, $\langle\langle i, j \rangle\rangle$ next-to-nearest pairs, and $\langle i, j, k \rangle$ nearest-neighbor triplets, etc.

Now divide the entire lattice into identical cubical blocks that cover the whole lattice, with b sites along each edge of the block. There are thus b^d spins in the block B, which we denote collectively as $\{s\}_B$. The block spin is

$$s_B' = f\{s\}_B \qquad (18.22)$$

where f is a mapping of $\{s\}_B$ into the set $\{1, -1\}$. For example, for the majority rule with tiebreaker, f is the function defined in (18.4). It is convenient to define

$$P_B = \delta_K(s', f\{s\}_B) \qquad (18.23)$$

where δ_K is the Kronecker δ. This function tells us whether a particular configuration $\{s\}_B$ gives $s_B' = 1$ or $s_B' = -1$. Taking the product of P_B over all blocks, we have a weight function

$$P\{s', s\} = \prod_B P_B \qquad (18.24)$$

which depends on the set of all block spins $\{s'\}$ and the set of all original spins $\{s\}$. It is equal to 1 if $\{s\}$ gives rise to $\{s'\}$, and 0 otherwise. Clearly,

$$P\{s', s\} \geq 0$$
$$\sum_{\{s'\}} P\{s', s\} = 1 \qquad (18.25)$$

These are the only properties of a block spin we shall use.

The partition function of the system can now be written as

$$Q = \sum_{\{s\}} e^{-E\{s\}} = \sum_{\{s'\}} \sum_{\{s\}} P\{s', s\} e^{-E\{s\}} \qquad (18.26)$$

where we have used (18.25). The block-spin Hamiltonian $E'\{s'\}$ is defined by

$$e^{-E'\{s'\}} \equiv e^{N\mu} \sum_{\{s\}} P\{s', s\} e^{-E\{s\}} \tag{18.27}$$

where the constant μ is to be so chosen that

$$\sum_{\{s'\}} E'\{s'\} = 0 \tag{18.28}$$

which conforms to (18.19). Thus, E' is again of the form (18.18), except that K_α is replaced by a new value K'_α.* We can now rewrite (18.24) as

$$Q = \sum_{\{s\}} e^{-E\{s\}} = e^{-N\mu} \sum_{\{s'\}} e^{-E'\{s'\}} \tag{18.29}$$

The transformation from E to E' is called a renormalization-group (RG) transformation, formally indicated by

$$E' = R(E) \tag{18.30}$$

In the limit of an infinite system, the sets $\{s\}$ and $\{s'\}$ become the same, and only the coupling constants K_α change in an RG transformation. Thus it is more appropriate to represent an RG transformation in the form

$$K'_\alpha = R_\alpha(K_1, K_2, \ldots) \tag{18.31}$$

Regarding K as components of a vector, and R a matrix operator, we can also write

$$K' = R(K) \tag{18.32}$$

The RG transformations are so named because they "renormalize" the coupling constants, and that they have group property: If $R_1(K)$ and $R_2(K)$ are RG transformations, so is $R_1 R_2(K)$. They do not form a group because block spins cannot be "unblocked," and thus there are no inverse transformations. (Mathematically one can invert the map R, except possibly at isolated singular points; but this has no physical relevance.)

Define the free energy per spin $g(K)$ (in units of kT) by

$$e^{-Ng(K)} = \sum_{\{s\}} e^{-E\{s\}} \tag{18.33}$$

In the block-spin system we have

$$e^{-N'g(K')} = \sum_{\{s'\}} e^{-E'\{s'\}} \tag{18.34}$$

where $N' = b^{-d}N$. The same function g appears in both (18.33) and (18.34) because $E\{s\}$ and $E'\{s'\}$ are the same functions except for the values of the

---

*The reason is that (18.18) is the most general form of the Hamiltonian whether we are dealing with the original spins or the block spins.

coupling constants. Using (18.29) we obtain

$$g(K) = \mu(K) + b^{-d}g(K')$$ (18.35)

This, together with (18.32), describes how the system behaves under an RG transformation, which increases the unit of length by a factor b.

18.4 FIXED POINTS AND SCALING FIELDS

Let $K^{(n)}$ denote the coupling constants resulting from n successive applications of a given RG transformation. These coupling constants are given by the recursion relation

$$K^{(n+1)} = R(K^{(n)})$$ (18.36)

It is important to note that R is independent of n, a fact that can be proven by constructing R formally through (18.22).

A fixed point K^* of the map R is defined by

$$K^* = R(K^*)$$ (18.37)

We assume that $K^{(n)}$ approaches a fixed point as $n \to \infty$.* The Hamiltonian E^* corresponding to K^* is called the fixed-point Hamiltonian. A fixed point could be physically significant, because it is a point at which the system becomes invariant under a change of length scale. That means the correlation length is either 0 or ∞. The latter corresponds to a critical point, which is the physically interesting case. The case with zero correlation length, as we have encountered in the one-dimensional Ising model, corresponds to infinite temperature, and usually can be recognized and rejected.

We now investigate the behavior of the system near a fixed point, which we assume to correspond to a critical point. Subtracting (18.37) from (18.36), we have

$$K^{(n+1)} - K^* = R(K^{(n)}) - K^*$$ (18.38)

Assuming n to be very large we can make the linear approximation

$$R(K^{(n)}) = R(K^*) + \mathbf{W}(K^{(n)} - K^*)$$ (18.39)

where $\mathbf{W}$ is the matrix whose elements are

$$\mathbf{W}_{\alpha\beta} = \left. \frac{\partial R_\alpha(K)}{\partial K_\beta} \right|_{K=K^*}$$ (18.40)

Substituting (18.39) into (18.38), using (18.37), we obtain

$$K^{(n+1)} - K^* = \mathbf{W}(K^{(n)} - K^*)$$ (18.41)

---

*In principle $K^{(n)}$ may approach a fixed point, go into a limit cycle, or exhibit ergodic behavior.

Now choose K^* as the origin in the coupling-constant space, and introduce new coordinate axes along the directions defined by the left eigenvectors of the matrix $\mathbf{W}$:

$$\phi\mathbf{W} = \lambda\phi \tag{18.42}$$

There are of course many different eigenvectors. We suppressed their labeling for simplicity. The vector $K^{(n)}$ can be represented by the coordinates

$$v^{(n)} = \sum_\alpha \phi_\alpha (K^{(n)} - K^*)_\alpha \tag{18.43}$$

which are called "scaling fields." Their usefulness lies in the fact that they do not mix with one another under the RG transformation, as the following calculation shows:

$$v^{(n+1)} = \sum_\alpha \phi_\alpha (K^{(n+1)} - K^*)_\alpha = \sum_\alpha \phi_\alpha \mathbf{W}_{\alpha\beta}(K^{(n)} - K^*)_\beta$$
$$= \lambda \sum_\alpha \phi_\alpha (K^{(n)} - K^*)_\alpha = \lambda v^{(n)} \tag{18.44}$$

Under the RG transformation, v is said to "scale" with a factor λ. It increases if $\lambda > 1$, and decreases if $\lambda < 1$. Since the RG transformation increases the unit of length by a factor b, we expect λ to have the form

$$\lambda = b^y \tag{18.45}$$

where y is the dimension of v.

In the neighborhood of a fixed point, it is convenient to use the scaling fields $\{v_1, v_2, \dots\}$ as independent variables, replacing the coupling constants $\{K_1, K_2, \dots\}$. The fixed point corresponds to all $v = 0$. We can rewrite (18.35) in the form

$$g(v_1, v_2, \dots) = \mu(v_1, v_2, \dots) + b^{-d}g(\lambda_1 v_1, \lambda_2 v_2, \dots) \tag{18.46}$$

There being no reason for μ to be singular at the fixed point, we shall assume it is regular. We identify the fixed point with a critical point, and identify the second term in (18.46) as the singular part of the free energy. It satisfies the homogeneity rule

$$g_{\text{sing}}(v_1, v_2, \dots) = b^{-d}g_{\text{sing}}(\lambda_1 v_1, \lambda_2 v_2, \dots) \tag{18.47}$$

A scaling field is called "irrelevant" if $\lambda < 1$, because it tends to 0 under repeated coarse-graining. In the neighborhood of a critical point the system behaves as if it had never existed. It is called "relevant" if $\lambda > 1$, for any nonzero initial value will be magnified under coarse-graining. To be at the critical point, we have to specially set it to zero. The case $\lambda = 1$ is called "marginal," which we shall not consider, for it depends on the details of the system.

In the coupling-constant space (K space), the fixed point lies on a hypersurface, called the "critical surface," defined by $v_i = 0$ for all the *relevant* scaling fields v_i. A point on the critical surface will approach the fixed point under successive RG transformations, while a point not on the surface will eventually veer away from the fixed point, as illustrated schematically in Fig. 18.4. Since

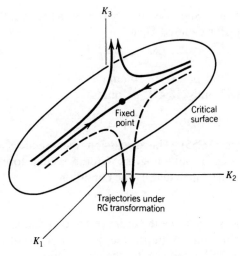

Fig. 18.4 The critical surface for a particular fixed point. It is a hypersurface in coupling-constant space obtained by setting all relevant variables to zero. Points on this surface correspond to systems in the same universality class, with the same critical exponents.

each point in K space represents a physical system, the critical surface contains different systems belonging to a universality class, sharing the same critical properties.

To illustrate how the critical exponents can be obtained from the eigenvalues λ, let us specialize to a familiar case by assuming that there are two relevant fields, v_1 and v_2, identified respectively with field h and temperature t:

$$v_1 = h, \qquad \lambda_1 = b^{D_h}$$
$$v_2 = t, \qquad \lambda_2 = b^{D_t} \tag{18.48}$$

The behavior of the correlation length ξ at $h = 0$ can be deduced as follows. Under an RG transformation the unit of length increases by a factor b. Hence $\xi' = \xi/b$. By definition $t' = b^{D_t}t$. Hence $\xi^{D_t}t$ is invariant under the RG, i.e., $\xi^{D_t}t \sim 1$. Therefore

$$\xi(t) \sim t^{-1/D_t}, \qquad \therefore \nu = \frac{1}{D_t} \tag{18.49}$$

a result we had assumed earlier in (16.56).

The argument above can be rephrased more physically. Under successive block-spin transformations, there will be come a point when the correlation length is equal to the size of a block, so that nearest-neighbor blocks are uncorrelated. This point must correspond to a definite t_0, which cannot depend on the initial temperature. Suppose one starts at some temperature t, and arrives at t_0 after n RG transformations. Then

$$t_0 = b^{nD_t}t \tag{18.50}$$

Noting that $\xi = b^n$ at this point, we obtain (18.49).

Substituting (18.48) into (18.47), we obtain

$$g_{\text{sing}}(h, t) = b^{-d}g_{\text{sing}}(b^{D_h}h, b^{D_t}t) \tag{18.51}$$

which is the same as (16.55). The discussion then reduces to that following (16.55). The crucial step in obtaining the Widom scaling form, from which the critical exponents can be calculated, is the condition

$$b = |t|^{-1/D_t}$$

which chooses the block size to be the order of the correlation length. This step is the "true" renormalization. With it, the block size b disappears from the problem, blotting out all reference to the microscopic structure. The critical exponents are given in terms of D_h and D_t in (16.59).

In conclusion, we have shown that

1. A calculation of the eigenvalues of the matrix **W**, which represents the RG transformation near a fixed point, will yield all the critical exponents corresponding to that fixed point.

2. The critical exponents are the same for all systems in the universality class defined by the critical surface containing the fixed point.

18.5 MOMENTUM-SPACE FORMULATION

We shall define the RG transformation for the physically more interesting Landau theory. In principle we have to consider the most general form of the Hamiltonian (in units of kT) in Landau theory:

$$E[m] = \int (dx)\psi(m(x))$$

$$\psi(m(x)) = \tfrac{1}{2}|\nabla m(x)|^2 + \sum_{n=1}^{\infty} K_n m^n(x) + \cdots \tag{18.52}$$

where terms not written out involve higher derivatives of $m(x)$, each with its own coupling constant. The coefficient of the first term is chosen to be $\tfrac{1}{2}$ to fix the scale of the order parameter. The coupling constants are functions of the cutoff Λ, which eventually tends to infinity. Our goal is to obtain finite physical answers in that limit.

To that end, we continue the coarse-graining, which introduced Λ in the first place, to ever longer wavelengths. As before, the RG transformation is defined as coarse-graining followed by rescaling, to make the system look like the original one. We shall define the procedure in "momentum space," the k space of Fourier

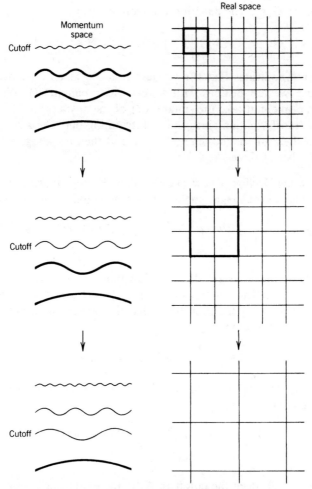

Fig. 18.5 Coarse-graining in momentum space and in real space. In the former, one effectively lowers the cutoff. In the latter, one blots out finer details, enlarging the effective lattice spacing.

analysis. Figure 18.5 shows a comparison between coarse-graining in momentum and in "real" space: Instead of expanding the unit cell in real space, we lower the effective cutoff in momentum space.

In terms of the Fourier transforms $\tilde{m}(k)$ of the order parameter defined in (17.8), the Hamiltonian takes the form

$$E[\tilde{m}] = \frac{1}{2} \int (dk)(k^2 + r_0)|\tilde{m}(k)|^2 + \cdots \qquad (18.53)$$

where we only display the Gaussian terms (putting $K_2 = r_0/2$) to establish the normalization of $\tilde{m}(k)$. The partition function is given by

$$Q \equiv e^{-G} = \mathcal{N} \prod_{|k| < \Lambda}' \int d\tilde{m}(k)\, d\tilde{m}*(k)\, e^{-E[\tilde{m}]} \tag{18.54}$$

where the prime indicates that the product is to be taken only over half of k space (because we are dealing with a real order parameter). We could also disregard the prime, but take the square root of the answer.

The RG transformation consists of "integrating out" the k values within a shell between the radii Λ and Λ/b ($b > 1$), and then rescaling. Specifically the procedure consists of three steps:

1. *Integration.* Define a new Hamiltonian E' by "integrating out" the k values whose magnitude lies between Λ and Λ/b ($b > 1$):

$$e^{-E'[\tilde{m}]} \equiv e^{\Omega} \prod_{\frac{\Lambda}{b} < |k| < \Lambda}' \int d\tilde{m}(k)\, d\tilde{m}*(k)\, e^{-E[\tilde{m}]} \tag{18.55}$$

where the constant Ω is a function of Λ and all the coupling constants. The new Hamiltonian $E'[\tilde{m}]$ depends only on $\tilde{m}(k)$ with $|k| < \Lambda/b$. Apart from that, it has the same form as (18.53) with new coupling constants:

$$E'[\tilde{m}] = \frac{1}{2} \int (dk)(Ak^2 + \tilde{r}_0)|\tilde{m}(k)|^2 + \cdots \tag{18.56}$$

Note that the coefficient of the k^2 term is changed. The partition function can be rewritten as

$$Q \equiv e^{-G} = e^{\Omega}\mathcal{N} \prod_{|k| < \frac{\Lambda}{b}}' \int d\tilde{m}(k)\, d\tilde{m}*(k)\, e^{-E'[\tilde{m}]} \tag{18.57}$$

2. *Rescaling.* Restore the cutoff to Λ by increasing the unit of length by a factor b, by changing the variable of integration to

$$k' = bk \tag{18.58}$$

The Hamiltonian now reads

$$E'[\tilde{m}] = \frac{b^{-d}}{2} \int (dk')\left(\frac{A}{b^2}k'^2 + \tilde{r}_0\right)\left|\tilde{m}\left(\frac{k'}{b}\right)\right|^2 + \cdots \tag{18.59}$$

3. *Normalization.* Restore the standard normalization of the order parameter, i.e., make the coefficient of the k'^2 term in (18.59) equal to $\frac{1}{2}$. This can be done by replacing $\tilde{m}(k)$ by

$$\tilde{m}'(k') \equiv \sqrt{\frac{A}{b^{d+2}}}\, \tilde{m}\left(\frac{k'}{b}\right) \tag{18.60}$$

which is the analog of the block-spin transformation. The final RG-transformed Hamiltonian is

$$E'[\tilde{m}'] = \frac{1}{2}\int (dk')(k'^2 + r_0')|\tilde{m}'(k')|^2 + \cdots, \qquad \left(r_0' = \frac{b^2}{A}\tilde{r}_0\right) \quad (18.61)$$

The partition function is, in terms of E',

$$Q = e^{\Omega}\mathcal{N}' \prod_{|k'| < \Lambda}' \int d\tilde{m}'(k')\, d\tilde{m}'^*(k')\, e^{-E'[\tilde{m}']} \quad (18.62)$$

where $\mathcal{N}'$ is a new normalization constant. (It is infinite both in the infinite-volume limit and the infinite-cutoff limit, but physically irrelevant.)

The net result is a mapping in coupling-constant space, of the same form as (18.32). The free energy also satisfies a relation like (18.35). Therefore the formal analysis in the last section can be taken over without change.

Because of the continuous nature of the model here, we can carry out the RG transformation in infinitesimal steps; i.e., we can take $b \sim 1$. The infinitesimal transformation contains all the information needed to obtain the critical properties of the system. In particular, all we need to know are the rate of change of the coupling constants with respect to b. Thus, the problem can be reduced to solving a set of differential equations. We shall not develop the general theory further, but merely illustrate the main points through examples.

18.6 THE GAUSSIAN MODEL

The Three Steps

We illustrate the recipe for momentum-space RG transformation in the Gaussian model. The Hamiltonian is

$$E[\tilde{m}] = \frac{1}{2}\int (dk)(k^2 + r_0)|\tilde{m}(k)|^2 - h\tilde{m}(0) \quad (18.63)$$

where $h = H/kT$ is a spatially constant external field.

Step 1 is trivial, because the k terms can be integrated out individually:

$$Q = \mathcal{N}' \prod_{|k| < \frac{\Lambda}{b}}' \int d\tilde{m}(k)\, d\tilde{m}^*(k)\, e^{-E'[\tilde{m}]} \quad (18.64)$$

Note that $E' = E$. The only change is the cutoff, and the normalization constant.

Step 2 restores the old cutoff by changing the variable of integration to $k' = bk$:

$$E'[\tilde{m}] = \frac{b^{-d}}{2}\int (dk')\left(\frac{k'^2}{b^2} + r_0\right)\left|\tilde{m}\left(\frac{k'}{b}\right)\right|^2 - h\tilde{m}(0) \quad (18.65)$$

Step 3 restores the normalization of $\tilde{m}(k)$:

$$\tilde{m}'(k') \equiv b^{-(d+2)/2}\tilde{m}\left(\frac{k'}{b}\right) \tag{18.66}$$

$$E'[\tilde{m}'] = \frac{1}{2}\int(dk')\left(k'^2 + r_0'\right)|\tilde{m}'(k')|^2 - h'\tilde{m}'(0) \tag{18.67}$$

where

$$r_0' = b^2 r_0, \qquad \therefore D_t = 2$$

$$h' = b^{(d+2)/2}h, \qquad \therefore D_h = \frac{d+2}{2} \tag{18.68}$$

The only fixed point is $r_0 = h = 0$, and both r_0 and h are relevant variables, (since $b > 1$). The scaling of these variables are just what naive dimensional analysis would predict.

The correlation length can be obtained by setting $\xi^2 r_0 = $ const., by the same reasoning as in the derivation of (18.48). Thus

$$\xi \sim r_0^{-1/2} \sim t^{-1/2} \tag{18.69}$$

The critical exponents can be obtained from (16.59) using D_t and D_h, with results as given earlier in (17.91).

Though trivial, the above example illustrates an important point: To obtain the critical exponents in a continuous system, it suffices to make an infinitesimal RG transformation, because all we need is the rate of change of the coupling constants.

Renormalization of the Partition Function

Since the partition function can be calculated exactly, we can show explicitly how one can renormalize it to obtain a finite free energy in the limit of infinite cutoff. For simplicity consider the case of no external field. The partition function has been explicitly calculated in (17.79):

$$Q(a_0 t, \Lambda) = \mathcal{N}\prod_{|k|<\Lambda}\left(k^2 + a_0 t\right)^{-1/2} \tag{18.70}$$

We see that Q would diverge as $\Lambda \to \infty$, if a_0 were fixed. But a_0 may depend on Λ. (It is an "unrenormalized" coupling constant.) Hence we can hope to make the above finite.

It is trivial to push the effective cutoff down, from Λ to Λ/b ($b > 1$):

$$Q(a_0 t, \Lambda) = \mathcal{N}\left\{\prod_{|k|<\Lambda/b}\left(k^2 + a_0 t\right)^{-1/2}\right\}\left\{\prod_{\Lambda/b<|k|<\Lambda}\left(k^2 + a_0 t\right)^{-1/2}\right\} \tag{18.71}$$

We can set $t = 0$ in the second factor, since it is a regular at $t = 0$. This makes it a constant, which can be absorbed into the normalization factor:

$$Q(a_0 t, \Lambda) = \mathcal{N}' \prod_{|k| < \Lambda/b} \left(k^2 + a_0 t\right)^{-1/2} \tag{18.72}$$

Now put $k = k'/b$ to restore the cutoff to Λ:

$$Q(a_0 t, \Lambda) = \mathcal{N}' b^{N(\Lambda)} \prod_{|k'| < \Lambda} \left(k'^2 + a_0 b^2 t\right)^{-1/2} \tag{18.73}$$

where $N(\Lambda)$ is the number of k values contained in a d-sphere of radius Λ. Equating the right sides of (18.72) and (18.73) yields the relation

$$Q(a_0 t, \Lambda/b) = b^{N(\Lambda)} Q(a_0 b^2 t, \Lambda) \tag{18.74}$$

which can be rewritten

$$Q(a_0 t, \Lambda) = b^{-N(\Lambda)} Q\left(\frac{a_0 t}{b^2}, \frac{\Lambda}{b}\right) \tag{18.75}$$

Choosing $b = \Lambda/\lambda$, where λ is arbitrary, we have

$$Q(a_0 t, \Lambda) = Z(\Lambda, \lambda) Q(at, \lambda)$$

$$a = \left(\frac{\lambda}{\Lambda}\right)^2 a_0 \tag{18.76}$$

where a is called the renormalized coupling constant. It is an arbitrary parameter of the theory, since a_0 is unknown and physically unknowable. The constant $Z = (\Lambda/\lambda)^{-N(\Lambda)}$ is physically irrelevant. Thus, the original cutoff disappears from the problem.

Dropping an irrelevant additive constant, we obtain a finite expression for the Gibbs free energy per unit volume:

$$\frac{G(t)}{kTV} = -\frac{1}{V} \log Q(at, \lambda) = \frac{1}{2} \int_0^\lambda dk \, S_d(k) \log \left(k^2 + at\right)$$

$$S_d(k) = \frac{d \pi^{d/2} k^{d-1}}{\Gamma\left(\frac{d}{2} + 1\right)} \tag{18.77}$$

where $S_d(k)$ is the surface area of a d-sphere of radius k. The arbitrary constant λ is called the renormalization point. Changing it only leads to a finite additive constant to the free energy density, without affecting the infrared singularities of the integral in the limit $t \to 0$. This explains why all dimensions are canonical in the Gaussian model. In more complicated models λ could be entangled with the infrared singularity at $t = 0$ and lead to anomalous dimensions. We can of course obtain the critical exponents by direct calculation of the thermodynamic functions, as was done in Section 17.7, except that there are no divergent constants anywhere.

18.7 THE LANDAU-WILSON MODEL

We now discuss m^4 theory, the simplest nontrivial case in Landau theory[*]:

$$E[m] = \int (dx)\{\tfrac{1}{2}|\nabla m(x)|^2 + \tfrac{1}{2}r_0 m^2(x) + u_0 m^4(x)\} \qquad (18.78)$$

Renormalization-Group Equations

We shall integrate out an infinitesimally thin shell of k values in momentum space, effectively lowering the cutoff to Λ/b ($b \approx 1$). In a large but finite volume V, the number of states in the shell is

$$N = \tfrac{1}{2}V S_d(\Lambda)\,\Delta k$$
$$\Delta k/\Lambda = \Delta(\log \Lambda) = \log b \qquad (18.79)$$

where $S_d(\Lambda)$ is the surface area of a d-sphere of radius Λ. We write $m(x) = \overline{m}(x) + \delta m(x)$, where $\delta m(x)$ contains only Fourier components to be integrated out, and $\overline{m}(x)$ is the average order parameter, a little bit more coarse-grained than $m(x)$.

Instead of Fourier analyzing $\delta m(x)$ we expand it in a set of N real wave packets $\phi_1(x), \ldots, \phi_N(x)$, which are formed by superposing plane waves whose k values lie in the thin shell, and their centers are so chosen that they cover the whole space of the system. They are rather extended objects with dimension $\Delta x \sim 1/\Delta k$, containing waves of extremely short wavelengths ($\sim 1/\Lambda$), and have the following properties:

$$\int (dx)[\phi_i(x)]^2 = 1$$

$$\int (dx)[\phi_i(x)]^n \approx 0 \qquad (n \text{ odd}) \qquad (18.80)$$

$$\int (dx)\phi_i(x)\phi_j(x) \approx 0 \qquad (i \neq j)$$

Only approximate orthogonality is required, to avoid having long tails in the wave packets. In view of the qualitative nature of subsequent calculations, there is no point in specifying these states in greater detail. Thus, in summary,

$$m(x) = \overline{m}(x) + \delta m(x)$$
$$\delta m(x) = \sum_{i=1}^{N} c_i \phi_i(x) \qquad (18.81)$$

To carry out step 1 in the RG recipe, substitute (18.81) into (18.78) and integrate out the terms containing $\delta m(x)$. We use (18.80) plus the following

---

[*]See K. G. Wilson and J. Kogut, *Phys. Rep. C* **12**, 75 (1974), for detailed review and original literature. The qualitative treatment given here follows K. G. Wilson, *Rev. Mod. Phys.* **55**, 583 (1983).

approximations:

(a) $\bar{m}(x)$ is assumed to be constant over a wave packet.

(b) $\int (dx)|\nabla \phi_i|^2 \approx \Lambda^2$.

(c) $(\delta m)^4$ terms are neglected.

Some estimates are given below:

$$\int (dx)\tfrac{1}{2}|\nabla [\bar{m}(x) + \delta m(x)]|^2 \approx \int (dx)|\nabla \bar{m}(x)|^2 + \Lambda^2 \sum_{i=1}^{N} c_i^2 \qquad (18.82)$$

$$\int (dx)[\bar{m}(x) + \delta m(x)]^2 \approx \int (dx)\bar{m}^2(x) + \sum_{i=1}^{N} c_i^2 \qquad (18.83)$$

$$\int (dx)[\bar{m}(x) + \delta m(x)]^4 \approx \int (dx)\bar{m}^4(x) + 6 \sum_{i=1}^{N} c_i^2 \bar{m}^2(x_i) \qquad (18.84)$$

where $\bar{m}(x_i)$ denotes $\bar{m}(x)$ at the center of the ith wave packet. Thus

$$E[m] = E[\bar{m}] + \sum_{i=1}^{N} \left[\tfrac{1}{2}(\Lambda^2 + r_0) + 6u_0\bar{m}^2(x_i)\right]c_i^2 \qquad (18.85)$$

The RG-transformed Hamiltonian E' is defined by

$$e^{E'[\bar{m}]} \equiv \prod_{i=1}^{N} \int_{-\infty}^{\infty} dc_i \, e^{-E[m]}$$

$$= e^{-E[\bar{m}]} \prod_{i=1}^{N} \sqrt{\pi/2} \left[\Lambda^2 + r_0 + 12u_0\bar{m}^2(x_i)\right]^{-1/2}$$

which gives

$$E'[\bar{m}] = E[\bar{m}] + \frac{1}{2} \sum_{i=1}^{N} \log\left\{\left(\frac{2\Lambda^2}{\pi}\right)\left[1 + \frac{r_0}{\Lambda^2} + \frac{12u_0}{\Lambda^2}\bar{m}^2(x_i)\right]\right\} \qquad (18.86)$$

Note that in this approximation the coefficient of $|\nabla m|^2$ is not changed by coarse-graining. This immediately tells us that the rescaling of $m(x)$ will be dictated by naive dimensional analysis, as in the Gaussian model. The same considerations as we went through in the Gaussian model will yield a canonical value for the dimension of the external field:

$$D_h = \frac{d + 2}{2} \qquad (18.87)$$

Returning to (18.86), we can replace the sum there by an integral, by noting (18.79):

$$\sum_{i=1}^{N} = S_d(\Lambda)\,\Delta k \int (dx) = C_d \Lambda^d (\log b) \int (dx)$$

$$C_d = \frac{\pi^{d/2}}{\Gamma\left(\dfrac{d}{2} + 1\right)} \qquad (18.88)$$

Expanding the logarithm in (18.86), keeping only terms up to second order in r_0 and u_0, and dropping an irrelevant constant, we obtain

$$E'[\overline{m}] = E[\overline{m}] + C_d \Lambda^d \log b \int (dx) \left[6\left(\frac{u_0}{\Lambda^2} - \frac{r_0 u_0}{\Lambda^4} \right) \overline{m}^2(x) - \frac{36 u_0^2}{\Lambda^4} \overline{m}^4(x) \right]$$

(18.89)

Using (18.78), we rewrite this in the form

$$E'[\overline{m}] = \int (dx) \left[\tfrac{1}{2} |\nabla \overline{m}(x)|^2 + \tfrac{1}{2} \tilde{r}_0 \overline{m}^2(x) + \tilde{u}_0 \overline{m}^4(x) \right]$$

$$\tilde{r}_0 = r_0 + 12(\log b) C_d \left(\Lambda^{d-2} u_0 - \Lambda^{d-4} r_0 u_0 \right)$$

(18.90)

$$\tilde{u}_0 = u_0 - 36(\log b) C_d \Lambda^{d-4} u_0^2$$

In step 2 of the RG transformation we increase the unit of length by a factor b, thus changing the variable of spatial integration from x to $x' = x/b$. In step 3, we restore the standard normalization of $m(x)$. The net results is to replace r_0 and u_0, respectively, by

$$r_0' = b^2 \tilde{r}_0, \qquad u' = b^{4-d} \tilde{u}_0$$

(18.91)

Since $b \approx 1$, we write $b^n \approx 1 + n \log b$. Substituting (18.90) into (18.91), keeping only terms to first order in $\log b$, we obtain

$$r_0' - r_0 = \left[2r_0 + 12 C_d \left(\Lambda^{d-2} u_0 - \Lambda^{d-4} r_0 u_0 \right) \right] \log b$$

$$u_0' - u_0 = \left[(4 - d) u_0 - 36 C_d \Lambda^{d-4} u_0^2 \right] \log b$$

(18.92)

The first terms on the right sides give the change expected under naive dimensional analysis. The additional terms give rise to non-Gaussian exponents.

The cutoff Λ is defined only up to a finite factor. The same is true of r, for we only require that it be proportional to t. Thus the factor C_d in (18.92) can be absorbed by changing Λ and r by suitable factors.

Since the RG transformations here proceeds by infinitesimal steps instead of finite steps, we regard the transformed coupling constants as continuous functions of $\log b$. Let us define

$$r(\tau) \equiv r_0'$$

$$u(\tau) \equiv u_0'$$

(18.93)

$$\tau \equiv \log b$$

with the "initial" values $r(0) = r_0$, $u(0) = u_0$. We then have the differential equations

$$\frac{dr}{d\tau} = 2r + 12 \Lambda^{d-2} u - 12 \Lambda^{d-4} ur$$

$$\frac{du}{d\tau} = (4 - d) u - 36 \Lambda^{d-4} u^2$$

(18.94)

These are called renormalization-group equations.

The development above is of the nature of a fable, it is not really true, but illustrates a point. Specifically, it is not true that $\bar{m}(x)$ is approximately constant over a wave packet, for it contains wavelengths only infinitesimally longer than those in the latter. But if we did not make believe it was true, the RG transformation would have generated arbitrary powers of $\bar{m}(x)$, which would have to be included from the beginning, making the problem hopeless. The simplification has enabled us to illustrate the spirit of the method in a simple way. The real reason we could be so bold is that the resulting RG equations turn out to be valid in the neighborhood of $d = 4$, namely, they are correct to first order in $\epsilon \equiv 4 - d$. Somehow the approximations work, at least in a limiting case.

Fixed Points and Trajectories

To analyze the implications of the RG equations (18.94), it is convenient to convert them into dimensionless forms, by introducing the dimensionless coupling constants

$$x \equiv \frac{r}{\Lambda^2}, \qquad y \equiv \frac{u}{\Lambda^{4-d}} \tag{18.95}$$

Then

$$\frac{dx}{d\tau} = 2x - 12xy + 12y$$
$$\frac{dy}{d\tau} = \epsilon y - 36y^2 \tag{18.96}$$

where

$$\epsilon \equiv 4 - d \tag{18.97}$$

which we shall consider a small quantity.

The fixed points x^* and y^* are defined to be points at which $dx/d\tau = dy/d\tau = 0$. There are two fixed points:

$$
\begin{array}{lll}
\text{Gaussian fixed point:} & x^* = 0, & y^* = 0 \\
\text{Nontrivial fixed point:} & x^* = -\epsilon/6, & y^* = \epsilon/36
\end{array} \tag{18.98}
$$

We see that the nontrivial fixed point approach the Gaussian fixed point as $\epsilon \to 0$. This is reason why one can solve this problem for small ϵ—it is close to the trivial Gaussian model.

In the neighborhood of a fixed point write

$$x = x^* + \delta x, \qquad y = y^* + \delta y \tag{18.99}$$

Then

$$\frac{d}{d\tau}\begin{pmatrix} \delta x \\ \delta y \end{pmatrix} = T\begin{pmatrix} \delta x \\ \delta y \end{pmatrix}, \qquad T = \begin{pmatrix} 2(1 - 6y^*) & 12(1 - x^*) \\ 0 & \epsilon - 72y^* \end{pmatrix} \tag{18.100}$$

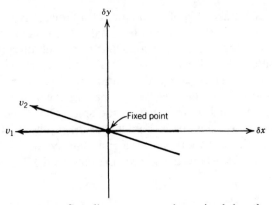

Fig. 18.6 Coordinate system determined by the scaling fields v_1 and v_2, in the neighborhood of a fixed point.

The left eigenvectors ϕ_i and eigenvalues λ_i of T are easily found to be

$$\phi_1 = \begin{pmatrix} 1 \\ \dfrac{12(1 - x^*)}{2 + 60y^* - \epsilon} \end{pmatrix}, \qquad \lambda_1 = 2(1 - 6y^*) \tag{18.101}$$

$$\phi_2 = \begin{pmatrix} 0 \\ 1 \end{pmatrix}, \qquad \lambda_2 = (\epsilon - 72y^*)$$

Now define scaling fields v_1, v_2 by

$$v_i = (\phi_i, q), \qquad q \equiv \begin{pmatrix} \delta x \\ \delta y \end{pmatrix} \tag{18.102}$$

Under the RG transformation we have $dv_i/d\tau = \lambda_i v_i$. Hence

$$v_i(\tau) = v_0 e^{\lambda_i \tau} = v_0 b^{\lambda_i} \tag{18.103}$$

Explicitly we have

$$v_1 = \delta x + \frac{12(1 - x^*)}{2 + 60y^* - \epsilon} \delta y \tag{18.104}$$

$$v_2 = \delta y$$

Figure 18.6 shows the direction along which each scaling field increases with the other held fixed, in the neighborhood of a fixed point.

The scaling fields and eigenvalues for the two fixed points are given below:

$$
\begin{aligned}
\text{Gaussian:} \quad & v_1 = \delta x + 6\,\delta y, & & \lambda_1 = 2 \\
& v_2 = \delta y, & & \lambda_2 = \epsilon \\
\text{Nontrivial:} \quad & v_1 = \delta x + (6 - \epsilon)\,\delta y, & & \lambda_1 = 2 - \tfrac{1}{3}\epsilon \\
& v_2 = \delta y, & & \lambda_2 = -\epsilon
\end{aligned} \tag{18.105}
$$

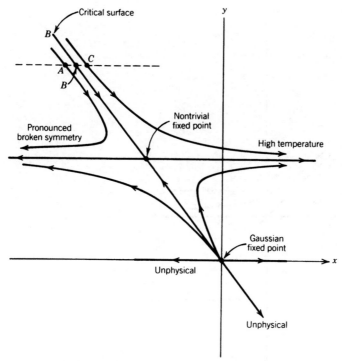

Fig. 18.7 Flow diagram of Landau-Wilson model for $d < 4$. The two fixed points are close to each other when the dimension of space is close to 4. This is why the model can be solved by expansion in powers of $\epsilon = 4 - d$.

The fixed points and the trajectories of the coupling constants under successive RG transformations are shown in Figs. 18.7 and 18.8. Note that the lower half plane including the negative x axis is unphysical. Figure 18.7 shows the case for $\epsilon > 0$ ($d < 4$). For the nontrivial fixed point, v_2 is an irrelevant scaling field and v_1 a relevant one. Hence the critical surface is the line $v_1 = 0$. Neglecting the irrelevant field means setting $\delta y = 0$, hence $v_1 = \delta x$. Hence v_1 is the temperature variable t. From (18.103) we have

$$D_t = \lambda_1 = 2 - \tfrac{1}{3}\epsilon \qquad (18.106)$$

For illustration, we call attention to the points A, B, C, which correspond to three different initial temperatures. The system at B has $t = 0$, and will tend toward the fixed point under RG transformations. The system at A has $t < 0$, and will veer off to a region of increasing negative r_0, with pronounced broken symmetry. The system at C has $t > 0$, and will run off to higher temperatures.

The correlation length can be obtained, as in the derivation of (18.48), by setting $\xi \sim b$ when $b^{\lambda_1} t \sim 1$. Thus we again have $\xi \sim t^{-1/D_t}$, or $\nu = 1/D_t$. The rest of the critical exponents can now be obtained from (16.59) using D_h and D_t.

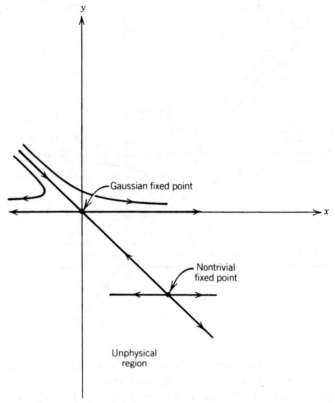

Fig. 18.8 Flow diagram of Landau-Wilson model for $d > 4$. The nontrivial fixed point is in an unphysical region. To stabilize the system, one needs an m^6 term, which probably will bring about a first-order phase transition, as indicated by mean-field theory.

The results to first order in ϵ are given below:

$$\alpha = 2 - \frac{d}{2}\left(1 + \frac{\epsilon}{12}\right) \qquad \delta = \frac{d+2}{d-2}$$

$$\beta = \frac{d-2}{4}\left(1 + \frac{\epsilon}{6}\right) \qquad \nu = \frac{1}{2} + \frac{\epsilon}{12}$$

$$\gamma = 1 + \frac{\epsilon}{6} \qquad\qquad \eta = 0$$

Figure 18.8 shows the fixed points for $\epsilon < 0$ $(d > 4)$. The nontrivial fixed point is unstable since both scaling fields are relevant. However, the fixed point is in an unphysical region, with negative u_0. The approximation breaks down here. We cannot neglect the m^6 term, for it is needed for stability of the system. Based on the results of mean-field theory we expect that trajectories starting at the unstable fixed point will run off to a region of first-order transition.

We conclude with two remarks. First, the results here can be demonstrated more rigorously* for $\epsilon \to 0^+$ through a systematic expansion in powers of ϵ. The fact that $\epsilon = 4 - d$ here has to do with the fact that $d = 4$ is the critical dimension for the fixed point associated with the m^4 interaction. For the fixed point associated with the m^6 interaction, for example, one would take $\epsilon = 3 - d$.

The case $d = 4$ is of special interest because it corresponds to a quantum-field theory with a self-interacting scalar field (usually referred to as ϕ^4 theory.) We see here that in the limit $\epsilon \to 0$, the nontrivial and Gaussian fixed points coincide, and the theory reduces to the Gaussian model—a free-field theory. This shows that the unrenormalized self-interaction in ϕ^4 theory disappears upon renormalization, and the theory is trivial.

PROBLEMS

18.1 Show that the RG transformation (18.36) that takes the coupling constants $K^{(n)}$ to $K^{(n+1)}$ is independent of n. Do this by finding a formal expression of $K^{(n+1)}$ in terms of $K^{(n)}$ with the help of (18.20).

18.2 Migdal-Kadanoff Approximation. Consider a two-dimensional Ising model on a square lattice without external field:

$$E(s) = -J \sum_{\langle i, j \rangle} s_i s_j$$

The Migdal-Kadanoff approximation consists of two steps, illustrated in the sketches. First, the sites marked by an $\times$ in sketch (a) are removed. To compensate for this, the remaining bonds are doubled in strength, so that the Hamiltonian becomes

$$E(s) = -2J \sum_{\langle i, j \rangle}{}' s_i s_j$$

where the prime indicates that this Hamiltonian is defined on the lattice shown in sketch

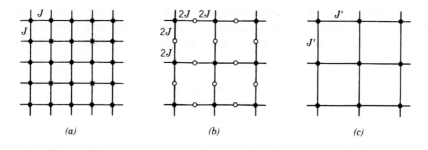

(a) $\qquad\qquad$ (b) $\qquad\qquad$ (c)

*See Wilson and Kogut, *op. cit.*

(b). This is a rather drastic approximation whose chief virtue is that it makes the renormalization of the model simple.

(a) Make a renormalization-group transformation by summing over the spins on the sites indicated by open circles in sketch (b), to end up with the lattice of sketch (c). Show that one obtains a nearest-neighbor Ising model on the new lattice, with Hamiltonian

$$E'(s) = -J' \sum_{\langle i,j \rangle} s_i s_j$$

such that the partition differs from the original one by only a constant factor. Put $kT = 1$. Defining $x = 2J$, $x' = 2J'$, derive the recursion relation

$$x' = \tfrac{1}{2}(x^2 + x^{-2})$$

(b) Recall that in the one-dimensional Ising model the first points are trivial ones at $x^* = 0$ and $x^* = 1$. Show that in addition to these, there is a new fixed point. Expanding x' in its neighborhood, show

$$x' - x^* = \left(x^* - \frac{1}{x^{*3}} \right)(x - x^*)$$

Show that the new fixed point is unstable.

(c) Calculate the exponent ν in terms of x^*.

18.3 Consider the following Landau Hamiltonian involving a two-component order parameter $m = (m_1, m_2)$

$$E[m] = \int (dx) \left\{ \tfrac{1}{2} \left[|\nabla m_1|^2 + |\nabla m_2|^2 \right] + \tfrac{1}{2} r_0 \left(m_1^2 + m_2^2 \right) \right.$$
$$\left. + g_1 \left(m_1^4 + m_2^4 \right) + g_2 m_1^2 m_2^2 \right\}$$

This model reduces to the XY model when $g_2 = 2g_1$. It has the appropriate symmetry to describe a structural phase transition in a crystal with a fourth order axis. The order parameters m_1 and m_2 may be interpreted as a projection of the atomic position onto a coordinate system normal to this axis.

(a) First treat the problem in the mean-field approximation. Assuming $g_1 > 0$ and putting $y \equiv g_2/g_1$, study the different phases as a function of r_0 and y. Specifically show that for $r_0 < 0$, spontaneous symmetry breaking occurs, such that

$$m_1 \neq 0, \ m_2 = 0 \quad \text{or} \quad m_2 \neq 0, \ m_1 = 0 \quad (y > 2)$$
$$m_1 = \pm m_2 \qquad\qquad\qquad\qquad (-2 < y < 2)$$

Show that the system is thermodynamically unstable for $y < -2$. Draw a phase diagram in the $r_0 y$ plane, indicating phase boundaries of first- and second-order phase transitions.

(b) Use the approximate method of Wilson, as discussed in the text, to derive the following RG equations:

$$\frac{dg_1}{d\tau} = \epsilon g_1 - A \left(36 g_1^2 + g_2^2 \right)$$

$$\frac{dg_2}{d\tau} = \epsilon g_2 - A \left(24 g_1 g_2 + 8 g_2^2 \right)$$

where A is a constant, $\epsilon = 4 - d$, and $\tau = \log b$. The unit of length is increased by a factor b in the RG transformation.

(*c*) Using the dimensionless parameters

$$x = r\Lambda^{-2}, \qquad y = g_2/g_1, \qquad z = g_1\Lambda^{-\epsilon}$$

show

$$dx/d\tau = 2x + 12(1 - x)(1 + y/6)z$$
$$dy/d\tau = y(y - 2)(y - 6)z$$
$$dz/d\tau = z\left[\epsilon - z(36 + y^2)\right]$$

(*d*) Consider ϵ to be small, and neglect $O(\epsilon^2)$. Investigate the fixed points and their stabilities.

N-BODY SYSTEM
OF IDENTICAL PARTICLES

A.1 THE TWO KINDS OF STATISTICS

An N-body system of identical particles is characterized by a Hamiltonian operator $\mathscr{H}$ that is invariant under the interchange of all the coordinates of any two particles. Any wave function for the system can be written as a linear superposition of eigenfunctions Ψ_n of the Hamiltonian:

$$\mathscr{H}\Psi_n(q_1,\ldots,q_N) = E_n\Psi_n(q_1,\ldots,q_N) \tag{A.1}$$

where q_i denotes the collection of all the coordinates of the ith particle, including the position coordinates, and the spin and other internal coordinates, if any. To study the general symmetry property of any wave function, it is sufficient to study the general symmetry property of Ψ_n.

Let $\mathscr{P}$ be an operator that, when applied to Ψ_n, interchanges the positions of q_i and q_j:

$$\mathscr{P}\Psi_n(\ldots,q_i,\ldots,q_j,\ldots) = \Psi_n(\ldots,q_j,\ldots,q_i,\ldots) \tag{A.2}$$

By definition we have

$$\mathscr{P}^{-1}\mathscr{H}\mathscr{P} = \mathscr{H} \tag{A.3}$$

Therefore

$$\mathscr{H}(\mathscr{P}\Psi_n) = E_n(\mathscr{P}\Psi_n) \tag{A.4}$$

i.e., if Ψ_n is an eigenfunction of $\mathscr{H}$ belonging to the eigenvalue E_n, then $\mathscr{P}\Psi_n$ is also an eigenfunction of $\mathscr{H}$ belonging to the same eigenvalue.

A possible property of Ψ_n is that $\mathscr{P}\Psi_n$ is proportional to Ψ_n. If this is so, then $\mathscr{P}^2\Psi_n = \Psi_n$. Hence the proportionality constant is either $+1$ or -1, and Ψ_n is either symmetric or antisymmetric under the interchange of two coordinates:

$$\mathscr{P}\Psi_n = \pm\Psi_n \tag{A.5}$$

Suppose that for a given system (A.5) is fulfilled. The question naturally arises whether we can have some Ψ_n symmetric while others are antisymmetric. The answer is as follows. There is nothing to forbid such a situation; but then the symmetric wave functions $\{\Psi_n^{(+)}\}$ and the antisymmetric ones $\{\Psi_n^{(-)}\}$ form two classes of wave functions that do not mix, in the sense of the "superselection rule"

$$\left(\Psi_m^{(+)}, \mathcal{O}\Psi_n^{(-)}\right) = 0 \tag{A.6}$$

Proof

$$\left(\Psi_m^{(+)}, \mathcal{O}\Psi_n^{(-)}\right) = \left(\Psi_m^{(+)}, \mathscr{P}^{-1}\mathcal{O}\mathscr{P}\Psi_n^{(-)}\right) = \left(\mathscr{P}\Psi_m^{(+)}, \mathcal{O}\mathscr{P}\Psi_n^{(-)}\right)$$
$$= -\left(\Psi_m^{(+)}, \mathcal{O}\Psi_n^{(-)}\right) \qquad \blacksquare$$

where $\mathcal{O}$ is any operator invariant under a permutation of particles. Since only such operators will ever be considered (otherwise the particles are not identical), the sets $\{\Psi_n^{(+)}\}$ and $\{\Psi_n^{(-)}\}$ form two disjoint systems of wave functions that do not influence each other. Hence it is sufficient to consider a system for which all Ψ_n are either $\Psi_n^{(+)}$ or $\Psi_n^{(-)}$.

According to (A.5), two wave functions differing only by an interchange of two coordinates correspond to one and the same state of the system. This has a direct bearing on the correct counting of states for a given energy. Hence (A.5) is said to define the statistics of the system. The plus sign refers to Bose statistics and the minus sign refers to Fermi statistics.

The property (A.5) is not the only one consistent with (A.4). In general $\mathscr{P}\Psi_n$ may be a linear combination of eigenfunctions $\Psi_1, \ldots, \Psi_g$, which must all have the eigenvalue E_n. If this is the case, the eigenvalue E_n has an intrinsic degeneracy that cannot be removed by the introduction of interactions among the particles, because any perturbation $\mathcal{O}$ added to the Hamiltonian must have the property $\mathscr{P}\mathcal{O}\mathscr{P}^{-1} = \mathcal{O}$. Nothing in what we have said so far rules out such intrinsic degeneracies. Nevertheless nature apparently abhors degeneracies, because it is an experimental fact that (A.5) is the only possibility so far observed. If future experiments discover a type of particle that does not obey (A.5), new types of symmetry in addition to those for bosons and fermions must be explored.

When the principle of Lorentz covariance is imposed on physical systems, it is possible to show,* in the formalism of relativistic quantum field theory, that particles of integer spin must obey Bose statistics and particles of half-integer spin must obey Fermi statistics.[†] These results are in accord with experimental observations.

We must note that although the connection between spin and statistic is both an experimental fact and a theoretical requirement for the consistency of

---

*W. Pauli, *Phys. Rev.* **58**, 716 (1940).

[†] The requirement that particles of spin $\frac{1}{2}$ must obey Fermi statistics can be easily seen to be a consequence of relativity. Relativity directly implies the Dirac equations for spin-$\frac{1}{2}$ particles. The Fermi statistics is required in order that a hole theory of these particles be possible.

quantum-field theory it is not particularly relevant to the subject matter of this book. For the sake of mathematical simplicity, we frequently consider spinless fermions, which are not mathematical absurdities as long as we are not concerned with the theory of elementary particles.

A.2 N-BODY WAVE FUNCTIONS

N-Body System

Consider a system of N identical particles of mass m contained in a cube of volume V. For simplicity we regard these particles as spinless. The positions of the N particles are denoted by $\{\mathbf{r}_1, \mathbf{r}_2, \ldots, \mathbf{r}_N\}$, which is sometimes abbreviated to $\{1, 2, \ldots, N\}$. The Hamiltonian for the system is the sum of the kinetic-energy operator K and the potential-energy operator Ω:

$$\mathcal{H} = K + \Omega$$
$$K = -\frac{\hbar^2}{2m}\left(\nabla_1^2 + \cdots + \nabla_N^2\right) \tag{A.7}$$
$$\Omega = \sum_{i<j} v_{ij}, \qquad v_{ij} \equiv v(\mathbf{r}_i, \mathbf{r}_j)$$

A stationary wave function $\Psi(1, \ldots, N)$ for the system satisfies the Schrödinger equation

$$\mathcal{H}\Psi(1, \ldots, N) = E\Psi(1, \ldots, N) \tag{A.8}$$

and is normalized to unity in the volume V:

$$(\Psi, \Psi) = \int d^{3N}r\, \Psi^*(1, \ldots, N)\Psi(1, \ldots, N) = 1 \tag{A.9}$$

where each $\mathbf{r}_i$ is integrated over the volume V. The wave function is symmetric or antisymmetric with respect to the interchange of any pair of coordinates $\mathbf{r}_i, \mathbf{r}_j$, depending on whether the particles are bosons or fermions. To determine Ψ from (A.8) it is necessary that the boundary conditions of Ψ be given. We may, for example, impose periodic boundary conditions, so that

$$\Psi(\ldots, \mathbf{r}_i, \ldots) = \Psi(\ldots, \mathbf{r}_i + \mathbf{n}L, \ldots) \qquad (i = 1, \ldots, N) \tag{A.10}$$

where

$$L = V^{\frac{1}{3}}$$
$$\mathbf{n} = \text{a vector whose components are } 0, \pm 1, \pm 2, \ldots \tag{A.11}$$

As $V \to \infty$ the energy per particle E/N determined from (A.8) is usually assumed to be independent of the boundary conditions.

Properties of the Ground State Wave Function

We shall show that the ground state wave function Ψ_0 can be chosen to be real, and that for Bose statistics it can be chosen positive everywhere (i.e., it has no nodes). It will follow from the latter fact that the ground state of a Bose system of particles is unique.

That Ψ_0 can be chosen to be real follows from the fact that $\mathcal{H}$ is a real operator. If Ψ is any eigenfunction of $\mathcal{H}$, then Ψ^* is an eigenfunction belonging to the same eigenvalue. It follows that $\Psi_0 + \Psi_0^*$ and $i(\Psi_0 - \Psi_0^*)$ are eigenfunctions belonging to the same eigenvalue. Hence all eigenfunctions of $\mathcal{H}$ can be chosen to be real.

To show that Ψ_0 can be chosen to be positive, we note that Ψ_0 minimizes the expression $(\Psi_0, \mathcal{H}\Psi_0)$. Let $\Phi_0 \equiv |\Psi_0|$, which is a symmetric function.* It is easily shown that $(\Phi_0, \mathcal{H}\Phi_0) = (\Psi_0, \mathcal{H}\Psi_0)$. Therefore Φ_0 minimizes the expression $(\Phi_0, \mathcal{H}\Phi_0)$. That is, if we replace Φ_0 by $\Phi_0 + \delta\Phi_0$, where $\delta\Phi_0$ is an arbitrarily small variation, then $\delta(\Phi_0, \mathcal{H}\Phi_0) = 0$. This leads to the conclusion that Φ_0 satisfies (A.8). It follows that the first derivatives of Φ_0 must be continuous because the potential is finite. Hence Ψ_0 never changes sign, for if it did there would be a nodal surface on which not only Ψ_0 but also all the derivatives of Ψ_0 would vanish. Since (A.8) is a second-order partial differential equation, Ψ_0 would be identically zero. Therefore Ψ_0 never changes its sign, which we can take to be positive.

That Ψ_0 is unique follows from the fact that a ground state wave function never changes sign. If Ψ_1 and Ψ_2 are two normalized ground state wave functions, $\Psi_1 - \Psi_2$ is also a ground state wave function, which never changes sign. This means that one of them is always greater than the other. Hence Ψ_1 and Ψ_2 cannot be both normalized unless $\Psi_1 = \Psi_2$.

Complete Set of Wave Functions

It is convenient to introduce a complete set of N-body wave functions in terms of which any N-body wave function can be obtained by linear superposition. We now describe such a set.

First we define an arbitrary complete set of single-particle wave functions $u_\alpha(\mathbf{r})$ such that

$$\int d^3 r\, u_\alpha^*(\mathbf{r}) u_\beta(\mathbf{r}) = \delta_{\alpha\beta} \tag{A.12}$$

where $\delta_{\alpha\beta}$ is the Kronecker symbol, defined by

$$\delta_{\alpha\beta} = \begin{cases} 1 & (\alpha = \beta) \\ 0 & (\alpha \neq \beta) \end{cases} \tag{A.13}$$

Furthermore, we require that $u_\alpha(\mathbf{r})$ satisfy periodic boundary conditions. We may construct an N-body wave function by symmetrizing or antisymmetrizing the product

$$u_{\alpha_1}(1) u_{\alpha_2}(2) \cdots u_{\alpha_N}(N)$$

where $u_\alpha(1) \equiv u_\alpha(\mathbf{r}_1)$ and where $\{\alpha_1, \alpha_2, \ldots, \alpha_N\}$ is a set of given indices serving

---

*Note that for Fermi statistics the proof fails at this point.

as quantum numbers labeling the N-body wave function. Let

$$\Phi_\alpha(1,\ldots,N) \equiv \Phi_{\alpha_1,\ldots,\alpha_N}(\mathbf{r}_1,\ldots,\mathbf{r}_N) = \frac{1}{\sqrt{N!}} \sum_P \delta_P \left[u_{\alpha_1}(P1) \cdots u_{\alpha_N}(PN) \right]$$

$$= \frac{1}{\sqrt{N!}} \sum_P \delta_P \left[u_{P\alpha_1}(1) \cdots u_{P\alpha_N}(N) \right] \tag{A.14}$$

where P is the permutation on N objects that sends the ordered set $\{1,2,\ldots,N\}$ to the ordered set $\{P1, P2,\ldots, PN\}$ and where $\delta_P = \pm 1$, as determined by the following rule:

$$\delta_P = 1 \qquad \text{(bosons)}$$

$$\delta_P = \begin{cases} +1 & (P \text{ even}) \\ -1 & (P \text{ odd}) \end{cases} \quad \text{(fermions)} \tag{A.15}$$

The permutation P is even or odd according to whether it is equivalent to an even or an odd number of successive interchanges. It is obvious that a permutation of $\alpha_1,\ldots,\alpha_N$ changes (A.14) by at most a sign and does not lead to a new independent wave function. For fermions, (A.14) is equivalent to the definition

$$\Phi_\alpha(1,\ldots,N) = \frac{1}{\sqrt{N!}} \begin{vmatrix} u_{\alpha_1}(1) & u_{\alpha_2}(1) & \cdots & u_{\alpha_N}(1) \\ \vdots & & & \vdots \\ u_{\alpha_1}(N) & U_{\alpha_2}(N) & \cdots & u_{\alpha_N}(N) \end{vmatrix} \tag{A.16}$$

It is obvious from this form that for fermions Φ_α vanishes unless $\alpha_1,\ldots,\alpha_N$ are distinct quantum numbers.

From the orthogonality of the functions $u_\alpha(\mathbf{r})$ it easily follows that

$$(\Phi_\alpha, \Phi_\beta) = \int d^{3N}r \, \Phi_\alpha^*(1,\ldots,N)\Phi_\beta(1,\ldots,N) = 0$$

$$\text{(if } \{\alpha_1,\ldots,\alpha_N\} \text{ and } \{\beta_1,\ldots,\beta_N\} \text{ are not the same)}$$

Therefore (A.14) defines an orthogonal set of wave functions. We now calculate the norm of Φ_α:

$$(\Phi_\alpha, \Phi_\alpha) = \int d^{3N}r \, \frac{1}{N!} \sum_P \sum_Q \delta_P \delta_Q \left[u_{P\alpha_1}^*(1) u_{Q\alpha_1}(1) \right] \cdots \left[u_{P\alpha_N}^*(N) u_{Q\alpha_N}(N) \right]$$

For fermions we must have $P = Q$, because the α_i are all distinct. Therefore

$$(\Phi_\alpha, \Phi_\alpha) = 1 \qquad \text{(fermions)} \tag{A.17}$$

For bosons the α_i are not necessarily all distinct. Suppose that among $\alpha_1,\ldots,\alpha_N$ there are n_α having the value α. Then

$$(\Phi_\alpha, \Phi_\alpha) = \prod_\alpha (n_\alpha!) \qquad \text{(bosons)} \tag{A.18}$$

which is not necessarily unity.

The integer n_α is called the occupation number of the single-particle level α. Obviously we have the conditions

$$\sum_\alpha n_\alpha = N$$

$$n_\alpha = 0, 1, \ldots, N \qquad \text{(boson)} \qquad\qquad \text{(A.19)}$$

$$n_\alpha = 0, 1 \qquad\qquad \text{(fermion)}$$

Instead of $\{\alpha_1, \ldots, \alpha_n\}$ we can equally well label the wave function by the occupation numbers $\{n_0, n_1, \ldots\}$. Thus we also introduce the notation Φ_n, defined by

$$\Phi_n \equiv \Phi_\alpha \qquad\qquad \text{(A.20)}$$

where n stands for the set $\{n_0, n_1, \ldots\}$.

If we wish, we may use in place of Φ_α the wave function

$$\Phi'_\alpha \equiv \frac{\Phi_\alpha}{\sqrt{\prod\limits_\alpha (n_\alpha!)}} \qquad\qquad \text{(A.21)}$$

which is normalized to unity for both bosons and fermions. It is, however, neither necessary nor convenient to do this. The reason is as follows. Suppose we are to calculate the trace of an operator. We may write

$$\text{Tr}\, \mathcal{O} = \sum_{\{\alpha\}} (\Phi'_\alpha, \mathcal{O}\Phi'_\alpha) \qquad\qquad \text{(A.22)}$$

where the sum extends over all distinct sets $\{\alpha_1, \ldots, \alpha_N\}$. A convenient way to calculate this sum is to sum over each α_i independently and to take into account the fact that a permutation of the α_i must not be counted as a new term in the sum. That is, we write

$$\text{Tr}\, \mathcal{O} = \sum_{\alpha_1, \ldots, \alpha_N} \frac{\prod\limits_\alpha (n_\alpha!)}{N!} (\Phi'_\alpha, \mathcal{O}\Phi'_\alpha)$$

By (A.21) we have

$$\text{Tr}\, \mathcal{O} = \frac{1}{N!} \sum_{\alpha_1, \ldots, \alpha_N} (\Phi_\alpha, \mathcal{O}\Phi_\alpha) \qquad\qquad \text{(A.23)}$$

Thus it is actually more convenient to use the wave functions Φ_α as defined in (A.14).

Free-Particle Wave Functions

A useful choice of $u_\alpha(\mathbf{r})$ is the single-particle wave function for a free particle of momentum $\mathbf{p}$. The quantum number α is now explicitly $\mathbf{p}$, and we have

$$u_\mathbf{p}(\mathbf{r}) \equiv \frac{1}{\sqrt{V}} e^{i\mathbf{p} \cdot \mathbf{r}/\hbar} \qquad\qquad \text{(A.24)}$$

The allowed values of $\mathbf{p}$ are determined by the periodic boundary conditions

$$u_{\mathbf{p}}(\mathbf{r} + \mathbf{n}L) = u_{\mathbf{p}}(\mathbf{r}) \tag{A.25}$$

where $\mathbf{n}$ and L are defined in (A.11). This implies that the allowed values of $\mathbf{p}$ are

$$\mathbf{p} = \frac{2\pi\hbar\mathbf{n}}{L} \tag{A.26}$$

These values form a cubic lattice in momentum space with the lattice constant $2\pi\hbar/L$, which approaches 0 as $V \to \infty$. In this limit a volume element of size d^3p in momentum space contains $(V/h^3)\,d^3p$ lattice points. Thus as $V \to \infty$ a sum over $\mathbf{p}$ may be replaced by an integration over $\mathbf{p}$ in the following manner:

$$\sum_{\mathbf{p}} \to \frac{V}{h^3}\int d^3p \tag{A.27}$$

The functions defined by (A.24) obviously form an orthogonal set. The completeness of this set follows from the fact that any function can be Fourier analyzed.

The N-body wave functions built up from $u_{\mathbf{p}}(\mathbf{r})$ according to (A.14) are the N-body free-particle wave functions. They are denoted by $\Phi_p(1,\ldots,N)$ and are eigenfunctions of the kinetic-energy operator K:

$$K\Phi_p(1,\ldots,N) = \frac{1}{2m}\left(p_1^2 + \cdots + p_N^2\right)\Phi_p(1,\ldots,N) \tag{A.28}$$

where $\mathbf{p}_1,\ldots,\mathbf{p}_N$ are the momenta of the N-single-particle wave functions contained in Φ_p.

Example of Calculation: System of Bosons

We calculate $(\Phi_\alpha, \Omega\Phi_\alpha)$ for a system of bosons:

$$(\Phi_\alpha, \Omega\Phi_\alpha) = \int d^{3N}r\, \Phi_\alpha^* \sum_{i<j} v_{ij}\Phi_\alpha = \tfrac{1}{2}N(N-1)\int d^{3N}r\, \Phi_\alpha^* v_{12}\Phi_\alpha$$

where the second equality is obtained by renaming the integration variables $\mathbf{r}_1,\ldots,\mathbf{r}_N$ in an appropriate fashion in each term of the sum $\sum v_{ij}$. Using (A.14) we have

$$(\Phi_\alpha, \Omega\Phi_\alpha)$$

$$= \frac{N(N-1)}{N!2}\sum_P\sum_Q\int d^{3N}r\left[u_{P\alpha_1}^*(1)\cdots u_{P\alpha_N}^*(N)\right]v_{12}\left[u_{Q\alpha_1}(1)\cdots u_{Q\alpha_N}(N)\right]$$

$$= \frac{N(N-1)}{N!2}\sum_P\sum_Q\langle P\alpha_1, P\alpha_2|v|Q\alpha_1, Q\alpha_2\rangle\left(\delta_{P\alpha_3, Q\alpha_3}\cdots \delta_{P\alpha_N, Q\alpha_N}\right) \tag{A.29}$$

where

$$\langle\alpha, \beta|v|\gamma, \lambda\rangle \equiv \int d^3r_1\, d^3r_2\, u_\alpha^*(1)u_\beta^*(2)v_{12}u_\gamma(1)u_\lambda(2) \tag{A.30}$$

In (A.29) only two terms in the sum $\sum\limits_{Q}$ are nonzero, namely the terms satisfying the conditions (a) or (b):

(a) $\quad Q\alpha_1 = P\alpha_1,$ $\quad Q\alpha_2 = P\alpha_2,$ $\quad Q\alpha_j = P\alpha_j$ $\quad (j = 3, \dots, N)$

(b) $\quad Q\alpha_1 = P\alpha_2,$ $\quad Q\alpha_2 = P\alpha_1,$ $\quad Q\alpha_j = P\alpha_j$ $\quad (j = 3, \dots, N)$ $\quad$ (A.31)

Hence

$$(\Phi_\alpha, \Omega\Phi_\alpha) = \frac{N(N-1)}{N!2} \prod_\alpha (n_\alpha!)$$

$$\times \sum_P (\langle P\alpha_1, P\alpha_2 | v | P\alpha_1, P\alpha_2 \rangle + \langle P\alpha_1, P\alpha_2 | v | P\alpha_2, P\alpha_1 \rangle) \quad \text{(A.32)}$$

As P ranges through all the $N!$ permutations of the set $\{\alpha_1, \dots, \alpha_N\}$, the pair $\{P\alpha_1, P\alpha_2\}$ takes on all possible pairs of values $\{\alpha, \beta\}$ chosen from the set $\{\alpha_1, \dots, \alpha_N\}$. Suppose the occupation numbers for the single-particle states α, β are respectively n_α, n_β. Then the number of ways in which the pair $\{\alpha, \beta\}$ can be chosen from the set $\{\alpha_1, \dots, \alpha_N\}$ is

$$f_{\alpha\beta} = \begin{cases} n_\alpha n_\beta & (\alpha \neq \beta) \\ \frac{1}{2} n_\alpha (n_\alpha - 1) & (\alpha = \beta) \end{cases}$$

or

$$f_{\alpha\beta} = (1 - \delta_{\alpha\beta}) n_\alpha n_\beta + \frac{1}{2} \delta_{\alpha\beta} n_\alpha (n_\alpha - 1)$$

Furthermore, there are $(N-2)!$ permutations that affect only the quantum numbers $\{\alpha_3, \dots, \alpha_N\}$ and thus leave $\{\alpha_1, \alpha_2\}$ unchanged. Changing the label of Φ_α to occupation numbers we obtain

$$(\Phi_n', \Omega\Phi_n') = \tfrac{1}{2} N(N-1) \frac{(N-2)!}{N!} \sum_{\alpha, \beta} f_{\alpha\beta} (\langle \alpha, \beta | v | \alpha, \beta \rangle + \langle \alpha, \beta | v | \beta, \alpha \rangle)$$

or

$$(\Phi_n', \Omega\Phi_n') = \tfrac{1}{2} \sum_{\alpha, \beta} \left[(1 - \delta_{\alpha\beta}) n_\alpha n_\beta + \tfrac{1}{2} \delta_{\alpha\beta} n_\alpha (n_\alpha - 1) \right]$$

$$\times (\langle \alpha, \beta | v | \alpha, \beta \rangle + \langle \alpha, \beta | v | \beta, \alpha \rangle) \quad \text{(A.33)}$$

This result can be derived without all the tedious counting if we use the method of quantized fields.

For the free-particle wave functions (A.24), and for $v_{12} = \delta(\mathbf{r}_1 - \mathbf{r}_2)$, (A.30) reduces to

$$\langle \mathbf{p}_1, \mathbf{p}_2 | \delta | \mathbf{p}_1', \mathbf{p}_2' \rangle = \frac{1}{V} \quad \text{(A.34)}$$

Therefore for free-particle wave functions we have

$$\left(\Phi_n', \sum_{i<j} \delta(\mathbf{r}_i - \mathbf{r}_j) \Phi_n' \right) = \frac{1}{V} \left[\sum_{\mathbf{p} \neq \mathbf{k}} n_\mathbf{p} n_\mathbf{k} + \tfrac{1}{2} \sum_\mathbf{p} n_\mathbf{p} (n_\mathbf{p} - 1) \right]$$

Since

$$\sum_{p \neq k} n_p n_k = \sum_p n_p \sum_k n_k - \sum_p n_p^2 = N^2 - \sum_p n_p^2 \qquad (A.35)$$

we have

$$\left(\Phi_n', \sum_{i<j} \delta(\mathbf{r}_i - \mathbf{r}_j)\Phi_n'\right) = \frac{1}{V}\left(N^2 - \tfrac{1}{2}N - \tfrac{1}{2}\sum_p n_p^2\right) \qquad (A.36)$$

Example of Calculation: System of Fermions

We calculate $(\Phi_\alpha, \Omega\Phi_\alpha)$ for a system of fermions. The formula (A.29) remains valid if we insert the factor $\delta_P \delta_Q$ before the summand in (A.29). The conditions in (A.31) remain pertinent. Noting that $\delta_P \delta_Q = 1$ under the condition (a) of (A.30), and that $\delta_P \delta_Q = -1$ under the condition (b) of (A.30), we obtain in place of (A.32) the formula

$$(\Phi_\alpha, \Omega\Phi_\alpha) = \frac{N(N-1)}{N!2}\sum_P (\langle P\alpha_1, P\alpha_2|v|P\alpha_1, P\alpha_2\rangle - \langle P\alpha_1, P\alpha_2|v|P\alpha_2, P\alpha_1\rangle)$$

$$(A.37)$$

Noting that $n_\alpha = 0, 1$ we obtain in place of (A.33) the formula

$$(\Phi_n, \Omega\Phi_n) = \tfrac{1}{2}\sum_{\alpha, \beta} n_\alpha n_\beta(\langle \alpha, \beta|v|\alpha, \beta\rangle - \langle \alpha, \beta|v|\beta, \alpha\rangle) \qquad (A.38)$$

To illustrate how the N-body wave functions may be generalized to include the spin coordinates of the particles, let us consider fermions of spin $\hbar/2$. In addition to the position coordinate $\mathbf{r}$, each particle now has a spin coordinate σ which can take on only the values ± 1. The free-particle wave function $u_{ps}(\mathbf{r}, \sigma)$ of a particle is now labeled by the momentum $\mathbf{p}$ and the spin quantum number s, which can assume only the values ± 1. When $s = +1$ the particle is said to be in a state of up spin, and when $s = -1$, in down spin. Explicitly we have

$$u_{ps}(\mathbf{r}, \sigma) = \frac{1}{\sqrt{V}} e^{i\mathbf{p}\cdot\mathbf{r}/\hbar} \delta(s, \sigma) \qquad (A.39)$$

where

$$\delta(s, \sigma) = \begin{cases} 1 & (s = \sigma) \\ 0 & (s \neq \sigma) \end{cases} \qquad (A.40)$$

Sometimes we write out the two values of $u_{ps}(\mathbf{r}, \sigma)$ for $\sigma = \pm 1$, as follows

$$\begin{pmatrix} u_{\mathbf{p}, +1}(\mathbf{r}, +1) \\ u_{\mathbf{p}, +1}(\mathbf{r}, -1) \end{pmatrix} = \frac{1}{\sqrt{V}} e^{i\mathbf{p}\cdot\mathbf{r}/\hbar}\begin{pmatrix} 1 \\ 0 \end{pmatrix}$$

$$\begin{pmatrix} u_{\mathbf{p}, -1}(\mathbf{r}, +1) \\ u_{\mathbf{p}, -1}(\mathbf{r}, -1) \end{pmatrix} = \frac{1}{\sqrt{V}} e^{i\mathbf{p}\cdot\mathbf{r}/\hbar}\begin{pmatrix} 0 \\ 1 \end{pmatrix}$$

We do not use such a representation here.

If we let α stand for the collection of quantum numbers $\{\mathbf{p}_\alpha, s_\alpha\}$ and let $u_\alpha(\mathbf{r}_1, \sigma_1)$ be abbreviated by $u_\alpha(1)$, then (A.14) defines a complete orthonormal set of wave functions for the N-fermion system and (A.38) continues to be valid. Let $v_{12} = \delta(\mathbf{r}_1 - \mathbf{r}_2)$, which is independent of spin coordinates. Then

$$\langle \alpha, \beta | \delta | \alpha, \beta \rangle = \frac{1}{V}$$

$$\langle \alpha, \beta | \delta | \beta, \alpha \rangle = \frac{1}{V}\delta(s_\alpha, s_\beta)$$

Let $n_{\mathbf{p}s}$ denote the occupation number of the single-particle state with momentum $\mathbf{p}$ and spin quantum number s. Then (A.38) becomes

$$\left(\Phi_n, \sum_{i<j} \delta(\mathbf{r}_i - \mathbf{r}_j)\Phi_n \right) = \frac{1}{2V} \sum_{s,s'} \sum_{\mathbf{p},\mathbf{k}} n_{\mathbf{p}s} n_{\mathbf{k}s'}[1 - \delta(s, s')]$$

$$= \frac{1}{2V}\left(N^2 - \sum_s \sum_{\mathbf{p},\mathbf{k}} n_{\mathbf{p}s} n_{\mathbf{k}s} \right)$$

Let

$$N_+ \equiv \sum_{\mathbf{p}} n_{\mathbf{p},+1}$$

$$N_- \equiv \sum_{\mathbf{p}} n_{\mathbf{p},-1} = N - N_+ \tag{A.41}$$

Then

$$\left(\Phi_n, \sum_{i<j} \delta(\mathbf{r}_i - \mathbf{r}_j)\Phi_n \right) = \frac{N_+ N_-}{V} \tag{A.42}$$

A.3 METHOD OF QUANTIZED FIELDS

A system of N particles is equivalent to a quantized field. This equivalence is often used to great advantage in the calculation of the energy levels and the partition function of an N-particle system.

A quantized field is a system characterized by field operators $\psi(\mathbf{r})$ that are defined for all values of the coordinate $\mathbf{r}$ and operate on a Hilbert space. A vector in this Hilbert space corresponds to a state of the quantized field. It is our purpose to show that a quantized field can be so defined that its Hilbert space contains the Hilbert space of a given N-particle system. For simplicity we consider the N particles to be either all identical spinless bosons or all identical spinless fermions.

First we define the quantized fields that corresponds to bosons and fermions. The field operators of the two cases are defined by the following commutation rules.

Bosons	*Fermions*	
$[\psi(\mathbf{r}), \psi^\dagger(\mathbf{r}')] = \delta(\mathbf{r} - \mathbf{r}')$	$\{\psi(\mathbf{r}), \psi^\dagger(\mathbf{r}')\} = \delta(\mathbf{r} - \mathbf{r}')$	
$[\psi(\mathbf{r}), \psi(\mathbf{r}')] = 0$	$\{\psi(\mathbf{r}), \psi(\mathbf{r}')\} = 0$	(A.43)
$[\psi^\dagger(\mathbf{r}), \psi^\dagger(\mathbf{r}')] = 0$	$\{\Psi^\dagger(\mathbf{r}), \psi^\dagger(\mathbf{r}')\} = 0$	

where $\psi^\dagger$ is the Hermitian conjugate of ψ and $[A, B] \equiv AB - BA$, $\{A, B\} \equiv AB + BA$.

The definition of the quantized field is completed by defining two Hermitian operators—the Hamiltonian operator $\mathscr{H}$ and the number operator N_{op}. The Hamiltonian operator is

$$\mathscr{H} \equiv K + \Omega$$

$$K = -\frac{\hbar^2}{2m} \int d^3r \, \psi^\dagger(\mathbf{r}) \nabla^2 \psi(\mathbf{r}) \tag{A.44}$$

$$\Omega = \tfrac{1}{2} \int d^3r_1 \, d^3r_2 \, \psi^\dagger(\mathbf{r}_1) \psi^\dagger(\mathbf{r}_2) v_{12} \psi(\mathbf{r}_2) \psi(\mathbf{r}_1)$$

where $v_{12} \equiv v(\mathbf{r}_1, \mathbf{r}_2)$. The number operator is

$$N_{op} \equiv \int d^3r \, \psi^\dagger(\mathbf{r}) \psi(\mathbf{r}) \tag{A.45}$$

These definitions hold for both bosons and fermions. We can easily verify that

$$\left[\mathscr{H}, N_{op} \right] = 0 \tag{A.46}$$

Therefore $\mathscr{H}$ and N_{op} can be simultaneously diagonalized. We show that a simultaneous eigenstate of $\mathscr{H}$ and N_{op} is an energy eigenstate of a system of a definite number of particles.

Let a complete orthonormal basis of the Hilbert space be so chosen that any vector $|\Phi_n\rangle$ of the basis is a simultaneous eigenstate of $\mathscr{H}$ and N_{op}. Let a particular member of the basis be denoted by $|\Psi_{EN}\rangle$, with the properties that

$$\begin{aligned} \langle \Psi_{EN} | \Psi_{EN} \rangle &= 1 \\ \mathscr{H} | \Psi_{EN} \rangle &= E | \Psi_{EN} \rangle \\ N_{op} | \Psi_{EN} \rangle &= N | \Psi_{EN} \rangle \end{aligned} \tag{A.47}$$

The state $|0\rangle \equiv |\Psi_{00}\rangle$, called the vacuum state, is assumed to be unique. Its properties are

$$\begin{aligned} \langle 0 | 0 \rangle &= 1 \\ \mathscr{H} | 0 \rangle &= 0 \\ N_{op} | 0 \rangle &= 0 \end{aligned} \tag{A.48}$$

From (A.45) and (A.43) it is easily verified that

$$\begin{aligned} \left[\psi(\mathbf{r}), N_{op} \right] &= \psi(\mathbf{r}) \\ \left[\psi^\dagger(\mathbf{r}), N_{op} \right] &= -\psi^\dagger(\mathbf{r}) \end{aligned} \tag{A.49}$$

Hence

$$\begin{aligned} N_{op} \psi(\mathbf{r}) | \Psi_{EN} \rangle &= (N - 1) \psi(\mathbf{r}) | \Psi_{EN} \rangle \\ N_{op} \psi^\dagger(\mathbf{r}) | \Psi_{EN} \rangle &= (N + 1) \psi^\dagger(\mathbf{r}) | \Psi_{EN} \rangle \end{aligned} \tag{A.50}$$

Thus $\psi(\mathbf{r})$ decreases N by 1, and $\psi^\dagger(\mathbf{r})$ increases N by 1. By repeated application of $\psi^\dagger(\mathbf{r})$ to $|0\rangle$, we prove that the eigenvalues of N_{op} are

$$N = 0, 1, 2, \dots \tag{A.51}$$

Since $\psi(\mathbf{r})$ decreases N by 1, and the state with $N = 0$ is assumed to be unique, we have the identity

$$\langle \Phi_n | \psi(1)\psi(2) \cdots \psi(N) | \Psi_{EN} \rangle = 0 \qquad \text{unless } |\Phi_n\rangle \equiv |0\rangle \qquad \text{(A.52)}$$

where $\psi(j) \equiv \psi(\mathbf{r}_j)$.

Let a function of the N position coordinates $\mathbf{r}_1, \ldots, \mathbf{r}_N$ be defined by

$$\Psi_{EN}(1, \ldots, N) \equiv \frac{1}{\sqrt{N!}} \langle 0 | \psi(1) \cdots \psi(N) | \Psi_{EN} \rangle \qquad \text{(A.53)}$$

By (A.43) this function is symmetric (antisymmetric) with respect to the exchange of any two coordinates for bosons (fermions). The norm of $\Psi_{EN}(1, \ldots, N)$ is unity, i.e.,

$$\int d^{3N}r \, \Psi_{EN}^*(1, \ldots, N) \Psi_{EN}(1, \ldots, N) = 1 \qquad \text{(A.54)}$$

Proof Let

$$I \equiv \int d^{3N}r \, \Psi_{EN}^*(1, \ldots, N) \Psi_{EN}(1, \ldots, N)$$

$$= \frac{1}{N!} \int d^{3N}r \langle \Psi_{EN} | \psi^\dagger(N) \cdots \psi^\dagger(1) | 0 \rangle \langle 0 | \psi(1) \cdots \psi(N) | \Psi \rangle$$

By (A.52) we can write

$$I = \frac{1}{N!} \int d^{3N}r \sum_n \langle \Psi_{EN} | \psi^\dagger(N) \cdots \psi^\dagger(1) | \Phi_n \rangle \langle \Phi_n | \psi(1) \cdots \psi(N) | \Psi_{EN} \rangle$$

$$= \frac{1}{N!} \int d^{3N}r \langle \Psi_{EN} | [\psi^\dagger(N) \cdots \psi^\dagger(1)] [\psi(1) \cdots \psi(N)] | \Psi_{EN} \rangle$$

Now carry out the integration over $\mathbf{r}_1$. The relevant factor is

$$\int d^3r_1 \, \psi^\dagger(1)\psi(1) = N_{op}$$

Next carry out the integration over $\mathbf{r}_2$. The relevant factor is

$$\int d^3r_2 \, \psi^\dagger(2) N_{op} \psi(2) = N_{op}(N_{op} - 1)$$

By induction we can show that

$$I = \frac{1}{N!} \langle \Psi_{EN} | N_{op}(N_{op} - 1)(N_{op} - 2) \cdots 1 | \Psi_{EN} \rangle = 1 \qquad \blacksquare$$

The connection between the quantized field and an N-body system is furnished by the following theorem.

THEOREM

$$\left(-\frac{\hbar^2}{2m}\sum_{j=1}^{N}\nabla_j^2 + \sum_{i<j}v_{ij}\right)\Psi_{EN}(1,\ldots,N) = E\Psi_{EN}(1,\ldots,N) \quad (A.55)$$

Proof By (A.47) and (A.53)

$$\frac{1}{\sqrt{N!}}\langle 0|[\psi(1)\cdots\psi(N)]\mathcal{H}|\Psi_{EN}\rangle = E\Psi_{EN}(1,\ldots,N) \quad (A.56)$$

Since $\mathcal{H}|0\rangle = 0$, and $\mathcal{H}$ is Hermitian, we also have $\langle 0|\mathcal{H} = 0$. Hence the left side of (A.56) has the form of a commutator:

$$J \equiv \frac{1}{\sqrt{N!}}\langle 0|[\psi(1)\cdots\psi(N)]\mathcal{H}|\Psi_{EN}\rangle$$

$$= \frac{1}{\sqrt{N!}}\langle 0|[\psi(1)\cdots\psi(N),\mathcal{H}]|\Psi_{EN}\rangle$$

$$= \frac{1}{\sqrt{N!}}\sum_{j=1}^{N}\langle 0|\psi(1)\cdots[\psi(j),\mathcal{H}]\cdots\psi(N)|\Psi_{EN}\rangle \quad (A.57)$$

where the last step is obtained through repeated use of the identity

$$[AB,C] = [A,C]B + A[B,C]$$

We explicitly calculate $[\psi(j),\mathcal{H}]$. From (A.44) we have

$$[\psi(i),\mathcal{H}] = [\psi(j),K] + [\psi(j),\Omega]$$

For Bosons

$$[\psi(j),K] = -\frac{\hbar^2}{2m}\int d^3r\,[\psi(j),\psi^\dagger(\mathbf{r})\nabla^2\psi(\mathbf{r})]$$

$$= -\frac{\hbar^2}{2m}\int d^3r\,[\psi(j),\psi^\dagger(\mathbf{r})]\nabla^2\psi(\mathbf{r})$$

$$= -\frac{\hbar^2}{2m}\nabla_j^2\psi(j)$$

$$[\psi(j),\Omega] = \tfrac{1}{2}\int d^3r_1\,d^3r_2\,[\psi(j),\psi^\dagger(1)\psi^\dagger(2)]v_{12}\psi(2)\psi(1)$$

$$= \tfrac{1}{2}\int d^3r_1\,d^3r_2\,\{[\psi(j),\psi^\dagger(1)]\psi^\dagger(2) + \psi^\dagger(1)[\psi(j),\psi^\dagger(2)]\}$$

$$\times v_{12}\psi(2)\psi(1)$$

$$= \left[\int d^3r\,\psi^\dagger(\mathbf{r})v(\mathbf{r},\mathbf{r}_j)\psi(\mathbf{r})\right]\psi(j)$$

For Fermions

$$[\psi(j), K] = -\frac{\hbar^2}{2m} \int d^3r \left[\psi(j), \psi^\dagger(\mathbf{r}) \nabla^2 \psi(\mathbf{r})\right]$$

$$= -\frac{\hbar^2}{2m} \int d^3r \left\{\psi(j), \psi^\dagger(\mathbf{r})\right\} \nabla^2 \psi(\mathbf{r})$$

$$= -\frac{\hbar^2}{2m} \nabla_j^2 \psi(j)$$

$$[\psi(j), \Omega] = \tfrac{1}{2} \int d^3r_1 d^3r_2 \left[\psi(j), \psi^\dagger(1)\psi^\dagger(2)\right] v_{12} \psi(2)\psi(1)$$

$$= \tfrac{1}{2} \int d^3r_1 d^3r_2 \left[\left\{\psi(j), \psi^\dagger(1)\right\}\psi^\dagger(2) - \psi^\dagger(1)\left\{\psi(j), \psi^\dagger(2)\right\}\right]$$
$$\times v_{12}\psi(2)\psi(1)$$

$$= \left[\int d^3r \, \psi^\dagger(\mathbf{r}) v(\mathbf{r}, \mathbf{r}_j)\psi(\mathbf{r})\right]\psi(j)$$

Hence for both bosons and fermions we have

$$[\psi(j), \mathscr{H}] = \left[-\frac{\hbar^2}{2m}\nabla_j^2 + X(j)\right]\psi(j) \tag{A.58}$$

where

$$X(j) = \int d^3r \, \psi^\dagger(\mathbf{r}) v(\mathbf{r}, \mathbf{r}_j)\psi(\mathbf{r}) \tag{A.59}$$

The following properties of $X(j)$ are trivial:

$$[\psi(i), X(j)] = v_{ij}\psi(i) \tag{A.60}$$

$$X(j)|0\rangle = 0, \qquad \langle 0|X(j) = 0 \tag{A.61}$$

Substitution of (A.59) into (A.57) yields

$$J = -\frac{\hbar^2}{2m}\sum_{j=1}^{N} \nabla_j^2 \Psi_{EN}(1, \ldots, N)$$

$$+ \frac{1}{\sqrt{N!}}\sum_{j=1}^{N} \langle 0|\psi(1)\cdots\psi(j-1)X(j)\psi(j)\cdots\psi(N)|\Psi_{EN}\rangle \tag{A.62}$$

We now commute $X(j)$ all the way to the left with the help of (A.60):

$$[\psi(1)\cdots\psi(j-1)X(j)\psi(j)\cdots\psi(N)]$$
$$= [\psi(1)\cdots\psi(j-2)X(j)\psi(j-1)\cdots\psi(N)]$$
$$\quad + v_{j-1,j}[\psi(1)\cdots\psi(N)]$$
$$= [\psi(1)\cdots\psi(j-3)X(j)\psi(j-2)\cdots\psi(N)]$$
$$\quad + (v_{j-2,j} + v_{j-1,j})[\psi(1)\cdots\psi(N)]$$
$$= \cdots$$
$$= \left[X(j) + \sum_{i=1}^{j-1} v_{ij}\right][\psi(1)\cdots\psi(N)] \tag{A.63}$$

Substituting this into (A.62) and using (A.61) we obtain

$$J = \left[-\frac{\hbar^2}{2m} \sum_{j=1}^{N} \nabla_j^2 + \sum_{i<j} v_{ij} \right] \Psi_{EN}(1, \ldots, N) \qquad \blacksquare$$

For convenience in actual calculations, let us introduce a complete orthonormal set of single-particle wave functions $\{u_\alpha(\mathbf{r})\}$ with

$$\int d^3r \, u_\alpha^*(\mathbf{r}) u_\beta(\mathbf{r}) = \delta_{\alpha\beta} \qquad (A.64)$$

Then we may expand the field operators $\psi(\mathbf{r})$ and $\psi^\dagger(\mathbf{r})$ in the following manner:

$$\psi(\mathbf{r}) = \sum_\alpha a_\alpha u_\alpha(\mathbf{r})$$
$$\psi^\dagger(\mathbf{r}) = \sum_\alpha a_\alpha^\dagger u_\alpha^*(\mathbf{r}) \qquad (A.65)$$

where, in accordance with (A.43), a_α and $a_\alpha^\dagger$ are operators satisfying the following commutation rules:

Bosons	Fermions	
$[a_\alpha, a_\beta^\dagger] = \delta_{\alpha\beta}$	$\{a_\alpha, a_\beta^\dagger\} = \delta_{\alpha\beta}$	
$[a_\alpha, a_\beta] = 0$	$\{a_\alpha, a_\beta\} = 0$	(A.66)
$[a_\alpha^\dagger, a_\beta^\dagger] = 0$	$\{a_\alpha^\dagger, a_\beta^\dagger\} = 0$	

It easily follows that the eigenvalues of $a_\alpha^\dagger a_\alpha$ are

$$n_\alpha \equiv a_\alpha^\dagger a_\alpha = \begin{cases} 0, 1, 2, \ldots & \text{(bosons)} \\ 0, 1 & \text{(fermions)} \end{cases} \qquad (A.67)$$

In terms of a_α and $a_\alpha^\dagger$ we have

$$N_{op} = \sum_\alpha a_\alpha^\dagger a_\alpha \qquad (A.68)$$

$$\mathcal{H} = \frac{\hbar^2}{2m} \sum_{\alpha, \beta} \langle \alpha | -\nabla^2 | \beta \rangle a_\alpha^\dagger a_\beta + \tfrac{1}{2} \sum_{\alpha, \beta, \gamma, \lambda} \langle \alpha\beta | v | \gamma\lambda \rangle (a_\alpha a_\beta)^\dagger (a_\lambda a_\gamma) \qquad (A.69)$$

where

$$\langle \alpha | -\nabla^2 | \beta \rangle = -\int d^3r \, u_\alpha^* \nabla^2 u_\beta$$

$$\langle \alpha\beta | v | \gamma\lambda \rangle = \int d^3r_1 \, d^3r_2 \, u_\alpha^*(1) u_\beta^*(2) v_{12} u_\gamma(1) u_\lambda(2) \qquad (A.70)$$

Let a set of integers $\{n_0, n_1, \ldots\}$ be given, such that each n_α is a possible value of $a_\alpha^\dagger a_\alpha$ as given by the rule (A.67). Define the state $|n\rangle$ by

$$|n\rangle \equiv |n_0, n_1, \ldots\rangle \equiv C_n \left[(a_0^\dagger)^{n_0} (a_1^\dagger)^{n_1} \cdots \right] |0\rangle \qquad (A.71)$$

where C_n is a normalization constant so chosen that $\langle n|n\rangle = 1$:

$$C_n = \frac{1}{\sqrt{\prod_\alpha (n_\alpha!)}} \qquad \text{(A.72)}$$

It can be verified that

For Bosons

$$a_\alpha |\ldots, n_\alpha, \ldots\rangle = \sqrt{n_\alpha} |\ldots, n_\alpha - 1, \ldots\rangle$$

$$a_\alpha^\dagger |\ldots, n_\alpha, \ldots\rangle = \sqrt{n_\alpha + 1} |\ldots, n_\alpha + 1, \ldots\rangle \qquad \text{(A.73)}$$

For Fermions*

$$a_\alpha |\ldots, n_\alpha, \ldots\rangle = \xi_\alpha \sqrt{n_\alpha} |\ldots, n_\alpha - 1, \ldots\rangle$$

$$a_\alpha^\dagger |\ldots, n_\alpha, \ldots\rangle = \xi_\alpha \sqrt{n_\alpha + 1} |\ldots, n_\alpha + 1, \ldots\rangle \qquad \text{(A.74)}$$

where $\xi_\alpha = \pm 1$ according to whether $\sum_{\beta < \alpha} n_\beta$ is an even or odd integer. The operator a_α is called the annihilation operator for the single-particle state α, and the operator $a_\alpha^\dagger$ is called the creation operator for the single-particle state α.

For both bosons and fermions we have

$$a_\alpha^\dagger a_\alpha |n\rangle = n_\alpha |n\rangle \qquad \text{(A.75)}$$

Therefore

$$N_{op} |n\rangle = \left(\sum_\alpha n_\alpha\right) |n\rangle \qquad \text{(A.76)}$$

By the use of (A.73) and (A.74) any matrix element $\langle n|\mathscr{H}|n'\rangle$ can be obtained trivially from (A.69).

By (A.55) and (A.54) the complete set of wave functions Φ_n defined in (A.14) can also be represented in the form

$$\frac{1}{\sqrt{\prod_\alpha (n_\alpha!)}} \Phi_n(1, \ldots, n) = \frac{1}{\sqrt{N!}} \langle 0|\psi(1) \ldots \psi(N)|n\rangle \qquad \text{(A.77)}$$

Therefore

$$\left(\Phi_n, \left[-\frac{\hbar^2}{2m} \sum_{j=1}^N \nabla_j^2 + \sum_{i<j} v_{ij}\right] \Phi_{n'}\right) = \langle n|\mathscr{H}|n'\rangle \qquad \text{(A.78)}$$

In particular, the results (A.33) and (A.38) can be trivially obtained through the use of this relation.

---

*Note that for fermions $|n\rangle \equiv 0$ if any $n_\alpha > 1$.

A.4 LONGITUDINAL SUM RULES

We derive sum rules relevant to density fluctuations (longitudinal sound waves) for either a Bose or a Fermi system of particles whose Hamiltonian is (A.44), except that we shall subtract from it the ground state energy E_0 for convenience, and we shall set $\hbar = 1$:

$$\mathcal{H} = \frac{1}{2m} \int d^3x \, |\nabla \psi(x)|^2 + \frac{1}{2} \int d^3x \, d^3y \, \psi^\dagger(y) \psi^\dagger(x) v(x-y) \psi(x) \psi(y) - E_0 \tag{A.79}$$

The density operator is

$$\rho(x) = \psi^\dagger(x) \psi(x) \tag{A.80}$$

with Fourier transform

$$\rho_k = V^{-1/2} \int d^3x \, e^{ik \cdot x} \rho(x), \qquad \rho_{-k} = \rho_k^\dagger \tag{A.81}$$

Let $|0\rangle$ denote the ground state* in the N-particle sector:

$$\mathcal{H}|0\rangle = 0 \tag{A.82}$$

The sum rules we shall derive are

$$n^{-1} \langle 0 | \rho_k^\dagger \mathcal{H} \rho_k | 0 \rangle = k^2/2m \tag{A.83}$$

$$\lim_{k \to 0} n^{-1} \langle 0 | \rho_k^\dagger \mathcal{H}^{-1} \rho_k | 0 \rangle = 1/2mc^2 \tag{A.84}$$

where n is the density, and c is the velocity of sound at absolute zero.

We shall use the creation and annihilation operators $a_k^\dagger$ and a_k as defined by

$$\psi(x) = V^{-1/2} \sum_k a_k e^{ik \cdot x} \tag{A.85}$$

with

$$\begin{aligned} \left[a_k, a_p^\dagger \right] &= \delta_{kp} \quad \text{(Bose)} \\ \left[a_k, a_p^\dagger \right] &= \delta_{kp} \quad \text{(Fermi)} \end{aligned} \tag{A.86}$$

Then

$$\rho_k = V^{-1/2} \sum_p a_{p+k}^\dagger a_p \tag{A.87}$$

Let

$$\xi_k \equiv -iV^{-1/2} \sum_p k \cdot p \, a_{k+p}^\dagger a_p \tag{A.88}$$

*One should not confuse this with the vacuum state, denoted by the same symbol in the last section.

We have the commutation relations

$$[\rho_\mathbf{k}, \rho_\mathbf{p}] = 0, \tag{A.89}$$

$$[\xi_\mathbf{k}, \rho_\mathbf{p}] = -iV^{-1/2}\mathbf{k} \cdot \mathbf{p}\rho_{\mathbf{k}+\mathbf{p}} \tag{A.90}$$

$$[\mathcal{H}, \rho_\mathbf{k}] = \frac{k^2}{2m}\rho_\mathbf{k} + \frac{i}{m}\xi_\mathbf{k} \tag{A.91}$$

To derive (A.83) note that since $\mathcal{H}|0\rangle = 0$,

$$\langle 0|\rho_{-\mathbf{k}}\mathcal{H}\rho_\mathbf{k}|0\rangle = \tfrac{1}{2}\{\langle 0|\rho_{-\mathbf{k}}[\mathcal{H}, \rho_\mathbf{k}]|0\rangle + \langle 0|[\rho_{-\mathbf{k}}, \mathcal{H}]\rho_\mathbf{k}|0\rangle\}$$
$$= (i/2m)\langle 0|[\rho_{-\mathbf{k}}\xi_\mathbf{k} - \xi_{-\mathbf{k}}\rho_\mathbf{k}]|0\rangle$$

Using the invariance of $|0\rangle$ under space inversion, we obtain

$$n^{-1}\langle 0|\rho_{-\mathbf{k}}\mathcal{H}\rho_\mathbf{k}|0\rangle = (i/2mn)\langle 0|[\rho_{-\mathbf{k}}, \xi_\mathbf{k}]|0\rangle = k^2/2m \qquad \blacksquare$$

To derive (A.84) we add a small perturbation to $\mathcal{H}$ so that the Hamiltonian becomes

$$\tilde{\mathcal{H}} = \mathcal{H} + \mathcal{H}' \tag{A.92}$$

$$\mathcal{H}' = \lambda V^{-1/2} \int d^3x \cos(\mathbf{k} \cdot \mathbf{x})\rho(\mathbf{x}) = \tfrac{1}{2}\lambda(\rho_\mathbf{k} + \rho_\mathbf{k}^\dagger) \tag{A.93}$$

Eventually we take the limit $\lambda \to 0$ and $\mathbf{k} \to 0$. Let the new ground state be $|\Psi_0\rangle$, and the new ground state energy be $E_0 + \Delta E_0$:

$$\tilde{\mathcal{H}}|\Psi_0\rangle = \Delta E_0|\Psi_0\rangle \tag{A.94}$$

$$\mathcal{H}|0\rangle = 0 \tag{A.95}$$

Let

$$X \equiv n^{-1}\langle 0|\rho_\mathbf{k}^\dagger \mathcal{H}^{-1}\rho_\mathbf{k}|0\rangle \tag{A.96}$$

We make three independent calculations.

(a) By applying second-order perturbation theory we obtain

$$\Delta E_0 = -\tfrac{1}{2}\lambda^2 nX \tag{A.97}$$

(b) The density in the new system is a function of position. We denote it by $n + \Delta n(\mathbf{x})$. By applying second-order perturbation theory we find

$$\Delta n(\mathbf{x}) = \langle \Psi_0|\rho(\mathbf{x})|\Psi_0\rangle - n = -\frac{22nX}{\sqrt{V}} \cos\mathbf{k} \cdot \mathbf{x} \tag{A.98}$$

(c) We calculate ΔE_0 another way by writing

$$\Delta E_0 = \langle \Psi_0|\mathcal{H}|\Psi_0\rangle + \langle \Psi_0|\mathcal{H}'|\Psi_0\rangle \tag{A.99}$$

The last term can be calculated by perturbation theory:

$$\langle \Psi_0|\mathcal{H}'|\Psi_0\rangle = -\lambda^2 nX. \tag{A.100}$$

To calculate the first term in (A.99) we make the assumption that as

$\lambda \to 0$ and $\mathbf{k} \to 0$, the wave function of the state $|\Psi_0\rangle$ is locally the same as that of $|0\rangle$ except that the density has a new value. Let $E_0 = V\epsilon(n)$. Then

$$\langle \Psi_0 | \mathcal{H} | \Psi_0 \rangle = \int d^3x \, \epsilon(n + \Delta n(\mathbf{x})) - E_0 = \lambda^2 X^2 n^2 (\partial^2\epsilon/\partial n^2) \quad \text{(A.101)}$$

where we have used (A.98). Hence as $\lambda \to 0$, $\mathbf{k} \to 0$,

$$\Delta E_0 \to \lambda^2 nX \left[Xn(\partial^2\epsilon/\partial n^2) - 1 \right] = \lambda^2 nX(mc^2 X - 1) \quad \text{(A.102)}$$

We obtain the final result by comparing (A.102) with (A.97):

$$X(mc^2 X - 1) \xrightarrow[k \to 0]{} \tfrac{1}{2}X \qquad \text{(A.103)}$$

Hence

$$X \xrightarrow[k \to 0]{} 1/2mc^2 \qquad \blacksquare$$

Note that throughout the derivations no use is made of the statistics of the particles, or the fact that $|0\rangle$ is an eigenstate of the number operator. Furthermore, the sum rules remain valid if $\mathcal{H}$ is replaced by the operator $\mathcal{H} - \mu N_{\text{op}}$, where μ is a fixed real number and N_{op} is the number operator.

INDEX

Abramovitz, M., 243
Absolute zero, 14
  unattainability of, 27
Accommodation, coefficient of, 110
Adiabatic process, 4
Aharanov, Y., 257
Aharanov-Bohm effect, 257
Als-Nielsen, J., 438
Anderson, P. W., 299, 322, 407
Anderson's equation, 322
Andronikashvili, experiment of, 312
Annihilation operator, 483
Antiferromagnetism, 349
Antiparticle, 156, 294
Arnold, G. P., 310
Assumption of molecular chaos:
  analog in BBGKY hierarchy, 69
  and Boltzmann's equation, 89
  and Boltzmann's H theorem, 86
Avogadro's number, 6

Baker-Campbell-Hausdorf theorem, 204
Baryon conservation, 154
BBGKY hierarchy, 65, 68
Bendt, P. J., 310
Bernoulli's equation, 104
Bessel, F. W., 248
Beta brass, 346
Beth, E., 224
Bethe-Peierls approximation, 357
Binary alloy, 346
Birgeneau, R. J., 349, 411, 438
Blackbody radiation, 278
  entropy per photon in, 302
Blatt, J. M., 233
Bloch, F., 409
Block spin:
  defined, 441
  transformation, 441, 446
Bogoliubov, N. N., 65, 299, 331
Bogoliubov transformation, 331
Bohm, D., 257
Bohr magneton, 260
Boiling point, effect of solute on, 46
Boltzmann's constant, 4
Boltzmann's H theorem:
  analysis of, 85
  and second law of thermodynamics, 79
  stated and proved, 74

Boltzmann transport equation:
  conservation theorems, 99
  derived, 62
    from BBGKY hierarchy, 71
  validity of, 89
Bose-Einstein condensate, 298
Bose-Einstein condensation, 189, 286
  criterion for, 299
  in imperfect gas, 297
  and lambda transition, 293, 310
Bosons, 179
  momentum-space attraction between, 295
  statistical attraction between, 203
Boyle's law, 4
Bragg, W. L., 353
Bragg-Williams approximation, 240, 353
Brownian motion, 50
Buckingham, M. J., 302, 309

Canonical ensemble:
  defined, 143
  derived with Darwin-Fowler method, 193
  energy fluctuations, 146
  equivalence with microcanonical, 146
  and thermodynamics, 145
Carnot engine, 10
Carnot's theorem, 12
Centigrade degree, 4
Chain relation, 20
Chandrasekhar, S., 253
Chandrasekhar limit, 253
Chapman, S., 105
Chemical potential, 32, 150, 154
Chiu, H. Y., 249
Circle theorem:
  electrostatic analog for, 212
  statement of, 210
Clapeyron equation, 35
Clausius statement, 10
Clausius' theorem, 14
Cluster expansion:
  classical, 213
  quantum mechanical, 220
Cluster integral:
  classical, 216
  for quantum ideal gases, 223
  quantum mechanical, 222
Coarse-graining, 453
Cohen, E. G. D., 189